THE SAUROPODS

THE SAUROPODS

Evolution and Paleobiology

Kristina A. Curry Rogers and Jeffrey A. Wilson

UNIVERSITY OF CALIFORNIA PRESS
Berkeley   Los Angeles   London

University of California Press, one of the most distinguished university presses in the United States, enriches lives around the world by advancing scholarship in the humanities, social sciences, and natural sciences. Its activities are supported by the UC Press Foundation and by philanthropic contributions from individuals and institutions. For more information, visit www.ucpress.edu.

University of California Press
Berkeley and Los Angeles, California

University of California Press, Ltd.
London, England

© 2005 by the Regents of the University of California

Library of Congress Cataloging-in-Publication Data

The Sauropods: Evolution and Paleobiology / [edited by] Kristina A. Curry Rogers, Jeffrey A. Wilson.
   p. cm.
Includes bibliographical references and index.
ISBN 0-520-24623-3 (cloth : alk. paper)
   1. Saurischia. 2. Saurischia—Evolution. 3. Saurischia—Morphology. 4. Saurischia—Anatomy. 5. Paleobiology. I. Curry Rogers, Kristina A., 1974- II. Wilson, Jeffrey A.
QE862.S3S28 2005
567.913—dc22
                                               2005010624

10 09 08 07 06 05
10 9 8 7 6 5 4 3 2 1

Cover: *Titanosaurid Nesting Grounds*, painting by Mark Hallett.

DEDICATION

This volume is dedicated to our friend John S. "Jack" McIntosh. Perhaps more than any other single person, Jack has influenced the course of sauropod studies since World War II. In addition to being one of the foremost students of sauropods (McIntosh and Berman, 1975; Berman and McIntosh, 1978; McIntosh, 1989, 1990a, b), Jack has devoted himself to archiving and interpreting original map and quarry data from the early history of North American dinosaur paleontology (Ostrom and McIntosh, 1966; Kohl and McIntosh, 1997). We close *The Sauropods: Evolution and Paleobiology* with an interview with Jack that records his thoughts on the past, present, and future of sauropod studies. Perhaps the greatest attribute of sauropod dinosaurs is that they attract students like Jack, whose devotion and contagious enthusiam continue after half a century.

Berman, D. S. and J. S. McIntosh. 1978. Skull and relationships of the Upper Jurassic sauropod *Apatosaurus* (Reptilia, Saurischia). Bulletin of the Carnegie Museum of Natural History 8: 1–35.

Kohl, M. F. and J. S. McIntosh. 1997. Discovering Dinosaurs in the Old West. The Field Journals of Arthur Lakes. Smithsonian Institution Press, Washington and London, 198 pp.

McIntosh, J. S. 1989. The sauropod dinosaurs: a brief survey. In: Padian, K., and Chure, D. J. (eds.). The Age of Dinosaurs. Short Courses in Paleontology No. 2. University of Tennessee, Knoxville. Pp. 85–99.

———. 1990a. Sauropoda. In: Weishampel, D. B., Dodson, P., and Osmólska, H. (eds.). The Dinosauria. University of California Press, Berkeley. Pp. 345–401.

———. 1990b. Species determination in sauropod dinosaurs with tentative suggestions for their classification. In: Carpenter, K., and Currie, P. J. (eds.) Dinosaur Systematics: Approaches and Perspectives. Cambridge University Press, Cambridge.

McIntosh, J. S. and D. S. Berman. 1975. Description of the palate and lower jaw of *Diplodocus* (Reptilia: Saurischia) with remarks on the nature of the skull of *Apatosaurus*. Journal of Paleontology 49: 187–199.

Ostrom, J. H. and J. S. McIntosh. 1966. Marsh's Dinosaurs. Yale University Press, New Haven, 388 pp.

CONTENTS

Acknowledgments / ix

Introduction: Monoliths of the Mesozoic / 1
Jeffrey A. Wilson and Kristina Curry Rogers

1 • **OVERVIEW OF SAUROPOD PHYLOGENY AND EVOLUTION** / 15
Jeffrey A. Wilson

2 • **TITANOSAURIA: A PHYLOGENETIC OVERVIEW** / 50
Kristina Curry Rogers

3 • **PHYLOGENETIC AND TAXIC PERSPECTIVES ON SAUROPOD DIVERSITY** / 104
Paul Upchurch and Paul M. Barrett

4 • **SAUROPODOMORPH DIVERSITY THROUGH TIME: MACROEVOLUTIONARY AND PALEOECOLOGICAL IMPLICATONS** / 125
Paul M. Barrett and Paul Upchurch

5 • **STRUCTURE AND EVOLUTION OF A SAUROPOD TOOTH BATTERY** / 157
Paul C. Sereno and Jeffrey A. Wilson

6 • **DIGITAL RECONSTRUCTIONS OF SAUROPOD DINOSAURS AND IMPLICATIONS FOR FEEDING** / 178
Kent A. Stevens and J. Michael Parrish

7 • **POSTCRANIAL PNEUMATICITY IN SAUROPODS AND ITS IMPLICATIONS FOR MASS ESTIMATES** / 201
Mathew J. Wedel

8 • **THE EVOLUTION OF SAUROPOD LOCOMOTION: MORPHOLOGICAL DIVERSITY OF A SECONDARILY QUADRUPEDAL RADIATION** / 229
Matthew T. Carrano

9 • **STEPS IN UNDERSTANDING SAUROPOD BIOLOGY: THE IMPORTANCE OF SAUROPOD TRACKS** / 252
Joanna L. Wright

10 • **NESTING TITANOSAURS FROM AUCA MAHUEVO AND ADJACENT SITES: UNDERSTANDING SAUROPOD REPRODUCTIVE BEHAVIOR AND EMBRYONIC DEVELOPMENT** / 285
Luis M. Chiappe, Frankie Jackson, Rodolfo A. Coria, and Lowell Dingus

11 • **SAUROPOD HISTOLOGY: MICROSCOPIC VIEWS ON THE LIVES OF GIANTS** / 303
Kristina Curry Rogers and Gregory M. Erickson

A Conversation with Jack McIntosh / 327
Jeffrey A. Wilson and Kristina Curry Rogers

List of Contributors / 335

Index / 337

ACKNOWLEDGMENTS

This volume was developed from a symposium dedicated to Jack McIntosh presented at the 2001 Society of Vertebrate Paleontology annual meeting, which we convered with Dan Chure. We thank Jack McIntosh for allowing us to organize a symposium in his honor. We also thank the participants of that symposium, most of whom authored chapters in this volume, for their excellent papers and patience throughout the process of bringing this book to publication. We especially thank Blake Edgar (UC Press) for initially supporting this project and his continual guidance throughout.

Most of all, we thank our partners, Ray Rogers and Monica Wilson, for their love and support.

INTRODUCTION

Monoliths of the Mesozoic

Jeffrey A. Wilson and Kristina Curry Rogers

GEORGE GAYLORD Simpson (1987:71) expressed his impressions of the well-known North American sauropod *Diplodocus* in the form of a poem to his mother, written while he was studying Mesozoic mammals at Oxford University:

> Oh! Thou imbecile reptile Diplodocus!
> Whoever created so odd a cuss?
> With a tail like a neck,
> And a neck like a tail—
> I wonder, by heck,
> If you ever do fail
> To remember your ends,
> And when danger impends
> Do stand still, which is bad,
> or still more, run tail first,
> Or indeed run both ways, which is rather worst!

Simpson adorned his poem with a caricature of *Diplodocus longus* (fig. I.1), which—it must be said—looks very sauropod-like. That the anatomy of a sauropod can be adequately conveyed in a humorous sketch attests to the relatively simple and recognizable body plan that characterizes the group. Sauropods have deep, barrel-shaped chests supported by four pillarlike legs. They have a relatively small skull that is perched at the end of an elongate neck, which in turn is balanced by a long tail that tapers tipward. Numerous synapomorphies reflecting this general body plan diagnose the basalmost sauropod nodes, and small and large differences in all regions of the skeleton allow recognition of 121 sauropod species (Upchurch et al. 2004) that were globally distributed during most of the Mesozoic Era.

Sauropods are paradoxical animals because they are built on an obvious and memorable body plan but are nonetheless one of the most taxonomically diverse dinosaur groups. This volume is dedicated to exploring sauropod systematics and paleobiology by presenting an up-to-date summary of our knowledge and remaining questions in these areas. While acknowledging and embracing the paradoxical nature of sauropods, we hope to explain how their body plan was constructed, explore its variations, and dispel the myth that it led to evolutionary stagnation and eventual replacement by more "advanced" herbivorous dinosaurs.

SAUROPODS AS MONOLITHS

Sauropods are the largest animals known to have walked the earth, as recorded in numerous

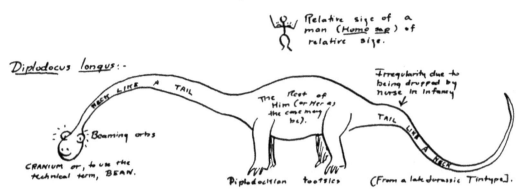

FIGURE I.1. Caricature of *Diplodocus longus* by George Gaylord Simpson (from Simpson 1987).

footprints found around the globe (e.g., Bird 1941, 1944; Hunt et al. 1994). "Monolithic" seems an apt descriptor of both their size and their tight adherence to the body plan described above, of which there are few reversals observed during the nearly 150 million years of sauropod evolution.

The recently published second edition of *The Dinosauria* (Weishampel et al. 2004) contains 22 chapters focused on "dinosaur systematics." The partitioning of chapters and number of pages devoted to each offer insight into specialists' perception of different dinosaur groups as well as the attention each has been given historically (table I.1). Not surprisingly, theropods have garnered the most attention, with their diversity (282 species) partitioned among nine chapters and 185 pages. Thyreophorans (68 species, 58 pages), ornithopods (107 species, 71 pages), and marginocephalians (56 species, 53 pages) are partitioned into three chapters each. Prosauropods (23 species, 27 pages) and sauropods (121 species, 64 pages) are the only major dinosaur groups represented by single chapters in *The Dinosauria 2*. This allotment may be justified for Prosauropoda, which is the smallest of the groups mentioned yet the only to receive more pages than its species count. In contrast, Sauropoda is the second most diverse dinosaur group, representing 18%, or nearly one-fifth, of the 661 recognized dinosaur species. Together, sauropods and theropods encompass >60% of dinosaur species diversity. That sauropods are lumped into a single chapter and given approximately half a page per species suggests that they are at least perceived as monolithic by dinosaur specialists. Is there any justification or explanation for this characterization?

SAUROPOD FOSSIL RECORD

The sauropod fossil record itself may be responsible for the monolithic perception of sauropods. Sauropods first appear in the fossil record during the Late Triassic, during which there are currently several candidate earliest-appearing sauropods. Together, these body fossils and ichnofossils suggest a late Carnian or Norian origin for the group (summarized in Wilson 2005). Possible Carnian sauropods include *Blikanasaurus* (Yates 2003, 2004; Yates and Kitching 2003; Upchurch et al. 2004) and the Portezuelo Formation trackmaker (Marsicano and Barredo 2004); probable Norian sauropods include *Antenonitrus* (Yates and Kitching 2003) and the *Tetrasauropus* trackmaker (Lockley et al. 2001).

The notion of Triassic sauropods is new to this century but was expected based on the first appearance of other saurischians (e.g., Upchurch, 1995; Wilson and Sereno 1998). Prior to this, a lengthy ghost lineage implied by these relationships preceded the exclusively post-Triassic sauropod record (fig. I.2), which began with the fragmentary remains of the Lower Jurassic sauropods *Vulcanodon* (Raath 1972) and *Barapasaurus* (Jain et al. 1975) and the complete remains of the Middle Jurassic *Shunosaurus* (Zhang 1988; Chatterjee and Zheng 2003).

TABLE I.1
Species Counts and Pages Devoted to Systematics in The Dinosauria, 2nd Edition

CHAPTER		PAGES	SPECIES	PAGES/SPECIES
1	Basal Saurischia	22	11	2.00
	Theropoda			
2	Ceratosauria	24	33	0.73
3	Basal Tetanurae	40	74	0.54
4	Tyrannosauroidea	26	19	1.37
5	Ornithomimosauria	14	12	1.17
6	Therizinosauroidea	14	12	1.17
7	Oviraptorosauria	19	18	1.06
8	Troodontidae	12	9	1.33
9	Dromaeosauridae	14	20	0.70
10	Basal Avialae	22	74	0.30
11	Prosauropoda	27	23	1.17
12	Sauropoda	64	121	0.53
13	Basal Ornithischia	10	4	2.50
	Thyreopoda			
14	Basal Thyreophora	8	5	1.60
15	Stegosauria	20	17	1.18
16	Ankylosauria	30	46	0.65
	Ornithopoda			
17	Basal Ornithopoda	20	24	0.83
18	Basal Iguanodontia	25	38	0.66
19	Hadrosauridae	26	45	0.58
	Marginocephalia			
20	Pachycephalosauria	14	17	0.82
21	Basal Ceratopsia	16	21	0.76
22	Ceratopsidae	23	18	1.28
	Total Dinosauria	490	661	—
	Sauropoda contribution	13.06%	18.31%	—

NOTE: Introductory chapters that contained no species descriptions (i.e., "Saurischia," "Ornithischia") are not included in the page tally.

Together, these taxa indicate that the sauropod body plan was constructed early in their evolutionary history and that the earliest sauropods resemble later sauropods more than they do sauropod outgroups (such as prosauropods). That is, until recently, few fossils have been available to document the transition between sauropods and their hypothesized sister-taxa. This fact of the fossil record was borne out in lower-level phylogenetic analyses of sauropod dinosaurs. Upchurch (1998) recorded 60 synapomorphies diagnosing nodes basal to Eusauropoda, and Wilson (2002) identified 74 synapomorphies arising at Sauropoda and Eusauropoda. These two analyses independently recognize that synapomorphies appearing at basal sauropod nodes represent 26% of *all* synapomorphies identified. Few, if any, dinosaur groups have such a base-heavy distribution of synapomorphies. Thus the sauropod fossil

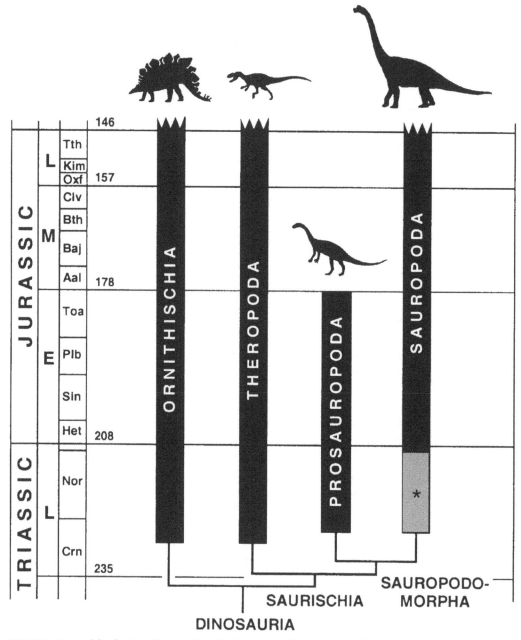

FIGURE I.2. Temporal distribution and relationships of major lineages of dinosaurs during the Triassic and Jurassic. The vertical bar representing sauropod lineage duration is divided into a black section (pre-2000) and a gray section with asterisk (post-2000). Icons from Wilson and Sereno (1998) and Sereno (1999); timescale based on Harland et al. (1990). (Modified from Wilson 2002:fig. 2.)

record has in part led the monolithic depiction of sauropods.

As new Triassic sauropods are discovered, many of the synapomorphies identified by Upchurch (1998) and Wilson (2002) will likely diffuse stemward and articulate the transition from bipedal, short-necked, nonspecialized herbivores to quadrupedal, long-necked, herbivorous monoliths.

SAUROPOD SYSTEMATIC RECORD

Romer (1968:137–138) bookended his discussion of sauropods in his *Notes and Comments on Vertebrate Paleontology* with the following laments: "A proper classification of the great amphibious sauropods has been the despair of every one working on the group," and "It will be a long time, if ever, before we obtain a valid, comprehensive picture of sauropod classification and phylogeny." The perception that sauropod interrelationships were intractable has contributed to their monolithic status, perhaps even more than do the circumstances of their fossil record. How can it be otherwise if subgroups are not identifiable?

Although a "valid" and "comprehensive" sauropod phylogeny was not available during Romer's time, steps had already been initiated to resolve the relationships of constituent taxa. Sauropod interrelationships were resolved in stages, beginning with early classifications by Marsh (1895, 1898), Janensch (1929), and Huene (1932), followed much later by Bonaparte's (1986a, 1986b) recognition of "eo" and "neo" sauropods, McIntosh's (1989, 1990a, 1990b) delineation of numerous sauropod families, and the relatively recent use of cladistic techniques (Russell and Zheng 1993; Calvo and Salgado 1995; Upchurch 1995, 1998; Wilson and Sereno 1998; Wilson 2002; Upchurch et al. 2004). Together, these and other studies have gained substantial consensus on the interrelationships of sauropods, although several contentious areas remain (see chapters by Wilson and Curry Rogers).

THE PACE OF DISCOVERY

Romer (1968:137–138) suggested that the fragmentary nature of many sauropod taxa contributed to their unresolved interrelationships: "The reasons for our difficulties are apparent. Few complete skeletons exist; feet and skulls are rare; many of the numerous described forms are based on fragmentary material." This was certainly the case, and the improvement in our understanding of sauropod phylogeny is the result of an improved sauropod fossil record. Recent key discoveries, in stratigraphic order, include the discovery of primitive sauropods in Africa (e.g., Charig et al. 1965; Raath 1972; Yates and Kitching 2003; Allain et al. 2004) and India (e.g., Jain et al. 1975; Yadagiri 2001; Middle Jurassic sauropods from China (Zhang 1988; He et al. 1988, 1998; Tang et al. 2001; Ouyang and Ye 2002) and Argentina (Bonaparte 1986b); and well preserved sauropod skeletons from the Cretaceous of Asia (Borsuk-Bialynicka 1977; Suteethorn et al. 1995), South America (Salgado and Bonaparte 1991; Calvo and Salgado 1995; González Riga 2003; Martínez et al. 2004), India (Jain and Bandyopadhyay 1997), Africa (Jacobs et al. 1993; Sereno et al. 1999), and Madagascar (Curry Rogers and Forster 2001).

The history of discovery of the 121 sauropod species recognized as valid by Upchurch et al. (2004) is summarized in figure 1.3. Prior to the "Dinosaur Renaissance," inaugurated with Ostrom's (1969) description of *Deinonychus*, sauropod discoveries rarely topped more than a handful every five years. A notable outlier, however, is the burst of sauropod discoveries and descriptions in the late 1870s that coincides with peak Cope–Marsh activity. Following 1969, however, sauropod discoveries always exceed five per five years, steadily increasing through the late 1990s, when 25 sauropods were named. The subsequent drop in sauropod discoveries in figure 3 is an artifact of only counting up to the year 2002. Projecting to 2005 based on these numbers suggests an excess of 10 sauropods for this interval. This relatively sudden surge in sauropod discoveries is more striking when we consider that 50 sauropod species were named since the first edition of *The Dinosauria*, and that one-third of all sauropod species have been named *since* the first cladistic analyses of Sauropoda (Calvo and Salgado 1995; Upchurch 1995).

Both the steady improvement of the sauropod fossil record and the intensified interest in and resolution of sauropod phylogeny are beginning to differentiate the group into more manageable pieces, such as basal sauropods, macronarians, and diplodocoids. Both of these advances are relatively recent—cladistic analyses and escalation in sauropod

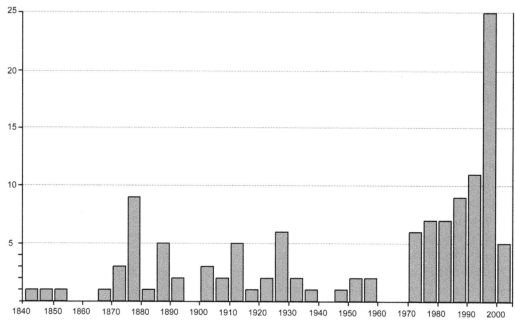

FIGURE I.3. Sauropod species named since the description of *Cetiosaurus* (Owen, 1841) and recognized as valid by Upchurch et al. (2004).

discoveries postdate the first edition of *The Dinosauria* (1990) but predate *The Dinosauria 2* (2004).

SAUROPODS LIVING LARGE

Body size is the most recognizable characteristic of sauropods, and can be expected to have influenced all aspects of their biology (Peters 1983; LaBarbera 1989). The largest sauropods are estimated to have reached adult body masses of 40 to 70 metric tons or more (Peczkis 1994:appendix), an upper bound reached independently within multiple sauropod lineages (Diplodocoidea, "*Seismosaurus*"; Macronaria, *Brachiosaurus*; Titanosauria, *Argentinosaurus*). The smallest sauropods (e.g., *Magyarosaurus*, *Saltasaurus*) may have weighed between 1.5 and 3 metric tons (Erickson et al. 2001) and represent one of the few phylogenetic decreases in body size among dinosaurs.

Sauropods appear to have attained large adult body size by rapid post-hatching growth (e.g., Rimblot-Baly et al. 1995; Curry 1999; Sander 2000; Erickson et al. 2001; Sander and Tückmantel 2003) rather than by the slow, prolonged growth strategy common in other reptiles (e.g., Enlow and Brown 1956, 1957; Case 1978a; Francillon-Vieillot et al. 1990). Interestingly, although other vertebrates attained comparable body sizes, sauropods may be the only giants whose young hatch from eggs. Even the largest of adult sauropods began as hatchlings measuring only one meter long and weighing less than 10 kg (Chiappe et al. 1998, 2001), a range of ontogenetic size exceeding that for any other dinosaur lineage. In contrast, other large-bodied vertebrates bore live young, including chondrichthyans (sharks [Dulvy and Reynolds 1997]), mammals (whales [Clapham et al. 1999]), and marine reptiles such as plesiosaurs (Cheng et al. 2004), ichthyosaurs (Böttcher 1990; Maxwell and Caldwell 2003), and mosasaurs (Caldwell and Lee 2001).

The enormity of sauropod dinosaurs has cast a shadow over studies of their paleobiology. Struggling to find adequate descriptors for sauropod size—"ponderous," "behemoth," "enormous," "stupendous," and "massive" being a few—many early paleontologists assumed that sauropods could not support their body weight on

FIGURE I.4. Oliver P. Hay's 1910 reconstruction of "The Form and Attitudes of *Diplodocus*." Hay's interpretation of sauropod paleobiology—lazing on beaches of ancient rivers, lizardlike in their stance and habit—was common until quite recently.

land (e.g., Owen 1875; Osborn 1899; Hatcher 1901). Although their anatomy was trumpeted as a "marvel of construction... a mechanical triumph for great size, lightness, and strength" (Osborn 1899:213), early life reconstructions depicted sauropods as large, lumbering, and near or up to their necks in ancient swamps (fig. I.4). Even the first-described sauropod footprints were initially interpreted as having been made in an aquatic environment (Bird 1941, 1944). This perception of sauropods as unwieldy, archaic herbivores relegated to evolutionary backwaters was prominent as recently as the early 1990s:

> Their large sizes, small heads, simple teeth, and tiny brains served them well for millions of years. But in the Cretaceous, more progressive, large-headed, larger-brained dinosaurs appeared (the ornithopods and marginocephalians) and vegetation changed.... The old giants retreated to southern continents, where the newcomers did not flourish.
>
> (Dodson 1991:34).

Consequently, Cretaceous Gondwanan sauropods were viewed as Jurassic relics rather than thriving lineages (Gilmore 1946; Lucas and Hunt 1989; Dodson 1991).

Ironically, some of these interpretations themselves can be looked on as holdovers from a previous era. Studies by Walter Coombs (1975, 1978) and Robert Bakker (1968, 1971a, 1971b, 1986) reinvented sauropods as dynamic, terrestrial vertebrates that might have been agile enough to feed tripodally, use their tails as weapons, and generate their own body heat via high food consumption capabilities (fig. I.5). These interpretations challenged the prevailing perception of sauropods as archaic dinosaurs destined for extinction and helped underscore how little we know about sauropod paleobiology. The recent surge in sauropod discoveries around the world, combined with the taxonomic revision of fragmentary genera and the first testable hypotheses of relationship, provide the requisite framework for delving deeper into these questions of sauropod paleobiology. In this volume, we attempt to gain new understanding of "nature's grandest extravagances" (Dodson 1991:34).

FIGURE I.5. John Gurche's dramatically posed *Barosaurus* protecting its young from an attacking *Allosaurus*. This depiction casts sauropods as active animals capable of rearing up on their hind legs.

SAUROPOD EVOLUTION AND PALEOBIOLOGY

The Sauropods: Evolution and Paleobiology opens with three chapters aimed at delimiting the evolutionary history and diversity of Sauropoda. These chapters provide a phylogenetic context for the subsequent chapters on sauropod paleobiology. Chapter 1 highlights our current view of sauropod phylogeny and concludes that sauropods were successful in terms of their geographic and temporal distributions, biomass, morphological complexity, and diversity at both higher and lower levels. Our understanding of sauropod phylogeny is particularly lucid with regard to the Diplodocoidea and Macronaria and the cranial specializations in each of these major groups. Wilson highlights the difficulties in tracing two major parts of the sauropod phylogeny—even the earliest known sauropods have the features of the axial and appendicular skeletons that characterize all sauropods, implying a 10 million- to 15 million-year ghost lineage. Similarly, the preponderance of partial material and relative conservatism of the postcranial skeleton has obfuscated the other end of sauropod evolution, that of Rebbachisauridae and Titanosauria. Wilson highlights the importance of these taxa for biogeography during the Late Mesozoic as the continents attained their current positions. Similarly, rebacchisaurids and titanosaurs bear dental specializations and were coeval with the burgeoning ornithischian populations in both the northern and the southern hemisphere. The connections between Late Cretaceous survivorship and narrow-crowned dentition provide an interesting opportunity for future work.

In chapter 2, Curry Rogers expands on the work of Wilson and examines the interrelationships among Titanosauria, a significant yet poorly understood sauropod clade. She outlines the current consensus with regard to titanosaur phylogeny and highlights the skeleton of *Rapetosaurus krausei*, the first titanosaur with associated cranial and postcranial remains, as a keystone taxon in current phylogenetic analyses. Her chapter concludes with one of the first detailed analyses of lower-level titanosaur relationships, which helps set the stage for a more detailed analysis of the paleobiology of this unique, derived group of macronarians.

Building on recent phylogenetic analyses, Upchurch and Barrett investigate sauropod "success rate" in chapter 3 by taking a closer look at taxic and phylogenetic estimates of sauropod diversity. Both estimates compare favorably with other estimates of sauropod diversity, indicating that important radiations occurred in the Middle and Late Jurassic, as well as at the end of the Cretaceous. Such results contradict the long-standing view that sauropod diversity reached its zenith in the Jurassic and its nadir in the Cretaceous, and highlights the high diversity of particular sauropod clades late in the history of sauropod evolution. Barrett and Upchurch expand on this analysis of sauropod diversity in chapter 4 by exploring the effects of macroevolutionary mechanisms potentially responsible for these patterns. The authors determine that proposed causes and their effects do not always fit the observed sauropod diversity pattern (i.e., competition between rebacchisaurids and titanosaurs; competition between sauropods and ornithopods). In addition, the authors highlight the potential for coevolution among titanosaurs and angiosperms.

The theme of complex dentitions and herbivory during the last radiation of sauropods is apparent in the description of the unique dental battery present in *Nigersaurus* presented in chapter 5. Sereno and Wilson compare the dental batteries of *Nigersaurus* to those of euornithopods and neoceratopsians, and conclude that divergent functions and uncorrelated progression signal independent causes for the evolution of dental batteries. This contrasts with the view that a single environmental cue (i.e., the evolution of angiosperms) prompted the evolution of complex dental batteries, and that sauropods and ornithischians were in direct competition for resources during the last stages of the Mesozoic.

In chapter 6, Stevens and Parrish present and implement their method for three-dimensional reconstructions of sauropod skeletons, focusing on the pose of the neck and its implications for sauropod herbivory. Their reconstructed feeding envelopes challenge the view that all sauropods were high browsers. In contrast to the giraffelike reconstructions of *Brachiosaurus*, or the tripodal stance that is so often depicted in the popular media, Stevens and Parrish identify most sauropods as medium to low browsers. In their view, sauropod feeding is constrained not only by the dentition, but also by the axial flexibility and forage availability and abundance.

Although sauropod vertebrae have long been touted as being specialized for reducing weight while providing strength, the relationship has not been systematically analyzed and the implications of potential weight-reducing methods qualified. In chapter 7, Wedel investigates the axial skeleton from the perspective of pneumaticity and "efficiency of design." Concluding that most eusauropod vertebrae were pneumatic, Wedel estimates that these vertebrae are in most cases between 50% and 60% air. The effects of pneumaticity might reduce sauropod mass estimates by as much as 10%, and provide some interesting new possibilities for interpreting sauropod physiology.

Chapters 8 and 9 focus on sauropod limbs and their signficance for interpreting locomotor capabilities in sauropods. In chapter 8, Carrano utilizes a phylogenetic framework to address the morphological and functional diversity observed in the secondarily quadrupedal sauropods. Without question, the appendicular morphology of sauropods is constrained by being big in a terrestrial environment. Carrano concludes that sauropods exhibit considerable morphological diversity associated with varying locomotor patterns. For example, titanosaurs can be characterized in part by their wide-gauge gaits and the coeval development of reduced body size in some clades, potential tripodality, and so forth. As outlined by Wright in chapter 9, sauropod trackways have provided widely divergent views of sauropod habits during the history of discovery, with early tracks thought to demonstrate that sauropods spent more time wading than walking in more terrestrial settings (Bird 1939, 1941, 1944). Wright outlines the history of discovery of sauropod tracks, summarizes our understanding of trackmakers, and critiques the ichnological evidence that sauropods traveled in organized herds.

Sauropod size and reproduction are discussed in chapter 10. Chiappe et al. document the first discoveries of identifiable sauropod embryos and nests—the fantastic accumulation of fossils from Auca Mahuevo, Argentina. This site provides definitive evidence that sauropods were egg-laying reptiles. Chiappe et al. analyze the taphonomy and morphology of this amazing assemblage of sauropod eggs, providing us with a clearer understanding of how sauropods reproduced and what hatchlings looked like, an example of nesting structure, and the first unspeculative interpretations of sauropod nesting behavior.

In chapter 11, Curry Rogers and Erickson survey sauropod growth rates from the perspectives of bone histology, long bone growth, and developmental mass. Instead of the century-long ontogenies predicted by early workers extrapolating from reptilian growth rates for the large-bodied sauropods (e.g., Case 1978a, 1978b), Curry Rogers and Erickson present a dramatically different view of the ontogeny for *Apatosaurus*. Growth rates on par with those of some of the largest living vertebrates were normal for sauropods, and development of a characteristic trend for dinosaur growth allows predictions of growth rates for the largest sauropods ever known. They suggest that sauropods reached their enormous adult sizes quickly—likely within 15 years. To compare—at age 5, an African elephant is only ~1 metric ton; at the same age, *Apatosaurus* might have been closer to 20 metric tons.

FUTURE DIRECTIONS

New sauropods will continue to be discovered in Mesozoic strata from around the world, despite the improbability of the preservation of their enormous bodies and fragile skulls and axial bones. These future discoveries will surely remedy some of the temporal and geographic gaps in our current record of Sauropoda and reveal anatomical surprises that will revise our estimates of their genealogy. *The Sauropods: Evolution and Paleobiology* not only documents what is currently known about sauropod evolutionary history, but also highlights deficiencies in our understanding. In that sense, we hope this volume sparks interest, provokes questions, and provides fresh ground for continued research on sauropod dinosaurs.

Several gaps in our understanding of sauropod paleobiology are not addressed in this volume. Many of these are directly related to the sauropod fossil record, which documents as few as 20 genera known from better than 90% of the skeleton (Upchurch et al. 2004). One of the most conspicuous deficiencies is the absence of transitional Triassic forms bridging the morphological gap between sauropods and their closest relatives, prosauropods, which obfuscates the origin of the sauropod body plan. Other poorly sampled horizons include the Middle Jurassic, during which nearly all the main neosauropod lineages are hypothesized to have evolved, and the Early Cretaceous, when rebbachisaurid and titanosaur sauropods underwent interesting changes in herbivory and locomotion that were manifest in later Cretaceous forms. Geographical biases in the sauropod fossil record also persist. Australia and Antarctica stand out as southern continental landmasses that have produced only a few, fragmentary sauropod taxa. Similarly, with the exception of a few well-preserved taxa from North Africa, the continent has only provided a glimpse of later-stage sauropod evolution. North America, despite its rich Late Jurassic sauropod record, hints of an as-yet-undiscovered Cretaceous sauropod fauna. These large-scale evolutionary questions are not the only ones impeded by an imperfect fossil record; straightforward anatomical questions also persist. Examples include the arrangement and diversity of body armor (osteoderms), the configuration and embryological identity of carpal bones, and the nature and arrangement of gastralia and clavicles.

Other questions are not necessarily dependent on discoveries of new and better fossils, but on our ability to interpret those already collected. Although we have made inroads, summarized in several chapters in this volume, sauropod life history stands out as a complex issue knotting physiology, development, ecology, and behavior. Basic questions remain, including the trajectory of body size change within the phylogenetic, temporal, and geographic distribution of sauropods; the sequence of fusion of skeletal elements (particularly in the vertebral column) throughout ontogeny; the masticatory forces producing wear facets on teeth; the composition of sauropod diets; and the relative roles of oral and gastric maceration of plant material. More challenging still are questions of sauropod thermal biology and energetics, reproductive behavior, and social interactions. These and other questions will propel future sauropod research.

LITERATURE CITED

Allain, R., Aquesbi, N., Dejax, J., Meyer, C., Monbaron, M., Montenat, C., Richir, P., Rochdy, M., Russell, D., and Taquet, P. 2004. A basal sauropod dinosaur from the Early Cretaceous of Morocco. *C. R. Palovol.* 3: 199–208.

Bakker, R.T. 1968. The superiority of dinosaurs. Discovery 3: 11–22.

———. 1971a. The ecology of brontosaurs. Nature 229: 172–174.

———. 1971b. Dinosaur physiology and the origin of mammals. Evolution 25: 636–658.

———. 1986. Dinosaur Heresies. William Morrow, New York.

Bird, R.T. 1939. Thunder in his footsteps. Nat. Hist. 43: 254–261.

———. 1941. A dinosaur walks into the museum. Nat. Hist. 47: 74–81.

———. 1944. Did *Brontosaurus* ever walk on land? Nat. Hist. 53: 61–67.

Bonaparte, J.F. 1986a. The early radiation and phylogenetic relationships of sauropod dinosaurs, based on vertebral anatomy. *In*: K. The Beginning of the Age of Dinosaurs. Cambridge University Press, Cambridge. Pp. 247–258.

———. 1986b. Les dinosaures (Carnosaures, Allosauridés, Sauropodes, Cétiosauridés) du Jurassique moyen de Cerro Cóndor (Chubut, Argentina). Ann. Paléontol. 72: 325–386.

Borsuk-Bialynicka, M. 1977. A new camarasaurid sauropod *Opisthocoelicaudia skarzynskii*, gen. n., sp. n. from the Upper Cretaceous of Mongolia. Palaeontol. Polonica 37: 1–64.

Böttcher, R. 1990. Neue Erkenntnisse über die Fortpflanzungsbiologie der Ichthyosaurier (Reptilia). Stuttgarter Beiträge Naturkunde (Ser B) 164: 13–51.

Caldwell, M.W., and Lee, M.S.Y. 2001. Live birth in Cretaceous marine lizards (mosasaurids). Proc. Roy. Soc. London B 268: 2397–2401.

Calvo, J.O., and Salgado, L. 1995. *Rebbachisaurus tessonei*, sp. nov. a new Sauropoda from the Albian-Cenomanian of Argentina; New evidence on the origin of Diplodocidae. GAIA 11: 13–33.

Case, T.J. 1978a. On the evolution and adaptive significance of postnatal growth rates in the terrestrial vertebrates. Q. Rev. Biol. 53: 243–282.

———. 1978b. Speculations on the growth rate and reproduction of some dinosaurs. Paleobiology 4: 320–328.

Charig, A.J., Attridge, J., and Crompton, A.W. 1965. On the origin of the sauropods and the classification of the Saurischia. Proc. Linn. Soc. London 176: 197–221.

Chatterjee, S., and Zheng, Z. 2003. Cranial anatomy of *Shunosaurus*, a basal sauropod dinosaur from the Middle Jurassic of China. Zool. J. Linn. Soc. 136: 145–169.

Cheng, Y.-N., Wu, X.-C., and Ji, Q. 2004. Triassic marine reptiles gave birth to live young. Nature 432: 383–386.

Chiappe, L.M., Coria, R.A., Dingus, L., Jackson, F., Chinsamy, A., and Fox, M. 1998. Sauropod dinosaur embryos from the Late Cretaceous of Patagonia. Nature 396: 258–261.

Chiappe, L.M., Salgado, L., and Coria, R.A. 2001. Embryonic skulls of titanosaur sauropod dinosaurs. Science 293: 2444–2446.

Clapham, P.J., Wetmore, S.E., Smith, T.D., and Mead, J.G. 1999. Length at birth and at independence in humpback whales. J. Cetacean Res. Manage. 1: 141–146.

Coombs, W.P., Jr. 1975. Sauropod habits and habitats. Palaeogeogr. Palaeoclimatol. Palaeoecol. 17: 1–33.

———. 1978. Theoretical aspects of cursorial adaptations in dinosaurs. Q. Rev. Biol. 53: 393–418.

Curry, K.A. 1999. Ontogenetic histology of *Apatosaurus* (Dinosauria: Sauropoda): New insights on growth rates and longevity. J. Vertebr. Paleontol. 19: 654–665.

Curry Rogers K.A., and Forster, C. 2001. The last of the dinosaur titans: a new sauropod from Madagascar. Nature 412: 530–534.

Dodson, P. 1991. Lifestyles of the huge and famous. Nat. Hist. 100: 30–34.

Dulvy, N.K., and Reynolds, J.D. 1997. Evolutionary transitions among egg-laying, live-bearing and maternal inputs in sharks and rays. Proc. Roy. Soc. Lond. B 264: 1309–1315.

Enlow, D.H., and Brown, S.O. 1956. A comparative histological study of fossil and recent bone tissue (Part 1). Tex. J. Sci. 8: 405–443.

———. 1957. A comparative histological study of fossil and recent bone tissue (Part 2).Tex. J. Sci. 9: 186–214.

Erickson, G., Curry Rogers, K., and Yerby, S. 2001. Dinosaurian growth patterns and rapid avian growth rates. Nature 412: 429–433.

Francillon-Vieillot, H., de Buffrénil, V., Castanet, J., Géraudie, J., Meunier, F.J., Sire, J.Y., Zylberberg, L., and de Ricqlès, A. 1990. Microstructure and mineralization of vertebrate skeletal tissues. In: Skeletal Biomineralization: Patterns, Processes, and Evolutionary Trends, Vol. 1. Van Nostrand Reinhold, New York. Pp. 473–530.

Gilmore, C.W. 1946. Reptilian fauna of the North Horn Formation of central Utah. U.S. Geol. Surv. Prof. Paper 210C: 1–52.

González Riga, B. 2003. A new titanosaur (Dinosauria, Sauropoda) from the Upper Cretaceous of Mendoza Province, Argentina. Ameghiniana 40: 155–172.

Harland, W.B., Armstrong, R.L., Cox, A.V., Craig, L.E., Smith, A.G., and Smith, D.G. 1990. A Geologic Time Scale 1989. Cambridge University Press, Cambridge, 263 pp.

Hatcher, J.B. 1901. *Diplodocus* (Marsh): Its osteology, taxonomy, and probable habits, with a restoration of the skeleton. Mem. Carnegie Mus. 1: 1–63.

He, X.-L., Li, K., and Cai, K.-J. 1988. [The Middle Jurassic dinosaur fauna from Dashanpu, Zigong, Sichuan. Vol. IV. Sauropod Dinosaurs (2) *Omeisaurus tianfuensis*]. Sichuan Scientific and Technological Publishing House, Chengdu. (In Chinese with English abstract.)

He, X., Wang, C., Liu, S., Zhou, F., Liu, T., Cai, and Dai, B. 1998. [A new sauropod dinosaur from the Early Jurassic in Gongxian County, South Sichuan]. Acta Geologica Sichuan 18: 1–6. (In Chinese with English abstract.)

Huene, F. von. 1932. Die fossil Reptil-Ordnung Saurischia, ihre Entwicklung und Geschichte. Monograph. Geol. Palaeontol. 4: 1–361.

Hunt, A.P., Lockley, M.G., Lucas, S.G., and Meyer, C.A. 1994. The global sauropod fossil record. GAIA 10: 261–279.

Jacobs, L.L., Winkler, D.A., Downs, W.R., and Gomani, E.M. 1993. New material of an Early Cretaceous titanosaurid sauropod dinosaur from Malawi. Palaeontology 36: 523–534.

Jain, S.L., and Bandyopadhyay, S. 1997. New titanosaurid (Dinosauria: Sauropoda) from the Late Cretaceous of central India. J. Vertebr. Paleontol. 17: 114–136.

Jain, S.L., Kutty, T.S., Roy-Chowdhury, T., and Chatterjee, S. 1975. The sauropod dinosaur from the Lower Jurassic Kota Formation of India. Proc. Roy. Soc. London B 188: 221–228.

Janensch, W. 1929. Material und Formengehalt der Sauropoden in der Ausbeute der Tendaguru-Expedition. Palaeontographica 2 (Suppl. 7): 1–34.

LaBarbera, M. 1989. Analyzing body size as a factor in ecology and evolution. Annu. Rev. Ecol. System. 20: 97–117.

Lockley, M.G., Wright, J.L., Hunt, A.G., and Lucas, S.G. 2001. The Late Triassic sauropod track record comes into focus: old legacies and new paradigms. New Mexico Geological Society Guidebook, 52nd Field Conference, Geology of the Llano Estacado. Pp. 181–190.

Lucas, S.G., and Hunt, A.P. 1989. *Alamosaurus* and the sauropod hiatus in the Cretaceous of the North American Western Interior. In: Farlow, J.O., (ed.). Paleobiology of the Dinosaurs. Geolog. Soc. Am. Special Paper 238: 75–85.

Marsh, O.C. 1895. On the affinities and classification of the Dinosaurian reptiles. Am. J. Sci. 21: 417–423.

———. 1898. On the families of sauropodous dinosaurs. Am. J. Sci. 6(4): 487–488.

Marsicano, C.A., and Barredo, S.P. 2004. A Triassic tetrapod footprint assemblage from southern South America: palaeobiogeographical and evolutionary implications. Palaeogeogr. Palaeoclimatol. Palaeoecol. 203: 313–335.

Martínez, R.D., Giménez, O., Rodríguez, J., Luna, M., and Lamanna, M.C. 2004. An articulated specimen of the basal titanosaurian (Dinosauria:

Sauropoda) *Epachthosaurus sciuttoi* from the early Late Cretaceous Bajo Barreal Formation of Chubut Province, Argentina. J. Vertebr. Paleontol. 24: 107–120.

Maxwell, E. E., and Caldwell, M. W. 2003. First record of live birth in Cretaceous ichthyosaurs: closing an 80 million year gap. Proc. Roy. Soc. London B 270: 104–107.

McIntosh, J. S. 1989. The sauropod dinosaurs: a brief survey. Pp. In: Padian, K., and Chure, D. J. (eds.). The Age of Dinosaurs. Short Courses in Paleontology No. 2. University of Tennessee, Knoxville. Pp. 85–99.

———. 1990a. Sauropoda. In: Weishampel, D. B., Dodson, P., and Osmólska, H. (eds.). The Dinosauria. University of California Press, Berkeley. Pp. 345–401.

———. 1990b. Species determination in sauropod dinosaurs with tentative suggestions for their classification. *In*: Carpenter, K., and Currie, P. J. (eds.). Dinosaur Systematics: Approaches and Perspectives. Cambridge University Press, Cambridge.

Osborn, H. F. 1899. A skeleton of *Diplodocus*. Mem. Am. Mus. Nat. Hist. 1: 191–214.

Ostrom J. H. 1969. Osteology of *Deinonychus antirrhopus*, an unusual theropod from the Lower Cretaceous of Montana. Bull. Peabody Mus. Nat. Hist. 30: 1–165.

Ouyang, H., and Ye, Y. 2002. [The first mamenchisaurian skeleton with complete skull, *Mamenchisaurus youngi*]. Sichuan Science and Technology Publishing House, Chengdu. (In Chinese with extended English abstract.)

Owen, R. 1875. Monographs of the fossil Reptilia of the Mesozoic formations (Pt. II) (genera *Bothriospondylus*, *Cetiosaurus*, *Omosaurus*). Palaeontogr. Soc. Monogr. 29: 15–93.

Peczkis, J. 1994. Implications of body-mass estimates for dinosaurs. Journal of Vertebrate Paleontology 14: 520–533.

Peters, R. H. 1983. The ecological implications of body size. Cambridge University Press, Cambridge/shill New York.

Raath, M. 1972. Fossil vertebrate studies in Rhodesia: a new dinosaur (Reptilia, Saurischia) from the near the Trias-Jurassic boundary. Arnoldia 30: 1–37.

Rimblot-Baly, F., de Ricqlès, A., and Zylberberg, L. 1995. Analyse paléohistologique d'une série de croissance partielle chez *Lapparentosaurus madagascariensis* (Jurassique Moyen): Essai sur la dynamique de croissance d'un dinosaure sauropode. Ann. Paléontolo. 81: 49–86.

Romer, A. S. 1968. Notes and Comments on Vertebrate Paleontology. University of Chicago Press, Chicago. 304 pp.

Russell, D. A., and Zheng, Z. 1993. A large mamenchisaurid from the Junggar Basin, Xinjiang, People's Republic of China. Can. J. Earth Sci. 30: 2082–2095.

Salgado, L. and Bonaparte, J. F. 1991. Un nuevo sauropodo dicraeosauridae, *Amargasaurus cazaui*, gen. et. sp. nov., de la Formacion La Amarga, Neocomiano de la Provincia del Neuquen, Argentina, Ameghiniana 28: 333–346.

Sander, P. M. 2000. Longbone histology of the Tendaguru sauropods: implications for growth and biology. Paleobiology 26: 466–488.

Sander, P. M., and Tückmantel, C. 2003. Bone lamina thickness, bone apposition rates, and age estimates in sauropod humeri and femora. Paläontol. Z. 77(1): 139–150.

Sereno, P. C. 1999. The evolution of dinosaurs. Science 284: 2137–2147.

Sereno P C., Beck, A. L., Dutheil, D. B., Larsson, H. C. E., Lyon, G. H., Moussa, B., Sadleir, R. W., Sidor, C. A., Varricchio, D. J., Wilson, G. P., and Wilson, J. A. 1999. Cretaceous sauropods from the Sahara and the uneven rate of skeletal evolution among dinosaurs. Science 286: 1342–1347.

Simpson, G. G. 1987. Simple Curiosity: Letters from George Gaylord Simpson to His Family, 1921–1970. Laporte, L. F. (ed.). University of California Press, Berkeley. 340 pp.

Suteethorn, V., Martin, V., Buffetaut, E., Triamwichanon, S., and Chaimanee. 1995. A new dinosaur locality in the Lower Cretaceous of northeastern Thailand. C. R. Acad. Sci. Paris (Ser. IIa) 321: 1041–1047.

Tang, F., Jin, X., Kang, X., and Zhang, G. 2001. *Omeisaurus maoianus*, a complete Sauropoda from Jingyan, Sichuan. Research works of Natural Museum of Zhejiang, China Ocean Press, Beijing. (In Chinese with extended English abstract).

Upchurch, P. 1995. The evolutionary history of sauropod dinosaurs. Philos. Trans. Roy. Soc. London B 349: 365–390.

———. 1998. The phylogenetic relationships of sauropod dinosaurs. Zool. J. Linn. Soc. 124: 43–103.

Upchurch, P., Barrett, P., (and) Dodson, P. 2004. Sauropoda. *In*: Weishampel, D. B., P. Dodson, and Osmólska, H. (eds.). The Dinosauria, 2nd ed. University of California Press, Berkeley. pp. 259–324.

Weishampel, D. B., Dodson, P., and Osmólska, H. (eds.). 2004. The Dinosauria, 2nd ed. University of California Press, Berkeley. 861 pp.

Wilson, J. A. 2002. Sauropod dinosaur phylogeny: critique and cladistic analysis. Zool. J. Linn. Soc. 136: 217–276.

Wilson, J. A. 2005. Integrating ichnofossil and body fossil records to estimate locomotor posture and spatiotemporal distribution of early sauropod dinosaurs: a stratocladistic approach. Paleobiology 31: 400–427.

Wilson, J.A., and Sereno, P.C. 1998. Early evolution and higher-level phylogeny of sauropod dinosaurs. Soc. Vertebr. Paleontol. Mem. 5: 1–68 (supplement to J. Vertebr. Paleontol. 18).

Yadagiri, P. 2001. The osteology of *Kotasaurus yamanpalliensis*, a sauropod dinosaur from the Early Jurassic Kota Formation of India. J. Vertebr. Paleontol. 21: 242–252.

Yates, A.M. 2003. A new species of the primitive dinosaur, *Thecodontosaurus* (Saurischia: Sauropodomorpha) and its implications for the systematics of early dinosaurs. J. Syst. Palaeontol. 1: 1–42.

———. 2004. *Anchisaurus polyzelus* Hitchcock: the smallest known sauropod dinosaur and the evolution of gigantism amongst sauropodomorph dinosaurs. Postilla 230: 1–58.

Yates, A.M., and Kitching, J.W. 2003. The earliest known sauropod dinosaur and the first steps towards sauropod evolution. Proc. Roy. Soc. London B 270: 1753–1758.

Zhang, Y. 1988. [The Middle Jurassic dinosaur fauna from Dashanpu, Zigong. China. Volume III. Sauropod dinosaur (I). *Shunosaurus*]. Sichuan Publishing House of Science and Technology, Chengdu. (In Chinese with extended English summary.)

ONE

Overview of Sauropod Phylogeny and Evolution

Jeffrey A. Wilson

SAUROPOD STUDIES FROM OWEN TO THE PRESENT

This year marks the one hundred sixty-fourth anniversary of Richard Owen's (1841) description of the first sauropod—*Cetiosaurus*, the "whale lizard"—on the basis of vertebrae and limb elements from localities across England. Although these remains "had been examined by Cuvier and pronounced to be cetaceous" (Buckland 1841:96), Owen (1841:458–459) demonstrated the saurian affinities of *Cetiosaurus* on the basis of several features, including the absence of epiphyses (growth plates) on caudal vertebrae (fig. 1.1). He differentiated *Cetiosaurus* from other extinct saurians on the basis of its large size and characteristics of its vertebrae (see Upchurch and Martin 2003:215). Owen (1841:462) concluded his initial description with this assessment: "The vertebræ, as well as the bones of the extremities, prove its marine habits . . . the surpassing bulk and strength of the *Cetiosaurus* were probably assigned to it with carnivorous habits, that it might keep in check the Crocodilians and Plesiosauri." He regarded *Cetiosaurus* as a crocodilian by the "form of the long bones" and "the toes being terminated by strong claws" (Owen 1842:102), but this assessment was based on limited anatomical evidence (Owen 1875:27). Key data emerged with the discovery of abundant *Cetiosaurus* bones in Oxfordshire by John Phillips. Thomas Huxley examined this "splendid series of remains" before the publication of Phillips' (1871) monograph and was the first to place *Cetiosaurus* within Dinosauria (Iguanodontidae [Huxley, 1869:35]). Phillips (1871) interpreted *Cetiosaurus* as a plant-eating dinosaur and hypothesized that its limb bones were "suited for walking." He could not rule out the possibility that it was amphibious, however, concluding that it was a "marsh-loving or riverside animal." Owen (1875:27) later acquiesced, referring *Cetiosaurus* to the Dinosauria because of its four sacral vertebrae. He admitted that it may have had some terrestrial capabilities but concluded that *Cetiosaurus* was an estuarine or marine animal based on its "organ of swimming," the tail (Owen 1875:41).

These early interpretations, based on somewhat limited samples, were followed by the discovery of abundant sauropod skeletons in western North America and eastern Africa during

FIGURE 1.1. Sagittally sectioned posterior caudal vertebra of *Cetiosaurus oxoniensis* (OUM-J13697) with label in Owen's hand. This sectioned vertebra was used to demonstrate the lack of epiphyses at either end of the caudal centrum. Scale equals 5 cm.

the late nineteenth and early twentieth centuries. O. C. Marsh and E. D. Cope described numerous new and well represented sauropod genera from the Morrison Formation of the western United States, including the first complete sauropod skull (*Diplodocus* [Marsh 1884]), reconstructions of the skeletons of *Brontosaurus* by Marsh (1883; fig. 1.2) and *Camarasaurus* by Cope (Osborn and Mook, 1921:pl. 82; fig. 1.2), and the first mount of a complete sauropod skeleton (*Diplodocus* [Anonymous 1905]). These discoveries provided the first examples of ontogenetic variation and phylogenetic diversity in sauropods. Later, German expeditions to East Africa (present-day Tanzania) produced sauropod material rivaling that from North America. Janensch and others led field crews at Tendaguru, where they collected more than 235,000 kg of fossils (Maier 2003:105) that represented many new genera described over the course of 50 years (e.g., Janensch, 1914, 1929a, 1935–36, 1950, 1961). The abundance and diversity of sauropod remains unearthed in North America and Africa not only answered many of the queries posed by early sauropod researchers (e.g., dinosaurian affinities and terrestrial habits of sauropods) but also posed new ones. One of the major controversies that extended across the Atlantic surrounded the posture of sauropods. American scientists favored an upright, columnar posture, whereas their German colleagues deemed a lacertilian pose more appropriate (Holland 1910; Desmond 1975). A second question, less controversial but farther-reaching, emerged from the study of these two large collections of sauropod material—How should sauropod diversity be classified?

TRADITIONAL CLASSIFICATION

When Marsh (1878) coined the suborder Sauropoda, it included only a single family, Atlantosauridae. Several of the features Marsh (1878:412) listed in that initial diagnosis of Sauropoda are now well-corroborated synapomorphies for the group or for more exclusive sauropod subgroups that were not identified at the time of Marsh's writing. Marsh invented new families to accommodate the increasing sauropod diversity revealed by new discoveries worldwide (e.g., Atlantosauridae, Morosauridae, Diplodocidae, Pleurocoelidae, Titanosauridae). The formal familial diagnoses for these groups (Marsh 1884, 1895) also recognized features currently considered synapomorphies for sauropod subclades. These diagnoses, however, did not resolve how these groups were interrelated; Marsh's ranked classifications did not function as hypotheses of evolutionary descent.

On the basis of his burgeoning Tendaguru collection, Janensch (1929a) produced a very different classification of Sauropoda that employed higher level groupings. He recognized two principal sauropod subgroups, one with broad, laterally facing nares and spatulate tooth crowns and the other with elevated, dorsally facing nares and narrow tooth crowns. Janensch named these two families Bothrosauropodidae and Homalosauropodidae, and recognized three and four subfamilies within each, respectively. Huene (1956) followed this dichotomous scheme, raising Janensch's subfamilies to familial rank and Janensch's families to "family-group" rank. In contrast to that of Marsh, Janensch's classification could be interpreted as an evolutionary

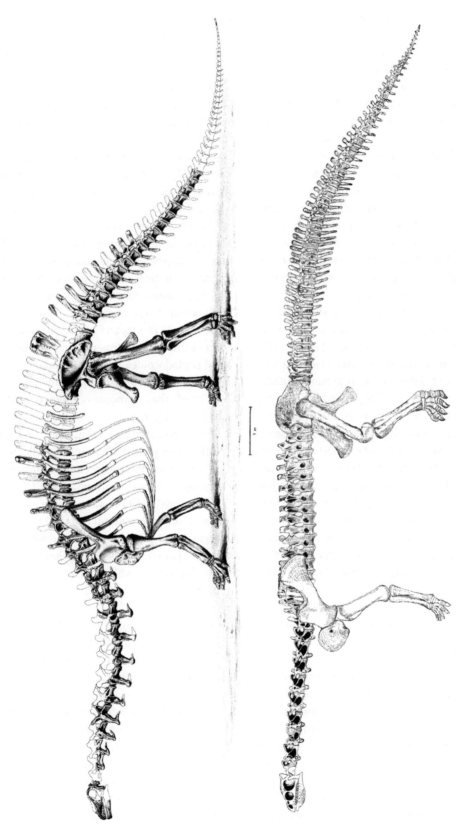

FIGURE 1.2. First reconstruction of sauropod dinosaurs. Top: "*Brontosaurus*" (=*Apatosaurus*) *excelsus* from Marsh (1883). From a lithograph later published in Ostrom and McIntosh (1966). Bottom: *Camarasaurus supremus*, as drawn by Ryder in 1877 under the direction of Cope. Later published in Osborn and Mook (1921:pl. 82).

hypothesis that involved divergence between two lineages differing in tooth morphology.

A dichotomous scheme for higher-level classification of sauropods based on tooth form and narial position became widely accepted, despite nomenclatural differences (Brachiosauridae versus Titanosauridae [Romer 1956, 1966]; Camarasauridae versus Atlantosauridae [Steel 1970]). Other traditional classifications of sauropods, however, follow Marsh in recognizing taxa of equivalent rank (usually families) with no higher-level hierarchical information (e.g., McIntosh 1990). Bonaparte (1986a) also utilized serially ranked families, but he regarded Late Jurassic and younger sauropod families ("Neosauropoda") as advanced relative to older forms ("Eosauropoda").

Numerical methods for assessing phylogenetic relationships in sauropod dinosaurs were first introduced by Gauthier (1986) in his analysis of saurischian dinosaurs. His character choice reflected those cited by previous authors (e.g., Romer 1956; Steel 1970) and his topology consequently conformed to the traditional dichotomy. Since then, more than a dozen cladistic analyses focusing on Sauropoda or its subgroups have appeared (Russell and Zheng 1993; Calvo and Salgado 1995; Upchurch 1995, 1998; Salgado et al. 1997; Wilson and Sereno 1998; Sanz et al. 1999; Curry 2001; Curry Rogers and Forster, 2001; Wilson 2002; Calvo and González Riga 2003; González Riga 2003; Upchurch et al. 2004). Together these analyses have scored 1,964 characters in 229 sauropod taxa, resulting in a variety of phylogenetic hypotheses that are discussed briefly below.

CLADISTIC HYPOTHESES

The main topological disagreement among early cladistic analyses of Sauropoda centered on the relationships of broad- and narrow-crowned sauropods. Upchurch (1995) presented the first large-scale cladistic analysis of sauropods, in which he proposed a slightly modified version of the traditional dichotomy that resolved broad tooth crowns as a primitive feature and narrow tooth crowns as a uniquely derived feature characterizing *Diplodocus*-like taxa (i.e., Diplodocoidea) and titanosaurs. Salgado et al. (1997) were the first to depart from this traditional dichotomy by providing character evidence linking narrow-crowned titanosaurs to the broad-crowned *Brachiosaurus*, rather than to the other narrow-crowned group (Diplodocoidea). This result was corroborated by Wilson and Sereno (1998). In a subsequent analysis, Upchurch (1998) produced a topology that agreed in many ways with those of Salgado et al. (1997) and Wilson and Sereno (1998) but also explored the relationships of genera not treated by either. These three analyses agree on several topological points, including the separation of early-appearing genera (e.g., *Vulcanodon, Shunosaurus, Barapasaurus, Omeisaurus*) from a derived clade called Neosauropoda (Bonaparte 1986a), the identification of the two constituent neosauropod lineages Diplodocoidea (e.g., *Apatosaurus*) and Macronaria (e.g., *Camarasaurus*), and the positioning of the titanosaur lineage within Macronaria (fig. 1.3).

Despite points of agreement, other topological differences persist. The most significant of these centers on the phylogenetic affinities of two groups of Asian sauropods: the Chinese "euhelopodids" (*Shunosaurus," Omeisaurus, Mamenchisaurus, Euhelopus*) and the Mongolian nemegtosaurids (*Nemegtosaurus, Quaesitosaurus*). Upchurch (1995) proposed "Euhelopodidae" as a clade that evolved while China was geographically isolated from Europe from Middle Jurassic until Early Cretaceous times (Russell 1993; Z. Luo 1999; Barrett et al. 2002; Upchurch et al. 2002; Zhou et al. 2003). It evolved independently of its sister-taxon Neosauropoda but was eventually replaced by it during the Cretaceous (Upchurch 1995, 1998). In contrast, Wilson and Sereno (1998) suggested that Chinese sauropods are paraphyletic, with *Omeisaurus* occupying the sister-taxon to Neosauropoda (as in Upchurch 1995, 1998), but *Shunosaurus* positioned basally and *Euhelopus* positioned apically. This result was corroborated by Wilson (2002), whose analysis

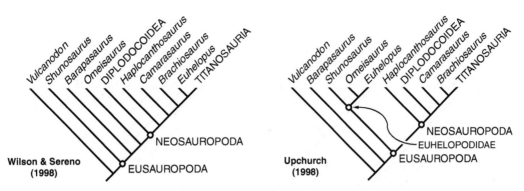

FIGURE 1.3. Hypotheses of the relationships of sauropod dinosaurs based on (left) Wilson and Sereno (1998) and (right) Upchurch (1998).

resolved some of Upchurch's (1998) "euhelopodid" characters as supporting the monophyly of *Omeisaurus* and *Mamenchisaurus* (Omeisauridae). A Templeton test (e.g., Larson 1994) showed that "euhelopodid" paraphyly could not be statistically rejected by the matrix of Upchurch (1998), but the "euhelopodid" monophyly could be rejected by the matrix of Wilson (2002). Thus far, no other analysis has specifically investigated the relationships of these Chinese sauropods, but Upchurch's most recent analysis supported paraphyly of some "euhelopodid" genera (see Upchurch et al. 2004; Barrett and Upchurch, chapter 4).

A second area of disagreement involves the relationships of the isolated skulls of the sauropods *Nemegtosaurus* and *Quaesitosaurus* from the Late Cretaceous of Mongolia. These slender-crowned taxa were originally described as *Dicraeosaurus*-like (Nowinski 1971), a designation consistent with the presumed diplodocid affinities of the Late Jurassic Chinese *Mamenchisaurus* (McIntosh 1990), as well as the conventional division of sauropods into narrow-crowned and broad-crowned groups. More recently, cladistic analyses have produced new hypotheses of relationships for *Nemegtosaurus* and *Quaesitosaurus*, including the monophyletic sister-taxon of diplodocoids (Yu 1993; Upchurch 1998, 1999; Upchurch et al. 2002), basal members of a clade including diplodocoids and titanosaurs (Upchurch 1995), and, most recently, titanosaurs (Salgado et al. 1997; Curry Rogers and Forster 2001; Wilson 2002, 2005a). Although the weight of the evidence is in favor of titanosaur affinities for *Nemegtosaurus* and *Quaesitosaurus*, convergences with diplodocoids are noteworthy (Upchurch 1999; Curry Rogers and Forster 2001; see below).

In addition to areas of disagreement, there are unresolved areas resulting from lack of information. Two such areas involve the origin of sauropods and the diversification of their latest surviving lineage, Titanosauria. Sauropods have long been absent from Triassic rocks, but their two saurischian sister-taxa (Prosauropoda, Theropoda) are found in lowermost Upper Triassic horizons. Recent discoveries of Triassic sauropod body fossils and ichnofossils (see below) have provided the first opportunity to resolve sauropod origins, but additional field and museum research is needed. Renewed interest in titanosaurs, whose interrelationships remain resolved, have been fueled by descriptions of many new discoveries in the field (Curry Rogers, chapter 2). These include the first titanosaur with associated cranial and cranial remains (*Rapetosaurus* Curry Rogers and Forster 2001, 2004), the first embryonic titanosaur remains (Chiappe et al. 1998, 2001; Salgado et al. 2005), and nearly complete associated or articulated postcranial skeletons from South America (*Mendozasaurus* González Riga 2003; *Epachthosaurus* Martínez et al. 2004; *Gondwanatitan* Kellner and Azevedo 1999), Asia (*Phuwiangosaurus* Martin et al. 1994;

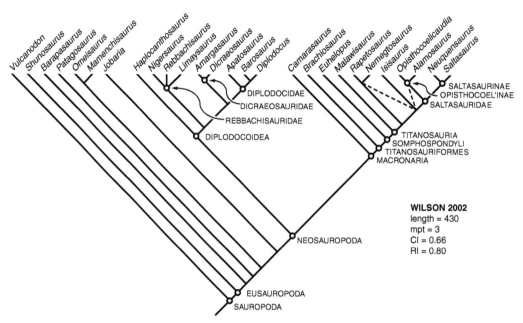

FIGURE 1.4. Phylogenetic relationships of sauropod genera based on Wilson (2002). Dashed lines indicate nodes that are lost in trees two steps longer than the most parsimonious tree. Taxonomy has been updated for *Rayososaurus* (=*Limaysaurus* [Salgado 2004]) and "*Titanosaurus*" *colberti* (=*Isisaurus* [Wilson and Upchurch 2003]).

Tangvayosaurus Allain et al. 1999), India (*Isisaurus* Jain and Bandyopadhyay 1997), Europe (*Lirainosaurus* Sanz et al. 1999; *Ampelosaurus* Le Loeuff 1995, 2003), and Africa (*Malawisaurus* Jacobs et al. 1993; *Paralatitan* Smith et al. 2001). Several analyses have investigated titanosaur phylogeny (most notably Curry [2001] and Curry Rogers and Forster [2001]), and there are several points of agreement among them (Wilson and Upchurch 2003). These preliminary analyses are the first step toward establishing a framework for titanosaur evolutionary history, but at least a dozen valid titanosaur genera have yet to be accommodated by a phylogenetic analysis, in addition to the many undescribed specimens uncovered in recent years.

The topology of Wilson's (2002) analysis of Sauropoda, based on 27 taxa scored for 234 characters, is assumed in this paper (fig. 1.4). Outgroup choice, character descriptions, character coding assumptions, character–taxon matrix, and tree statistics are given by Wilson (2002). Below, the evolutionary events diagnosing several major sauropod clades are discussed. For each event, a set of synapomorphies is presented that has been identified in various analyses (table 1.1). Appendix 1.1 lists each character and its states.

MAJOR EVOLUTIONARY EVENTS IN SAUROPODA AND ITS SUBGROUPS

Sauropoda is a monophyletic group whose body plan (fig. 1.2) is supported by more than 40 synapomorphies, many of which were not lost within the 150 million-year history of the group (McIntosh 1990; Upchurch 1995, 1998; Wilson and Sereno 1998). Modification of this basic architecture, as it pertains to the evolution of herbivory, neck elongation, and locomotion within five clades (Sauropoda, Eusauropoda, Neosauropoda, Diplodocoidea, Macronaria) is explored here. Important to this discussion is the presumed ancestry of Sauropoda, which is not yet agreed on. Whereas most researchers favor a monophyletic Prosauropoda (Sereno 1989; Galton 1990; Wilson and Sereno 1998; Galton and Upchurch 2000, 2004; Benton et al. 2000), recent analyses of sauropodomorph relationships (Yates 2001 2003, 2004; Yates and Kitching

TABLE 1.1.
Synapomorphies for the Five Sauropod Clades Discussed

	CHARACTER NUMBER, BY CLADISTIC ANALYSIS			
	SALGADO ET AL. (1997)	UPCHURCH (1998)	WILSON & SERENO (1998)	WILSON (2002)
Columnar, quadrupedal posture (Sauropoda)				
elongate forelimbs	—	158	1	172
elongate metatarsal V	—	—	15	225
straight limb elements	4	186	1	149
reduction of olecranon	—	161	4	167
femur with eccentric cross-section	—	191	10	198
unossified limb articular surfaces	—	—	—	—
unossified distal carpals	—	164	—	173
unossified distal tarsals (3 & 4)	—	197	13	216
Herbivorous specializations (Eusauropoda)				
tooth rows shortened	—	73	67	66
precise occlusion	—	—	35	67
tooth rows arched	—	59	31	65
teeth overlap	—	—	34	69
enamel wrinkling	—	—	33	71
broad crowns	—	71	32	70
dentary deepens anteriorly	—	57	30	55
Neck elongation (Eusauropoda)				
number of neck vertebrae	5	76	37	80
number of dorsal vertebrae	—	95	70	91
Hindfoot posture (Eusauropoda)				
pes shortened relative to tibia	—	—	50	223
spreading metatarsus	—	—	52	217
metatarsal I broader than II–V	—	—	51	221
pedal phalangeal count reduced	—	200	57	233
metatarsal II broader than III–IV*	—	—	73	224
pedal unguals directed laterally*	—	—	64	228
Reduced ossification of wrist & ankle (Neosauropoda)				
reduction to two carpals	—	163	79	173
astragalus reduced	—	195	85	210
Forefoot posture (*Jobaria* + Neosauropoda)				
bound metacarpus	—	169	80	175
tightly arched metacarpus	—	169	81	176
Herbivorous specializations (Diplodocoidea)				
tooth row restricted anteriorly	—	74	—	66
mandible squared in dorsal view	—	59	—	65
jaw articulation shifted forward	—	?27	—	46, 53
pterygoid flange and adductor fossa shifted forward	—	—	—	37
loss of crown overlap	—	—	—	69
cylindrical tooth crowns	—	70	—	70
enhanced tooth replacement rate	—	—	—	74

TABLE 1.1. (continued)

	CHARACTER NUMBER, BY CLADISTIC ANALYSIS			
	SALGADO ET AL. (1997)	UPCHURCH (1998)	WILSON & SERENO (1998)	WILSON (2002)
Presacral specializations (Flagellicaudata)				
forked neural spines	—	92	106	85, 89
elongate neural spines*	—	—	—	93
number of cervical vertebrae*	—	77	37	80
number of dorsal vertebrae*	—	95	70	91
Tail specializations (Diplodocoidea)				
elongate caudal centra	—	134	—	137
biconvex caudal centra	—	134	—	136
30 or more archless caudal centra*	—	128	—	138
Wide-gauge limb posture (Saltasauridae)				
femur distal condyles beveled	—	—	—	201
eccentric femoral midshaft	—	—	—	198
coracoid quadrangular†	29	153	—	156
scapular blade deflected dorsally†	—	—	—	151
crescentic sternal plates†	26	154	—	158
humeral distal condyles exposed anteriorly	—	—	—	163
humeral distal condyles divided	—	—	—	164
humeral deltopectoral crest expanded	—	—	3	161
prominent olecranon†	—	161	4	167
distal radius expanded transversely†	—	—	—	170
distal tibia expanded transversely†	7	—	—	205
iliac blade directed laterally†	28	172	—	187
femur deflected medially†	19	187	100	199
carpus unossified*	—	165	79	173
manual phalanges absent*	27	—	43	181

NOTE: Character numbers are those employed in four major cladistic analyses of sauropod relationships. Asterisks (*) denote synapomorphies that apply at slightly *less* inclusive nodes; daggers (†) denote synapomorphies that apply at slightly *more* inclusive nodes (see text for details).

2003) resolve taxa considered "prosauropods" to be paraphyletic. Although the earliest of these analyses supports a fully pectinate arrangement of "prosauropods" (Yates 2001, 2003:fig. 22), the most recent analyses resolve a monophyletic core of prosauropods flanked basally by primitive forms and apically by sauropod-like forms (Yates and Kitching 2003:fig. 4; Yates 2004:fig. 13). Sereno (1998) specified phylogenetic definitions that designate Prosauropoda and Sauropoda reflexive stem-based clades that comprise the node-based Sauropodomorpha. Applying this phylogenetic definition to the Yates and Kitching (2003:fig. 13) topology, the monophyletic core should be called Prosauropoda, the derived sauropod-like forms should be included in Sauropoda, and taxa resolved as outgroups to those clades are non-sauropodomorph saurischians. The phylogenetic definitions for this node–stem triplet are as follows (Sereno 1998:table 4) (boldface type indicates node-based definitions; regular type indicates stem-based definitions):

Sauropodomorpha Huene 1932—
Plateosaurus engelhardti, Saltasaurus loricatus, their most recent common ancestor and all descendants.

Prosauropoda Huene 1920—All sauropodomorphs closer to *Plateosaurus engelhardti* than to *Saltasaurus loricatus*.

Sauropoda Marsh 1878—All sauropodomorphs closer to *Saltasaurus loricatus* than to *Plateosaurus engelhardti*.

Below, I summarize the major specializations relating to herbivory, neck elongation, and locomotion for each of five major sauropod clades. The synapomorphies discussed are listed in table 1.1, alongside their usage in various cladistic analyses of sauropod relationships. Appendix 1.1 gives a full character list with primitive and derived states.

SAUROPODA

Probable sauropod body fossils and ichnofossils are present in Upper Triassic (Carnian) sediments, but their referrals require confirmation (summarized in Wilson 2005b). The partial hindlimb of *Blikanasaurus* is proportioned similarly to those of later sauropods (Yates 2003, 2004; Yates and Kitching 2003; Upchurch et al. 2004), but correlation with body size cannot yet be ruled out. Likewise, trackways from the Upper Triassic (Carnian) Portezuelo Formation of West–Central Argentina resemble those of later sauropods, but their identification remains tentative (Marsicano and Barredo 2004). The oldest definitive sauropod fossils are the *Tetrasauropus* trackways preserved in the Chinle Group of western North America, which are Norian–Rhaetian in age (ca. 210 mya [Lockley et al. 2001; see Wright, chapter 9]). Slightly younger or coeval ?Rhaetian strata in Thailand preserve the fragmentary remains of *Isanosaurus* (Buffetaut et al. 2000). *Isanosaurus* may be more derived than the slightly younger *Vulcanodon* (Raath 1972), which is generally considered the most primitive sauropod (Wilson 2002: fig.13, table 13). Because the basalmost sauropods *Vulcanodon*, *Isanosaurus*, and *Gongxianosaurus* (fig. 1.5) lack complete cranial remains and much of the vertebral column, the majority of the features diagnosing Sauropoda are appendicular synapomorphies. Of these, many are related to the adoption of a columnar, graviportal posture, which involved independent changes in limb proportions, posture, and ossification.

COLUMNAR, QUADRUPEDAL POSTURE

Outgroups to Sauropoda are primitively bipedal and characterized by relatively short forelimbs that generally represent less than half the length of the hindlimb. In these forms, the proximal hindlimb is shorter than the distal hindlimb, nearly half of whose length is provided by the metatarsus. Sauropoda is characterized by modifications of proportions both within and between the fore- and the hindlimbs, a modification related to quadrupedalism. Sauropods have elongate forelimbs that are at least 70% of the hindlimb length, nearly twice that of their outgroups (table 1.2, fig. 1.5). This change was accommodated by an overall lengthening of the forelimb, especially the distal elements (fig. 1.6), and an overall shortening of distal hindlimb elements relative to the proximal element (fig. 1.7). Reduction of the distal hindlimb did not include metatarsal V, which attains at least 70% of the length of metatarsal IV in all sauropods, effecting a more symmetrical pes with five weight-bearing digits. Although lengthening of the forelimb and relative shortening of the distal hindlimb characterize all sauropods, future discoveries may suggest that these features are not correlated. Both early sauropod body fossils and ichnofossils suggest that quadrupedalism evolved in sauropods sometime prior to the Late Triassic (Wilson 2005b). Adoption of a quadrupedal pose within Sauropoda represents one of four such acquisitions within Dinosauria, each of which is associated with body size increase (Carrano 2000, 2005; see Carrano, chapter 8).

Associated with the proportional changes that facilitate a quadrupedal pose are specializations that allow a columnar, rather than flexed, limb posture. In basal dinosaurs, the disposition of limb articular surfaces and shaft curvature suggest a slightly flexed resting pose for the hip, knee, shoulder, and elbow joints. In sauropod outgroups, for example, the anteroposterior curvature of the femur offsets proximal and distal

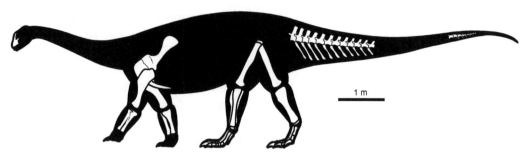

FIGURE 1.5. Silhouette skeletal reconstruction of *Gongxianosaurus shibeiensis* in left lateral view. Reconstruction based on unnumbered specimens pertaining to three individuals described by He et al. (1998). The majority of the skeleton pertains to a possibly subadult individual represented by an articulated pectoral girdle and forelimb and an articulated hindlimb that were discovered in association (He et al. 1998:1). The two series of articulated caudal vertebrae likely pertain to a distinct, adult individual, as does the premaxilla. Both the caudal series and the premaxilla have been scaled to the size of the appendicular elements. The relative size of missing elements (i.e., skull, neck, trunk, manus) was based on the basal sauropods *Vulcanodon* and *Shunosaurus*. Additional elements attributed to *Gongxianosaurus* (Luo and Wang 2000) are not included in this reconstruction because they have not yet been figured or described in detail.

condyles of the femur approximately 20° from horizontal (fig. 1.8A, B). Likewise, bony extensor processes on the ulna (olecranon) and tibia (cnemial crest) are prominent in immediate sauropod outgroups but do not project above the dorsal surface of the ulna and tibia, respectively, in *Vulcanodon*, *Gongxianosaurus* (fig. 1.5), and most other sauropods (fig. 1.7). Reduction of these processes suggests a more columnar alignment of the elbow and knee joints. In addition, the

TABLE 1.2
Limb Proportions in Selected Saurischian Genera

	FORE:HIND	MT III:TIBIA	REFERENCE(S)
Theropoda			
Eoraptor	0.43	0.43	Sereno (pers. comm.)
Herrerasaurus	0.47	0.52	Sereno (1993), Novas (1993)
Prosauropoda			
Jingshanosaurus	0.42	0.57	Zhang & Yang (1994)
Lufengosaurus	0.50	0.57	Young (1941)
Plateosaurus	0.52	0.48	Huene (1926)
?Sauropoda			
Blikanasaurus	—	0.36	Galton & van Heerden (1985)
Antenonitrus	0.81	0.38	Yates & Kitching (2003)
Sauropoda			
Vulcanodon	0.78	0.37, 0.32	Raath (1972), Cooper (1984)
Gongxianosaurus	0.62	0.38	He et al. (1988)
Shunosaurus	0.67	0.27	Zhang (1988)
Omeisaurus	0.90	0.28	He et al. (1988)
Jobaria	0.88	0.28	Sereno et al. (1999)
Apatosaurus	0.72	0.21	Gilmore (1936)
Camarasaurus*	0.83	0.24	Gilmore (1925)
Camarasaurus	(0.85)	0.24	McIntosh & al. (1996)
Opisthocoelicaudia	0.79	0.25	Borsuk-Bialynicka (1977)

NOTE: Forelimb length equals the sum of the lengths of the humerus, radius, and longest metacarpal; hindlimb length equals the sum of the lengths of the femur, tibia, and longest metatarsal. Asterisk(*) indicates measurement of a juvenile individual; parentheses indicate an estimated value. Abbreviation: mt, metatarsal.

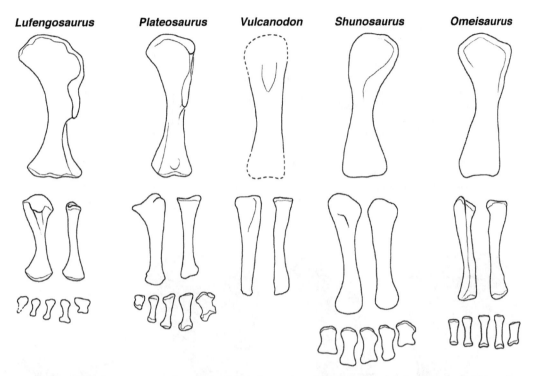

FIGURE 1.6. Forelimb proportions in the prosauropods *Lufengosaurus* and *Plateosaurus* and the basal sauropods *Vulcanodon*, *Shunosaurus*, and *Omeisaurus*. Forelimbs have been scaled to the same humeral length. Based on Young (1947), Huene (1926), Raath (1972), Zhang (1988), and He et al. (1998), respectively.

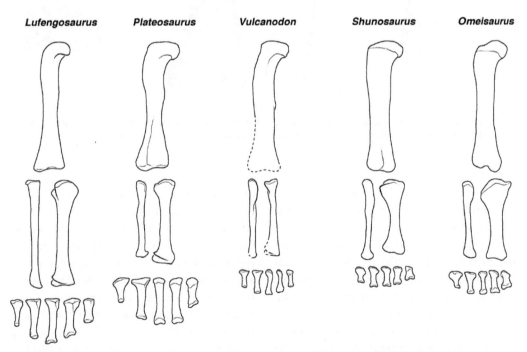

FIGURE 1.7. Hindlimb proportions in the prosauropods *Lufengosaurus* and *Plateosaurus* and the basal sauropods *Vulcanodon*, *Shunosaurus*, and *Omeisaurus*. Hindlimbs have been scaled to the same femoral length. Based on Young (1947), Huene (1926), Raath (1972), Zhang (1988), and He et al. (1998), respectively.

FIGURE 1.8. Femoral curvature in the saurischian dinosaurs *Herrerasaurus* (A), *Massospondylus* (B), *Vulcanodon* (C), and *Isanosaurus* (D). Femora are figured in right medial view and have been scaled to the same length to facilitate comparison. The left femur of *Isanosaurus* has been reversed. Based on Novas (1993), Cooper (1981), Cooper (1984), and Buffetaut et al. (2000), respectively.

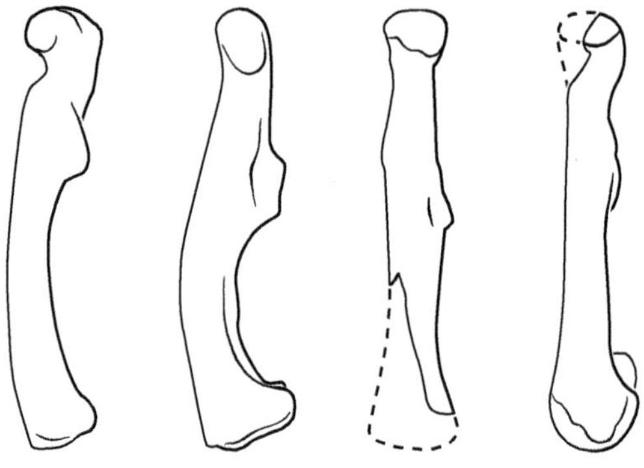

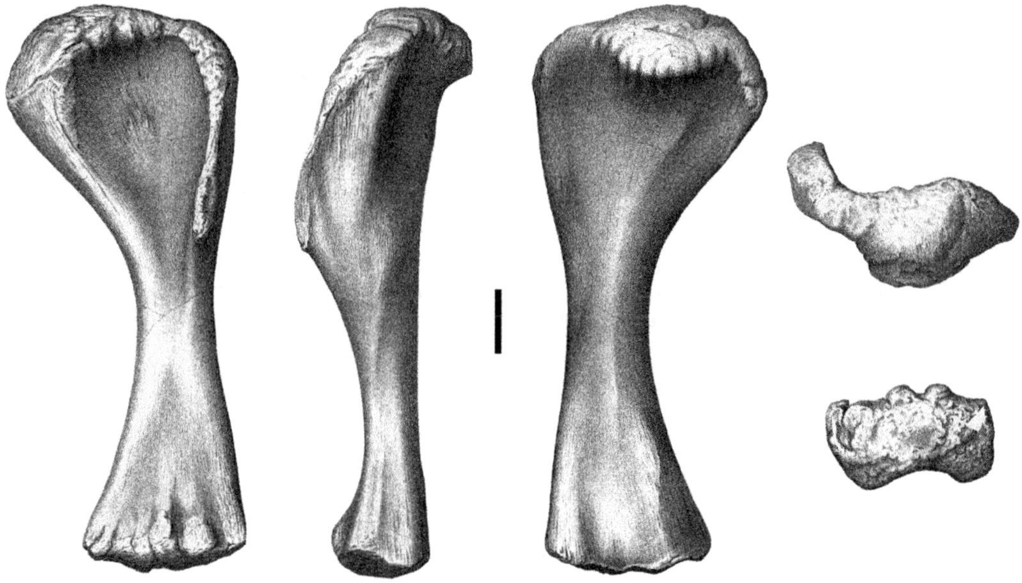

FIGURE 1.9. Left humerus of *Camarasaurus grandis* (YPM 1901) in anterior (left), lateral (middle), posterior (right), proximal (top), and distal (bottom) views (from Ostrom and McIntosh 1966:pl. 49). Scale bar equals 30 cm.

longest weight-bearing elements in the skeleton (humerus, femur) have eccentric midshaft cross sections that are broader mediolaterally than anteroposteriorly. Distal limb elements (radius/ulna, tibia/fibula) do not share this cross-sectional geometry, but they bear weight in tandem and are together broader mediolaterally than anteroposteriorly.

Reduced ossification of limb elements represents the third major appendicular specialization characterizing Sauropoda. A conspicuous feature of sauropod limb elements is that their artic-ular ends have a rugose, irregular surface, whereas their shafts are smooth. Owen (1841:461) recognized this feature in *Cetiosaurus*, noting that "the articular surfaces which are preserved are covered with large tubercles for the attachment of thick cartilage." Similarly, Marsh (1878:413) described the humerus of *Camarasaurus* as "rough, and well covered with cartilage" (fig. 1.9). The thickness of this cartilage cap has not yet been estimated but is implied in articulated skeletons by the difference in volumes of the acetabulum and femoral head. The proximal carpal and tarsal elements have few to no nonarticular surfaces and are completely made up of rough, rugose bone. Consequently, the configuration of sauropod wrist and ankle elements relative to adjacent elements is difficult to determine because little of the articular surfaces remains. Distal carpals have not been identified in any sauropod skeleton, and distal tarsals have only been recovered for *Gongxianosaurus*, in which discoidal ossifications are preserved atop metatarsal III and between metatarsal IV and metatarsal V (He et al. 1998:fig. 4C; fig. 1.5). Retention of ossified distal tarsals may suggest that *Gongxianosaurus* is the most primitive sauropod, but their absence in other basal sauropods (e.g., *Vulcanodon*) has not yet been confirmed by articulated material.

EUSAUROPODA

Eusauropoda is the node-based group including *Shunosaurus lii, Saltasaurus loricatus*, their most recent common ancestor, and all descendants (fig. 1.4). This definition specifies all named sauropods except *Vulcanodon, Gongxianosaurus*, and *Rhoetosaurus*, as well as the possible early sauropods *Blikanasaurus* and *Antenonitrus*. The oldest well-preserved eusauropod is the Middle Jurassic *Shunosaurus*, which is known from several complete skeletons (Zhang 1988). Consequently, many of the synapomorphies diagnosing Eusauropoda are ambiguous and may obtain a broader distribution once basal forms are known more completely. Eusauropod synapomorphies greatly outnumber those of any other node within Sauropoda—Wilson (2002) reported 53, fewer than half of which could be scored in more basal taxa. Thus, the ambiguous (i.e., cranial and axial) synapomorphies may have evolved as early as the divergence of Sauropoda from Prosauropoda in earliest Late Triassic (Carnian, 220 mya [Flynn et al. 1999]). The unambiguous (i.e., hindlimb) synapomorphies, on the other hand, signal more recent modifications since the divergence of Eusauropoda from Sauropoda in the Late Triassic (Rhaetian, 210 mya [Buffetaut et al. 2000]).

HERBIVOROUS SPECIALIZATIONS

The Middle Jurassic *Shunosaurus* is the earliest-appearing sauropod known from well-preserved cranial remains. All cranial synapomorphies of Eusauropoda are ambiguous and may later be shown to characterize more inclusive groups. *Shunosaurus* possessed a sophisticated dental apparatus that is highly modified relative to that of prosauropods, indicating that eusauropods modified the shape of the crowns as well as their arrangement along the tooth row. Principal among these changes is the acquisition of precisely occluding dentition, a feature that is unknown elsewhere in Saurischia.

Prosauropods and theropods primitively have lower tooth rows that extend the length of the dentary but upper tooth rows that extend farther posteriorly to midorbit. With different lengths and numbers of teeth, upper and lower teeth have mismatched occlusion that generates no regular wear pattern. Additionally, prosauropods and theropods have tooth rows that are relatively straight in dorsal or ventral view. Right and left sides meet at an acute angle, and none of the teeth are oriented transversely (fig. 1.10A, B). Sauropods differ in all of these respects. Nearly all sauropods known by cranial remains have tooth rows that are of even length and contain similar numbers of teeth. The upper tooth row terminates at or in front of the antorbital fenestra, and the dentary always has an edentulous region posterior to the last tooth. In dorsal view, the tooth rows are curved rather than straight, and at least two teeth are oriented transversely (fig. 1.10C, D). Together, these changes signal precise occlusion in sauropods, as evidenced by crown wear facets generated by tooth-to-tooth wear (Calvo 1994). In dorsal view, the tooth rows are outwardly arched rather than straight, and at least two teeth are oriented transversely (fig. 1.10C, D). The entire tooth row is transversely oriented in some sauropods (see "Diplodocoidea," below). Most sauropods develop an imbricate arrangement of teeth in which the mesial edge of each tooth is overlapped by the distal edge of the preceding tooth.

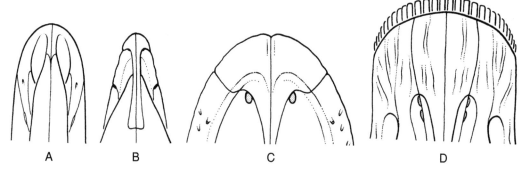

FIGURE 1.10. Snouts of the theropod *Herrerasaurus* (A), the prosauropod *Plateosaurus* (B), and the sauropods *Brachiosaurus* (C) and *Diplodocus* (D) in dorsal view. Based on reconstructions from Sereno (1993) and Wilson and Sereno (1998)

Eusauropod tooth crowns also have distinctive shape and texture. All teeth have a characteristically wrinkled enamel texture whose function is unknown. Coarseness of enamel wrinkling varies to some extent within sauropods, with narrow-crowned teeth usually exhibiting much finer wrinkling than broad tooth crowns. Sauropod tooth crowns are primitively spatulate, with a D-shaped cross section.

Precise tooth-to-tooth occlusion is not lost within Sauropoda, but many of the other herbivorous innovations are modified in later lineages, principally Diplodocoidea (see Sereno and Wilson, chapter 5). Because all known sauropod skulls share these features, their sequence of acquisition is not yet known.

NECK ELONGATION

The primitive saurischian precaudal vertebral count is 27, although the relative number of cervical, dorsal, and sacral vertebrae vary in Theropoda (9-15-3, respectively) and Prosauropoda (10-14-3). Vertebral counts are not known for non-eusauropods, but *Vulcanodon* has a sacrum with four coosified vertebrae (Raath 1972). The fourth sacral vertebra in sauropods is a caudosacral, based on osteological and developmental evidence (Wilson and Sereno 1998). The eusauropod *Shunosaurus* (13-13-4) is the basalmost sauropod genus for which the vertebral count is known. Compared to outgroups, eusauropods are characterized by two neck elongation events: (1) incorporation of one dorsal vertebra into the cervical series and (2) duplication of two cervical vertebrae to achieve the primitive eusauropod precaudal count of 13-13-4.

Later, *Patagosaurus*, *Omeisaurus*, and more derived sauropods acquire a fifth sacral without changing the precaudal count (13-12-5), which most likely represents the incorporation of a dorsal vertebra into the sacrum (rather than the addition of a cervical and loss of a dorsal). Nearly all of the dozen subsequent neck-lengthening events characterize individual neosauropod genera and are not synapomorphies of larger clades. The exception is Diplodocidae (15-10-5), which incorporated two dorsal vertebrae into the cervical series. Thus, there is no progressive increase in neck length within Sauropoda; rather, individual genera were specialized for their neck length. All three means of neck lengthening (incorporation, duplication, elongation) were employed within Sauropoda.

HINDFOOT POSTURE

Theropods and prosauropods are interpreted as having a digitigrade pes, a posture in which the heel and proximal metatarsals were held off the ground, and the distal metatarsals and phalanges contacted the substrate (Carrano 1997). Eusauropods are characterized by several changes that together result in a unique hindfoot posture that is easily recognized in footprints (fig. 1.11). These include the independent modification of the length, arrangement, and robustness of the metatarsus, as well as the reduction in the number and size of the pedal phalanges.

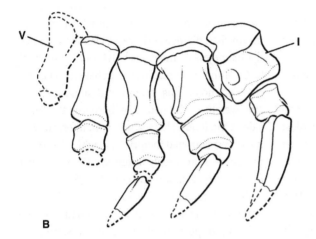

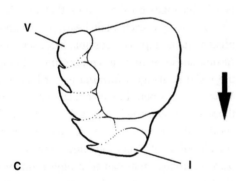

FIGURE 1.11. Right pes of *Apatosaurus* in proximal (A) and dorsal (B) views; C, right pes print of ?*Brontopodus* oriented relative to the trackway midline (arrow). *Apatosaurus* modified from Gilmore (1936:figs. 25, 27, 28); ?*Brontopodus* modified from Thulborn (1990:fig. 6.16f). Abbreviations: I-V, digits I-V.

Sauropod outgroups have long distal hindlimbs, in which the metatarsus accounts for 40% to 50 of the tibial length (table 1.2). In contrast, the eusauropod metatarsus is markedly abbreviated and comprises less than 25% of the tibial length. The proportions of the basal sauropods *Vulcanodon* and *Gongxianosaurus*, as well as those of *Blikanasaurus* and *Antenonitrus*, are intermediate between sauropod outgroups and eusauropods such as *Shunosaurus* (figs. 1.5, 1.7, table 1.2). In addition to these proportional changes, the eusauropod metatarsus attains a spreading configuration in which the proximal ends are not in mutual contact, as they are in sauropod outgroups. In dorsal view, for example, the metatarsal shafts are separated by intervening spaces (fig. 1.11B). These changes effect a more spreading hindfoot posture in which the metatarsus was held in a subhorizontal, rather than subvertical, orientation. Wilson and Sereno (1998:41) recognized this as a "semi-digitigrade" foot posture. Carrano (1997:fig. 1B) termed the inferred foot posture in sauropods "sub-unguligrade," referring to the specialized foot posture of hippopotamids, rhinoceratids, and proboscideans, in which the metatarsus is held vertically, a fleshy pad supports the foot, and the penultimate and ungual phalanges contact the substrate. Although the hypothesized sauropod hindfoot posture is similar to "sub-unguligrady" (viz. the fleshy heel pad), the metatarsus is thought to have been held in a nearly horizontal rather then a vertical orientation, and the nonungual phalanges are hypothesized to have contacted the substrate. The term *semi-digitigrady* is used here to refer to the foot posture hypothesized for Eusauropoda.

Although the earliest sauropods are interpreted as having a digitigrade posture, there are no footprints attributed to sauropods that indicate such a foot posture. Rather, the earliest sauropod trackways bear elongate pes prints that indicate a semidigitigrade hindfoot posture (e.g., Portezuelo trackways [Marsicano and Barredo 2003], *Tetrasauropus* [Lockley et al. 2001]). This 22 million- to 44 million-year discrepancy between the first appearance of semi-digitigrade pedal posture in the body fossil and that in the ichnofossil records may indicate early appearance of eusauropods, homoplasy, or that hindfoot posture has been erroneously interpreted in early sauropods. Based on a stratocladistic analysis of ichnological and body fossil data, Wilson (2005b) suggested that semi-digitigrady evolved in the Late Triassic and was either reversed or misinterpreted in the early sauropods *Vulcanodon* and *Gongxianosaurus*. A preliminary study of skeletal remains referred to *Plateosaurus* has inferred a less digitigrade posture than traditionally posited for prosauropods (Sullivan et al. 2003), which underscores difficulties in determining locomotor posture from osteology.

In addition to revising the temporal origin of semi-digitigrade hindfoot posture in early sauropods, ichnofossils indicate that the subhorizontal foot was supported by a fleshy heel (fig. 1.11C).

Eusauropods are also characterized by a departure from the within-pes proportions that characterize other saurischians. Body weight in theropods and prosauropods is accommodated by three and four pedal digits, respectively. In these taxa, shaft breadth varies little across the metatarsus (table 1.2), implying that body weight was borne subequally by its constituent elements. The eusauropod pes, in contrast, displays marked asymmetry of metatarsal shaft diameters in which metatarsal I is broader at than all others. The disparity among metatarsals II–V becomes more pronounced in more derived sauropods. *Omeisaurus*, *Mamenchisaurus*, and Neosauropoda are diagnosed by a metatarsus in which the minimum shaft diameters decrease laterally such that the diameters of metatarsals III and IV are 50% to 60% that of metatarsal II (table 1.3). Hatcher (1901:51) noted this pattern and suggested that "the weight of the body was borne by the inner side of the foot." This feature is manifest in well-preserved sauropod footprints, in which the inner margin is more deeply impressed than the outer margin (e.g., Pittman and Gillette 1989:322).

The acquisition of a semidigitigrade hindfoot posture is accompanied by reduction of the phalangeal portion of the pes. Prosauropods and basal theropods retain a full complement of pedal phalanges on digits I–IV that invariably number 2-3-4-5, each digit bearing an ungual phalanx (table 1.4). The possible basal sauropod *Blikanasaurus* retains the same count. Although the pedal phalangeal formula is not known in *Vulcanodon*, its penultimate phalanges resemble those of prosauropods and are not drastically shortened (Cooper 1984:figs. 34, 35). The articulated hindfoot of *Gongxianosaurus* confirms that basal sauropods maintained a high number of phalanges (2-3-4-5), which themselves were longer than broad (fig. 1.5). The pes of eusauropods is reduced in both the number and the size of phalangeal elements. The penultimate phalanx in digits II–IV is reduced to a plate-shaped disc or lost in *Shunosaurus*, *Omeisaurus*, and various neosauropods. The greatest number of phalanges retained in eusauropod digit IV is three (e.g., *Shunosaurus*, *Omeisaurus*, *Camarasaurus*), two fewer than the outgroup condition of five. Despite the loss of two phalanges, an ungual is maintained on digit IV. Eusauropods clearly demonstrate non-terminal phalangeal reduction, which maintains the size and functionality of the unguals amid substantial

TABLE 1.3
Pedal Proportions in Select Saurischian Genera

	MT I:II:III:IV	MT V:IV
	(MINIMUM BREADTH)	(LENGTH)
Theropoda		
Eoraptor	—	0.56
Herrerasaurus	0.60:0.90:0.95:1	0.59
Prosauropoda		
Jingshanosaurus	1.0:1.0:1.0:1.0	0.55
Lufengosaurus	0.84:0.84:0.84:1	0.56
Plateosaurus	1.0:1.0:1.0:1.0	0.61
?Sauropoda		
Blikanasaurus	1:1:0.89:0.65	0.53
Antenonitrus	0.62:0.86:1?	—
Sauropoda		
Gongxianosaurus	—	0.64
Vulcanodon	0.73:0.64:0.82:1	0.75
Shunosaurus	1:0.92:0.85:0.85	0.70
Omeisaurus	1:0.87:0.56:0.62	0.90
Apatosaurus	1:0.75:0.50:0.55	—
Camarasaurus*	—	0.75
Camarasaurus	1:0.69:0.44:0.50	0.81
Opisthocoelicaudia	1:0.78:0.67:0.44	0.78
"*Barosaurus*"	1:0.68:0.41:0.36	0.95

NOTE: References as in Table 1.2; "*Barosaurus*" data from Janensch (1961). Asterisk (*) indicates measurement of a juvenile individual.

digital shortening. Non-terminal phalangeal reduction may have also produced a mammal-like phalangeal count in cynodont-grade synapsids (Hopson 1995).

Further modification of the pedal configuration described above diagnoses sauropods more derived than *Shunosaurus*. In *Barapasaurus*, *Omeisaurus*, and all neosauropods, the four pedal unguals are directed laterally with respect to the digit axis. This reorientation of the pedal unguals is accomplished by a beveled proximal articular surface and twisting of the axis of the ungual. Wilson and Sereno (1998) scored the basal sauropods *Vulcanodon* and *Shunosaurus* with the primitive condition (i.e., anteriorly directed unguals), and the primitive condition appears to characterize *Blikanasaurus*, *Antenonitrus*, and *Gongxianosaurus*. Although laterally directed pedal unguals first appear in the body fossil record in the Lower Jurassic *Barapasaurus*, they appear 13 million to 35 million years earlier in the ichnofossil record. Upper Triassic *Tetrasauropus* trackways (fig. 1.11) clearly preserve impressions of unguals deflected laterally relative to the axis of the pes (Lockley et al. 2001), indicating that this feature evolved earlier than implied by body fossils alone (Wilson 2005b).

NEOSAUROPODA

Neosauropoda is the node-based group including *Diplodocus longus*, *Saltasaurus loricatus*, and all descendants of their most recent common ancestor (Wilson and Sereno 1998; fig. 1.4). Within this node-based group, the two reflexive stem-groups (Diplodocoidea, Macronaria) form a stable node–stem triplet (boldface type indicates node-based definitions; regular type indicates stem-based definitions).

TABLE 1.4
Manual and Pedal Phalangeal Counts in Select Saurischian Genera

	MANUS	PES
Theropoda		
Eoraptor	2*-3*-4*-1-0	2*-3*-4*-5*-1
Herrerasaurus	2*-3*-4*-1-0	2*-3*-4*-?-1
Prosauropoda		
Jingshanosaurus	2*-3*-4*-3*-1	2*-3*-4*-5*-1
Lufengosaurus	2*-3*-4*-3-1	2*-3*-4*-5*-1
Plateosaurus	2*-3*-4*-3-2	2*-3*-4*-5*-2
?Sauropoda		
Blikanasaurus	—	2*-3*-4*-?5*-?
Antenonitrus	—	—
Sauropoda		
Vulcanodon	—	—
Gongxianosaurus	—	2*-3*-4*-5*-?
Shunosaurus	2*-2-2-2-2	2*-3*-3*-3*-2
Omeisaurus	2*-2-?-?-1	2*-3*-3*-3*-2
Diplodocus	—	2*-3*-3?-2-0
Camarasaurus	2*-1-1-1-1	2*-3*-4*-2*-?
Brachiosaurus	2*-1-1-1-1	—
Opisthocoelicaudia	0-0-0-0-0	2*-2*-2*-1?-?

NOTE: Asterisk (*) indicates a clawed digit. References as in Table 1.2; *Diplodocus* data from Hatcher (1901).

Neosauropoda Bonaparte 1986b—
Diplodocus longus, *Saltasaurus loricatus*, their most recent common ancestor, and all descendants.

Diplodocoidea Upchurch 1995—All neosauropods more closely related to *Diplodocus longus* than to *Saltasaurus loricatus*.

Macronaria Wilson and Sereno 1998—All neosauropods more closely related to *Saltasaurus loricatus* than to *Diplodocus longus*.

Bonaparte (1986b:369) originally referred to neosauropods as the "end-Jurassic" sauropods—members of Dicraeosauridae, Diplodocidae, Camarasauridae, and Brachiosauridae. Although no definitive skeletal remains referable to this group have been recorded prior to the "end Jurassic," the phylogenetic definition of Neosauropoda is not temporally bounded. Cretaceous neosauropods include rebbachisaurid diplodocoids and Titanosauria (left out of Bonaparte's definition), and the near-simultaneous appearance of the principal neosauropod lineages in the Late Jurassic implies that one or more of them were present in the Middle Jurassic. Neosauropoda accommodates the majority of sauropod genera and encompasses most of its morphological diversity.

The recently described *Jobaria* has been resolved as the outgroup of Neosauropoda (Sereno et al. 1999; Wilson 2002; fig. 1.4) on the basis of a number of advanced features they share. *Jobaria* has been alternatively resolved within Neosauropoda as a basal macronarian (Upchurch et al. 2004), but the evidence supporting this hypothesis is presently outweighed by the retention of several primitive characters. The relevant synapomorphies of *Jobaria* + Neosauropoda is discussed alongside those distinguishing Neosauropoda. *Jobaria* and

Neosauropoda can be distinguished by a novel forefoot posture in which the manus is arranged into a tight semicircle and held vertically. Neosauropoda, in turn, is distinguished by marked reduction in the number and ossification of carpal and tarsal elements.

REDUCED OSSIFICATION OF CARPALS AND TARSALS

The evolutionary history of sauropod dinosaurs documents reduced ossification of carpal and tarsal elements. This tendency may be related to the reduced ossification that characterizes all sauropod weight-bearing elements (fig. 1.9; see Sauropoda, above). In the carpus of *Herrerasaurus* (Sereno 1993:fig. 15) and *Eoraptor* (P. Sereno, pers. comm.), a large radiale, ulnare, and series of four distal carpals are present. In prosauropods, the proximal carpals are very reduced or absent (unossified), but the medial three distal carpals (dc 1–3) are present and articulate with the proximal ends of metacarpals I–III, respectively (e.g., *Massospondylus* [Cooper 1981:figs. 35, 36], *Lufengosaurus*, [Young 1941:fig.15]). The earliest sauropod for which a carpus is known preserves only three block-shaped carpals that decrease in size laterally, based on their presumed position (*Shunosaurus* [Zhang 1988:figs. 2, 48]). Because they were found closely associated with metacarpals I–III, rather than with the radius and ulna, Wilson and Sereno (1998:47) regarded them as distal carpals. Other nonneosauropods show a similar pattern: *Omeisaurus* has three carpals of decreasing size (He et al. 1988), as does *Jobaria* (Sereno et al. 1999). In *Jobaria*, one surface of the largest carpal has two triangular facets that match the proximal surfaces of metacarpals I and II, suggesting that it is a distal carpal; the other surface bears no discernible articular surface. Thus prosauropods and basal sauropods retain only three ossified distal carpals and lack ossified proximal carpals. Neosauropods further reduce the number of ossified carpals to two or fewer. In *Camarasaurus*, two block-shaped carpals are present and fitted to the metacarpals. As Osborn (1904:182) noted, the fitted articulation between the carpals and the metacarpals suggests that the primary axis of the wrist joint was positioned more proximally, between these carpals and the bones of the forearm. Other neosauropods have a single carpal element positioned above metacarpals II and III (e.g., *Apatosaurus* [Hatcher 1902; Gilmore 1936]). One individual of *Apatosaurus*, however, preserves a carpal element hypothesized to articulate with metacarpals IV and V (Filla and Redman 1994). As discussed later, some sauropods lack ossified carpus altogether (see "Macronaria" below).

In prosauropods and basal theropods, the body of the astragalus (i.e., the portion below the ascending process) is trapezoidal. In the prosauropod *Massospondylus* and the basal theropod *Herrerasaurus*, the proximodistal and anteroposterior depth of the medial side of the astragalus equals or exceeds that of the lateral side (Cooper 1981:fig. 71f; Novas 1989:figs. 2.5–10). The basal sauropod *Shunosaurus* appears to retain the primitive condition, based on the only available view (anterior) of the astragalus (Zhang 1988:fig. 54). In contrast, the astragalus in *Jobaria* and Neosauropoda appears wedge-shaped in both proximal and anterior views. In *Jobaria*, for example, the proximodistal and anteroposterior depth of the astragalus diminishes markedly toward its medial side. In addition, in proximal and distal views, the primitive posteromedial corner of the astragalus is absent, and the astragalus has a subtriangular, rather than subrectangular, shape.

FOREFOOT POSTURE

Prosauropods and basal ornithischians retain the primitive dinosaur condition in which the proximal ends of the metacarpals are not closely appressed and are only slightly arched in articulation. The metacarpals are subrectangular in proximal view, and their intermetacarpal articular surfaces do not extend down the shaft. For example, the articulated manus of *Massospondylus* is cupped approximately 90° between metacarpal I and metacarpal V, and most of this arch occurs in metacarpals I and III, whose lateral articular surfaces form an acute angle with the anterior surface (Cooper 1981:fig. 37). A similar condition is present in basal ornithischians (e.g.,

Lesothosaurus [Sereno 1991]). The condition in theropods, however, differs from that of prosauropods and basal ornithischians. Although the manus is bound proximally in theropods, the metacarpus retains the same 90° proximal curvature present in prosauropods and ornithischians (*Herrerasaurus* [Sereno 1993,fig. 15], *Deinonychus*, [Ostrom 1969:fig.62]). The configuration and pose of the metacarpus of basal sauropods such as *Shunosaurus* and *Omeisaurus* are not agreed on. Whereas Wilson and Sereno (1998) suggested that their forefoot posture resembles that of prosauropods and basal ornithischians, in which the manus is spreading and only slightly arched ventrally, both Upchurch (1998) and Bonnan (2003) interpreted them as having the derived, digitigrade forefoot posture that characterizes *Jobaria* and neosauropods.

The metacarpus of *Jobaria* and neosauropods is arranged into a tightly bound, digitigrade structure that is hypothesized to have contacted the substrate at the metacarpal–phalangeal joints. The metacarpals arranged into a vertical cylinder in which all are subequal in length and have well-developed intermetacarpal articular surfaces that extend distally to midshaft (e.g., *Camarasaurus* [Ostrom and McIntosh 1966:figs. 55–59]). Proximally, the metacarpal heads are wedge-shaped and articulate in a tight arc of approximately 270°. This tubular arrangement of the metacarpals is due to their medial and lateral articular surfaces meeting the external (anterior) aspect of the metacarpal at an acute angle. A tightly curled, digitigrade manus, defined by these osteological features, is present without exception in Neosauropoda.

Although cladistic studies have regarded digitigrade forefoot posture as diagnostic of Eusauropoda (Upchurch 1998), Neosauropoda (Upchurch 1995; Wilson and Sereno 1998), or *Jobaria* + Neosauropoda (Wilson 2002), ichnofossils suggest a much earlier origin. Upper Triassic trackways from North America (Lockley et al. 2001) and South America (Marsicano and Barredo 2004), as well as Lower Jurassic trackways from Italy (Dalla Vecchia 1994), Poland (Gierlinski 1997), and Morocco (Ishigaki 1988), document sauropod trackmakers with a digitigrade manus (fig. 1.12). These trackways record the appearance of a digitigrade forefoot posture 22 million to 44 million years earlier than predicted by Upchurch (1998) and 57 million years earlier than predicted by Wilson and Sereno (57 my). This discrepancy can be interpreted as the early appearance of Neosauropoda or the early appearance of digitigrade foot posture in non-neosauropods. Assessment of forefoot posture in non-neosauropods was based solely on the published illustrations of the only basal taxa preserving manual remains, *Shunosaurus* (Zhang 1988:fig. 49, pl. 14) and *Omeisaurus* (He et al. 1988:figs. 47, 48; pl. 14, figs. 4–6). On the basis of these illustrations, Wilson and Sereno (1998:48) argued that *Shunosaurus* and *Omeisaurus* lacked a digitigrade forefoot posture because their metacarpals have poorly defined intermetacarpal articular surfaces. Other dinosaurs with a vertically oriented, digitigrade foot posture have metapodials that are tightly appressed (bound) proximally and have well-marked intermetapodial facets that extend down their shafts (e.g., *Herrerasaurus* pes, *Iguanodon* manus). Additionally, the metacarpals of *Shunosaurus* and *Omeisaurus* are subrectangular proximally, implying that they were only slightly arched proximally (~90°) in articulation, unlike *Jobaria* and neosauropods (Wilson and Sereno 1998:fig. 40). Upchurch (1998:68), however, argued that despite these considerations, *Shunosaurus* and *Omeisaurus* had forefeet that were both digitigrade and U-shaped proximally, features he regarded as a single character (table 1.1). Trackways from Italy, Poland, and Morocco preserve a digitigrade manus that is not tightly arched (fig. 1.12), suggesting that the bound metatarsus and its tightly arched configuration are independent characters. The trackways further suggest that the bound metacarpus was acquired earlier in sauropod history than was the tubular metacarpus.

DIPLODOCOIDEA

One of two reflexive neosauropod stem-groups, Diplodocoidea includes all neosauropods more

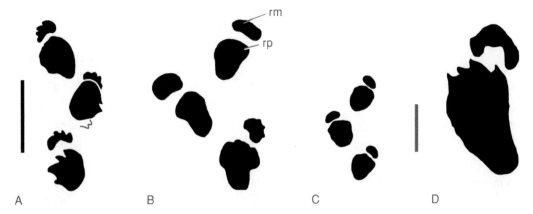

FIGURE 1.12. Early sauropod footprints. (A), *Tetrasauropus* trackway from the Upper Triassic (Chinle Group) of Cub Creek (based on Lockley et al. (2001); (B), trackway from the Lower Jurassic (Pliensbachian) of the Atlas Mountains, Morocco (based on Farlow 1992:fig. 2a, b); (C), trackway (ROLM 28) from the Lower Jurassic (Hettangian–Pliensbachian) of Lavini di Marco, Italy (based on Dalla Vecchia 1994:fig. 2); (D) manus–pes pair of ?*Parabrontopodopus*, from the Hettangian of the Holy Cross Mountains, Poland (based on Gierlinski 1997:fig. 1b). (A–C) are oriented relative to trackway midline; trackway midline cannot be determined for (D). Scale equals 50 cm for (A), 10 cm for (B–D).

closely related to *Diplodocus longus* than to *Saltasaurus loricatus* (Wilson and Sereno 1998). By this definition, Diplodocoidea unites *Haplocanthosaurus*, Rebbachisauridae, Dicraeosauridae, and Diplodocidae. The position of *Haplocanthosaurus* as the basalmost diplodocoid is weakly supported and awaits further confirmation by additional material. Apart from the position of *Haplocanthosaurus*, the relationships within and between diplodocoid families are stable. The three families have stem-based definitions specifying all taxa more closely related to their namesake genus than to either of the other two namesake genera. Sereno (1998) formally defined Dicraeosauridae and Diplodocidae; Rebbachisauridae is phylogenetically defined for the first time here. A revised phylogenetic nomenclature for Diplodocoidea and its subgroups is proposed below (boldface type indicates node-based definitions; regular type indicates stem-based definitions):

Diplodocoidea Upchurch 1995—All neosauropods more closely related to *Diplodocus longus* than to *Saltasaurus loricatus*.

Rebbachisauridae Bonaparte 1997—All diplodocoids more closely related to *Rebbachisaurus garasbae* than to *Diplodocus longus*.

Flagellicaudata Harris and Dodson 2004—*Diplodocus longus*, *Dicraeosaurus hansemanni*, their most recent common ancestor, and all descendants.

Dicraeosauridae Janensch 1929b—All diplodocoids more closely related to *Dicraeosaurus hansemanni* than to *Diplodocus longus*.

Diplodocidae Marsh 1884—All diplodocoids more closely related to *Diplodocus longus* than to *Dicraeosaurus hansemanni*.

This arrangement of taxon names affords a node–stem triplet within Diplodocoidea that unites two well known stem-based groups (Diplodocidae, Dicraeosauridae) whose sister-taxon relationship has been long recognized.

Like all neosauropod lineages, earliest diplodocoids are found in Upper Jurassic rocks. Diplodocidae is currently restricted to the Late Jurassic of North America (*Diplodocus*, *Apatosaurus*, *Barosaurus*, *Seismosaurus*) and Africa ("*Barosaurus*" *africanus*). Dicraeosauridae is also known from the Late Jurassic of Africa (*Dicraeosaurus*) but survives into the Early Cretaceous of South America (*Amargasaurus*). Rebbachisauridae is the latest surviving diplodocoid clade and is restricted to the Cretaceous of Africa (*Nigersaurus*, *Rebbachisaurus*),

TABLE 1.5

Data Support in the Two Neosauropod Lineages Macronaria and Diplodocoidea

	MACRONARIA	DIPLODOCOIDEA
number of taxa	11	9
cranial		
% missing data	58	55
% character support	30	39.5
axial		
% missing data	35	33
% character support	37	45.5
appendicular		
% missing data	41	55
% character support	33	15

NOTE: The relative proportions of cranial, axial, and appendicular characters supporting the interrelationships of these clades are compared below. Missing data scores were based on Wilson (2002:table 8). Total percentage missing data was higher in Diplodocoidea (48%) than Macronaria (44%).

South America (*Limaysaurus*), and Europe (*Histriasaurus*, Salas rebbachisaurid).

The features supporting the relationships within Diplodocoidea and differentiating its composite genera are supported by a predominance of cranial and axial synapomorphies (table 1.5). These include a major transformation in skull shape and a highly modified vertebral column, discussed below.

HERBIVOROUS SPECIALIZATIONS

As discussed above, the basic sauropod skull plan is quite distinct from those of basal saurischians (see "Eusauropoda" below). The set of features comprising this plan evolved sometime prior to the first appearance of Eusauropoda (Middle Jurassic) and, with few exceptions, was retained until their last appearance (latest Cretaceous). The diplodocoid skull is perhaps the most unique among Sauropoda, and may be thought of as the result of exaggeration of several eusauropod features combined with novelties that evolved stepwise within Diplodocoidea.

The broadening of the snout and shortening of the tooth row that characterizes Eusauropoda is exaggerated in diplodocoids, which evolved upper and lower tooth rows that are restricted anterior to the antorbital fenestra and arranged in jaws that are rectangular in dorsal view (fig. 1.10). In dicraeosaurids and diplodocoids, most teeth are positioned on the transverse portion of the jaw ramus. Rebbachisaurids further this trend by restricting all teeth to the transverse portion of the jaw, which extends lateral to the ramus (see Sereno and Wilson, chapter 5). Transversely oriented tooth rows are unknown elsewhere in Dinosauria.

Other modifications of the diplodocoid skull are novelties that have no precedent in sauropod evolution. One set of such features that characterizes Diplodocoidea is the reorientation of the braincase and part of the palate relative to the dermal skull. In sauropod outgroups and in most sauropods, the jaw articulation lies at the posterior extreme of the skull, behind the orbit. Likewise, the basipterygoid processes are short and point ventrally, and the adductor fossa is positioned on the posterior half of the lower jaw, just below the orbit. The diplodocoid skull differs in each of these respects, due to a reorientation of the dermal skull relative to the braincase. In diplodocoids the quadrate is oriented anteriorly such that the jaw joint is positioned below the orbit in lateral view. The pterygoid and its connection to the braincase via the basipterygoid processes are shifted forward. These two changes effectively shorten the lower jaw and shift anteriorly the adductor fossa. Although they diagnose the same node, these features are not considered to be correlated. Some but not all of these changes are manifest in other taxa, such as pachycephalosaurs, which have anteriorly oriented basipterygoid processes with otherwise typical quadrate articulation and lower jaw length. This change in the shape of the diplodocoid skull effects an inclined line of action for the principal jaw closing musculature, which may have resulted in more fore–aft motion than present in other sauropods (Upchurch and Barrett 2000).

One aspect of diplodocoid skulls represents a reversal from the basic eusauropod condition. Relative to their outgroups, eusauropods have broad crowns that overlap one another along the tooth row. Diplodocoids, in contrast, reduce crown size and lose the crown overlap diagnostic of Eusauropoda. These features are correlated, because crown overlap requires some expansion of the crown relative to the root. Reduction of crown size may also be correlated with the relative shortening of the tooth row, if tooth number remains constant. A significant consequence of crown reduction is that additional replacement teeth can pack the jaw ramus. Up to five replacement teeth fill a given position in *Diplodocus* (Holland 1924:fig. 3), whereas up to seven are present in *Nigersaurus* (Sereno et al. 1999:fig. 2D). This specialization may allow enhanced tooth replacement rates (see Sereno and Wilson, chapter 5).

PRESACRAL SPECIALIZATIONS

Neural spines vary in length, shape, and orientation throughout Dinosauria, but only within sauropods are they completely divided. Forked neural spines appear several times in Sauropoda, usually as an autapomorphies for genera (e.g., *Camarasaurus, Euhelopus, Opisthocoelicaudia*). Flagellicaudata (Diplodocidae + Dicraeosauridae) is the only suprageneric group characterized by forked neural spines. These usually extend from the anterior cervical neural spines to those of the mid-dorsal region, but they may extend to the anterior caudal neural spines in some taxa (e.g., *Diplodocus*). The forked neural spines are longest in *Amargasaurus* and *Dicraeosaurus*, in which they are more than four times the centrum height, and shortest in *Apatosaurus*, in which they are shorter than the centrum height. A median tubercle may occasionally be present between the two rami of the neural spines of the pectoral region, but this feature is not present in all diplodocoids. Forked neural spines may have been a specialization that allowed passage of elastic ligaments (Janensch 1929b; Alexander 1985), such as the ligamentum nuchae and ligamentum elasticum interspinale (Tsuihiji 2004). The presence of forked neural spines and implied ligaments in some taxa but not in others remains unexplained.

As mentioned earlier, more than a dozen neck lengthening events appear within Neosauropoda, whose basic complement of cervical, dorsal, and sacral vertebrae is 13-12-5. All are autapomorphies for genera, except for one event that characterizes Diplodocidae. The diplodocids *Apatosaurus* and *Diplodocus* incorporated two dorsal vertebrae into the neck to obtain the precaudal count of 15-10-5. Although precaudal counts are not known for other diplodocids, dicraeosaurids and rebbachisaurids do not appear to share this feature.

TAIL SPECIALIZATIONS

The number of caudal vertebrae comprising the tail is fairly similar in outgroups to Sauropoda. Caudal counts are known for the prosauropods *Jingshanosaurus* (44), *Lufengosaurus* (43), and *Plateosaurus* (41), as well as the basal theropods *Eoraptor* (45; Sereno, pers. comm.) and *Herrerasaurus* (43–45). This number is retained in the basal sauropod *Shunosaurus* (43) and slightly increased in more derived sauropods such as *Omeisaurus* (50–55) and *Camarasaurus* (53). A marked increase in the number of caudal vertebrae characterizes Diplodocidae, which nearly doubles the primitive count (*Diplodocus*, 80 + ; *Apatosaurus*, 82). Tail elongation in diplodocids is the result of supernumerary distal caudal vertebrae, of which there are more than 30. These archless distal caudal centra are not only numerous, but also distinctly biconvex and elongate. Although biconvex distal caudal centra are known in other neosauropods, none are as elongate or numerous as in diplodocids (Wilson et al. 1999). Together, this series of 30 or more elongate, biconvex centra constitute a "whiplash" tail, which has been interpreted as a defensive (e.g., Holland 1915) or noisemaking (Myhrvold and Currie 1997) specialization. Although their caudal counts are unknown, the presence of elongate, biconvex caudal centra in the rebbachisaurids *Limaysaurus* (Calvo and Salgado 1995; Salgado 2004) and *Nigersaurus* (personal observation), as well as the dicraeosaurid *Dicraeosaurus* (Janensch 1929b), suggests that the whiplash tail may have been a general diplodocoid feature. However, this can only be confirmed by future discoveries of articulated remains.

MACRONARIA

The second of the two reflexive neosauropod stem-groups, Macronaria includes all neosauropods more closely related to *Saltasaurus loricatus* than to *Diplodocus longus* (Wilson and Sereno 1998). By this definition, Macronaria unites *Camarasaurus, Brachiosaurus, Euhelopus,* and Titanosauria (fig. 1.4). Two node–stem triplets are recognized within Macronaria, one for Titanosauriformes (Wilson and Sereno 1998) and the other for Saltasauridae Sereno (1998), a titanosaur subgroup. Phylogenetic definitions within Macronaria are as follows (boldface type indicates node-based definitions; regular type indicates stem-based definitions):

> Macronaria Wilson and Sereno 1998—All neosauropods more closely related to *Saltasaurus loricatus* than to *Diplodocus longus.*
>
> **Titanosauriformes** Salgado et al. 1997—*Brachiosaurus brancai, Saltasaurus loricatus,* their most recent common ancestor, and all descendants.
>
> Brachiosauridae Riggs 1904—All titanosauriforms more closely related to *Brachiosaurus brancai* than to *Saltasaurus loricatus.*
>
> Somphospondyli Wilson and Sereno 1998—All titanosauriforms more closely related to *Saltasaurus loricatus* than to *Brachiosaurus brancai.*
>
> **Titanosauria** Bonaparte and Coria 1993—*Andesaurus delgadoi, Saltasaurus loricatus,* their most recent common ancestor, and all descendants.
>
> **Saltasauridae** Powell 1992—*Opisthocoelicaudia skarzynskii, Saltasaurus loricatus,* their most recent common ancestor, and all descendants.
>
> Opisthocoelicaudiinae McIntosh 1990—All saltasaurids more closely related to *Opisthocoelicaudia skarzynskii* than to *Saltasaurus loricatus.*
>
> Saltasaurinae Powell 1992—All saltasaurids more closely related to *Saltasaurus loricatus* than to *Opisthocoelicaudia skarzynskii.*

Macronaria is more taxonomically diverse and widespread than its neosauropod counterpart Diplodocoidea. Like other neosauropod lineages, macronarians first appear in the Late Jurassic. However, the simultaneous appearance of *Camarasaurus, Brachiosaurus,* and the possible titanosaur *Janenschia* suggests an earlier origin for the group. Furthermore, trackway evidence may suggest a Middle Jurassic origin for titanosaurs (Wilson and Carrano 1999; Day et al. 2002, 2004; see below) and thus all neosauropod lineages. Macronarians are the only sauropod subgroup to persist until the end of the Cretaceous, represented as titanosaurs in the Maastrichtian in North America (*Alamosaurus* [Gilmore 1946]), India (*Isisaurus* [Jain and Bandyopadhyay 1997]), Europe (*Magyarosaurus* [Huene 1932], *Ampelosaurus* [LeLoeuff 1995]), Asia (*Nemegtosaurus* [Nowinski 1971], *Opisthocoelicaudia* [Borsuk-Bialynicka 1977]), Madagascar (*Rapetosaurus* [Curry Rogers and Forster 2001]), Africa (cf. Titanosauria [Rauhut and Werner 1999]), and South America (*Gondwanatitan* [Kellner and Azevedo

1999]). Titanosauria includes several extremely large forms (e.g, *Antarctosaurus giganteus, Argyrosaurus, Argentinosaurus*; fig. 1.13, left), but also genera diminuitive by sauropod standards (e.g., *Saltasaurus, Neuquensaurus*; fig. 1.13, right). The body size range in Titanosauria exceeds that in other sauropod subgroups and provides an opportunity to evaluate temporal and morphological patterns of body size change within the group, once a genus-level phylogeny is established.

Appendicular synapomorphies comprise a substantial proportion of character support within Macronaria, particularly within its latest-surviving clade, Titanosauria (table 1.5). Although several appendicular synapomorphies apply at basal macronarian nodes, perhaps the most striking changes occur within Titanosauria and are related to the acquisition of a wide-gauge limb posture.

WIDE-GAUGE LIMB POSTURE

Animals with parasagittal limb stance walk or run on land with their limbs held close to the body midline. In these forms, the supporting elements swing anteroposteriorly and contact the substrate near the body midline. As the animal reaches higher speeds, these contacts approach and sometimes touch or cross the midline. A parasagittal limb stance can be observed directly in living therian mammals and in birds (e.g., Muybridge 1957). Squamates and crocodylians, in contrast, have a sprawling gait in which the proximal limb elements are oriented close to the horizontal plane and the limbs contact the substrate at some distance from the body midline (e.g., Blob 2001). Fossilized trackways provide indirect evidence for parasagittal locomotion in extinct dinosaurs (e.g., Thulborn 1982). As observed in living therians and birds, theropod and ornithopod fore- and hindfoot impressions are quite close to or overlap the trackway midline. Likewise, sauropod trackways evidence a parasagittal limb stance, although the placement of the fore- and hindfeet relative to the midline varies within the clade. "Narrow-gauge" sauropod tracks are

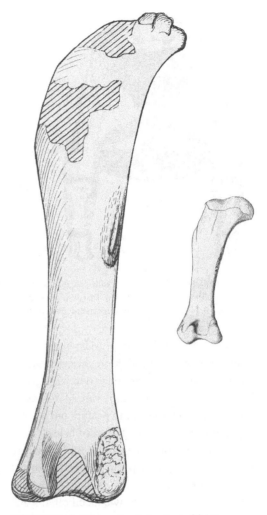

FIGURE 1.13. Left femora (posterior view) of the titanosaurs *Antarctosaurus giganteus* (length, 231 cm) and *Neuquensaurus australis* (length, 70 cm). To scale (from Huene 1929:pl. 20, 36). Both femora correspond to adult individuals.

defined as those in which manus and pes impressions are "close [to] or even intersect the trackway midline," whereas "wide-gauge" trackways are "well away from the trackway midline" (Farlow 1992:108, 109). Variation in gauge width has been inferred to be taxonomic, with narrow-gauge stance presumed to be primitive and wide-gauge stance derived (Wilson and Carrano 1999). Further, the presence of certain morphological characteristics of saltasaurid titanosaurs has suggested that they are the wide-gauge trackmakers. Wilson and Carrano (1999) recognized three hindlimb features that support

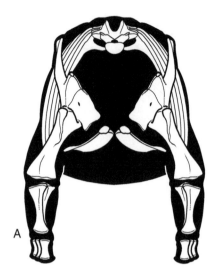 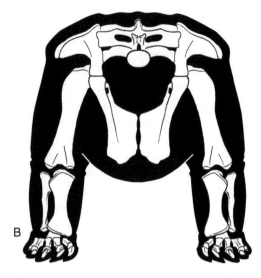

FIGURE 1.14. Limb skeleton of the wide-gauge saltasaurid *Opisthocoelicaudia*. Pectoral girdle and forelimb (A) and pelvic girdle and hindlimb (B) in anterior view. Forelimb reconstruction based on illustrations and photographs in Borsuk-Bialynicka (1977:fig. 9B, pl. 7–9, 11); hindlimb reconstruction modified from Wilson and Carrano (1999).

the hypothesis that saltasaurids were wide-gauge trackmakers. In addition, they recognized forelimb features that are related to wide-gauge locomotion. Still other features are merely associated with wide-gauge limb posture but are not required by it. These are discussed below.

Acquisition of wide-gauge limb posture requires the manus and pes to contact the ground at some distance from the midline. This was achieved in saltasaurids by two modifications that allowed the femur to angle outward from the body wall. First, the proximal third of the femur is canted inward relative to the rest of the shaft (fig. 1.14B). A similar characteristic is present in forest bovids that walk with their femora more abducted than do their closest relatives (Kappelman 1988). Second, the distal condyles are not aligned orthogonal to the long axis of the femur, as in other sauropods. Instead, the distal femoral condyles of saltasaurids are beveled 10° dorsomedially. As shown in figure 1.14, this conformation orients the axis of the knee parallel to the ground and perpendicular to the ground reaction force when the limb is angled away from the body. So far, this feature is restricted to saltasaurids. A third feature that may facilitate a wide-gauge limb posture in saltasaurids is the highly eccentric femoral midshaft cross section. It has already

been mentioned that all sauropod femora (except some diplodocines) are broader mediolaterally that anteroposteriorly. This shape provides greater resistance to mediolateral bending. Saltasaurids, however, exaggerate this feature well beyond that of typical sauropods. This increased femoral eccentricity may have offered greater resistance to the increased bending moment imposed by a wide-gauge limb posture. The distal tibia, whose distal end is diagnostically broader in titanosaurs than in other sauropods, may also be specialized to counter mediolateral bending.

A series of pectoral girdle and forelimb features is related to the acquisition of wide-gauge limb posture in saltasaurids. The anterior thorax and the shoulder girdle are broader in saltasaurids than in other sauropods, owing to the combined effects of the elongate coracoids and the enlarged, crescentic sternal plates (fig. 1.14A). However, because the pectoral girdle has no bony connection to the vertebral column, the absolute distance between the glenoid and the midline cannot be determined. The forelimb is characterized by several reversals of early sauropod synapomorphies associated with the evolution of a columnar, graviportal posture. The humerus in saltasaurids is unique in that it bears a prominent deltopectoral crest, and its distal condyles are both divided and exposed anteriorly. These features are not present in other sauropods. Likewise, the ulna is characterized by a prominent olecranon process that projects above the articular surface of the ulna, as it does in sauropod outgroups but not in other sauropods. These reversals—particularly extension of the ulnar articular surface onto the anterior surface of the humerus and the prominent olecranon process—suggest a more habitually flexed forelimb posture in saltasaurids. Other features are consistent with this interpretation, including the increased transverse diameter of the distal radius, which is a shape that better resists mediolateral bending moments.

A third set of features is novelties of wide-gauge limb trackmakers that do not signal a modified limb posture but they may offer insight into the function of this novel locomotory specialization. Saltasaurids are characterized by a short tail that consists of approximately 35 stout caudals, many fewer than in sauropods primitively and less than half the number in diplodocids. The articular surfaces of all titanosaur caudal centra are concavo-convex; in all but *Opisthocoelicaudia* the anterior face of the centrum is concave (procoelous). Borsuk-Bialynicka (1977) and Wilson and Carrano (1999) suggested that this shortened tail may have functioned as a third support when saltasaurids reared during feeding or mating. A second saltasaurid synapomorphy may also be related to occasional bipedal stance or tripodal rearing. The preacetabular processes or saltasaurid ilia are flared laterally such that they are oriented nearly perpendicular to the body axis. Wilson and Carrano (1999) suggested that, among other effects, flared ilia move the origination site of the femoral protractor muscles laterally, bringing them into anteroposterior alignment with the direction of travel. Finally, one feature peculiar to saltasaurids and their subgroups is a pronounced reduction in the ossification of the carpus, manus, and tarsus. Carpal elements have not been found associated with manual elements in any titanosaur and are not present among the articulated forelimb elements of *Alamosaurus* and *Opisthocoelicaudia* (Gilmore 1946:pl. 4; Borsuk-Bialynicka 1977:29). In both cases, radius, ulna, and metacarpals were all preserved in articulation, but no intervening carpal elements were found. Manual phalanges have been reported only rarely in association with titanosaur skeletons, and never has an ungual been reported. The manual phalanges that have been reported are extremely reduced (e.g., Borsuk-Bialynicka 1977), and it is likely that the manus had no fleshy digits. Like the carpus and manus, the ankle is extremely reduced in the saltasaurids. The saltasaurid astragalus is distinct among dinosaurs in the extreme reduction of its mediolateral diameter, which is subequal to the anteroposterior and proximodistal diameters. In articulation, the pyramidal astragalus of the saltasaurid *Opisthocoelicaudia* contacts the fibula and the lateral aspect of the tibia but does not reach the medial extreme of the distal tibia (fig. 1.14B).

CONCLUSIONS

Sauropods were "successful" dinosaurs by virtue of their geographic distribution, temporal survivorship, biomass, generic diversity, higher-level diversity, and morphological complexity. Although historically studies of sauropod systematics have lagged behind those of other dinosaur subgroups, a burst of analyses in the last decade has begun to elucidate the evolutionary history of the group.

The stratigraphic distribution of the first representatives of Sauropoda and of their sister-taxon Prosauropoda implies a 10 million- to 15-million year missing lineage during which the score of features diagnosing sauropods evolved. Synapomorphies related to precisely occlusion, neck elongation, and columnar posture evolved during the Late Triassic and Early Jurassic and characterize all sauropods. Although all main neosauropod lineages appear simultaneously around the globe in the Late Jurassic, it is probable that neosauropods were present in the Middle Jurassic and possible that they were present in the Early Jurassic. Future sampling of these poorly sampled intervals will better illuminate early neosauropod evolution. The two principal neosauropod subgroups, Macronaria and Diplodocoidea, are the predominant sauropods during the Late Jurassic and Cretaceous. The descent and diversification within these two groups were shaped by changes in different regions of the skeleton. Diplodocoids are characterized by cranial and axial synapomorphies that led to the evolution of a dental battery in rebbachisaurids and a whiplash tail in flagellicaudatans. In contrast, a series of appendicular changes led to the evolution of a wide-gauge limb posture in the macronarian subgroup Titanosauria, which was the dominant sauropod lineage of the Cretaceous, represented on nearly all continental landmasses by more than 40 species.

Despite advances in understanding of the group, substantial gaps in our knowledge of sauropod history still exist. Like the animals themselves, our understanding of sauropod history is deepest in the middle but somewhat thinner on both ends. The sequence of changes leading to the sauropod body plan from the primitive saurischian condition is still poorly understood. For instance, it is not known whether herbivorous specializations preceded large body size and quadrupedality. New discoveries of basal sauropod taxa are needed to address this question. At the other end of sauropod history, phylogenetic understanding of the two latest-surviving sauropod groups, Rebbachisauridae and Titanosauria, are as yet unknown, but new discoveries have already begun to bring clarity to this problem. These two lineages are important biogeographically during the end of sauropod history and may signal an interesting survivorship pattern. Despite the fact that both broad- and narrow-crowned sauropod taxa were present on most continental landmasses, only narrow-crowned taxa survived into the Late Cretaceous each independent case (see Barrett and Upchurch, chapter 4). No broad-crowned sauropod teeth have been reported from Late Cretaceous sediments. Future discoveries and analyses are required to better understand the relationship, if any, between Late Cretaceous survivorship and narrow-crowned dentition.

ACKNOWLEDGMENTS

I thank Jack McIntosh for the advice, encouragement, criticism, and sauropod details he has shared with me during the past decade but, also, for his single-handed championing of sauropods for so many years. I thank Paul Sereno for discussion of many of the ideas presented in this chapter. I also profited from correspondence with Ricardo Martínez, Leo Salgado, and Paul Upchurch. I thank Adam Yates for sharing with me his in press manuscript on *Anchisaurus*. Chris Sidor provided useful information on *Blikanasaurus*. I am grateful for detailed reviews by Cathy Forster and Thomas Holtz, Jr. Figures 1.5–1.8, 1.10, and 1.14 were skillfully completed by B. Miljour from illustrations by J.A.W. This research was supported by grants from The Dinosaur Society, the American Institute for Indian Studies, The Hinds Fund, and the Scott Turner Fund. I obtained translations of several papers from the Polyglot Paleontologist (www.informatics.sunysb.edu/anatomicalsci/paleo/): Bonaparte (1986a, 1997), Bonaparte and Coria (1993), He et al. (1998), and Powell (1992). C. Yu translated excerpts from He et al. (1988) and Zhang (1988). Janensch (1929b) was translated by S. Klutzny through a Jurassic Foundation grant (to J.A.W. and M. Carrano).

LITERATURE CITED

Alexander, R. M. 1985. Mechanics of posture and gait of some large dinosaurs. Zool. J. Linn. Soc. 83:1–25.

Allain, R., Taquet, P., Battail, B., Dejax, J., Richir, P., Véran, M., Limon-Duparcmeur, F., Vacant, R., Mateus, O., Sayarath, P., Khenthavong, B., and Phouyavong, S. 1999. Un nouveau genere de dinosaure sauropode de la formation des Grés supérieurs (Aptien-Albien) du Laos. C.R. Acad. Sci. Paris Sci. Terre Planétes 329: 609–616.

Anonymous. 1905. The presentation of a reproduction of *Diplodocus carnegiei* to the trustees of the

British Museum. Annals of the Carnegie Museum 3: 443–452.

Barrett, P.M., Hasegawa, Y., Manabe, M., Isaji, S., and Matsuoka, H. 2002. Sauropod dinosaurs from the Lower Cretaceous of eastern Asia: taxonomic and biogeographical implications. Palaeontology 45: 1197–1217.

Benton, M.J., Juul, L., Storrs, G.W., and Galton, P.M. 2000. Anatomy and systematics of the prosauropod dinosaur *Thecodontosaurus antiquus* from the Upper Triassic of southwest England. J. Vertebr. Paleontol. 20: 77–108.

Blob, R.W. 2001. Evolution of hindlimb posture in nonmammalian therapsids: biomechanical tests of paleontological hypotheses. Paleobiology 27: 14–38.

Bonaparte, J.F. 1986a. The early radiation and phylogenetic relationships of sauropod dinosaurs, based on vertebral anatomy. In: Padian, K.(eds.). The Beginning of the Age of Dinosaurs. Cambridge University Press, Cambridge. pp. 247–258.

———. 1986b. Les dinosaures (Carnosaures, Allosauridés, Sauropodes, Cétiosaurides) du Jurassique moyen de Cerro Cóndor (Chubut, Argentine). Ann. Paléontol. 72: 325–386.

———. 1997. *Rayososaurus agrioensis* Bonaparte 1995. Ameghiniana 34: 116.

Bonaparte, J.F., and Coria, R.A. 1993. Un nuevo y gigantesco saurópodo titanosaurio de la Formación Río Limay (Albiano-Cenomanio) de la Provincia del Neuquén, Argentina. Ameghiniana 30: 271–282.

Bonnan, M.F. 2003. The evolution of manus shape in sauropod dinosaurs: implications for functional morphology, limb orientation, and phylogeny. J. Vertebr. Paleontol. 23: 595–613.

Borsuk-Bialynicka, M. 1977. A new camarasaurid sauropod *Opisthocoelicaudia skarzynskii*, gen. n., sp. n. from the Upper Cretaceous of Mongolia. Palaeontol. Polonica 37: 1–64.

Buckland, W. 1841. Geology and Minerology Considered with Reference to Natural Theology, 2nd ed. Lea & Blanchard, Philadelphia. 468 pp.

Buffetaut, E., Suteethorn, V., Cuny, G., Tong, H., Le Loeuff, J., Khansubha, S., and Jongautchariyakul., S. 2000. The earliest known sauropod dinosaur. Nature 407: 72–74.

Calvo, J.O. 1994. Jaw mechanics in sauropod dinosaurs. GAIA 10: 183–193.

Calvo, J.O. and González Riga, B.J. 2003. *Rinconsaurus caudamirus* gen. et sp. nov., a new titanosaurid (Dinosauria, Sauropoda) from the Late Cretaceous of Patagonia, Argentina. Revista Geologica de Chile 30: 333–353.

Calvo, J.O., and Salgado, L. 1995. *Rebbachisaurus tessonei*, sp. nov. A new Sauropoda from the Albian-Cenomanian of Argentina; new evidence on the origin of Diplodocidae. GAIA 11: 13–33.

Carrano, M.T. 1997. Morphological indicators of foot posture in mammals: a statistical and biomechanical analysis. Zool. Linn. Soc. 121: 77–104.

———. 2000. Homoplasy and the evolution of dinosaur locomotion. Paleobiology 26: 489–512.

———. 2005. Body-size evolution in the Dinosauria. In: Carrano, M.T., Gaudin, T.J., Blob, R.W., and Wible, J.R. (eds.). Amniote Paleobiology: Phylogenetic and Functional Perspectives on the Evolution of Mammals, Birds and Reptiles. Chicago: University of Chicago Press (in press).

Chiappe, L.M., Coria, R.A., Dingus, L., Jackson, F., Chinsamy, A., and Fox, M. 1998. Sauropod dinosaur embryos from the Late Cretaceous of Patagonia. Nature 396: 258–261.

Chiappe, L.M., Salgado, L., and Coria, R.A., 2001. Embryonic skulls of titanosaur sauropod dinosaurs. Science 293: 2444–2446.

Cooper, M.R. 1981. The prosauropod dinosaur *Massospondylus carinatus* Owen from Zimbabwe: its biology, mode of life and phylogenetic significance. Occas. Papers Natl. Mus. Rhodesia B Nat. Sci. 6: 689–840.

———. 1984. Reassessment of *Vulcanodon* and the origin of the Sauropoda. Palaeontol. Africana 25: 203–231.

Curry, K.A. 2001. The Evolutionary History of the Titanosauria. Ph.D. thesis, State University of New York, Stony Brook.

Curry Rogers, K.A., and Forster, C. 2001. The last of the dinosaur titans: a new sauropod from Madagascar. Nature 412: 530–534.

Curry Rogers, K.A., and Forster, C. 2004. The skull of *Rapetosaurus krausei* (Sauropoda: Titanosauria) from the Late Cretaceous of Madagascar. J. Verteb. Paleontol. 24:121–144.

Dalla Vecchia, F.M. 1994. Jurassic and Cretaceous sauropod evidence in the Mesozoic carbonate platforms of the southern Alps and Dinards. GAIA 10: 65–73.

Day, J.J., Upchurch, P., Norman, D.B., Gale, A.S., and Powell, H.P. 2002. Sauropod trackways, evolution, and behavior. Science 296: 1659.

Day, J.J., Norman, D.B., Gale, A.S., Upchurch, P., and Powell, H.P. 2004. A Middle Jurassic dinosaur trackway site from Oxfordshire, U.K. Palaeontology 47: 319–348.

Desmond, A.J. 1975. The Hot-Blooded Dinosaurs. Blond and Briggs, London. 238 pp.

Farlow, J.O. 1992. Sauropod tracks and trackmakers: integrating the ichnological and skeletal records. Zubía 10: 89–138.

Filla, B. J., and Redman, P. D. 1994. *Apatosaurus yahnapin*: a preliminary description of a new species of diplodocid dinosaur from the Late Jurassic Morrison Formation of southern Wyoming, the first sauropod found with a complete set of "belly ribs." Forty-fourth Annual Field Conference—1994 Wyoming Geological Association Guidebook 44: 159–178.

Flynn, J. J., Parrish, J. M., Rakotosamimanana, B., Simpson, W. F., Whatley, R. L., and Wyss, A. R. 1999. A Triassic fauna from Madagascar, including early dinosaurs. Science 286: 763–765.

Galton, P. M. 1990. Prosauropoda. In: Wieshampel, D. B., Dodson, P., and Osmólska, H. (eds.). The Dinosauria. University of California Press, Berkeley. Pp. 320–345.

Galton, P. M. and Upchurch, P. 2000. Prosauropod dinosaurs: homeotic transformations ("frame shifts") with third sacral as a caudosacral or a dorsosacral. J. Vertebr. Paleontol. 20: 43A.

———. 2004. Prosauropoda. In: Weishampel, D. B., Dodson, P., and Osmólska, H. (eds.), The Dinosauria, 2nd ed. University of California Press, Berkeley. Pp. 232–258.

Galton, P. M., and van Heerden, J. 1985. Partial hindlimb of *Blikanasaurus cromptoni* n. gen. and n. sp., representing a new family of prosauropod dinosaurs from the Upper Triassic of South Africa. Géobios 18: 509–516.

Gauthier, J. A. 1986. Saurischian monophyly and the origin of birds. In: Padian K., ed. The origin of birds and the evolution of flight. Memoirs of the California Academy of Sciences 8: 1–55.

Gierlinski, G. 1997. Sauropod tracks in the Early Jurassic of Poland. Acta Palaeontol. Polonica 42: 533–538.

Gierlinski, G., Pienkowski, G., and Niedzwiedzki, G. 2004. Tetrapod track assemblage in the Hettangian of Soltyków, Poland, and its paleoenvironmental background. Ichnos 11: 195–213.

Gilmore, C. W. 1925. A nearly complete articulated skeleton of *Camarasaurus*, a saurischian dinosaur from the Dinosaur National Monument, Utah. Mem. Carnegie Mus. 10: 347–384.

Gilmore, C. W. 1936. Osteology of *Apatosaurus* with special reference to specimens in the Carnegie Museum. Mem. Carnegie Mus. 11: 175–300.

———. 1946. Reptilian fauna of the North Horn Formation of central Utah. U. S. Geol. Surv. Prof. Paper 210C: 1–52.

González Riga, B. J. 2003. A new titanosaur (Dinosauria, Sauropoda) from the Upper Cretaceous of Mendoza Province, Argentina. Ameghiniana 40: 155–172.

Harris, J. D., and Dodson, P. 2004. A new diplodocoid sauropod dinosaur from the Upper Jurassic Morrison Formation of Montana, USA. Acta Palaeontol. Polonica 49: 197–210.

Hatcher, J. B. 1901. *Diplodocus* (Marsh): its osteology, taxonomy, and probable habits, with a restoration of the skeleton. Mem. Carnegie Mus. 1: 1–63.

———. 1902. Structure of the forelimb and manus of *Brontosaurus*. Ann. Carnegie Mus. 1: 356–376.

He, X.-L., Li, K., and Cai, K.-J. 1988. [The Middle Jurassic dinosaur fauna from Dashanpu, Zigong, Sichuan. Vol. IV. Sauropod Dinosaurs (2) *Omeisaurus tianfuensis*]. Sichuan Scientific and Technological Publishing House, Chengdu. 143 pp.

He, X.-L., Wang, C., Liu, S., Zhou, F., Liu, T., Cai, K., and Dai., B. 1998. [A new sauropod dinosaur from the Early Jurassic in Gongxian County, South Sichuan]. Acta Geol. Sichuan 18: 1–6.

Holland, W. J. 1910. A review of some recent criticisms of the restorations of sauropod dinosaurs existing in the museums of the United States, with special reference to that of *Diplodocus carnegiei* in the Carnegie Museum. Am. Nat. 44: 259–283.

———. 1915. Heads and tails: a few notes relating to the structure of the sauropod dinosaurs. Annals of the Carnegie Museum 9: 273–278.

———. 1924. The skull of *Diplodocus*. Mem. Carnegie Mus. 9: 379–403.

Hopson, J. A. 1995. Patterns in evolution of the manus and pes of non-mammalian synapsids. J. Vertebr. Paleontol. 15: 615–639.

Huene, F. V. 1920. Bemerkungen zur Systematik und Stammesgeschicht einiger Reptilien. Z. Induktive Abstammungslehre Vererbungslehre 22: 209–212.

———. 1926. Vollständige Osteologie eines Plateosauriden aus dem schwäbischen Keuper. Geol. Palaeontol. Abhandlungen 15: 129–179.

———. 1932. Die fossil Reptil-Ordnung Saurischia, ihre Entwicklung und Geschichte. Monograph. Geol. Palaeontol. 4: 1–361.

———. 1956. Paläontologie und Phylogenie der Niederen Tetrapoden. VEB Gustav Fischer Verlag, Jena. 716 pp.

Huxley, T. H. 1869. On the classification of the Dinosauria, with observations on the Dinosauria of the Trias. Q. Rev. Geol. Soc. London 26: 32–51.

Ishigaki, S. 1988. Les empreintes de dinosaures du Jurassique inférieur du Haut Atlas central marocain. Notes Serv. Géol. Maroc 44: 79–86.

Jacobs, L. L., D. A., Winkler, Downs, W. R., and Gomani, E. M. 1993. New material of an Early Cretaceous titanosaurid sauropod dinosaur from Malawi. Palaeontology 36: 523–534.

Jain, S. L., and Bandyopadhyay, S. 1997. New titanosaurid (Dinosauria: Sauropoda) from the

Late Cretaceous of central India. J. Vertebr. Paleontol. 17: 114–136.

Janensch, W. 1914. Ubersicht uber die Wirbeltierfauna der Tendaguru-Schichten, nebst einer kurzen Charakterisierung der neu aufgefuhrten Arten von Sauropoden. Arch. Biontol. 3: 81–110.

———. 1929a. Material und Formengehalt der Sauropoden in der Ausbeute der Tendaguru Expedition. Palaeontographica (Suppl. 7) 2: 1–34.

———. 1929b. Die Wirbelsäule der gattung *Dicraeosaurus*. Palaeontographica (Suppl. 7) 7: 39–133.

———. 1935–1936. Die Schädel der Sauropoden *Brachiosaurus, Barosaurus,* und *Dicraeosaurus* aus den Tendaguru-Schichten Deutsch-Ostafrikas. Palaeontographica (Suppl. 7) 3: 147–298.

———. 1950. Die wirbelsäule von *Brachiosaurus brancai*. Palaeontographica (Suppl. 7) 3: 27–93.

———. 1961. Die Gliedmaszen und Gliedmaszengürtel der Sauropoden der Tendaguru-Schichten. Palaeontographica (Suppl. 7) 3: 177–235.

Kappelman, J. 1988. Morphology and locomotor adaptations of the bovid femur in relation to habitat. J. Morphol. 198: 119–130.

Kellner, A.W.A., and Azevedo, S.A.K. de 1999. A new sauropod dinosaur (Titanosauria) from the Late Cretaceous of Brazil. Proc. Second Gondwanan Dinosaur Symp. 15: 111–142.

Larson, A. 1994. The comparison of morphological and molecular data in phylogenetic systematics. In: Schierwater, B. S. B., Wagner, G. P., and DeSalle, R. (eds.). Molecular Biology and Evolution: Approaches and Applications. Birkhauser Verlag, Basel. Pp. 371–390.

Le Loeuff, J. 1995. *Ampelosaurus atacis* (nov. gen., nov. sp.), un nouveau Titanosauridae (Dinosauria, Sauropoda) du Crétacé supérieur de la Haute Vallée de l'Aude (France). C.R. Acad. Sci. Paris (Ser. II) 321: 693–699.

Le Loeuff, J. 2003. The first articulated titanosaurid skeleton in Europe. In: Meyer, C.A. (ed.). First European Association of Vertebrate Paleontologists Meeting. Natural History Museum, Basel.

Lockley, M.G., Wright, J.L., Hunt, A.G., and Lucas, S.G. 2001. The Late Triassic sauropod track record comes into focus: old legacies and new paradigms. New Mexico Geological Society Guidebook, 52nd Field Conference, Geology of the Llano Estacado. Pp. 181–190

Luo, Y., and Wang, C. 2000. A new sauropod, *Gongxianosaurus*, from the Lower Jurassic of Sichuan, China. Acta Geol. Sinica 74: 132–136.

Luo, Z. 1999. A refugium for relicts. Nature 400: 23–25.

Maier, G. 2003. African Dinosaurs Unearthed. Indiana University Press, Bloomington. 380 pp.

Marsh, O.C. 1878. Principal characters of American Jurassic dinosaurs. Pt. I. Am. J. of Sci. (Ser. 3) 16: 411–416.

———. 1883. Principal characters of American Jurassic dinosaurs: Part VI. Restoration of *Brontosaurus.* Am. J. Sci. (Ser. 3) 26: 81–86.

———. 1884. Principal characters of American Jurassic dinosaurs. Part VII. On the Diplodocidae, a new family of the Sauropoda. Am. J. Sci. (Ser. 3) 27: 161–168.

———. 1895. On the affinities and classification of the dinosaurian reptiles. Am. J. Sci. (Ser. 3) 50: 413–423.

Marsicano, C.A., and Barredo, S.P. 2004. A Triassic tetrapod footprint assemblage from southern South America: palaeobiogeographical and evolutionary implications. Palaeogeogr. Palaeoclimatol. Palaeoecol. 203: 313–335.

Martin, V., Buffetaut, E., and Suteethorn, V. 1994. A new genus of sauropod dinosaur from the Sao Khua Formation (Late Jurassic or Early Cretaceous) of northeastern Thailand. C.R. Acad. Sci. Paris (Ser. II) 319: 1085–1092.

Martínez, R., Giménez, O., Rodríguez, J., Luna, M., and Lamanna, M.C. 2004. An articulated specimen of the basal Titanosaurian (Dinosauria: Sauropoda) *Epachthosaurus sciuttoi* from the early Late Cretaceous Bajo Barreal Formation of Chubut Province, Argentina. J. Vertebr. Paleontol. 24: 107–120.

McIntosh, J.S. 1990. Sauropoda. In: Weishampel, D.B., Dodson, P., and Osmólska, H. (eds.). The Dinosauria. University of California Press, Berkeley. Pp.345–401.

McIntosh, J.S., Miles, C.A., Cloward, K.C., and Parker, J.R. 1996. A new nearly complete skeleton of *Camarasaurus.* Bull. Gunma Mus. Nat. Hist. 1: 1–87.

Myhrvold, N.P., and Currie, P.J. 1997. Supersonic sauropods? Tail dynamics in the diplodocids. Paleobiology 23: 393–409.

Muybridge, E. 1957. Animals in Motion. Brown, L.S.(ed.). Dover, New York.

Novas, F.E. 1989. The tibia and tarsus in Herrerasauridae (Dinosauria, incertae sedis) and the origin and evolution of the dinosaurian tarsus. J. Paleontol. 63: 677–690.

———. 1993. New information on the systematics and postcranial skeleton of *Herrerasaurus ischigualastensis* (Theropoda: Herrerasauridae) from the Ischigualasto Formation (Upper Triassic) of Argentina. J. Vertebr. Paleontol. 13: 400–423.

Nowinski, A. 1971. *Nemegtosaurus mongoliensis* n. gen., n. sp., (Sauropoda) from the uppermost Cretaceous of Mongolia. Palaeontol. Polonica 25:57–81.

Osborn, H. F. 1904. The manus, sacrum and forelimb of Sauropoda. Bull. Am. Mus. Nat. Hist. 20: 181–190.

Osborn, H. F., and Mook, C. C. 1921. *Camarasaurus, Amphicoelias*, and other sauropods of Cope. Mem. Am. Mus. Nat. Hist. 3: 247–387.

Ostrom, J. H. 1969. Osteology of *Deinonychus antirrhopus*, an unusual theropod from the Lower Cretaceous of Montana. Bull. Yale Peabody Mus. Nat. Hist. 30: 1–165.

Ostrom, J. H., and McIntosh, J. S. 1966. Marsh's Dinosaurs, the Collections from Como Bluff. Yale University Press, New Haven, CT, London. 388 pp.

Owen, R 1841. A description of a portion of the skeleton of *Cetiosaurus*, a gigantic extinct saurian occurring in the Oolitic Formation of different parts of England. Proc. Geol. Soc. London 3: 457–462.

———. 1842. Report on British fossil reptiles, Part II. Reptiles. Rep. Br. Assoc. Adv. Sci. 1841: 60–204.

———. 1875. Monographs on the British fossil Reptilia of the Mesozoic formations. Part II. (Genera *Bothriospondylus, Cetiosaurus, Omosaurus*). Palaeontol. Soc. Monogr. 29: 15–93.

Phillips, J. 1871. Geology of Oxford and the Valley of the Thames. Clarendon Press, Oxford. 529 pp.

Pittman, J. G., and Gillette, D. D. 1989. The Briar Site: a new sauropod dinosaur tracksite in Lower Cretaceous beds of Arkansas. In: Gillette, D. D., and Lockley, M. G. (eds.). Dinosaur Tracks and Traces. Cambridge University Press, Cambridge, New York. Pp. 313–332.

Powell, J.E. 1992. Osteologia de *Saltasaurus loricatus* (Sauropoda-Titanosauridae) del Cretácico Superior del Noroeste argentino. In Sanz, J. L. and Buscalioni, A. D., (eds.). Los Dinosaurios y su Entorno Biotico. Instituto "Juan de Valdes", Cuenca. Pp. 165–230.

Raath, M. 1972. Fossil vertebrate studies in Rhodesia: a new dinosaur (Reptilia, Saurischia) from the near the Trias-Jurassic boundary. Arnoldia 30: 1–37.

Rauhut, O. W. M., and Werner, C. 1997. First record of a Maastrichtian sauropod dinosaur from Egypt. Palaeontol. Africana 34: 63–67.

Riggs, E. S. 1904. Structure and relationships of opisthocoelian dinosaurs. Part II: The Brachiosauridae. Field Columbian Museum Geological Series 2: 229–248.

Romer, A. S. 1956. Osteology of the Reptiles. University of Chicago Press, Chicago, 772 pp.

———. 1966. Vertebrate Paleontology. University of Chicago Press, Chicago. 368 pp.

Russell, D. A. 1993. The role of central Asia in dinosaurian biogeography. Can. J. Earth Sci. 30: 2002–2012.

Russell, D.A., and Zheng, Z. 1993. A large mamenchisaurid from the Junggar Basin, Xinjiang, People's Republic of China. Can. J. Earth Sci. 30: 2082–2095.

Salgado, L. 2004. Lower Cretaceous rebbachisaurid sauropods from the Cerro Aguada del León (Lohan Cura Formation), Neuquén Province, northwestern Patagonia, Argentina. J. Vertebr. Paleontol. 24: 903–912.

Salgado, L., Coria, R. A., and Calvo, J. O. 1997. Evolution of titanosaurid sauropods. I: Phylogenetic analysis based on the postcranial evidence. Ameghiniana 34: 3–32.

Salgado, L., Coria, R. A., Chiappe, L. M. 2005. Osteology of the sauropod embryos from the Upper Cretaceous of Patagonia. Acta Paleontol. Polonica 50: 79–92.

Sanz, J. L., Powell, J. E., Le Loeuff, J., Martínez, R., and Pereda-Suberbiola, X. 1999. Sauropod remains from the Upper Cretaceous of Laño (northcentral Spain). Titanosaur phylogenetic relationships. Estud. Mus. Ci. Nat. de Álava 14: 235–255.

Sereno, P. C. 1989. Prosauropod monophyly and basal sauropodomorph phylogeny. J. Vertebr. Paleontol. 9: 38A.

———. 1991. *Lesothosaurus*, "fabrosaurids," and the early evolution of Ornithischia. J. Vertebr. Paleontol. 11: 168–197.

———. 1993. The pectoral girdle and forelimb of the basal theropod *Herrerasaurus ischigualastensis*. J. Vertebr. Paleontol. 13: 425–450.

———. 1998. A rationale for phylogenetic definitions, with application to the higher-level phylogeny of Dinosauria. Neues Jahrb. Geol. Paläontol. Abhandlungen 210: 41–83.

Sereno, P. C., and Novas, F. E. 1993. The skull and neck of the basal theropod *Herrerasaurus ischigualastensis*. J. Vertebr. Paleontol. 13: 451–476.

Sereno, P. C., Beck, A. L., Dutheil, D. B., Larsson, H. C. E., Lyon, G. H., Moussa, B., Sadleir, R. W., Sidor, C. A., Varricchio, D. J., Wilson, G. P., and Wilson, J. A. 1999. Cretaceous sauropods from the Sahara and the uneven rate of skeletal evolution among dinosaurs. Science 286: 1342–1347.

Smith, J. B., Lamanna, M. C., Lacovara, K. J., Dodson, P., Smith, J. R., Poole, J. C., Giengengack R., and Attia, Y. 2001. A giant sauropod dinosaur from an Upper Cretaceous mangrove deposit in Egypt. Science 292: 1704–1706.

Steel, R. 1970. Saurischia. Handb. Paläoherpetol. 13: 1–88.

Sullivan, C., Jenkins, F.A., Gatesy, S.M., and Shubin, N.H. 2003. A functional asssessment of hind foot posture in the prosauropod dinosaur *Plateosaurus*. J. Vertebr. Paleontol. 23: 102A.

Thulborn, R.A. 1982. Speeds and gaits of dinosaurs. Palaeogeogr. Palaeoclimatol. Palaeoecol. 38: 227–256.

———. 1990. Dinosaur Tracks. Chapman and Hall, London. 410 pp.

Tsuihiji, T. 2004. The ligament system in the neck of *Rhea Americana* and its implication for the bifurcated neural spines of sauropod dinosaurs. J. Vertebr. Paleontol. 24: 165–172.

Upchurch, P. 1995. The evolutionary history of sauropod dinosaurs. Philos. Trans. Roy. Soc. London B 349: 365–390.

———. 1998. The phylogenetic relationships of sauropod dinosaurs. Zool. J. Linn. Soc. 124: 43–103.

———. 1999. The phylogenetic relationships of the Nemegtosauridae (Saurischia, Sauropoda). Journal of Vertebrate Paleontology 19: 106–125.

Upchurch, P., and Barrett, P.M. 2000. The evolution of sauropod feeding. In: Sues, H.-D. (ed.). Evolution of Feeding in Terrestrial Vertebrates. Cambridge University Press, Cambridge. pp.79–122.

Upchurch, P., and Martin, J. 2003. The anatomy and taxonomy of *Cetiosaurus* (Saurischia, Sauropoda) from the Middle Jurassic of England. J. Vertebr. Paleontol. 23: 218–231.

Upchurch, P., Hunn, C.A., and Norman, D.B. 2002. An analysis of dinosaurian biogeography: evidence for the existence of vicariance and dispersal patterns caused by geological events. Proc. Roy. Soc. London B 269: 613–621.

Upchurch, P., Barrett, P.M., and Dodson, P. 2004. Sauropoda. In: Weishampel, D.B., Dodson, P., and Osmólska, H. (eds.). The Dinosauria, 2nd ed. University of California Press, Berkeley. Pp. 259–322.

Wilson, J.A. 2002. Sauropod dinosaur phylogeny: critique and cladistic analysis. Zool. J. Linn. Soc. 136: 217–276.

———. 2005a. Redescription of the Mongolian sauropod *Nemegtosaurus mongoliensis* Nowinski (Dinosauria: Saurischia) and comments on Late Cretaceous sauropod diversity. J. Syst. Palaeontol. 3: 283–318.

———. 2005b. Integrating ichnofossil and body fossil records to estimate locomotor posture and spatiotemporal distribution of early sauropod dinosaurs: a stratocladistic approach. Paleobiology 31: 400–423.

Wilson, J.A., and Carrano, M.T. 1999. Titanosaurs and the origin of "wide-gauge" trackways: a biomechanical and systematic perspective on sauropod locomotion. Paleobiology 25: 252–267.

Wilson, J.A., and Sereno, P.C. 1998. Early evolution and higher-level phylogeny of sauropod dinosaurs. Soc. Vertebr. Paleontol. Mem. 5: 1–68 (supplement to J. Vertebr. Paleontol. 18).

Wilson, J.A., Martinez, R.A., and Alcobar, O.A. 1999. Distal tail of a titanosaurian sauropod from the Upper Cretaceous of Mendoza, Argentina. Journal of Vertebrate Paleontology 19: 592–595.

Wilson, J.A., and Upchurch, P. 2003. A revision of *Titanosaurus* Lydekker (Dinosauria–Sauropoda), the first dinosaur genus with a 'Gondwanan' distribution. Journal of Systematic Palaeontology 1: 125–160.

Yates, A.M. 2001. A new look at *Thecodontosaurus* and the origin of sauropod dinosaurs. J. Vertebr. Paleontol. 21: 116A.

———. 2003. A new species of the primitive dinosaur, *Thecodontosaurus* (Saurischia: Sauropodomorpha) and its implications for the systematics of early dinosaurs. J. Syst. Palaeontol. 1: 1–42.

———. 2004. *Anchisaurus polyzelus* Hitchcock: the smallest known sauropod dinosaur and the evolution of gigantism amongst sauropodomorph dinosaurs. Postilla. Pp 1–58.

Yates, A.M., and Kitching, J.W. 2003. The earliest known sauropod dinosaur and the first steps towards sauropod evolution. Proc. Roy. Soc. London B 270: 1753–1758.

Young, C.C. 1941. A complete osteology of *Lufengosaurus huenei* Young (gen. et sp. nov.). Palaeontol. Sinica (C) 7: 1–53.

———. 1947. On *Lufengosaurus magnus* (sp. nov.) and additional finds of *Lufengosaurus huenei* Young. Palaeontol. Sinica (C) 12: 1–53.

Yu, C. 1993. The skull of *Diplodocus* and the phylogeny of the Diplodocidae. PhD Thesis: University of Chicago.

Zhang, Y. 1988. The Middle Jurassic dinosaur fauna from Dashanpu, Zigong, Sichuan. J. Chengdu Coll. Geol. 3: 1–87.

Zhang, Y., and Yang, Z. 1994. [A New Complete Osteology of Prosauropoda in Lufeng Basin, Yunnan, China; *Jingshanosaurus*]. Yunnan Publishing House of Science and Technology, Kunming. 100 pp.

Zhou, Z., Barrett, P.M., and Hilton, J. 2003. An exceptionally preserved Lower Cretaceous ecosystem. Nature 421: 807–814.

APPENDIX 1.1. CHARACTER DESCRIPTIONS

Note: Complete character descriptions for synapomorphies discussed in text and listed in Table 1.1 (from Wilson 2002). Primitive state is indicated as 0; 1–5 represent derived states.

37. Pterygoid, transverse flange (i.e., ectopterygoid process) position: posterior of orbit (0); between orbit and antorbital fenestra (1); anterior to antorbital fenestra (2).
46. Basipterygoid processes, length: short, approximately twice (0); or elongate, at least four times (1) basal diameter.
53. Basipterygoid processes, orientation: perpendicular to (0) or angled approximately 45° to (1) skull roof.
55. Dentary, depth of anterior end of ramus: slightly less than that of dentary at midlength (0); 150% minimum depth (1).
65. Tooth rows, shape of anterior portions: narrowly arched, anterior portion of tooth rows V-shaped (0); broadly arched, anterior portion of tooth rows U-shaped (1); rectangular, tooth-bearing portion of jaw perpendicular to jaw rami (2).
66. Tooth rows, length: extending to orbit (0); restricted anterior to orbit (1); restricted anterior to subnarial foramen (2).
67. Crown-to-crown occlusion: absent (0); present (1).
69. Tooth crowns, orientation; aligned along jaw axis, crowns do not overlap (0); aligned slightly anterolingually, tooth crowns overlap (1).
70. Tooth crowns, cross-sectional shape at midcrown: elliptical (0); D-shaped (1); cylindrical (2).
71. Enamel surface texture: smooth (0); wrinkled (1).
74. Replacement teeth per alveolus, number: two or fewer (0); more than four (1).
80. Cervical vertebrae, number: 9 or fewer (0); 10 (1); 12 (2); 13 (3); 15 or greater (4).
85. Anterior cervical neural spines, shape: single (0); bifid (1).
89. Posterior cervical and anterior dorsal neural spines, shape: single (0); bifid (1).
91. Dorsal vertebrae, number: 15 (0); 14 (1); 13 (2); 12 (3); 11 (4); 10 (5).
93. Dorsal neural spines, length: approximately twice (0) or approximately four times (1) centrum length.
136. Distalmost caudal centra, articular face shape: platycoelous (0); biconvex (1).
137. Distalmost biconvex caudal centra, length-to-height ratio: <4(0); >5(1).
138. Distalmost biconvex caudal centra, number: 10 or fewer (0); more than 30 (1).
149. Posture: bipedal (0); columnar, obligately quadrupedal posture (1).
151. Scapular blade, orientation: perpendicular to (0) or forming a 45° angle with (1) coracoid articulation.
156. Caracoid, anteroventral margin shape: rounded (0); rectangular (1).
158. Sternal plate, shape: oval (0); crescentic (1).
161. Humeral deltopectoral crest, shape: relatively narrow throughout length (0); markedly expanded distally (1).
163. Humeral distal condyles, articular surface shape: restricted to distal portion of humerus (0), exposed on anterior portion of humeral shaft (1).
164. Humeral distal condyle, shape: divided (0); flat (1).
167. Ulnar olecranon process, development: prominent, projecting above proximal articulation (0); rudimentary, level with proximal articulation (1).
170. Radius, distal breadth: slightly larger than (0) or approximately twice (1) midshaft breadth.
172. Humerus-to-femur ratio: <0.60 (0); ≥0.60 (1).
173. Carpal bones, number: three or more (0); two or fewer (1).
175. Metacarpus, shape: spreading (0); bound, with subparallel shafts and articular surfaces that extend half their length (1).
176. Metacarpals, shape of proximal surface in articulation: gently curving, forming a 90° arc (0); U-shaped, subtending a 270° arc (1).
181. Manual digits II and III, phalangeal number: 2-3-4-3-2 or more (0); reduced, 2-2-2-2-2 or less (1); absent or unossified (2).
187. Iliac preacetabular process, orientation: anterolateral to (0) or perpendicular to (1) body axis.

198. Femoral midshaft, transverse diameter: subequal to (0), 125%–150%, or (1) at least 185% (2) anteroposterior diameter.

199. Femoral shaft, lateral margin shape: straight (0); proximal one-third deflected medially (1).

201. Femoral distal condyles, orientation: perpendicular or slightly beveled dorsolaterally (0) or beveled dorsomedially approximately 10° (1) relative to femoral shaft.

205. Tibia, distal breadth: approximately 125% (0) or more than twice (1) midshaft breadth.

210. Astragalus, shape: rectangular (0); wedge-shaped, with reduced anteromedial corner (1).

216. Distal tarsals 3 and 4: present (0); absent or unossified (1).

217. Metatarsus, posture: bound (0); spreading (1).

221. Metatarsal I, minimum shaft width: less than (0) or greater than (1) that of metatarsals II–IV.

223. Metatarsal III length: more than 30% (0) or less than 25% (1) that of tibia.

224. Metatarsals III and IV, minimum transverse shaft diameters: subequal to (0) or less than 65% (1) that of metatarsals I or II (1).

225. Metatarsal V, length: shorter than (0) or at least 70% (1) length of metatarsal IV.

228. Pedal unguals, orientation: aligned with (0) or deflected lateral to (1) digit axis.

233. Pedal digit IV ungual, development: subequal in size to unguals of pedal digits II and III (0); rudimentary or absent (1).

TWO

Titanosauria

A PHYLOGENETIC OVERVIEW

Kristina Curry Rogers

TITANOSAUR BODY FOSSILS HAVE been recovered from every landmass except Antarctica and are present in Upper Jurassic to Upper Cretaceous strata. Their unique, wide-gauge trackways extend their record back still farther, to the Middle Jurassic (Santos et al. 1994; Wilson and Carrano 1999; Day et al., 2002; Wilson and Upchurch 2003). In spite of their extensive temporal and geographic ranges, their fossil record remains relatively poor, with most genera based only on fragmentary and/or disassociated postcranial remains. Presently, Titanosauria includes over 30 currently accepted genera representing more than one-third of sauropod diversity, and new titanosaur genera are being erected more prolifically than for any other group of sauropods (see Wilson and Curry Rogers, Introduction).

New titanosaur finds from around the globe paint a picture of a dramatically diverse group, which includes bizarre armored forms (e.g., *Agustinia*, [Bonaparte 1998], *Magyarosaurus* [Csiki 1999], *Neuquensaurus* [Powell 2003], *Saltasaurus* [Bonaparte and Powell 1980]) and imposing giants (*Argentinosaurus* [Bonaparte and Coria 1993], *Paralititan* [Smith et al. 2001]).

The discovery of preserved titanosaur eggs and embryos at Auca Maheuvo in Argentina illuminates aspects of their ontogeny and reproductive biology (Chiappe et al. 2001; see Chiappe et al., chapter 10). New bone histological research promises to provide important new insight into the growth strategies of various titanosaur taxa (e.g., Erickson et al. 2001; see Curry Rogers and Erickson, chapter 11), including the largest terrestrial vertebrates of all time (e.g., *Argentinosaurus*). There is even a hint of dwarfism among some titanosaur taxa (e.g., *Magyarosaurus* [Jianu and Weishampel 1999], *Saltasaurus* and *Neuquensaurus* [Coiki 1999; Powell 2003]).

In spite of these advances, Titanosauria has remained a phylogenetic puzzle. Although recent consensus regarding the macronarian origin of titanosaurs has been attained, included taxa and their interrelationships are still debated (e.g., Salgado et al. 1997; Wilson and Sereno 1998; Upchurch 1998; Sanz et al. 1999; Smith et al. 2001; Curry Rogers and Forster 2001, Wilson 2002; Wilson and Upchurch 2003, Martínez et al. 2004; Apesteguía 2004). Even the type genus and species, '*Titanosaurus indicus*,' has been invalidated (Wilson and Upchurch 2003). The only means of improving phylogenetic resolution for Titanosauria is through more detailed analyses and addition of well-preserved specimens. Taxa that are known from both cranial and postcranial data are essential if we are to bridge the gaps in our current understanding of titanosaur anatomy. This "whole-skeleton" view of Titanosauria allows controversial points in existing phylogenetic hypotheses to be tested with a more robust data set.

In this chapter I consider some of the contentious aspects of titanosaur phylogeny, with a focus on within-group relationships. The discussion includes an historical primer and brief description of *Rapetosaurus krausei*, a nearly complete titanosaur from the Upper Cretaceous Maevarano Formation of Madagascar. The chapter concludes with an analysis of titanosaur phylogeny that utilizes cranial and postcranial data. This phylogenetic treatment is intended to serve as a framework and catalyst for continued analyses of titanosaur biology, evolution, and paleobiogeography.

BACKGROUND ON TITANOSAUR SYSTEMATICS

TAXONOMIC BACKGROUND

The first titanosaur specimens were discovered in the late 1800s in India (*'Titanosaurus indicus'* Falconer 1868; Lydekker 1877; *'T.' blanfordi* Lydekker 1879), Europe (*Macruosaurus* Seeley 1869, 1876), Argentina (*Argyrosaurus* Lydekker 1893), and Madagascar (*'T.' madagascariensis* Depéret 1896a, 1896b). The definition and position of Titanosauridae have remained ambiguous since its original usage in 1893, and the taxon has long been used as a receptacle for enigmatic and fragmentary Cretaceous sauropods, with little agreement on included group members (Gilmore 1946; Romer 1956, 1966, 1968; Steel 1970; McIntosh 1990a, 1990b). For example, Romer (1956, 1966, 1968) united the Cretaceous taxa with procoelous caudal vertebrae (Titanosauria *sensu stricto* Bonaparte and Coria 1993) and placed diplodocoids within Titanosauridae. Similarly, Steel (1970) included Titanosauria *sensu stricto* and diplodocoids within a new group, Atlantosauridae. More recent workers following traditional methods for higher-level sauropod classification either restored Titanosauridae to family level (e.g., Berman and McIntosh 1978; Powell 2003; McIntosh 1989, 1990a, 1990b) or ignored titanosaurs altogether (e.g., Bonaparte 1986).

Titanosauria was established by Bonaparte and Coria (1993), who recognized the need for a higher taxon to subsume two distinct titanosaur groups: (1) Andesauridae, whose members are characterized by the presence of hyposphene-hypantra and amphiplatyan caudals, and (2) Titanosauridae, whose members lack hyposphene-hypantra and exhibit procoelous caudals. Andesauridae, also coined by Bonaparte and Coria (1993), is based only on primitive characters that specify a paraphyletic group and must be considered an informal name until taxa are identified that share synapomorphies with *Andesaurus* (Wilson and Upchurch 2003).

In a recent revision of 14 *Titanosaurus* species, including *'T. indicus'* (the type species) Wilson and Upchurch (2003) recognized only five valid species and deemed the type species *'T. indicus'* a *nomen dubium* due to its basis on obsolete characters. Consequently, they proposed the abandonment of all coordinated rank-taxa, including Titanosaurinae, Titanosauridae, and Titanosauroidea, and proposed and defined a standardized node and stem-based nomenclature for Titanosauria (table 2.1). This nomenclatural system is particularly significant given the inconsistent usage of names and taxa employed in the cladistic analyses outlined below (table 2.2, fig. 2.1).

PHYLOGENETIC BACKGROUND

The phylogenetic relationships of Titanosauria have received relatively little attention, in part because of the frustratingly fragmentary nature of most genera. In the past decade, only a handful of phylogenetic analyses have examined

TABLE 2.1
Phylogenetic Definitions for Titanosauria and Its Subclades

Titanosauria (Bonaparte and Coria 1993)	*Andesaurus delgadoi* (Calvo and Bonaparte 1991), *Saltasaurus loricatus* (Bonaparte and Powell 1980), their most recent common ancestor, and all descendants.
Lithostrotia (Upchurch et al. 2004)	*Malawisaurus dixeyi* (Haughton 1928), *Saltasaurus loricatus* (Bonaparte and Powell 1980), their most recent common ancestor, and all descendants.
Saltasauridae (Bonaparte and Powell 1980)	*Opisthocoelicaudia skarzynskii* (Borsuk-Bialynicka 1977), *Saltasaurus loricatus* (Bonaparte and Powell 1980), their most recent common ancestor, and all descendants.
Saltasaurinae (Bonaparte and Powell 1980)	All saltasaurids more closely related to *Saltasaurus loricatus* (Bonaparte and Powell 1980) than to *Opisthocoelicaudia skarzynskii* (Borsuk-Bialynicka 1977).
Opisthocoelicaudiinae (McIntosh 1990)	All saltasaurids more closely related to *Opisthocoelicaudia skarzynskii* (Borsuk-Bialynicka 1977) than to *Saltasaurus loricatus* (Bonaparte and Powell 1980).

NOTE: As outlined by Wilson and Upchurch (2003) and as employed in phylogenetic analysis, node-based definitions are in boldface type; stem-based definitions are in regular type.

Titanosauria in detail (Salgado et al. 1997; Curry 2001; Curry Rogers and Forster 2001; Curry Rogers 2001; Wilson 2002; Upchurch et al. 2004). Each of these analyses employed a different array of taxa, though a core group of relatively well-known titanosaurs was common to all (table 2.2). Most other cladistic studies have focused on higher-level sauropod phylogeny (Gauthier 1986; Yu 1993; Wilson and Sereno 1994, 1998; Calvo and Salgado 1995; Upchurch 1995, 1998, 1999; Wilson 1999; Sanz et al., 1999). Several of these studies have included one or more titanosaurs or have used the higher-level grouping Titanosauria as a terminal taxon. However, relationships within Titanosauria in these analyses tend to be poorly supported (e.g., Salgado et al. 1997; Upchurch 1998; Sanz et al. 1999; Curry Rogers and Forster 2001) or utilize only a small fraction of available characters. Several of these previous analyses are summarized below (table 2.2, figs. 2.1, 2.2).

Salgado et al. (1997) provided the first treatment of titanosaur ingroup relationships. They analyzed 38 postcranial characters (20 axial, 18 appendicular) in 10 titanosaurs and six basal sauropod taxa. Two most parsimonious trees (length, 54 steps; CI = 0.81) were generated, differing only in the placement of *Malawisaurus* and *Epacthosaurus*. Their consensus hypothesis was well resolved at the base of the tree, particularly with regard to all nontitanosaurs (fig. 2.1A). Salgado et al. (1997) were first to formally recognize the close relationship between *Brachiosaurus* and Titanosauria within Titanosauriformes and diagnosed Titanosauria with three unambiguous postcranial synapomorphies. Though character support for nodes within Titanosauria was fairly poor, Salgado et al. (1997) refuted Bonaparte and Coria's (1993) hypothesis of two distinct families within Titanosauria (Andesauridae and Titanosauridae) and, instead, only supported Titanosauridae monophyly. They included all titanosaurs except *Andesaurus* within this group based on the presence of proximal caudals with strongly procoelous "ball and socket" articulations. Saltasaurinae constituted the only other named taxon in the phylogenetic hypothesis of Salgado et al. (1997). It was supported by two

TABLE 2.2
Genera Employed for Seven Analyses of Titanosaur Relationships

	TITANOSAUR GENERA		TITANOSAUR GENERA
Salgado et al. 1997	*Aeolosaurus*	Curry 2001	**Augustinia**
	Alamosaurus		**Epacthosaurus**
	Andesaurus		'Jainosaurus'
	Argentinosaurus		Janenschia
	Epacthosaurus		Lirainosaurus
	Malawisaurus		**Magyarosaurus**
	Neuquensaurus		Malawisaurus
	Opisthocoelicaudia		**Malagasy Taxon B**
	Saltasaurus		Nemegtosaurus
	'Titanosaurinae indet.'		Neuquensaurus
			Opisthocoelicaudia
Upchurch 1998	Alamosaurus		**Paralititan**
	Andesaurus		Phuwiangosaurus
	Malawisaurus		Quaesitosaurus
	Opisthocoelicaudia		Rapetosaurus
	Phuwiangosaurus		**Rocasaurus**
	Saltasaurus		Saltasaurus
Sanz et al. 1999	Andesaurus		**Santa Rosa indet.**
	Argyrosaurus		Isisaurus
	Epacthosaurus		(=Titanosaurus)
	Lirainosaurus		colberti
	Opisthocoelicaudia		
	Peirópolis form	Wilson 2002	Alamosaurus
	Saltasaurus		Malawisaurus
			Nemegtosaurus
Curry Rogers			Neuquensaurus
and Forster 2001	Alamosaurus		Opisthocoelicaudia
	Antarctosaurus		Rapetosaurus
	Malawisaurus		Saltasaurus
	Nemegtosaurus		Isisaurus
	Neuquensaurus		(=Titanosaurus)
	Opisthocoelicaudia		colberti
	Quaesitosaurus		
	Rapetosaurus	Upchurch et al. 2004	Alamosaurus
	Saltasaurus		Andesaurus
	Isisaurus		Argentinosaurus
	(=Titanosaurus)		**Austrosaurus**
	colberti		**Gondwanatitan**
			Lirainosaurus
Curry 2001	**Aegyptosaurus**		Malawisaurus
	Aeolosaurus		Opisthocoelicaudia
	Alamosaurus		**Pellegrinisaurus**
	Ampelosaurus		Phuwiangosaurus
	Andesaurus		Saltasaurus
	Antarctosaurus		Isisaurus
	Argentinosaurus		(=Titanosaurus)
	Argyrosaurus		colberti

NOTE: Boldface type indicates genera included in only one analysis.

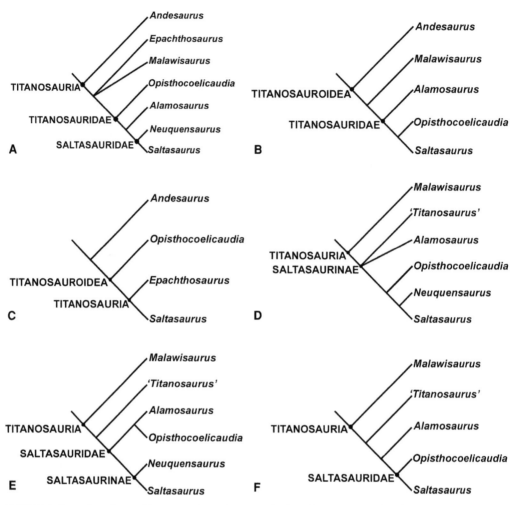

FIGURE 2.1. Recent hypotheses of titanosaur relationships. (A) Salgado et al. (1997); (B) Upchurch (1998); (C) Sanz et al. (1999); (D) Curry Rogers and Forster (2001); (E) Wilson (2002); (F) Upchurch et al. (2004). Topologies have been simplified to reflect genera included in more than one analysis (see table 2.2). (Modified from Wilson and Upchurch 2003.)

unambiguous synapomorphies related to caudal vertebral morphology and included the South American *Neuquensaurus* and *Saltasaurus*.

Upchurch (1998) analyzed 205 characters (74 cranial, 76 axial, 68 appendicular) among 26 sauropods, including 5 titanosaurs (*Alamosaurus, Andesaurus, Malawisaurus, Opisthocoelicaudia, Saltasaurus*) and several other controversial taxa (*Euhelopus, Phuwiangosaurus, Nemegtosaurus, Quaesitosaurus*) (fig. 2.1B). Although Upchurch (1995) originally supported the monophyly of a group including titanosaurs and diplodocoids as sister-taxa, the 1998 revision based on additional character data and taxa supported the monophyly of Titanosauriformes. Within Titanosauriformes,

Upchurch defined Titanosauroidea as all taxa more closely related to "true titanosaurids" (e.g., *Saltasaurus*) than they are to brachiosaurids. In terms of lower-level relationships, Upchurch provided the first evidence for inclusion of *Phuwiangosaurus* as a basal titanosaur. As in the analysis by Salgado et al. (1997), *Andesaurus* and *Malawisaurus* were successive sister-taxa to other titanosaurs. In spite of the recognized paraphyly of Andesauridae (Salgado et al. 1997), Upchurch (1998) retained the name in his systematic classification of the Sauropoda to designate a group of basal titanosaurs including *Phuwiangosaurus, Andesaurus,* and *Malawisaurus*. Upchurch (1998) restricted the Titanosauridae to the clade

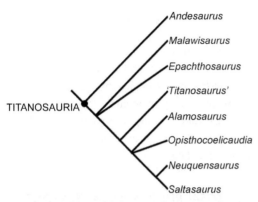

FIGURE 2.2. Adams consensus tree of titanosaur relationships based on the six analyses depicted in figure 2.1. (Modified from Wilson and Upchurch 2003.)

including *Alamosaurus*, *Opisthocoelicaudia*, and *Saltasaurus*. Upchurch (1998, 1999) did not support the inclusion of *Nemegtosaurus* and *Quaesitosaurus* within Titanosauroidea.

Wilson and Sereno (1998) analyzed 109 characters (32 cranial, 24 vertebral, 53 appendicular) for 10 sauropod taxa and obtained a single most parsimonious tree with 153 steps (fig. 2.1C). Titanosauria was included as a higher-level terminal taxon. Wilson and Sereno (1998) corroborated Titanosauriformes monophyly, but within Titanosauriformes they hypothesized a sister-group relationship between *Euhelopus* and Titanosauria within Somphospondylii. Relationships among titanosaurs were not discussed in a phylogenetic context in this analysis of higher-level sauropod relationships, but a few points of note were mentioned regarding Titanosauria as a terminal taxon (Wilson and Sereno 1998). They suggested that *Andesaurus* and *Malawisaurus* are basal titanosaurs that have an important role in highlighting derived features among the clade and confirmed that there is no phylogenetic support for Andesauridae as defined by Bonaparte and Coria (1993). In addition, Wilson and Sereno (1998) included both *Nemegtosaurus* and *Quaesitosaurus* within Titanosauria.

Sanz et al. (1999) conducted a phylogenetic analysis of 43 characters (1 cranial, 27 vertebral, 15 appendicular) derived from seven titanosaurs and the two neosauropods. They obtained a single most parsimonious tree of 70 steps (fig. 2.1D). A clade containing *Haplocanthosaurus* and *Andesaurus* was the sister-taxon to Titanosauroidea, thus revising Upchurch's (1998) use of the term to exclude *Andesaurus*. Within Titanosauroidea, Sanz et al. (1999) supported the monophyly of Titanosauria and proposed Eutitanosauria, for a more derived clade of titanosaurs diagnosed, in part, by the presence of dermal armor.

Curry Rogers and Forster (2001) analyzed 228 characters (74 cranial and 154 postcranial) among 16 ingroup taxa and supported the monophyly of Titanosauriformes (fig. 2.1E). Sixteen additional steps were required to move Titanosauria into its more traditional position as the sister-taxon to Diplodocoidea. The monophyly of Titanosauria was strongly supported by 20 characters, 9 of which were unambiguous, and two main titanosaur clades were distinguished by the analysis. One clade included *Rapetosaurus*, *Nemegtosaurus*, *Malawisaurus*, *Quaesitosaurus*, and *Antarctosaurus* and was diagnosed by four unambiguous cranial synapomorphies. *Malawisaurus* was the sister-taxon of *Quaesitosaurus* (*Rapetosaurus* + *Nemegtosaurus*). *Rapetosaurus* and *Nemegtosaurus* were united by seven ambiguous synapomorphies, five of which occur in one or more diplodocoids. A second clade (Saltasaurinae in their analysis) included *Alamosaurus*, '*T.*' *colberti*, Neuquensaurus, Saltasaurus, and *Opisthocoelicaudia*.

Wilson (2002) scored 234 morphological characters (76 cranial, 72 axial, 85 appendicular, 1 dermal) for 27 taxa. Three equally parsimonious trees resulted, each supporting 26 internal nodes (fig. 2.1F). In this analysis Titanosauriformes, Somphospondylii, and monophyly of Titanosauria were well supported. Titanosauria united *Malawisaurus*, Nemegtosauridae (=*Nemegtosaurus* + *Rapetosaurus*), '*T.*' *colberti* (=*Isisaurus colberti* [Wilson and Upchurch 2003]), and Saltasauridae (=Opisthocoelicaudiinae + Saltasaurinae). Wilson (2002) identified several problematic areas of the phylogeny, including the weak resolution of '*T.*' *colberti* (Wilson and Upchurch 2003) and *Nemegtosaurus* relative to Saltasauridae. The problems in resolution were hypothesized to reflect the incompletely known skeletons of these taxa and the nonoverlap of preserved anatomy among titanosaurs.

Upchurch et al. (2004) conducted a phylogenetic analysis of 309 characters among 41 sauropod genera. Characters in this analysis were synthesized from Wilson and Sereno (1998), Upchurch (1995, 1998, 1999), and Curry Rogers and Forster (2001). They obtained 1,056 most parsimonious trees and applied reduced consensus methods, which indicated that 36 of the 41 sampled sauropod genera have stable positions across all 1,056 most parsimonious trees. Deletion of five taxa, including the titanosaurs *Andesaurus* and *Argentinosaurus*, resulted in a single fully resolved topology (fig. 2.1G). A bootstrap analysis produced low support values (<50%) for all resolved clades within Titanosauria. Their analysis supported Titanosauriformes and Titanosauria monophyly, as well as a monophyletic Lithostrotia (diagnosed by the presence of osteoderms) and Saltasauridae (the least inclusive clade containing *Opisthocoelicaudia* and *Saltasaurus*). Notably, they did not include *Nemegtosaurus* or *Quaesitosaurus* within Titanosauria but, instead, placed them within Diplodocoidea. Though bootstrap values for this inclusion were low, the clade uniting these two Mongolian sauropod skulls was high (60%).

CONSENSUS AND CONTROVERSY

In spite of the diversity of hypotheses in each of the analyses outlined above, some resolution on titanosaur relationships is apparent (figs. 2.1, 2.2). Most analyses agree that *Andesaurus* is a titanosaur, as are *Malawisaurus*, and *Isisaurus* (= '*T.*' *colberti* [Wilson and Upchurch 2003]). More derived titanosaurs include *Opisthocoelicaudia*, *Alamosaurus*, and a group most commonly uniting *Saltasaurus* and *Neuquensaurus*. That said, Titanosauria is among the most generic-rich of any sauropod clade, and few of the individual specimens include more than isolated (and often poorly preserved) postcranial elements. Given the disparity of data and taxa considered, it is no wonder that general agreement and resolution on titanosaur ingroup phylogeny are lacking. Though new taxa continue to be named and described at a rapid pace (see Introduction), few provide the cranial and postcranial character data required if we are to address the most pertinent questions of titanosaur phylogeny. The only means of improved phylogenetic resolution for Titanosauria is through more detailed analyses and addition of key specimens with cranial and postcranial data.

RAPETOSAURUS KRAUSEI, A KEYSTONE TITANOSAUR

The recent discovery and description of *Rapetosaurus krausei* from the Upper Cretaceous Maevarano Formation of Madagascar (Curry Rogers and Forster 2001, 2004; Curry, 2001) provide the cranial and postcranial data from juvenile and adult skeletons required for more detailed, total-evidence analyses of titanosaur phylogeny.

HISTORICAL BACKGROUND

Fragmentary sauropod remains have been known from the Late Cretaceous of the Mahajanga Basin (Maevarano Formation, Campanian(?)–Maastrichtian [Rogers et al. 2000]) for over a century. In 1896, a shipment of fragmentary dinosaurian remains from the northwest coast of Madagascar arrived in the office of French paleontologist Charles Depéret. The sauropod remains—namely, two weathered procoelous caudal vertebrae (UCB [Université Claude Bernard, Lyon] 92829, UCB 92305; fig. 2.3A, B), a partial humeral diaphysis (UCB 92831; fig. 2.3C), and a large osteoderm (UCB 92827; fig. 2.3D) (Depéret 1896a, 1896b)—were recovered from several distinct localities (fig. 2.4A).

Depéret (1896a) assigned the four elements from Madagascar to a new species of titanosaur, *Titanosaurus madagascariensis*. At that early date, two titanosaurs were already known from India (*T. indicus* Lydekker 1877, *T. blanfordi* Lydekker 1879), and one taxon was recognized from the Wealden Sandstone of the Isle of Wight (*Macrurosaurus semnus* Seeley 1869, 1876). The

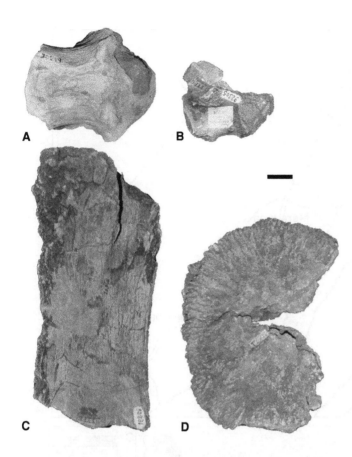

FIGURE 2.3. Syntype elements of 'Titanosaurus madagascariensis' (Depéret 1896a). (A) UCB (Universíte Claude Bernard) 92829, proximal caudal centrum in ventral view. This centrum is now referred to *Rapetosaurus krausei*. (B) UCB 92305, midcaudal centrum in left lateral view. This centrum is now referred to Malagasy Taxon B (Curry Rogers 2001). (C) UCB 92831 partial humeral diaphysis in anterior view. (D) UCB 92827 osteoderm in dorsal view. Scale bar equals 3 cm.

extremely procoelous caudal centra and proximally positioned neural arches of the Malagasy caudal vertebrae indicated a close relationship with the Indian titanosaurs, but several distinctions, including the deeply excavated ventral centrum and a double chevron articulation, warranted assignment to a new species. Depéret (1896a) was prescient in attributing the large dermal ossification to the new species of titanosaur. He cited the unique pattern of vascularization to rule out other dinosaurian forms characterized by dermal ossifications (e.g., ankylosaurs, stegosaurs), but this interpretation was subsequently called into question (e.g., Russell et al., 1976; Hoffstetter 1957). In more recent years, dermal ossifications have been recognized from titanosaurs around the world (e.g., Csiki 1999; Munyikwa et al., 1998), including new discoveries from the same strata in the Mahajanga Basin of Madagascar (Dodson et al. 1998).

Collecting efforts continued sporadically in the early 1900s (e.g., Thévenin 1907), and all the collected titanosaur bones, including proximal and midcaudals and fragmentary limb elements from the Mahajanga Basin, were referred to '*T.*' *madagascariensis*. Huene (1929) later subsumed the '*T.*' *madagascariensis* syntype material within the South American genus *Laplatasaurus*, and additional remains were reported from India (cf. *Laplatasaurus madagascariensis* Huene and Matley 1932). This reassessment stemmed from the more robust "shorter and wider" caudal centrum (UCB 92305) included in the '*T.*' *madagascariensis* syntype. More recent authors questioned Huene's assessment and reinstated the name '*T.*' *madagascariensis* for all Late Cretaceous sauropod material from Madagascar (e.g., Lavocat 1955 Besairie 1972; Russell et al. 1976; McIntosh 1990a, 1990b; Ravoavy 1991).

A REVISED PERSPECTIVE

Over the last decade, a diverse Late Cretaceous vertebrate assemblage that includes nonavian and avian theropod dinosaurs, fish, crocodiles,

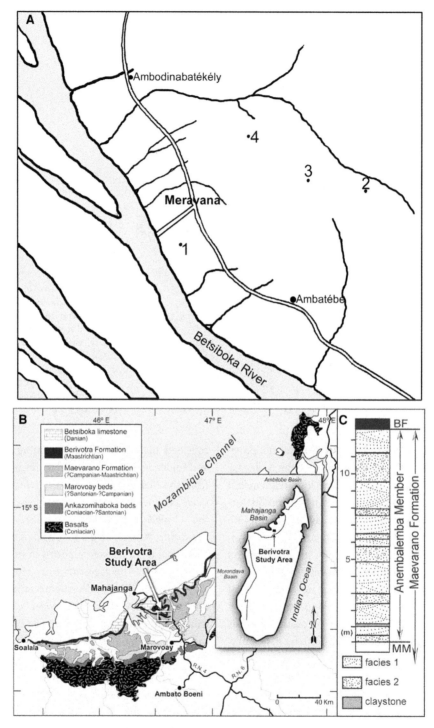

FIGURE 2.4. (A) Map modified from Depéret's (1896) report of fossil vertebrates from the Late Cretaceous of the Mahajanga Basin of northwestern Madagascar. Four localities were recorded by collectors as indicated. (B) Outcrop map of Upper Cretaceous and Tertiary strata in the Mahajanga Basin of northwestern Madagascar. Berivotra study area is highlighted. (C) Schematic profile of the Anembalemba Member (modified from Rogers 2004.) The Anembalema member in the Berivotra area is underlain by the Masorobe member of the Maevarano Formation (MM) and overlain by the marine Berivotra Formation (BF).

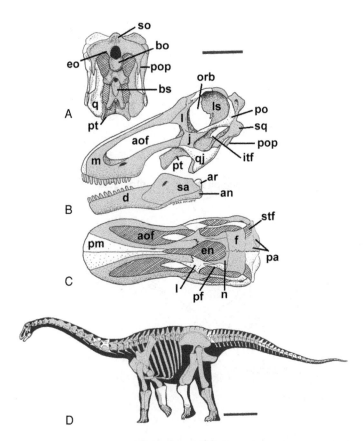

FIGURE 2.5. *Rapetosaurus krausei* from the Upper Cretaceous Maevarano Formation of Madagascar. This skull reconstruction is a composite based on holotype (UA 8698) and referred specimens (FMNH PR 2184–2192, 2194, 2196–2197) and shown in (A) posteroventral view, (B) left lateral view, and (C) dorsal view. Shaded bones are preserved. an, angular; aof, antorbital fenestra; ar, articular; bo, basioccipital; bs, basisphenoid; d, dentary; en, external nares; eo, exoccipital; f, frontal; itf, infratemporal fenestra; j, jugal; l, lacrimal; ls, laterosphenoid–orbitosphenoid; m, maxilla; n, nasal; orb, orbit; pa, parietal; pf, prefrontal; pm, premaxilla; po, postorbital; pop, paroccipital process; pt, pterygoid; q, quadrate; qj, quadratojugal; sa, surangular; so, supraoccipital; sq, squamosal; stf, supratemporal fenestra. Scale bar equals 10 cm. (D) *Rapetosaurus krausei* skeletal reconstruction based on a referred juvenile skeleton (FMNH PR 2209) recovered from a bonebed 2 in MAD93-18. Scale bar equals 1 m. Bones in gray are preserved. (Skull and skeletal reconstructions by Mark Hallett; modified from Curry Rogers and Forster 2001.)

mammals, frogs, turtles, and snakes has been recovered from the Anembalemba Member of the Maevarano Formation (fig. 2.4B) (e.g., Krause et al. 1994, 1997, 1999; Forster et al. 1996, 1998; Sampson et al. 1996, 1998, 2001; Curry 1997; Asher and Krause 1998; Gottfried and Krause 1998; Buckley and Brochu 1999; Buckley et al. 2000; Rogers et al. 2000; Krause 2001). All told, hundreds of new sauropod fossils have been recovered including associated and articulated remains. These new finds indicate the presence of two distinct titanosaur taxa in the Upper Cretaceous strata of Madagascar (Ravoavy 1991; Curry 2001; Curry Rogers and Forster 1999a, 1999b, 2001). The new taxa are most readily distinguished on the basis of their caudal vertebral morphologies along with other postcranial variations. Both caudal morphologies are included in the '*Titanosaurus*' *madagascariensis* syntype, which calls the validity of '*T.*' *madagascariensis* into question. Pending further study of the sauropod fauna in Madagascar (namely, Malagasy Taxon B), '*T.*' *madagascariensis* must be considered a *nomen dubium* (Curry 2001; Curry Rogers and Forster 2001; Wilson and Upchurch 2003).

RAPETOSAURUS KRAUSEI

The holotypic adult *Rapetosaurus krausei* skull, along with a well-preserved, associated juvenile postcranial skeleton with a partial skull, provides our first look at titanosaur anatomy from head to tail (fig. 2.5) (Curry 2001; Curry Rogers and Forster, 2001). The disarticulated bones of the holotype skull were recovered from a single stratum over an area of ~1 m² at locality MAD96-02 (fig. 2.6). Holotypic and the referred juvenile skeleton show no duplication of elements and exhibit exact sutural articulations. Additional skull material from a juvenile specimen was found in direct association with a nearly complete juvenile postcranial skeleton at locality MAD93-18 (fig. 2.7). As in the adult skull, juvenile material also exhibits sutural articulation and

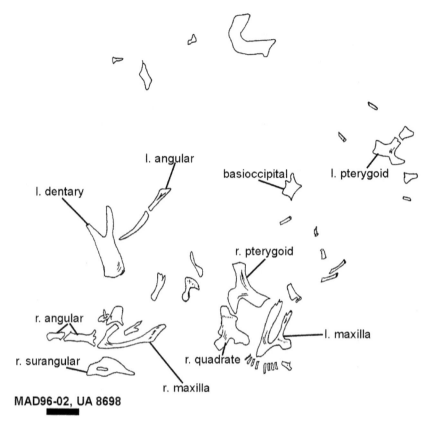

FIGURE 2.6. Map of quarry MAD96-02, which yielded the holotype skull of *Rapetosaurus krausei* (UA; Université d'Antananarivo 8698). Preserved bones exhibit exact sutural articulations and included the following: right maxilla with 8 teeth, left maxilla, right lacrimal, left jugal, right and left nasals, right quadrate, right and left pterygoids, partial basioccipital, right paroccipital process, left dentary with 11 teeth, right and left angulars, right surangular, and 5 additional teeth. Representative bones are labeled; scale bar equals 10 cm.

essentially no duplication of elements. Several of the juvenile skull elements from MAD93-18 share autapomorphies with the adult holotype (fig. 2.8), and thus the associated juvenile skeleton and skull were both assigned to a new titanosaur genus and species: *Rapetosaurus krausei* (Curry 2001; Curry Rogers and Forster 2001, 2004; Curry Rogers 2001).

Rapetosaurus is diagnosed by an enlarged antorbital fenestra extending along and anterior to the tooth row; a preantorbital fenestra positioned caudal to the antorbital fenestra; a rostrally located, dorsoventrally elongated subnarial foramen; a narrow and caudodorsally elongate maxillary jugal process; a median dome on the frontal; a V-shaped quadratojugal articulation on the quadrate; a supraoccipital with two rostrally directed median parietal processes; a pterygoid with a dorsoventrally expanded rostral process and an extremely shallow basipterygoid articulation; basiptyergoid processes that diverge only at their distal ends; and a dentary with 11 alveoli extending for two-thirds of the length of this element (Curry 2001, Curry Rogers and Forster 2001, 2004). Several of the appendicular skeletal characters reported as autapomorphies for *Rapetosaurus* in Curry Rogers and Forster (2001) are now interpreted to have a wider distribution among titanosaurs and are excluded from the diagnosis here as in Upchurch et al. (2004).

The skull of *Rapetosaurus* is superficially similar to skulls of diplodocoid taxa in its elongated, narrow morphology, with external nares retracted to the level of the orbits (figs. 2.5A–C, 2.8). The

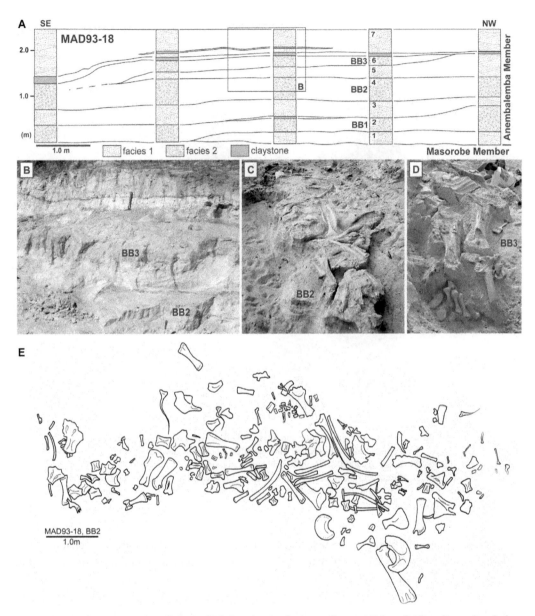

FIGURE 2.7. (A) Cross section through the multiple bone-bearing horizons of quarry MAD93-18. Three discrete bonebeds (BB1, BB2, BB3) each yield bones referable to *Rapetosaurus*. (B) BB2 yields the juvenile skull and skeleton described in the text. The skeleton was closely associated, but only a few vertebrae were articulated. BB1 and BB3 yielded bones of subadult and adult *Rapetosaurus*. (C) Juvenile *Rapetosaurus* skeleton from BB2; cervical vertebrae and ribs, radius in the foreground, scapulae, metacarpals, and femora in the distance. (D) Subadult *Rapetosaurus* skeleton from BB3; ribs are articulated in distance, and forelimb and hindlimb (including left foot) are also closely associated in BB3. (E) Map of BB2, quarry MAD93-18. Map highlights the skeleton of *Rapetosaurus krausei*. Scale bar equals 1 m. (Modified from Rogers 2005.)

basicranium of *Rapetosaurus* is tipped only slightly caudoventrally, and its teeth are peglike with high-angled wear facets. In spite of this general similarity to the skulls of taxa like *Diplodocus* and *Apatosaurus*, *Rapetosaurus* differs from these taxa in the presence of platelike pterygoids and external nares that are undivided and bound, in part, by the ascending process of the maxilla. The *Rapetosaurus* jugal has an elongate maxillary process that lengthens the antorbital region of the skull, and the infraorbital margin of the skull is shortened as the jugal

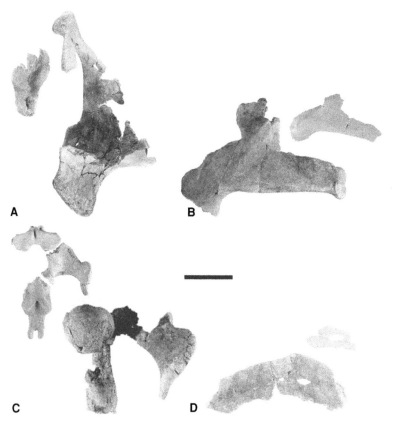

FIGURE 2.8. *Rapetosaurus krausei* adult and juvenile skull elements compared. (A) Juvenile (FMNH PR 2190) and adult (UA 8698) right quadrates in posterior view; (B) juvenile (FMNH PR 2191) and adult (UA 8698) left pterygoids in medial view, anterior to right; (C) adult (UA8698) and juvenile (FMNH PR 2184, 2197) basicrania (in posterior view; and (D) juvenile (FMNH PR 2187) and adult (UA8698) right surangulars in lateral view, anterior to left. Scale bar equals 3 cm.

contribution to the antorbital fenestra boundary is reduced. *Rapetosaurus* frontals are characterized by prominent midline elevations, and the caudal crest of the parietal is well defined. The basicranium includes elongated basipterygoid processes with a narrow, U-shaped division, and elongate, topographically low basal tubera. Teeth extend the full the length of the upper and lower jaws and have precise occlusion along their lingual surfaces. Comparisons of *Rapetosaurus* skull material to that of other known titanosaurs and neosauropods have provided cranial synapomorphies for Titanosauriformes and Somphospondylii, which were previously diagnosed only by postcranial features. In addition to new data for phylogeny reconstruction, the *Rapetosaurus* skull material provides new insights on cranial character correlation (e.g., narial retraction and braincase rotation) and indicates that these characters are decoupled from one another during this phase of sauropod skull evolution, as hypothesized by Chiappe et al. (2001).

The axial skeleton of *Rapetosaurus* (Figs. 2.5D, 2.9A–E) is characterized by at least 15 cervical vertebrae, 10 dorsal vertebrae, and 6 sacral vertebrae. The total number of caudals is unknown, but 18 middle and posterior caudals were preserved with FMNH PR 2209 from MAD93-18. The anterior cervical vertebrae are characterized by elongated centra (elongation index = centrum length/height of posterior face >4.0) with shallow, poorly defined lateral pneumatic fossae and butterfly-shaped neurocentral

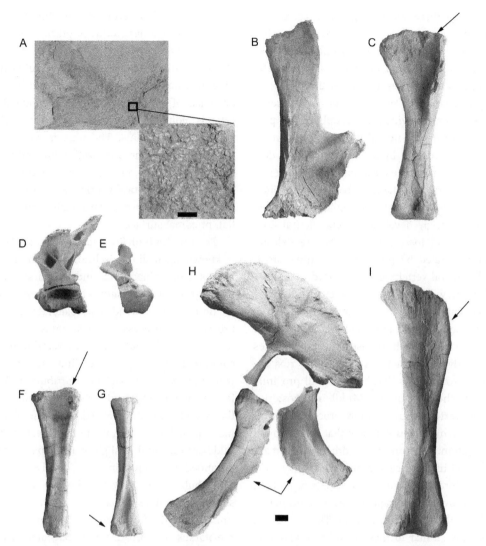

FIGURE 2.9. Representative postcranial elements from a juvenile specimen of *Rapetosaurus krausei* (FMNH PR 2209). (A) Spongy bone in presacral vertebrae. Middorsal vertebra is shown in right lateral view. Inset details somphospondylus bone texture in a middorsal centrum). (B) Right scapula in lateral view. Note the unexpanded distal blade and medially deflected glenoid. (C) Left humerus in anterior view. Arrow outlines rectangular proximal end. (D) Anterior dorsal vertebra in left lateral view with posteriorly inclined dorsal neural spines with rudimentary fossae and lamination. (E) Anterior caudal vertebra in left lateral view is procoelous with an anteriorly positioned neural arch. (F) Left ulna in anterior view. Arrow points to prominent olecranon process. (G) Left radius in posterior view. Arrow points to the medially beveled distal end. (H) left pubis, ischium, and ilium in lateral view. Arrows point to reduced ischium length relative to that of pubis. (I) Left femur in anterior view. Arrow points to the prominent proximolateral flange. Scale bar equals 3 cm.

articulations (resulting in constriction at the midpoint of the neural canal). Anterior cervical neural arches are relatively low and proximally divided (though single in their distal extremity), with strongly developed lateral lamination and fossae. The cervicodorsal transition is marked by a dramatic departure in standard cervical morphology, with centra and neural arches becoming short and broad, while retaining the butterfly-shaped neurocentral articulation. All cervical neural arches have lateral fossa containing deep pneumatic foramina with sharply demarcated borders. Cervical ribs are elongate and gracile, and extend posteriorly the length of at least three succeeding vertebrae. Dorsal vertebrae are characterized by strongly opisthocoelous centra with straight, parallel neurocentral articulations and deep lateral pneumatic fossae. Neural arches are low, with posteriorly tipped triangular neural spines and transversely oriented, winglike diapophyses. In posterior dorsals the neural spine is more vertically oriented, and by the most posterior dorsals the neural spine does not project posterior to the neural arch facets. Pre- and postspinal laminae are prominent throughout the series, as are strongly developed lateral neural arch laminae and fossae. Dorsal ribs are delicate and perforated by pneumatic formina proximally. Sacral vertebrae are unfused in FMNH PR 2209, with opisthocoelous centra anteriorly and slight procoely in the posteriormost sacral. Neural arches are more than twice the height of the centra, with rugose, sutural surfaces anteirorly and posteriorly. The sacral ribs are short, broad wings that bear a small proximal crest on the anterior face. Middle and posterior caudal vertebrae are strongly procoelous, with relatively high neural spines that retain spinal lamination. Centrum face width and height are subequal in all preserved caudals.

The forelimb of *Rapetosaurus* (figs. 2.9B, C, F, G) is known from scapula, coracoid, humerus, radius, ulna, and metacarpals. The scapula is generally similar to those of other titanosaurs but has a uniform scapular blade width, the scapular blade articulates with the body of the scapula at an acute angle, and the acromion process extends anterior to the coracoid articulation. The coracoid has a rounded outline and subequal dimensions. The lateral surface of the coracoid is convex, and it has a larger glenoid surface than it has for articulation with the scapula. The sternal plates are laminar and semilunate with strongly concave lateral borders and a highly convex medial border. A broad, subrectangular deltopectoral crest characterizes the proximal humerus of *Rapetosaurus*. The humerus has a straight diaphysis, with an elliptical cross section at its midpoint. The distal humerus has subequal transverse and anteroposterior dimensions. The radius of *Rapetosaurus* is gracile and distinctively antieorly bowed, with equidimensional proximal and distal ends. The radius and ulna articulation is clearly demarcated by the presence of a strong, interosseus ridge that extends along the full length of the posterior surface of the radius. The ulna is similar to those of other titanosaurs, with a low but well-defined olecranon process. The *Rapetosaurus* metacarpus is vertically oriented, with metacarpals bound to form a semilunate arch with a flat, proximal surface.

The hindlimb of *Rapetosaurus* (fig. 2.9H, I) is known from ilium, ischium, and pubis, as well as femur, tibia, fibula, and metatarsus. The pelvic girdle of *Rapetosaurus* is characterized by an ilium with a broad, anterolaterally flaring preacetabular process. The ischium is only 54% the length of the pubis, and the acetabulum is composed primarily of the ilium and pubis, with only a minor ischial contribution. The femur is characterized by a proximolateral flange and medially deflected head. The tibia is incomplete but has a short cnemial crest, tapers middiaphysis, and is slightly anteroposteriorly expanded in both proximal and distal extremes. The fibula is relatively simple with a slightly sigmoidal outline in lateral view and flared proximal and distal ends. The articulated metatarsus is a broad, nearly flat, arched pes strikingly different from the tightly bound manus. The length of the metatarsals increases from I to III, with metatarsal IV slightly shorter than metatarsal III, and metatarsal V short and triangular.

Rapetosaurus exhibits a suite of features that confirm its status as a titanosaur, and it is the only titanosaur with a wealth of both cranial and postcranial data. To date, no other titanosaur preserves as much high-quality skeletal data as *Rapetosaurus krausei*. As such, *Rapetosaurus* has significant implications for phylogenetic work within Titanosauria and is of great significance for delimiting variation within the clade. New data derived from the *Rapetosaurus* skeleton and skull highlight the fact that Titanosauria exhibited a wide range of morphologies and were not limited to a narrow window of variation.

THE INTERRELATIONSHIPS OF TITANOSAURIA

Now, with *Rapetosaurus* as a key, we can begin to delve deeper into titanosaur phylogeny. A generic-level phylogeny of Titanosauria is presented here. The analysis incorporates characters utilized by other workers (Calvo and Salgado 1995; Upchurch 1995, 1998, 1999; Salgado et al. 1997; Wilson and Sereno 1998; Wilson 1999; Curry Rogers and Forster 2001; Upchurch et al. 2004), as well as novel characters generated from research in museum collections and the primary literature. Details and implications of the analysis are summarized below. The appendixes contain a character–taxon matrix (appendix 2.1), a list of characters and character states (appendix 2.2), and two synapomorphy lists (appendixes 2.3 and 2.4).

ANALYSIS

Thirty-five terminal taxa (including 29 purported titanosaurs; table 2.3) were scored for 364 morphological characters (109 cranial, 139 axial, 113 appendicular, and 2 dermal characters) (appendixes 2.1 and 2.2) in MacClade (Maddison and Maddison 2000) and analyzed in PAUP* (Swofford 1999). Scoring was based on personal observations for all taxa except *Euhelopus* and *Phuwiangosaurus*, which were scored from published illustrations, photographs, and descriptions. The remains used to score certain genera deserve additional comment. Scoring of *Isisaurus colberti* was based on the type specimen (='*T. indicus*' Jain and Bandyopadhyay 1997). Several Indian titanosaur skulls were subsumed within *Jainosaurus septentrionalis* (Huene and Matley 1932; Berman and Jain 1983; Hunt et al. 1994), and the skull recently described as *Antarctosaurus septentrionalis* by Chatterjee and Rudra (1996) is here designated "Jabalpur Titanosaur indet." and is not definitively associated with postcranial remains.

Antarctosaurus was coded on the basis of the holotypic cranium and referred postcrania only.

Character polarity was determined by outgroup comparison with outgroup taxa regarded as paraphyletic with respect to the ingroup (*Brachiosaurus, Camarasaurus, Dicraeosaurus, Diplodocus, Euhelopus*). Most characters were binary, and multistate characters were left unordered and unweighted. The high number of terminal taxa and abundant missing data (table 2.4), particularly in the titanosaur sample precluded exact treebuilding methods, so heuristic searches were conducted with TBR branch swapping, and branches with maximum length zero were collapsed to yield polytomies. Characters were optimized under ACCTRAN and trees were rooted with *Camarasaurus*.

The strict consensus tree for 200,000 equally parsimonious trees (tree length, 989; CI = 0.56; RI = 0.70; RCI = 0.39) is shown in fig. 2.10. Many of the nodes have low decay indices, suggesting weak support for the hypothesized relationship (fig. 2.10). Specific phylogenetic hypotheses were more thoroughly investigated in MacClade (Maddison and Maddison 2000) by "fitting" data to alternative hypotheses (i.e., *Nemegtosaurus* nested within Diplodocoidea rather than Titanosauria). In all cases, these alternatives were far less parsimonious than the topologies presented in this analysis.

Titanosauriformes monophyly was unambiguously supported (appendix 2.3). No unambiguous characters support the monophyly of Somphospondylii or Titanosauria in the strict consensus summary of the data, and most titanosaurs collapse into an unwieldy polytomy. However, two clades of note are retained in the strict consensus tree: (1) a clade including *Lirainosaurus* as the sister-taxon to a polytomy comprised of typical saltasaurines (e.g., *Neuquensaurus, Saltasaurus, Rocasaurus*), *Quaesitosaurus*, Malagasy Taxon B, and the unnamed titanosaur from Jabalpur, India (fig. 2.10; nodes A, B); and (2) a clade uniting *Rapetosaurus* and *Nemegtosaurus* as sister taxa (fig. 2.10; node C). No unambiguous support

TABLE 2.3
Geological Age and Geographical Range for 27 Titanosaur Terminal Taxa Analyzed

	AGE (STAGE)	CONTINENT (COUNTRY)	MATERIAL	ORIGINAL REFERENCE
Aegyptosaurus baharijensis	Late Cretaceous (Cenomanian)	Africa (Egypt)	Axial, appendicular	Stromer 1932
Aeolosaurus rionegrus	Late Cretaceous (Campanian)	South America (Argentina)	Axial, appendicular	Powell 1986
Agustinia ligabuei	Early Cretaceous (Aptian)	South America (Argentina)	Axial, appendicular	Bonaparte 1998
Alamosaurus sanjuanensis	Late Cretaceous (Maastrichtian)	North America (USA)	Axial, appendicular	Gilmore 1922
Ampelosaurus atacis	Late Cretaceous (Maastrichtian)	Europe (France)	Cranial, axial, appendicular dermal	Le Loueff 1995
Andesaurus delgadoi	Early Cretaceous (Albian–Cenomanian)	South America (Argentina)	Axial, appendicular	Calvo and Bonaparte 1991
Antartosaurus wichmannianus	Late Cretaceous (Campanian–Maastrictian)	South America (Argentina)	Cranial, appendicular	Huene 1929
Argentinosaurus huinculensis	Early Cretaceous (Aptian–Coniacian)	South America (Argentina)	Axial, appendicular	Bonaparte and Coria 1993
Argyrosaurus superbus	Early Cretaceous (Senonian)	South America (Argentina)	Axial, appendicular	Lydekker 1893
Brazil Series B	Early Cretaceous (Senonian)	South America (Brazil)	Axial	Powell 1986
Epacthosaurus scuittoi	Early Cretaceous (Senonian)	South America (Argentina)	Axial, appendicular	Powell 1986
Isisaurus (= *Titanosaurus*) *colberti*	Late Cretaceous (Maastrichtian)	Asia (India)	Axial, appendicular	Jain and Bandyopadhyay 1997
Jainosaurus septentrionalis	Late Cretaceous (Campanian–Maastrichtian)	Asia (India)	Cranial	Hunt et al. 1994
Janenschia robusta	Late Jurassic (Kimmeridgian)	Africa (Tanzania)	appendicular	Wild 1991
Japalpur indet.	Late Cretaceous (Maastrictian)	Asia (India)	Cranial	Unpublished
Lirainosaurus astibiae	Late Cretaceous (Campanian–Maastrichtian)	Europe (Spain)	Cranial, axial, appendicular, dermal	Sanz et al. 1999
Magyarosaurus dacus	Late Cretaceous (Maastrichtian)	Europe (Romania)	Cranial, axial, appendicular, dermal	Huene 1932

TABLE 2.3 (continued)

	AGE (STAGE)	CONTINENT (COUNTRY)	MATERIAL	ORIGINAL REFERENCE
Malagasy Taxon B	Late Cretaceous (Campanian–Maastrichtian)	Africa (Madagas-car)	Axial, appendicular	Curry Rogers 2001
Malawisaurus dixeyi	Early Cretaceous (Aptian)	Africa (Malawi)	Cranial, axial, appendicular, dermal	Jacobs et al. 1993
Nemegtosaurus mongoliensis	Late Cretaceous (Maastrichtian)	Asia (Mongolia)	cranial	Nowinski 1971
Neuquensaurus	Late Cretaceous (Campanian–Maastrichtian)	South America (Argentina)	Axial, appendicular	Powell 1986
Opisthocoelicaudia skarzynskii	Late Cretaceous (Maastrichtian)	Asia (Mongolia)	Axial, appendicular	Borsuk-Bialynicka 1977
Paralititan stromeri	Late Cretaceous (Cenomanian)	Africa (Egypt)	Axial, appendicular	Smith et al. 2001
Phuwiangosaurus sirindhornae	Early Cretaceous	Asia (Thailand)	Axial, appendicular	Martin et al. 1994
Quaesitosaurus orientalis	Late Cretaceous (Maastrichtian)	Asia (Mongolia)	Cranial	Kurzanov and Bannikov 1983
Rapetosaurus krausei	Late Cretaceous (Campanian–Maastrichtian)	Africa (Madagascar)	Cranial, axial, appendicular, dermal	Curry Rogers and Forster 2001
Rocasaurus muniozi	Late Cretaceous (Campanian–Maastrichtian)	South America (Argentina)	Axial, appendicular	Salgado and Azpilicueta 2000
Saltasaurus loricatus	Late Cretaceous (Campanian–Maastrichtian)	South America (Argentina)	Cranial, axial, appendicular, dermal	Bonaparte and Powell 1980
Santa Rosa indet.	Late Cretaceous (Campanian–Maastrichtian)	South America (Argentina)	Cranial	Unpublished

exists for the saltasaurine polytomy, and the strict consensus tree does not resolve unambiguous support for components of the basal polytomy.

An Adams consensus tree highlights more resolved branches and reveals that several consistent nestings of taxa are among the most parsimonious trees (fig. 2.11). Nine additional unambiguous characters from the cranial and postcranial skeleton support Titanosauriformes monophyly, and 22 additional steps are required to move Titanosauria into its more traditional position as the sister-group to Diplodocoidea. Several sauropods generally regarded as close titanosaur relatives are at the base of the Adams consensus, each in an unresolved dichotomy with another taxon. *Euhelopus* and *Jainosaurus* are basal to the dichotomy including *Phuwiangosaurus* and *Andesaurus* within Titanosauria. More derived titanosaur relationships are not unambiguously supported in this analysis, but six clades of note are conserved in the Adams consensus: (1) Titanosauria (fig. 2.11; node A); (2) Saltasauridae (fig. 2.11; node J); (3) Lithostrotia (fig. 2.11; node B); and (4) Saltasaurinae (fig. 2.11; node G), (5) Opisthocoelicaudiinae (fig. 2.11; node I); and (6) an unnamed clade including *Rapetosaurus*,

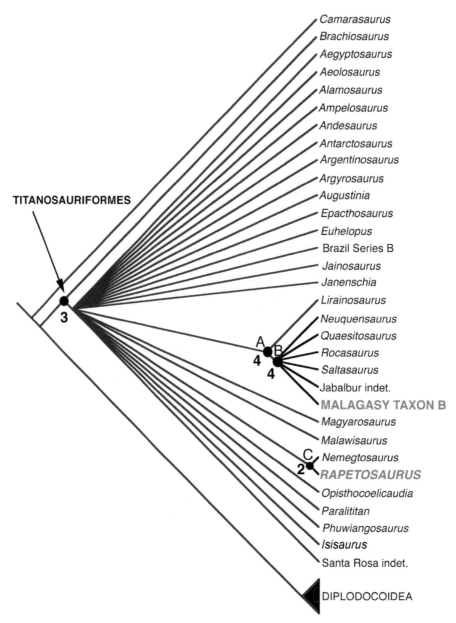

FIGURE 2.10. Strict consensus of 200,000 most parsimonious trees (989 steps, CI = 0.56, RI = 0.70) generated by a heuristic search in PAUP*. Trees were rooted with *Camarasaurus* as the outgroup and optimized with accelerated transformations, and all characters were unordered. Bremer support values are listed at labeled nodes. Labeled nodes (A–C) indicate monophyletic groups that are discussed more fully in text. No formal names are assigned to these clades at present.

Nemegtosaurus, *Malawisaurus*, and several other derived lithostrotians (fig. 2.11; node E). Unambiguous support for some of the nodes within these clades is listed in appendix 2.3. Other clades recognized in the Adam's consensus are polytomous, and unambiguous support is not calculated for these nodes (e.g., the polytomy of *Antarctosaurus*, *Argentinosaurus*, and *Opisthocoelicaudia*).

This maximum parsimony analysis of the total-evidence data set resulted in poor ingroup resolution in strict consensus summaries, as

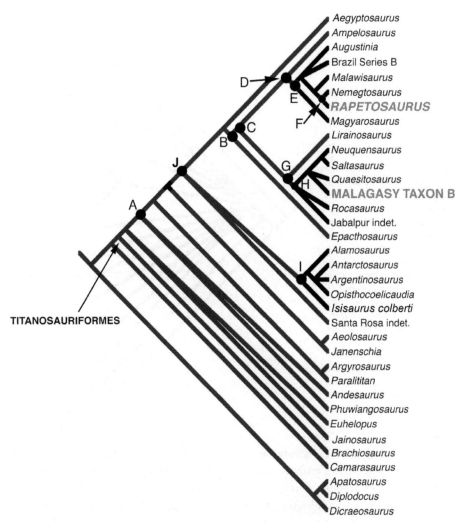

FIGURE 2.11. Adams consensus produced from analysis of 200,000 most parsimonious trees. Node A indicates Titanosauria, which includes a basal polytomy of *Phuwiangosaurus* and *Andesaurus*. Node B indicates Lithostrotia. Node E marks the *Rapetosaurus* clade. Node G indicates Saltasaurinae. Node I indicates Opisthocoelicaudiinae. Node J indicates Saltasauridae. Other labeled nodes demarcate monophyletic groups but are not formally named at present. Unambiguous upport for these unnamed clades, as well as for Titanosauridae, Saltasauridae, and the *Rapetosaurus* clade, is listed in appendix 2.3.

well as relatively low Bremer support values. MacClade's search for redundant taxa identified none in the matrix, and no taxa could be deleted using Wilkinson's (1995) rules for "safe" taxonomic deletion. Given these constraints, I followed the precedent set by several recent authors (Gauthier 1986; Rowe 1988; Benton 1990; Novacek 1992; Grande and Bemis 1998), and removed taxa that were 15% or less completely coded, and ran the matrix a second time. Fourteen titanosaurs were removed from the matrix (tables 2.4, 2.5), allowing more resolution in the Adams consensus of the data and a reduction in most parsimonious trees. Though the reduction in tree number allowed for detailed study of the 44 trees, it did not result in a better-resolved strict consensus tree (fig. 2.12). Exploration of each of the 44 equally parsimonious trees indicates that relationships among three major titanosaur clades are stable, with only *Quaesitosaurus* migrating throughout the cladogram. In fact, *Quaesitosaurus* can be equally

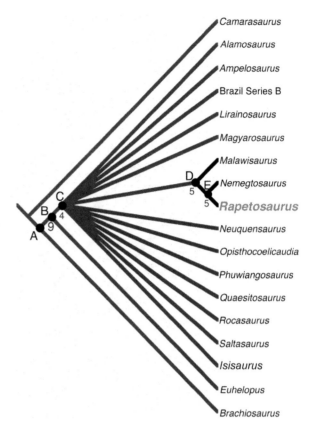

FIGURE 2.12. Cladogram showing phylogenetic hypothesis of titanosaur relationships, based on the strict consensus of 44 most parsimonious trees (704 steps, CI = 0.72, RI = 0.74) generated by a heuristic search in PAUP*. Trees were rooted with *Camarasaurus* as the outgroup, optimized with accelerated transformations, and all characters were unordered. Bremer support values are listed at labeled nodes. Labeled nodes (A–E) indicate monophyletic groups that are discussed more fully in text. No formal names are assigned to these clades at present.

parsimoniously placed as the sister-taxon to almost every titanosaur in the analysis and is responsible for the poor resolution in the strict consensus tree for the taxonomically reduced data set. That said, the strict consensus of the data does support the higher-level groupings of Titanosauriformes and Somphospondylii (fig. 2.12; nodes A, B), as well as a portion of Titanosauria (fig. 2.12; nodes C, D).

The Adams consensus tree (fig. 2.13) of the reduced data set indicates that *Phuwiangosaurus* occupies a basal position within Titanosauria. Unambiguous support for Titanosauria is not calculated in the Adams consensus tree. However, in individual trees several unambiguous postcranial characters support Titanosauria (appendix 2.4). Within Titanosauria, *Ampelosaurus* and *Lirainosaurus* form successive sister-taxa to several major titanosaur clades. These clades were also recovered in the Adams consensus for the complete data set discussed above. (1) Saltasauridae (fig. 2.13; node C), (2) Lithostrotia (fig. 2.13; node D), (3) Saltasaurinae (fig. 2.13; node E), (4) the *Rapetosaurus* clade (fig. 2.13; node G), and (5) Opisthocoelicaudiinae (fig. 2.13; node K). Where calculated, unambiguous support for these nodes is outlined in appendix 2.4.

DISCUSSION

The analysis presented here has several important implications for titanosaur phylogeny. First and foremost, in spite of an enormous amount of missing data (14 titanosaurs included in this study are coded for less than 15% of their data; tables 2.4, 2.5), several clades are well-supported. Titanosauriformes monophyly is apparent when all known titanosaurs are considered, in support of the dismantled grouping of titanosaurs and diplodocoids (e.g., Huene 1929; Romer 1956; Steel 1970). Titanosauriform monophyly is supported with cranial characters, as is the monophyly of Somphospondylii (titanosauriforms more closely related to *Saltasaurus* than to

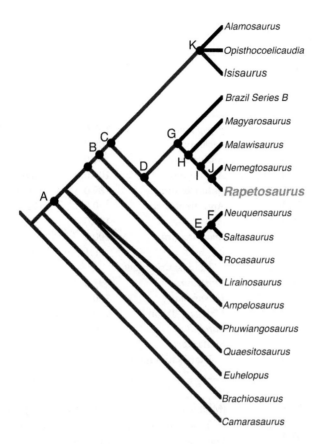

FIGURE 2.13. Adams consensus produced from analysis of 44 MPTs resulting from reduced analysis. Node A indicates Somphospondylii. An unlabeled node demarcates Titanosauria, which includes a basal polytomy of *Phuwiangosaurus* and *Andesaurus* with more derived titanosaurs. Node C indicates Saltasauridae. Node D indicates Lithostrotia. Node E marks Saltasaurinae. Node G indicates the *Rapetosaurus* clade. Node K indicates Opisthocoelicaudiinae. Other labeled nodes demarcate monophyletic groups but are not formally named at present. Unambiguous upport for these unnamed clades, as well as those outlined here, is listed in appendix 2.4.

Brachiosaurus [Wilson and Sereno 1998]). As such, the new, more comprehensive titanosaur sample adds strength to hypotheses formerly based only on postcranial data.

In addition to firm establishment of the appropriate titanosaur outgroup, this analysis also highlights new lower-level resolution. In individual trees and in Adams consensus summaries of the data, several titanosaur clades are sustained, all of which have been supported in some form in previous analyses (e.g., Salgado et al. 1997; Upchurch 1998; Wilson et al. 1999; Curry Rogers and Forster 2001). This analysis confirms that *Epachthosaurus*, *Andesaurus*, and *Argentinosaurus* are not closely related (contra Bonaparte and Coria 1993). Andesauridae (Bonaparte and Coria, 1993) is paraphyletic in this analysis, and forcing its monophyly requires an addition of 12 steps. While *Andesaurus* retains its position as the most basal member of Titanosauria, *Epachthosaurus* and *Argentinosaurus* are more nested within the group. *Argentinosaurus* is in a polytomy with *Antarctosaurus* and *Opisthocoelicaudia* within Titanosauridae. Salgado et al. (1997) also hypothesized a close relationship between *Argentinosaurus* and *Opisthocoelicaudia*. *Epachthosaurus* is the sister-taxon of *Lirainosaurus* + Saltasauridae, instead of being closely allied with *Malawisaurus* (Salgado et al. 1997).

The monophyly of Lithostrotia and Saltasauridae are strongly supported in this analysis, though *Malawisaurus* is not considered a basal titanosaur (contra Wilson 2002). Within each of these major groups, additional resolution is possible, and is in general agreement with the recent analyses by Wilson (2002) and Upchurch et al. (2004).

Opistocoelicaudiinae is supported in individual trees, though it is not apparent in the strict consensus of the data, and in this analysis it includes *Isisaurus* as the sister-taxon of *Alamosaurus* + an unresolved polytomy (*Opisthocoelicaudia*, *Antarctosaurus*, and *Argentinosaurus*). When poorly preserved taxa are excluded from

TABLE 2.4
Missing Data for Titanosaur Cranial and Postcranial Skeleton

	PERCENTAGE MISSING DATA	
	CRANIAL	POSTCRANIAL
Aegyptosaurus	100.0	89.6
Aeolosaurus	100.0	85.7
Augustinia	100.0	89.0
Alamosaurus	100.0	33.3
Ampelosaurus	72.4	46.6
Andesaurus	100.0	86.3
Antarctosaurus	89.0	100.0
Argentinosaurus	100.0	85.2
Argyrosaurus	100.0	92.0
Brazil Series B	100.0	66.6
Epacthosaurus	100.0	94.2
Isisaurus (= '*Titanosaurus*')	100.0	36.1
Jabalpur Titanosauria indet.	96.1	100.0
Jainosaurus	97.0	100.0
Janenschia	100.0	89.3
Lirainosaurus	93.6	55.3
Magyarosaurus	100.0	47.1
Malagasy Taxon B	100.0	96.4
Malawisaurus	58.7	51.4
Nemegtosaurus	31.2	100.0
Neuquensaurus	100.0	19.6
Opisthocoelicaudia	100.0	21.6
Paralititan	100.0	93.7
Phuwiangosaurus	100.0	43.1
Quaesitosaurus	38.5	100.0
Rapetosaurus	18.5	7.7
Rocasaurus	100.0	69.8
Saltasaurus	72.5	19.2
Santa Rosa Titanosaura indet.	92.6	100.0

TABLE 2.5
Taxa Excluded from Parsimony Analysis

	PERCENTAGE MISSING DATA	
	CRANIAL	POSTCRANIAL
Aegyptosaurus	100.0	89.6
Aeolosaurus	100.0	85.7
Andesaurus	100.0	86.3
Augustinia	100.0	89.0
Antarctosaurus	89.0	100.0
Argentinosaurus	100.0	85.2
Argyrosaurus	100.0	92.0
Epacthosaurus	100.0	94.2
Jabalpur Titanosauria indet.	96.1	100.0
Jainosaurus	97.0	100.0
Janenschia	100.0	89.3
Malagasy Taxon B	100.0	96.4
Paralititan	100.0	93.7
Santa Rosa Titanosaura indet.	92.6	100.0

the analysis, the relationship of *Opisthocoelicaudia* remains unresolved, and it is equally parsimoniously placed as the sister- taxon to *Alamosaurus* or *Isisaurus*.

Saltasaurinae is strongly supported in all analyses and contains traditional members of a longstanding group of titanosaurs (*Saltasaurus*, *Neuquensaurus*). *Lirainosaurus* is the sister-taxon of Saltasauridae. Saltasauridae includes traditional members of the group (e.g., *Saltasaurus*, *Neuquensaurus*), does not include others

(Mendoza titanosaur [Wilson et al. 1999]), and includes several other taxa (*Quaesitosaurus*, Malagasy Taxon B, and Jabalpur indet.), though support for inclusion of *Quaesitosaurus* and Jabalpur indet. is weak.

The *Rapetosaurus* clade is the last major group within Titanosauria recognized by this analysis. Although only the sister-taxon relationship between *Nemegtosaurus* and *Rapetosaurus* is retained in the strict consensus trees, in Adams consensus trees the *Rapetosaurus* clade includes a polytomy of the South American titanosaurs *Agustinia* and Brazil Series B as the sister-taxon to *Malawisaurus* (*Nemegtosaurus* + *Rapetosaurus*). In previous analyses (e.g., Salgado et al. 1997; Smith et al. 2001; Wilson 2002) *Malawisaurus* has occupied a more basal position within Titanosauria. Here (as in Curry Rogers and Forster 2001) *Malawisaurus* is firmly nested within the *Rapetosaurus* clade, and removal requires 11 additional steps.

The positions of several controversial sauropods are also better understood. Notably, as hypothesized by Upchurch (1998), *Phuwiangosaurus* is the sister-taxon of Titanosauria. Even more significantly, Mongolian sauropods are members of Titanosauria, in conflict with the hypothesis presented by Upchurch (1998, 1999) and Upchurch et al. (2004), who concluded that *Nemegtosaurus* and *Quaesitosaurus* were members of Diplodocoidea. Forcing either of the Mongolian sauropods into a sister-group relationship with Diplodocoidea results in an increase of at least 26 steps. This analysis also suggests that *Nemegtosaurus, Opisthocoelicaudia,* and *Quaesitosaurus* are likely not conspecific. Similarly, *Paralititan* and *Aegyptosaurus* are also not likely to be conspecific, or even closely related at the generic level. Finally, a survey of a sample of the 200,000 most parsimonious trees demonstrates that the primary lack of resolution in the consensus trees results from pervasive migration of several very incomplete titanosaurs (e.g., *Quaesitosaurus, Jainosaurus*).

CONCLUSION

Over the last several years, a handful of rigorous looks at the anatomy and phylogeny of Titanosauria have clarified our view of this enigmatic sauropod group. Revision of the type genus and species, *Titanosaurus indicus,* has provided us with a working definition of what a titanosaur is, and we are better able to diagnose titanosaurs. Discoveries of new fossils from around the globe extend the titanosaur reign geographically and temporally, and give us a clearer image of how titanosaurs made their living. We have even reached some resolution on titanosaur phylogeny, particularly with regard to higher-level relationships. Titanosauria is no longer shrouded in so deep a mystery, but we are far from a complete understanding of the group, particularly with regard to lower-level relationships. All analyses of titanosaur interrelationships have lacked a full body view for a diverse taxonomic sample due to the sheer lack of skeletal data for almost all titanosaur genera. *Rapetosaurus krausei* is a keystone taxon that is most significant because it preserves cranial and postcranial material that allows across-the-board comparisons with other titanosaurs, as well as a wealth of new character data from all parts of the skeleton.

Through the addition of *Rapetosaurus* and recent revision of several other key titanosaur genera (e.g., *Nemegtosaurus, Isisaurus*), the following key aspects of titanosaur phylogeny are addressed and more fully resolved.

1. Monophyly of Titanosauriformes can now be confirmed with unambiguous support from the cranial and postcranial skeleton.

2. Monophyly of Somphospondylii is also confirmed with unambiguous support from the cranial and postcranial skeleton.

3. Lithostrotia, Saltasauridae, Saltasaurinae, and Opisthocoelicaudiinae are monophyletic, though this analysis differs from other recent workers' (Wilson 2002; Upchurch et al. 2004) in the placement of several well-known titanosaur genera.

4. An unnamed clade uniting *Malawisaurus, Rapetosaurus krausei,* and *Nemegtosaurus* is monophyletic and confirms the inclusion of *Nemegtosaurus* in Titanosauria.

5. Resolution within the clades mentioned above is greatly improved in this analysis, though including the titanosaurs with a preponderance of missing data complicates resolution in strict consensus trees.

6. *Phuwiangosaurus* and *Andesaurus* are basal titanosaurs, but Andesauridae is not monophyletic. The other purported "andesaurids," *Argentinosaurus* and *Epachthosaurus* are nested within the clades mentioned above.

ACKNOWLEDGMENTS

Thanks go to Jack McIntosh for inspiring all of us to search for new sauropods, new approaches, and new answers. Thanks are also due to C. Forster, D. Krause, S. Sampson, and J. Gauthier, as well as R. Rogers, J. A. Wilson, M. Carrano, and M. O'Leary for their many comments and advice on preparing this chapter. D. B. Weishampel, J. A.

Wilson, and one anonymous reviewer greatly improved the quality of this chapter.

LITERATURE CITED

Asher, R. J., and Krause, D. W. 1998. The first pre-Holocene (Cretaceous) record of Anura from Madagascar. J. Vertebr. Paleontol. 18: 696–699.

Benton, M. J. 1990. The species of *Rhyncosaurus*, a rhyncosaur (Reptila, Diaprida) from the Middle Triassic of England. Philosophical Transactions of the Royal Society of London B. 328: 213–306.

Berman, D. S., and Jain, S. L. 1983. The braincase of a small sauropod dinosaur (Reptilia: Saurischia) from the Upper Cretaceous Lameta Group, Central India, with a review of Lameta Group Localities. Ann. Carnegie Mus. 51: 405–422.

Berman, D. S., and McIntosh, J. S. 1978. Skull and relationships of the Upper Jurassic sauropod *Apatosaurus* (Reptilia, Saurischia). Bull. Carnegie Mus. Nat. Hist. 8: 1–35.

Besairie, H. 1972. Géologie de Madagascar. I. Les terrains sédimentaires. Ann. Geol. Madagascar 35: 1–463.

Bonaparte, J. F. 1986. The early radiation and phylogenetic relationships of sauropod dinosaurs, based on vertebral anatomy. In: Padian, K. (ed.). The Beginning of the Age of Dinosaurs. Cambridge University Press, Cambridge, New York. Pp. 247–258.

Bonaparte, J. F. 1998. An armored Sauropod from the Aptian of northern Patagonia, Argentina. In: Second Symposium on Gondwana Dinosaurs, Tokyo. P. 10.

Bonaparte, J. F. and Coria, R. 1993. Un neuvo y gigantesco sauropodo titanosaurio de la Formación Río Limay (Albiano-Cenomanio) del a Provincia del Neuquén, Argentina. Ameghiniana 30: 271–282.

Bonaparte, J. F. and Powell, J. 1980. A continental assemblage of tetrapods from the Upper Cretaceous beds of El Brete, Northwestern Argentina (Sauropoda-Coelurosauria-Carnosauria-Aves). Mem. Soc. Geol. Fr. N. S. 139: 19–28.

Borsuk-Bialynicka, M. 1977. A new camarasaurid sauropod, *Opisthocoelicaudia skarzynskii*, gen. n., sp. n. from the Upper Cretaceous of Mongolia. Palaeontol. Polonica 37: 1–64.

Buckley, G. A., and Brochu, C. 1999. An enigmatic new crocodiles from the Upper Cretaceous of Madagascar. In: Unwin, D. (ed.). Cretaceous Fossil Vertebrates: Special Papers in Palaeontology No. 60. Paleontological Association, London. Pp. 149–175.

Buckley, G. A., Brochu, C., Krause, D. W., and Pol, D. 2000. A pug-nosed crocodyliform from the Late Cretaceous of Madagascar Nature 405: 941–944.

Calvo, J. O., and Bonaparte, J. F. 1991. *Andesaurus delgadoi* gen. et sp. nov. (Saurischia-Sauropoda), dinosaurio Titanosauridae de la Formación Río Limay (Albiano-Cenomanio), Neuquén, Argentina. Ameghiniana 28: 303–310.

Calvo, J. O., and Salgado, L. 1995. *Rebbachisaurus tesonei* sp. nov. A new Sauropoda from the Albian-Cenomanian of Argentina; new evidence of the origin of the Diplodocidae. GAIA 111: 13–33.

Chatterjee, S., and Rudra, D. K. 1996. KT events in India: impact, rifting, volcanism and dinosaur extinction. Mem. Queensland Mus. 39: 489–532.

Chiappe, L. M., Salgado, L., and Coria, R. A. 2001. Embryonic skulls of titanosaur sauropod dinosaurs. Science 293: 2444–2446.

Csiki, Z. 1999. New evidence of armoured titanosaurids in the Late Cretaceous—*Magyarosaurus dacus* from the Hateg Basin (Romania). Oryctos 2: 93–99.

Curry, K. A. 1997. Vertebrate fossils from the Upper Cretaceous Ankazomihaboka Sandstones, Mahajanga Basin, Madagascar. J. Vertebr. Paleontol. 17: 40A.

Curry, K. A. 2001. The Evolutionary History of the Titanosauria. Unpublished Ph.D. thesis, Stony Brook University, Stony Brook, NY. 552 pp.

Curry Rogers, K. 2001. A new sauropod from Madagascar: implications for lower level titanosaur phylogeny. J. Vertebr. Paleontol. 21: 43A.

Curry Rogers, K., and Forster, C. A. 1999a. New evidence of titanosaurian sauropods in the Late Cretaceous of Madagascar. In: Abstracts, VII International Symposium on Mesozoic Terrestrial Ecosystems, Buenos Aires. P. 20.

———. 1999b. New sauropods from Madagascar: a glimpse into titanosaur cranial morphology and evolution. J. Vertebr. Paleontol. 19: 40A.

———. 2001. The last of the dinosaur titans: a new sauropod from Madagascar. Nature 412: 530–534.

———. 2004. The skull of *Rapetosaurus krausei* (Sauropoda: Titanosauria) from the Late Cretaceous of Madagascar. J. Vertebr. Paleontol. 24: 121–144.

Day, J. J., Upchurch, P., Norman, D. B., Gale, A. S., and Powell, H. P. 2002. Sauropod trackways, evolution, and behavior. Science 296: 1659.

Depéret, C. 1896a. Note sur les Dinosauriens Sauropodes et Théropodes dans le Crétacé supérieur de Madagascar. Bull. Soc. Geol. France 21: 176–194.

———. 1896b. Sur L'existence de Dinosauriens, Sauropodes et Théropodes dans le Crétacé. supérieur de Madagascar. C.R. Hebdomadiares Sea. Acad. Sci. (Paris) 122: 483–485.

Dodson, P., Krause, D. W., Forster, C. A., Sampson, S. D., and Ravoavy, F. 1998. Titanosaurid (Sauropoda) osteoderms from the Late Cretaceous of Madagascar. J. Vertebr. Paleontol. 18: 563–568.

Erickson, G. M., Curry Rogers, K., and Yerby, S. 2001. Dinosaurian growth patterns and rapid avian growth rates. Nature 412: 429–433.

Falconer, H. 1868. Memorandum on two remarkable vertebrae sent by Dr. Oldham from Jubbulpore-Spilsbury bed. In: Paleontological Memoirs and Notes of the Late Hugh Falconer, Vol. 1. Pp. 418–419.

Forster, C. A., Chiappe, L. M., Krause, D. W., and Sampson, S. D. 1996. The first Cretaceous bird from Madagascar. Nature 382: 532–534.

Forster, C. A., Sampson, S. D., and Chiappe, L. M. 1998. The theropod ancestry of birds: new evidence from the Late Cretaceous of Madagascar. Science 279: 1915–1919.

Gauthier, J. 1986. Saurischian monophyly and the origin of birds. In: Padian, K. (ed.). The Origin of Birds and the Evolution of Flight. Memoirs of the California Academy of Sciences. Pp. 1–55.

Gilmore, C. W. 1922. A new sauropod dinosaur from the Ojo Alamo Formation of New Mexico. Smithson. Misc. Collect. 72: 1–11.

———. 1946. Reptilian fauna of the North Horn Formation of central Utah. U.S. Geol. Surv. Prof. Paper 210C: 1–52.

Gottfried, M. D., and Krause, D. W. 1998. First record of gars (Ginglymodi, Actinopterygii) on Madagascar: Late Cretaceous remains from the Mahajanga Basis. J. Vertebr. Paleontol. 18: 275–279.

Grande, L., and Bemis, W. E. 1998. A comprehensive phylogenetic study of amiid fishes (Amiidae) based on comparative skeletal anatomy. An empirical search for interconnected patterns of natural history. J. Vertebr. Paleontol. Mem. 4, 18: 1–690.

Hoffstetter, R. 1957. Quelques observations sur le stégosaurinés. Bulletin de Musée d'Histoire Naturelle, Paris 29: 537–547.

Huene, F. von. 1929. Los Saurisquios y Ornithisquios del Cretaceo Argentino. An. Mus. La Plata 2: 1–196.

Huene, F. von, and Matley, C. A. 1932. The Cretaceous Saurischia and Ornithischia of the central provinces of India. Mem. Geol. Surv. India 21: 1–74.

Hunt, A., Lockley, M. D., Lucas, S. G., and Meyer, C. A. 1994. The global sauropod fossil record. GAIA 10: 261–279.

Jacobs, L. L., Winkler, D. A., Downs, W. R., and Gomani, E. M. 1993. New material of an Early Cretaceous titanosaurid sauropod dinosaur from Malawi. Paleontology 36: 523–534.

Jain, S., and Bandyopadhyay, S. 1997. New titanosaurid (Dinosauria: Sauropoda) from the Late Cretaceous of central India. J. Vertebr. Paleontol. 17: 114–136.

Jianu, C. M., and Weishampel, D. B. 1999. The smallest of the largest, a new look at possible dwarfing in sauropod dinosaurs. Geol. Mijnbouw 78: 335–343.

Krause, D. W. 2001. Fossil molar from a Madagascan marsupial. Nature 412: 497.

Krause, D. W., Hartman, J. H., and Wells, N. A. 1997. Late Cretaceous vertebrates from Madagascar: implications for biotic change in deep time. In: Goodman, S. D., and Patterson, B. D. (eds.). Natural Change and Human Impact in Madagascar. Smithsonian Institution, Washington, DC. Pp. 3–43.

Krause, D. W., Hartman, J. H., Wells, N. A., Buckley, G. A., Lockwood, C. A., Wall, C. E., Wunderlich, R. E., Rabarison, J. A., and Randriamaramanana, L. 1994. Late Cretaceous mammals. Nature 368: 298.

Krause, D. W., Rogers, R. R., Forster, C. A., Hartman, J. H., Buckley, G. A., and Sampson, S. D. 1999. The Late Cretaceous vertebrate fauna of Madagascar: Implications for Gondwanan paleobiogeography. GSA Today 9: 1–7.

Kurzanov, S. M., and Bannikov, A. F. 1983. A new sauropod from the Upper Cretaceous of Mongolia. Paleontol. J. 2: 90–96.

Lavocat, R. 1955. Etude des gisements de Dinosauriens de la region de Majunga (Madagascar). Travaux Bur. Geol. 69: 1–19.

Le Loueff, J. 1995. *Ampelosaurus atacis* (nov. gen., nov. sp.), un nouveau Titanosauridae (Dinosauria, Sauropoda) du Crétacé superieur de la Haute Vallee de L'Aude (France). C. R. Acad. Sci. Paris (Ser. IIa) 321: 693–699.

Lydekker, R. 1877. Notices of new and other Vertebrata from Indian Tertiary and Secondary Rocks. Rec. Geol. Soc. India 10: 30–43.

———. 1879. Indian Pre-Tertiary Vertebrata. Part 3. Fossil Reptilia and Batrachia. Paleontol. Indica I: 1–36.

———. 1893. Contributions to the study of the fossil vertebrates of Argentina. I. The dinosaurs of Patagonia. An. Mus. La Plata Sec. Paleontol. 2: 1–14.

Maddison. W. P., and Maddison, D. R. 2000. MacClade, Version 4.0. Sinauer Associates, Sunderland, MA.

Martin, V., Buffetaut, E., and Suteethorn, V. 1994. A new genus of sauropod dinosaur from the Sao Khua Formation (Late Jurassic to Early Cretaceous) of northeastern Thailand. C. R. Acad. Sci. Paris 319: 1085–1092.

Martínez, R. D., Gimenez, O., Rodríguez, J., Luna, M., and Lamanna, M. C. 2004. An articulated specimen of the basal titanosaurian *Epachthosaurus sciuttoi* (Dinosauria: Sauropoda) from the early Late Cretaceous Bajo Barreal Formation of Chubut Province Argentina. J. Vertebr. Paleontol. 24: 107–120.

McIntosh, J. S. 1989. The sauropod dinosaurs: a brief survey. In: Padian, K., and Chure, D. (eds.). The Age of Dinosaurs. Short Courses in Paleontol N. 2. University of Knoxville, Knoxville, TN. Pp. 85–99.

———. 1990a. Sauropoda. In: Weishampel, D. B., Dodson, P., and Osmolska, H. (eds.). The Dinosauria. University of California Press, Berkeley. Pp. 345–401.

———. 1990b. Species determination in sauropod dinosaurs with tentative suggestions for their classification. In: Carpenter, K. and Curry, P. J. (eds.). Dinosaur Systematics: Approaches and Perspectives. Cambridge University Press, Cambridge, New York. Pp. 53–71.

Munyikwa, D., Sampson, S. D., Rogers, R. R., Forster, C. A., Curry, K. A., and Curtice, B. D. 1998. Vertebrate paleontology and geology of the Gokwe Formation, Zimbabwe. J. Afr. Earth Sci. 27: 142–143.

Novacek, M. J. 1992. Fossils, topologies, missing data, and the higher level phylogeny of eutherian mammals. Syst. Biol. 41: 58–73.

Nowinski, A. 1971. *Nemegtosaurus mongoliensis* n. gen. n. sp. (Sauropoda) from the uppermost Cretaceous of Mongolia. Palaeontol. Polonica 25: 57–81.

Powell, J. E. 2003. Revision of South American Titanosaurid dinosaurs: palaeobiological, palaeobiogeographical, and phylogenetic aspects. Rec. Queen Victoria Mus. III: 1–173.

Ravoavy, F. 1991. Identification et mise en catalogue des vertebras fossils récoltes dans le Crétace supérieur continental de la region de Berivotra (Majunga) fouille 1987. Univ. Antananarivo Mem. Recherche II: 55–104.

Rogers, R. R. 2005, Fine-gravel debris flows and extraordinary vertebrate burials in the Late Cretaceous of Madagascar. Geology 33 (4): 297–300.

Rogers, R. R., Hartman, J. H., and Krause, D. W. 2000. Stratigraphic analysis of Upper Cretaceous rocks in the Mahajanga Basin, northwestern Madagascar: Implications for ancient and modern faunas. J. Geol. 108: 275–301.

Romer, A. S. 1956. Osteology of the Reptiles. University of Chicago Press, Chicago. 772 pp.

———. 1966. Vertebrate Paleontology. University of Chicago Press, Chicago. 468 pp.

———. 1968. Notes and Comments on Vertebrate Paleontology. University of Chicago Press, Chicago. 304 pp.

Rowe, T. 1988. Definition, diagnosis, and origin of Mammalia. J. Vertebr. Paleontol. 8: 241–264.

Russell, D. A., Russell, D., Taquet, P., and Thomas, H. 1976. Nouvelles récoltes de vertebras dans les terrains continentaux du Crétacé supérieur de la region de Majunga (Madagascar). Soc. Geol. France C. R. Sommaires 5: 204–207.

Salgado, L., and Azpilicueta, C. 2000. Un nuevo saltasaurino (Sauropoda, Titanosauridae) de la provincia de Río Negro (Formacíon Allen, Cretácico Superior), Patagonia, Argentina. Ameghiniana 37: 259–264.

Salgado, L., Calvo, J. O., and Coria, R. A. 1997. Evolution of the titanosaurid sauropods. I. Phylogenetic analysis based on the postcranial evidence. Ameghiniana 34: 3–32.

Sampson, S. D., Carrano, M. T., and Forster, C. A. 2001. A bizarre predatory dinosaur from the Late Cretaceous of Madagascar. Nature 409: 504–506.

Sampson, S. D., Krause, D. W., Dodson, P., and Forster, C. A. 1996. The premaxilla of *Majungasaurus* (Dinosauria: Theropoda) with implications for Gondwanan paleobiogeography. J. Vertebr. Paleontol. 16: 601–605.

Sampson, S. D., Witmer, L. M., Forster, C. A., Krause, D. W., O'Connor, P. M., Dodson, P., and Ravoavy, F. 1998. Predatory dinosaur remains from Madagascar: Implications for the Cretaceous biogeography of Gondwana. Science 280: 1048–1051.

Santos, V. F., Lockley, M. G., Meyer, C. A., Carvalho, J., Galopim, A. M., and Moratalla, J. J. 1994. A new sauropod tracksite from the Middle Jurassic of Portugal. GAIA 10: 5–14.

Sanz, J. L., Powell, J. E., Le, Loueff, J. Martinez, R., and Suberbiola, X. P. 1999. Sauropod remains from the upper Cretaceous of Laño (northcentral Spain). Titanosaur phylogenetic relationships. In: Astiba, H., Corral, J. C., Murelaga, X., Oue-Extebarria, X., and Pereda-Suberbiola, X. (eds.). Geology and Palaeontology of the Upper Cretaceous Vertebrate-Bearing Beds of the Laño Quarry (Basque-Cantabrrian Region, Iberian Peninsula). Estud. Mus. Cie. Nat. Alava 14 (num. espec. 1): 235–255.

Seeley, H. G. 1869. Index to the fossil remains of Aves, Ornithsauria and Reptilia, from the Secondary system of strata arranged in the Woodwardian Museum of the University of Cambridge, Cambridge. 143 pp.

———. 1876. On *Macrurosaurus semnus* (Seeley), a long-tailed animal with procoelous vertebrae from the Cambridge Upper Greensand, preserved in the Woodwardian Museum of the University of Cambridge. Q. J. Geol. Soc. London 32: 440–444.

Smith, J. B., Lamanna, M. C., Lacovara, K. J., Dodson, P., Smith, J. R., Poole, J. C., Giegengack, R., and Attia, Y. 2001. A giant sauropod dinosaur from an Upper Cretaceous mangrove deposit in Egypt. Science 292: 1704–1706.

Steel, R. 1970. Handbuch der Paläoherpetology, Saurischia. Gustav Fischer Verlag, Jena. 87 pp.

Stromer, E. 1932. Wibeltierreste der Baharije-Stufe (untestes Cenoman). 11. Sauropoda. Abh. Bayer. Akad. Wissensch. Math.-Naturwiss. Abt. 10: 1–21.

Swofford, D. L. 1999. PAUP*: Phylogenetic Analysis Using Parsimony (*and Other Methods), Version 4.0b2a. Sinauer Associates, Sunderland, MA.

Thévenin, A. 1907. Dinosauriens (Paléontologie de Madagascar IV). Ann. Paleontol. 2: 121–136.

Upchurch, P. 1995. Evolutionary history of sauropod dinosaurs. Philos. Trans. Roy. Soc. London B 349: 365–390.

———. 1998. The phylogenetic relationships of sauropod dinosaurs. Zool. J. Linn. Soc. 124: 43–103.

———. 1999. The phylogenetic relationships of the Nemegtosauridae (Saurischia, Sauropoda). J. Vertebr. Paleontol. 19: 106–125.

Upchurch, P., Barrett, P. M., and Dodson, P. 2004. Sauropoda. In: Weishampel, D. B., Dodson, P., and Osmólska, H. (eds.). The Dinosauria, 2nd ed. University of California Press, Berkeley. Pp. 259–324.

Wild, R. 1991. *Janenschia* n. g. *robusta* (E. Fraas 1908) pro *Tornieria robusta* (E. Fraas, 1908) (Reptilia, Saurischia, Sauropodomorpha). Stuttgarter eitr. Naturk. B 173: 1–4.

Wilkinson, M. 1995. Coping with abundant missing entries in phylogenetic inference using parsimony. Syst. Biol. 44: 501–514.

Wilson, J. A. 1999. The Evolution and Phylogeny of Sauropod Dinosaurs. Unpublished Ph.D. dissertation, University of Chicago, Chicago. 384 pp.

———. 2002. Sauropod dinosaur phylogeny: critique and cladistic analysis. Zool. J. Linn. Soc. 136: 217–276.

Wilson, J. A., and Carrano, M. T. 1999. Titanosaur locomotion and the origin of "wide-gauge" trackways: a biomechanical and systematic perspective on sauropod locomotion. Paleobiology 25: 252–267.

Wilson, J. A., and Sereno, P. C. 1994. Higher-level phylogeny of sauropod dinosaurs. J. Vertebr. Paleontol. 14: 42A.

———. 1998. Early evolution and higher-level phylogeny of sauropod dinosaurs. J. Vertebr. Paleontol. (Suppl. to No. 2): 1–68.

Wilson, J. A., and Upchurch, P. 2003. A revision of *Titanosaurus* Lydekker (Dinosauria-Saurpoda), the first dinosaur genus with a 'Gondwanan' distribution. J. Syst. Palaeont. 1 (3): 125–160.

Wilson, J. A., Martinez, R. N., and Alcober, O. 1999. Distal tail segment of a titanosaur (Dinosauria: Sauropoda) from the Upper Cretaceous of Mendoza, Argentina. J. Vertebr. Paleontol. 19: 591–594.

Yu, C. 1993. The Skull of *Diplodocus* and the Phylogeny of the Diplodocidae. Unpublished Ph.D. dissertation, University of Chicago, Chicago. 150 pp.

APPENDIX 2.1. CHARACTER–TAXON MATRIX

```
                         1                   10                  20                  30
Aegyptosaurus     ? ? ? ? ? ? ? ? ? ? ? ? ? ? ? ? ? ? ? ? ? ? ? ? ? ? ? ? ? ? ? ? ? ?
Aeolosaurus       ? ? ? ? ? ? ? ? ? ? ? ? ? ? ? ? ? ? ? ? ? ? ? ? ? ? ? ? ? ? ? ? ? ?
Alamosaurus       ? ? ? ? ? ? ? ? ? ? ? ? ? ? ? ? ? ? ? ? ? ? ? ? ? ? ? ? ? ? ? ? ? ?
Ampelosaurus      ? ? ? ? ? ? ? ? ? ? ? ? ? ? ? ? ? ? ? ? ? ? ? ? ? ? ? ? ? ? 1 1 1 1
Andesaurus        ? ? ? ? ? ? ? ? ? ? ? ? ? ? ? ? ? ? ? ? ? ? ? ? ? ? ? ? ? ? ? ? ? ?
Antarctosaurus    ? ? ? ? ? ? ? ? ? ? ? ? ? ? ? ? ? ? ? ? ? ? ? ? ? ? ? ? ? 0 1 1 1 1
Apatosaurus       0 0 1 2 1 0 0 1 0 0 1 1 1 1 1 1 0 1 1 0 1 0 1 ? 0 1 0 0 0 1 0 0 0 0
Argentinosaurus   ? ? ? ? ? ? ? ? ? ? ? ? ? ? ? ? ? ? ? ? ? ? ? ? ? ? ? ? ? ? ? ? ? ?
Argyrosaurus      ? ? ? ? ? ? ? ? ? ? ? ? ? ? ? ? ? ? ? ? ? ? ? ? ? ? ? ? ? ? ? ? ? ?
Augustinia        ? ? ? ? ? ? ? ? ? ? ? ? ? ? ? ? ? ? ? ? ? ? ? ? ? ? ? ? ? ? ? ? ? ?
Brachiosaurus     1 1 0 1 1 1 0 1 0 0 0 0 1 0 0 0 1 0 0 2 0 0 1 0 1 0 0 1 0 0 0 0 1 0
Camarasaurus      1 1 0 1 1 0 0 1 0 0 0 1 0 1 0 0 0 1 0 0 0 0 0 1 0 0 0 1 0 0 0 0 0 0
Dicraeosaurus     0 ? 1 ? ? ? ? 1 ? ? 1 ? ? 1 ? 0 ? ? ? ? ? ? ? ? ? ? ? ? 1 0 1 0 0 1
Diplodocus        0 0 1 2 1 0 0 1 0 0 1 1 1 1 1 1 0 1 1 0 1 0 1 1 0 1 0 0 0 1 0 0 0 0
Epacthosaurus     ? ? ? ? ? ? ? ? ? ? ? ? ? ? ? ? ? ? ? ? ? ? ? ? ? ? ? ? ? ? ? ? ? ?
Euhelopus         0 0 0 0 0 0 0 ? 1 0 0 0 ? 0 0 ? 0 ? ? 0 ? 0 ? 0 ? 0 ? ? ? ? ? ? ? ?
Brazil Series B   ? ? ? ? ? ? ? ? ? ? ? ? ? ? ? ? ? ? ? ? ? ? ? ? ? ? ? ? ? ? ? ? ? ?
Jainosaurus       ? ? ? ? ? ? ? ? ? ? ? ? ? ? ? ? ? ? ? ? ? ? ? ? ? ? ? ? ? 1 ? ? ? ?
Janenschia        ? ? ? ? ? ? ? ? ? ? ? ? ? ? ? ? ? ? ? ? ? ? ? ? ? ? ? ? ? ? ? ? ? ?
Lirainosaurus     ? ? ? ? ? ? ? ? ? ? ? ? ? ? ? ? ? ? ? ? ? ? ? ? ? ? ? ? ? ? ? ? ? ?
Magyarosaurus     ? ? ? ? ? ? ? ? ? ? ? ? ? ? ? ? ? ? ? ? ? ? ? ? ? ? ? ? ? ? ? ? ? ?
Malawisaurus      1 ? 0 1 0 ? ? ? ? ? ? ? ? 0 0 ? 0 0 ? ? ? ? ? ? ? ? ? ? ? ? ? ? ? 2
Nemegtosaurus     0 1 1 1 ? 1 ? 0 0 1 0 0 1 1 ? 1 1 ? 1 ? ? 0 0 1 ? 1 1 0 0 1 0 1 1 1
Neuquensaurus     ? ? ? ? ? ? ? ? ? ? ? ? ? ? ? ? ? ? ? ? ? ? ? ? ? ? ? ? ? ? ? ? ? ?
Opisthocoelicaudia ? ? ? ? ? ? ? ? ? ? ? ? ? ? ? ? ? ? ? ? ? ? ? ? ? ? ? ? ? ? ? ? ? ?
Paralititan       ? ? ? ? ? ? ? ? ? ? ? ? ? ? ? ? ? ? ? ? ? ? ? ? ? ? ? ? ? ? ? ? ? ?
Phuwiangosaurus   ? ? ? ? ? ? ? ? ? ? ? ? ? ? ? ? ? ? ? ? ? ? ? ? ? ? ? ? ? ? ? ? ? ?
Quaesitosaurus    1 1 1 0 ? ? 0 0 ? ? 0 1 0 ? ? ? ? ? ? ? 0 1 1 ? ? 0 ? 0 1 0 0 1
Rapetosaurus      0 0 1 0 ? 1 1 ? 1 1 1 1 ? 0 2 1 1 0 0 1 1 0 1 1 ? 1 1 1 ? 1 0 0 1 1 1
Rocasaurus        ? ? ? ? ? ? ? ? ? ? ? ? ? ? ? ? ? ? ? ? ? ? ? ? ? ? ? ? ? ? ? ? ? ?
Saltasaurus       ? ? ? ? ? ? ? ? ? ? ? ? ? ? ? ? ? ? ? ? ? ? ? ? ? ? ? ? ? 0 1 1 0 1
Titanosaurus      ? ? ? ? ? ? ? ? ? ? ? ? ? ? ? ? ? ? ? ? ? ? ? ? ? ? ? ? ? ? ? ? ? ?
Santa Rosa indet. ? ? ? ? ? ? ? ? ? ? ? ? ? 0 1 ? ? ? ? ? ? ? ? ? ? ? ? ? ? 0 1 1 1 1
Jabalpur indet.   ? ? ? ? ? ? ? ? ? ? ? ? ? ? ? ? ? ? ? ? ? ? ? ? ? ? ? ? ? 1 ? ? ? ?
Malagasy Taxon B  ? ? ? ? ? ? ? ? ? ? ? ? ? ? ? ? ? ? ? ? ? ? ? ? ? ? ? ? ? ? ? ? ? ?
```

APPENDIX 2.1. (continued)

```
                      40                  50                  60                  70
Aegyptosaurus      ? ? ? ? ? ? ? ? ? ? ? ? ? ? ? ? ? ? ? ? ? ? ? ? ? ? ? ? ? ? ?
Aeolosaurus        ? ? ? ? ? ? ? ? ? ? ? ? ? ? ? ? ? ? ? ? ? ? ? ? ? ? ? ? ? ? ?
Alamosaurus        ? ? ? ? ? ? ? ? ? ? ? ? ? ? ? ? ? ? ? ? ? ? ? ? ? ? ? ? ? ? ?
Ampelosaurus       1 1 1 1 0 ? 0 ? ? ? ? ? ? ? ? ? ? ? 0 1 0 0 ? ? ? ? ? ? ? ? ?
Andesaurus         ? ? ? ? ? ? ? ? ? ? ? ? ? ? ? ? ? ? ? ? ? ? ? ? ? ? ? ? ? ? ?
Antarctosaurus     1 ? 1 2 1 0 2 0 ? ? ? ? ? ? ? ? ? ? 0 0 0 0 0 1 0 1 2 0 0 ? 1 1 0
Apatosaurus        0 0 1 1 ? 0 0 0 1 1 0 1 1 0 1 0 0 0 1 1 0 1 ? 1 1 0 0 0 0 0 1 0
Argentinosaurus    ? ? ? ? ? ? ? ? ? ? ? ? ? ? ? ? ? ? ? ? ? ? ? ? ? ? ? ? ? ? ?
Argyrosaurus       ? ? ? ? ? ? ? ? ? ? ? ? ? ? ? ? ? ? ? ? ? ? ? ? ? ? ? ? ? ? ?
Augustinia         ? ? ? ? ? ? ? ? ? ? ? ? ? ? ? ? ? ? ? ? ? ? ? ? ? ? ? ? ? ? ?
Brachiosaurus      1 0 0 0 1 0 1 0 0 0 1 0 1 0 0 0 0 0 ? 0 1 0 0 0 ? 0 0 0 0 0 1 1 0 1
Camarasaurus       0 0 0 1 0 0 0 0 1 0 0 0 1 0 0 1 0 0 0 0 0 1 0 0 0 0 0 0 0 0 2 1 1 0 0
Dicraeosaurus      0 0 0 1 1 1 1 0 ? ? ? ? 0 0 ? ? ? ? 0 0 0 1 1 1 ? 1 1 1 1 0 1 0 ? 0
Diplodocus         0 0 1 1 0 0 0 0 1 1 0 1 0 1 0 0 1 0 0 0 0 1 1 0 1 ? 1 1 0 0 0 0 0 1 0
Epachthosaurus     ? ? ? ? ? ? ? ? ? ? ? ? ? ? ? ? ? ? ? ? ? ? ? ? ? ? ? ? ? ? ?
Euhelopus          ? ? ? ? ? ? 0 ? ? 0 0 0 0 0 ? 0 0 0 ? ? ? ? 0 ? ? ? ? ? ? ? ? ? ?
Brazil Series B    ? ? ? ? ? ? ? ? ? ? ? ? ? ? ? ? ? ? ? ? ? ? ? ? ? ? ? ? ? ? ?
Jainosaurus        ? ? ? ? 0 1 0 ? ? ? ? ? ? ? ? ? ? ? ? 0 0 1 0 ? 0 ? ? ? ? ? ? ? ?
Janenschia         ? ? ? ? ? ? ? ? ? ? ? ? ? ? ? ? ? ? ? ? ? ? ? ? ? ? ? ? ? ? ?
Lirainosaurus      ? ? ? ? ? ? ? ? ? ? ? ? ? ? ? ? ? ? ? 0 0 ? ? ? ? ? ? ? ? ? 0
Magyarosaurus      ? ? ? ? ? ? ? ? ? ? ? ? ? ? ? ? ? ? ? ? ? ? ? ? ? ? ? ? ? ? ?
Malawisaurus       1 1 1 1 0 1 0 ? ? ? 0 1 ? ? ? 1 ? ? ? ? ? 0 0 0 0 ? 1 1 0 0 ? ? 1 ? 0
Nemegtosaurus      1 1 1 1 0 2 0 1 1 1 0 1 1 1 1 ? 1 1 ? 1 0 0 1 ? 0 ? 1 1 1 0 1 0 1 ? 1
Neuquensaurus      ? ? ? ? ? ? ? ? ? ? ? ? ? ? ? ? ? ? ? ? ? ? ? ? ? ? ? ? ? ? ?
Opisthocoelicaudia ? ? ? ? ? ? ? ? ? ? ? ? ? ? ? ? ? ? ? ? ? ? ? ? ? ? ? ? ? ? ? ?
Paralititan        ? ? ? ? ? ? ? ? ? ? ? ? ? ? ? ? ? ? ? ? ? ? ? ? ? ? ? ? ? ? ?
Phuwiangosaurus    ? ? ? ? ? ? ? ? ? ? ? ? ? ? ? ? ? ? ? ? ? ? ? ? ? ? ? ? ? ? ?
Quaesitosaurus     0 0 1 1 0 2 0 1 ? 1 0 0 0 1 1 1 1 0 1 1 1 0 0 ? 0 ? 1 1 1 0 1 1 1 ? 1
Rapetosaurus       1 1 1 1 0 2 0 0 ? ? ? ? 1 1 1 1 1 1 1 1 0 0 0 0 1 0 1 1 1 2 0 ? ? 1
Rocasaurus         ? ? ? ? ? ? ? ? ? ? ? ? ? ? ? ? ? ? ? ? ? ? ? ? ? ? ? ? ? ? ?
Saltasaurus        1 ? 0 1 1 2 2 1 ? ? ? ? ? ? ? ? ? ? ? 1 0 0 0 0 1 0 1 1 0 ? ? ? ? 1
Titanosaurus       ? ? ? ? ? ? ? ? ? ? ? ? ? ? ? ? ? ? ? ? ? ? ? ? ? ? ? ? ? ? ?
Santa Rosa indet.  1 1 0 2 0 0 2 1 ? ? ? ? ? ? ? ? ? ? 1 1 0 0 0 0 1 0 1 ? 0 0 1 1 1 0
Jabalpur indet.    ? ? 1 ? ? ? ? ? ? ? ? ? ? ? ? ? ? ? 1 1 0 ? 0 0 ? 1 0 0 ? 0 1 ? ? 1
Malagasy Taxon B   ? ? ? ? ? ? ? ? ? ? ? ? ? ? ? ? ? ? ? ? ? ? ? ? ? ? ? ? ? ? ?
```

APPENDIX 2.1. (continued)

```
                                      80                    90                    100
Aegyptosaurus      ? ? ? ? ? ? ? ? ? ? ? ? ? ? ? ? ? ? ? ? ? ? ? ? ? ? ? ? ? ? ? ? ? ?
Aeolosaurus        ? ? ? ? ? ? ? ? ? ? ? ? ? ? ? ? ? ? ? ? ? ? ? ? ? ? ? ? ? ? ? ? ? ?
Alamosaurus        ? ? ? ? ? ? ? ? ? ? ? ? ? ? ? ? ? ? ? ? ? ? ? ? ? ? ? ? ? ? ? ? ? ?
Ampelosaurus       ? ? ? ? ? ? ? ? ? ? ? ? 1 0 ? 0 0 0 0 ? ? ? ? ? ? ? ? ? 0 1 1 0 2 0
Andesaurus         ? ? ? ? ? ? ? ? ? ? ? ? ? ? ? ? ? ? ? ? ? ? ? ? ? ? ? ? ? ? ? ? ? ?
Antarctosaurus     0 0 0 ? 0 ? ? ? ? ? ? ? ? ? ? ? ? ? ? ? ? ? ? ? ? ? ? ? ? ? ? ? ? ?
Apatosaurus        0 0 1 1 0 2 1 ? 1 1 1 1 ? ? 1 1 0 ? ? ? ? 1 0 0 1 1 ? ? ? 0 1 1 2 3 2
Argentinosaurus    ? ? ? ? ? ? ? ? ? ? ? ? ? ? ? ? ? ? ? ? ? ? ? ? ? ? ? ? ? ? ? ? ? ?
Argyrosaurus       ? ? ? ? ? ? ? ? ? ? ? ? ? ? ? ? ? ? ? ? ? ? ? ? ? ? ? ? ? ? ? ? ? ?
Augustinia         ? ? ? ? ? ? ? ? ? ? ? ? ? ? ? ? ? ? ? ? ? ? ? ? ? ? ? ? ? ? ? ? ? ?
Brachiosaurus      1 1 1 0 1 1 0 2 0 1 0 0 0 0 0 1 0 0 0 1 0 1 1 0 0 1 1 1 1 0 1 0 0 0
Camarasaurus       0 1 0 0 1 1 0 0 0 ? 0 0 1 0 0 0 0 0 0 0 1 1 1 0 1 0 1 1 0 0 0 0 0 0 1
Dicraeosaurus      0 0 1 0 1 ? ? 0 ? ? ? ? ? 0 0 0 ? 1 0 0 ? ? ? ? ? ? ? ? ? ? 0 0 ? 2 3 ?
Diplodocus         0 0 1 1 0 2 1 ? 1 1 1 1 1 0 1 1 0 1 0 0 0 1 0 0 1 1 0 1 1 0 1 1 2 3 2
Epacthosaurus      ? ? ? ? ? ? ? ? ? ? ? ? ? ? ? ? ? ? ? ? ? ? ? ? ? ? ? ? ? ? ? ? ? ?
Euhelopus          ? ? ? ? ? 1 0 0 0 ? 0 ? ? 1 0 0 0 0 0 0 1 ? ? ? ? ? ? ? ? ? 0 1 0 0 0 1
Brazil Series B    ? ? ? ? ? ? ? ? ? ? ? ? ? ? ? ? ? ? ? ? ? ? ? ? ? ? ? ? ? ? ? ? ? ?
Jainosaurus        0 0 1 2 1 ? ? ? ? ? ? ? ? ? ? ? ? ? ? ? ? ? ? ? ? ? ? ? ? ? ? ? ? ?
Janenschia         ? ? ? ? ? ? ? ? ? ? ? ? ? ? ? ? ? ? ? ? ? ? ? ? ? ? ? ? ? ? ? ? ? ?
Lirainosaurus      ? ? ? ? ? ? ? ? ? ? ? ? ? ? ? ? ? ? ? ? ? ? ? ? ? ? ? ? ? ? ? ? 0 2 2
Magyarosaurus      ? ? ? ? ? ? ? ? ? ? ? ? ? ? ? ? ? ? ? ? ? ? ? ? ? ? ? ? ? ? ? ? ? ?
Malawisaurus       ? 0 ? 2 ? ? ? ? ? ? ? ? ? 0 ? 0 ? 0 1 0 ? ? ? ? ? ? ? ? ? 0 0 ? 0 1 1
Nemegtosaurus      0 1 1 2 1 ? ? 1 1 0 1 1 0 0 1 0 1 0 0 1 1 0 1 ? 0 0 0 0 0 0 1 0 0 1 2
Neuquensaurus      ? ? ? ? ? ? ? ? ? ? ? ? ? ? ? ? ? ? ? ? ? ? ? ? ? ? ? ? ? ? ? ? ? ?
Opisthocoelicaudia ? ? ? ? ? ? ? ? ? ? ? ? ? ? ? ? ? ? ? ? ? ? ? ? ? ? ? ? ? ? ? ? ? ?
Paralititan        ? ? ? ? ? ? ? ? ? ? ? ? ? ? ? ? ? ? ? ? ? ? ? ? ? ? ? ? ? ? ? ? ? ?
Phuwiangosaurus    ? ? ? ? ? ? ? ? ? ? ? ? ? ? ? ? ? ? ? ? ? ? ? ? ? ? ? ? ? ? ? ? ? ?
Quaesitosaurus     1 1 0 0 1 ? ? 1 ? ? ? ? 1 0 0 ? 0 0 1 0 1 1 ? 1 0 0 0 0 1 1 0 ? 2 2
Rapetosaurus       1 0 1 2 1 1 1 1 0 1 0 ? ? 0 0 1 ? ? 1 ? 1 1 1 1 0 ? ? ? 0 1 1 1 1 2
Rocasaurus         ? ? ? ? ? ? ? ? ? ? ? ? ? ? ? ? ? ? ? ? ? ? ? ? ? ? ? ? ? ? ? ? ? ?
Saltasaurus        1 1 0 2 1 ? ? ? ? ? ? ? ? ? ? ? ? ? ? ? ? ? ? ? ? ? ? ? ? ? ? ? ? ?
Titanosaurus       ? ? ? ? ? ? ? ? ? ? ? ? ? ? ? ? ? ? ? ? ? ? ? ? ? ? ? ? ? ? ? ? ? ?
Santa Rosa indet.  0 0 1 2 1 ? ? ? ? ? ? ? ? ? ? ? ? ? ? ? ? ? ? ? ? ? ? ? ? ? ? ? ? ?
Jabalpur indet.    0 1 1 ? 1 ? ? ? ? ? ? ? ? ? ? ? ? ? ? ? ? ? ? ? ? ? ? ? ? ? ? ? ? ?
Malagasy Taxon B   ? ? ? ? ? ? ? ? ? ? ? ? ? ? ? ? ? ? ? ? ? ? ? ? ? ? ? ? ? ? ? ? ? ?
```

APPENDIX 2.1. (continued)

	110									120									130									140				
Aegyptosaurus	?	?	?	?	?	?	?	?	?	?	?	?	?	?	?	?	?	?	?	?	?	?	?	?	?	?	?	?	?	?	?
Aeolosaurus	?	?	?	?	?	?	?	?	?	?	?	?	?	?	?	?	?	?	?	?	?	?	?	?	?	?	?	?	?	?	?
Alamosaurus	?	?	?	?	?	?	?	1	?	?	?	?	?	?	?	1	0	0	0	?	1	1	0	?	?	?	?	?	?	?	?
Ampelosaurus	1	0	0	0	?	0	?	?	0	?	1	?	0	1	2	?	0	0	?	?	1	1	1	0	1	1	1	0	0	0	0	3	1	0
Andesaurus	?	?	?	?	?	?	?	?	?	?	?	?	?	?	?	?	?	?	?	?	?	?	?	?	?	?	?	?	?	?	?
Antarctosaurus	?	?	?	?	?	?	?	?	?	?	?	?	?	?	?	?	?	?	?	?	?	?	?	?	?	?	?	?	?	?	?
Apatosaurus	1	1	1	1	0	0	1	2	0	1	1	0	0	1	0	0	0	1	0	1	1	1	1	1	1	1	0	1	0	0	0	1	3	0	0
Argentinosaurus	?	?	?	?	1	0	?	?	?	?	?	?	?	?	?	?	?	?	?	?	?	?	?	?	?	?	?	?	?	?	?
Argyrosaurus	?	?	?	?	?	?	?	?	?	?	?	?	?	?	?	?	?	?	?	?	?	?	?	?	?	?	?	?	?	?	?
Augustinia	?	?	?	?	1	1	?	?	1	?	0	?	?	?	2	?	?	0	?	?	?	?	?	?	1	0	0	0	1	0	0	0	3	1	1
Brachiosaurus	0	0	0	0	0	0	0	1	0	1	1	0	0	1	0	1	0	0	1	1	1	1	1	0	1	0	1	0	0	0	0	2	0	0
Camarasaurus	1	0	0	0	0	0	0	0	1	1	0	1	1	0	1	0	1	0	0	1	0	1	1	1	1	0	0	1	1	3	0	0
Dicraeosaurus	0	0	1	?	0	0	1	0	0	0	1	0	1	1	0	1	0	1	0	0	0	0	1	1	1	0	0	0	0	0	1	3	0	0
Diplodocus	1	1	1	1	0	0	1	2	0	1	1	1	0	1	0	1	0	1	1	0	1	1	1	1	0	1	1	1	0	0	0	1	3	0	0
Epacthosaurus	?	?	?	?	?	?	?	?	?	?	?	?	?	?	?	?	?	?	?	?	?	?	?	?	?	?	?	?	?	?	?
Euhelopus	1	0	0	0	1	0	?	2	1	?	0	1	0	1	0	1	0	0	0	1	1	1	1	?	1	0	1	0	0	0	0	0	3	0	0
Brazil Series B	?	?	?	?	1	1	?	1	1	?	1	1	0	1	2	1	?	0	1	0	1	?	?	0	?	1	1	0	?	0	0	0	3	1	1
Jainosaurus	?	?	?	?	?	?	?	?	?	?	?	?	?	?	?	?	?	?	?	?	?	?	?	?	?	?	?	?	?	?	?
Janenschia	?	?	?	?	?	?	?	?	?	?	?	?	?	?	?	?	?	?	?	?	?	?	?	?	?	?	?	?	?	?	?
Lirainosaurus	1	?	?	?	1	0	?	?	?	?	?	?	?	?	?	?	?	?	?	?	?	?	?	?	1	1	?	?	?	0	0	?	?
Magyarosaurus	?	?	?	?	1	1	?	?	1	?	1	?	?	?	2	?	?	0	?	?	?	?	?	0	?	1	1	0	0	0	0	3	?	?
Malawisaurus	1	1	0	0	1	0	?	?	1	0	0	?	?	?	1	?	?	0	?	?	?	?	?	0	0	0	1	0	?	0	0	0	0	0	0
Nemegtosaurus	0	1	1	0	?	?	?	?	?	?	?	?	?	?	?	?	?	?	?	?	?	?	?	?	?	?	?	?	?	?	?
Neuquensaurus	?	?	?	?	1	0	?	?	1	0	1	0	0	1	2	?	?	2	?	1	?	1	1	0	1	1	1	0	?	0	0	0	1	1	0
Opisthocoelicaudia	?	?	?	?	?	0	?	?	?	?	?	?	?	?	?	?	?	?	?	?	?	?	?	?	?	?	?	?	?	?	?
Paralititan	?	?	?	?	?	?	?	?	?	?	?	?	?	?	?	?	?	?	?	?	?	?	?	?	?	?	?	?	?	?	?
Phuwiangosaurus	?	?	?	?	1	?	?	1	?	1	1	0	0	2	?	?	0	?	?	1	1	1	0	1	1	1	1	?	0	0	1	3	0	0
Quaesitosaurus	0	0	?	0	?	?	?	?	?	?	?	?	1	?	?	?	?	?	?	?	?	?	?	?	?	?	?	?	?	?	?
Rapetosaurus	1	1	1	0	1	1	?	2	1	0	1	?	?	?	1	1	1	0	0	1	1	1	1	1	0	1	1	1	1	0	1	2	0	0
Rocasaurus	?	?	?	?	1	?	?	?	0	0	?	?	0	1	2	1	0	0	0	1	?	1	1	?	?	?	?	?	?	?	?	?	?
Saltasaurus	?	?	?	?	1	1	?	?	1	0	0	1	0	0	1	0	?	2	0	?	0	1	1	0	1	1	1	1	?	0	0	0	1	0	0
Titanosaurus	?	?	?	?	0	?	?	1	?	1	0	0	1	1	0	0	0	1	0	1	1	1	0	1	1	1	0	0	0	0	0	3	1	0
Santa Rosa indet.	?	?	?	?	?	?	?	?	?	?	?	?	?	?	?	?	?	?	?	?	?	?	?	?	?	?	?	?	?	?	?
Jabalpur indet.	?	?	?	?	?	?	?	?	?	?	?	?	?	?	?	?	?	?	?	?	?	?	?	?	?	?	?	?	?	?	?
Malagasy Taxon B	?	?	?	?	?	?	?	?	?	?	?	?	?	?	?	?	?	?	?	?	?	?	?	?	?	?	?	?	?	?	?

APPENDIX 2.1. (continued)

```
                            150                 160                 170
Aegyptosaurus     ? ? ? ? ? ? ? ? ? ? ? ? ? ? ? ? ? ? ? ? ? ? ? ? ? ? ? ? ? ? ? ? ? ? ?
Aeolosaurus       ? ? ? ? ? ? ? ? ? ? ? ? ? ? ? ? ? ? ? ? ? ? ? ? ? ? ? ? ? ? ? ? ? ? ?
Alamosaurus       0 0 ? ? 0 ? ? 0 0 0 1 1 ? 0 0 ? 0 0 1 3 1 0 0 0 ? ? ? 0 0 1 1 ? 1 1 0
Ampelosaurus      0 0 ? 0 0 0 0 1 0 0 0 0 1 1 0 0 0 ? 1 0 0 0 1 1 0 ? 1 1 1 0 1 1 1
Andesaurus        ? ? ? 1 ? 0 0 ? ? ? ? ? 0 0 ? ? ? ? ? ? ? ? ? 1 0 0 0 ? 0 0 ? ? ?
Antarctosaurus    ? ? ? ? ? ? ? ? ? ? ? ? ? ? ? ? ? ? ? ? ? ? ? ? ? ? ? ? ? ? ? ? ? ? ?
Apatosaurus       1 1 3 1 0 0 1 1 2 1 1 1 0 1 0 0 1 1 0 1 0 1 0 0 1 0 0 1 0 0 0 0 1 0 0
Argentinosaurus   0 0 ? 0 0 0 0 ? 0 0 1 1 0 1 0 1 0 0 1 3 0 1 0 1 ? 1 0 1 0 0 1 ? 0 1 0
Argyrosaurus      ? ? ? ? ? ? ? ? ? ? ? ? ? ? ? ? ? ? ? ? ? ? ? ? ? ? ? ? ? ? ? ? ? ? ?
Augustinia        0 0 ? ? 0 0 0 1 ? 0 ? ? 0 ? ? ? ? ? ? ? 3 ? 0 ? ? ? 1 ? ? 1 1 ? ? ? ?
Brachiosaurus     0 ? 1 1 0 0 1 1 0 0 1 1 0 1 1 0 0 0 1 0 0 1 0 0 1 1 0 0 0 0 0 1 1 0 0
Camarasaurus     1 0 1 0 0 0 1 0 2 1 0 1 0 1 0 0 0 0 0 0 0 0 1 0 0 1 1 0 1 0 0 0 1 0 0 0
Dicraeosaurus     1 1 1 0 0 1 0 1 ? 1 0 1 1 0 1 0 0 1 ? 0 0 ? 0 0 1 0 0 0 0 0 0 0 0 1 0
Diplodocus        1 1 3 0 0 0 1 1 2 1 0 1 0 1 0 0 1 1 0 0 0 1 1 0 1 0 0 0 0 0 0 0 0 0 0
Epacthosaurus     ? ? ? ? ? ? ? ? ? ? ? 0 ? ? ? ? ? ? ? ? ? 0 0 0 0 ? ? 0 ? ? 0 ?
Euhelopus         1 1 0 0 0 0 ? 1 0 1 1 0 0 1 0 ? ? 0 3 2 0 0 0 0 1 1 0 1 0 0 1 1 0 0
Brazil Series B   0 ? 3 ? 0 0 0 1 ? 0 1 1 0 0 1 0 ? ? ? ? ? 0 ? ? 1 1 1 0 1 1 1 0 ? 1 1
Jainosaurus       ? ? ? ? ? ? ? ? ? ? ? ? ? ? ? ? ? ? ? ? ? ? ? ? ? ? ? ? ? ? ? ? ? ? ?
Janenschia        ? ? ? ? ? ? ? ? ? ? ? ? ? ? ? ? ? ? ? ? ? ? ? ? ? ? ? ? ? ? ? ? ? ? ?
Lirainosaurus     0 0 ? 1 1 0 0 1 1 0 1 ? 0 0 0 0 0 0 0 1 ? 0 0 0 1 1 0 0 0 1 0 ? 1 1 0
Magyarosaurus     ? ? ? 0 0 0 0 1 0 0 1 ? ? 0 1 0 0 0 1 1 1 0 1 ? 1 1 ? ? 1 1 ? 0 1 ? ?
Malawisaurus      0 0 ? 1 1 0 0 1 2 0 1 1 2 1 0 0 0 0 0 2 1 0 1 0 ? 1 ? ? 1 1 1 ? ? 0 ?
Nemegtosaurus     ? ? ? ? ? ? ? ? ? ? ? ? ? ? ? ? ? ? ? ? ? ? ? ? ? ? ? ? ? ? ? ? ? ? ?
Neuquensaurus     0 0 ? 1 0 0 1 0 2 0 1 1 0 0 1 0 0 0 0 3 2 0 0 0 1 1 0 0 0 1 0 0 0 1 0
Opisthocoelicaudia 1 0 2 1 0 0 ? 0 0 1 0 1 0 1 0 0 0 0 1 1 1 0 0 1 0 1 1 1 1 1 0 1 1 0 1
Paralititan       ? ? ? ? ? ? ? ? ? ? ? ? ? ? ? ? ? ? ? ? ? ? ? ? ? ? ? ? ? ? ? ? ? ? ?
Phuwiangosaurus   1 0 ? 0 0 0 0 ? ? ? ? ? 0 0 1 0 ? ? ? ? ? ? ? ? ? 0 1 0 0 0 1 0 ? ? 0 ?
Quaesitosaurus    ? ? ? ? ? ? ? ? ? ? ? ? ? ? ? ? ? ? ? ? ? ? ? ? ? ? ? ? ? ? ? ? ? ? ?
Rapetosaurus      0 0 3 0 0 0 0 1 2 0 1 0 0 1 0 0 0 0 0 3 2 0 1 0 1 1 1 1 1 1 1 1 0 1 1
Rocasaurus        0 0 ? 1 ? 0 0 0 0 0 ? 1 ? 0 1 ? 0 0 1 3 2 0 0 0 ? ? ? 0 1 ? 0 ? ? 0
Saltasaurus       0 0 ? 1 0 0 0 1 2 0 1 1 0 0 1 0 0 0 0 3 2 0 0 0 1 1 0 0 1 1 0 0 0 1 ?
Titanosaurus      0 ? ? 0 0 0 0 1 2 0 0 1 0 0 0 0 0 0 0 3 2 0 0 0 1 1 0 0 0 1 0 0 1 0 ?
Santa Rosa indet. ? ? ? ? ? ? ? ? ? ? ? ? ? ? ? ? ? ? ? ? ? ? ? ? ? ? ? ? ? ? ? ? ? ? ?
Jabalpur indet.   ? ? ? ? ? ? ? ? ? ? ? ? ? ? ? ? ? ? ? ? ? ? ? ? ? ? ? ? ? ? ? ? ? ? ?
Malagasy Taxon B  ? ? ? ? ? ? ? ? ? ? ? ? ? ? ? ? ? ? ? ? ? ? ? ? ? ? ? ? ? ? ? ? ? ? ?
```

APPENDIX 2.1. (continued)

	180										190										200										210
Aegyptosaurus	?	?	?	?	?	?	?	?	?	?	?	?	?	?	?	?	?	?	?	?	?	?	?	?	?	?	?	?	?	?	?
Aeolosaurus	?	?	?	?	?	?	?	?	?	?	?	?	?	?	?	?	?	?	?	?	?	?	?	?	?	?	1	?	?	?	?
Alamosaurus	1	1	0	0	0	1	1	1	0	3	2	0	1	0	0	0	0	?	?	?	?	?	?	?	0	0	1	1	?	?	?
Ampelosaurus	0	0	0	?	?	0	0	0	0	?	0	1	0	0	0	0	0	1	?	?	0	?	0	?	?	?	?	?	0	?	?
Andesaurus	?	0	?	?	?	1	?	?	0	?	?	0	?	?	?	?	?	?	0	?	?	?	?	?	?	?	?	?	?	0	?
Antarctosaurus	?	?	?	?	?	?	?	?	?	?	?	?	?	?	?	?	?	?	?	?	?	?	?	?	?	?	?	?	?	?	?
Apatosaurus	1	0	0	0	1	0	1	1	0	3	3	0	0	0	0	0	0	0	?	0	3	1	1	1	1	0	0	0	0	0	0
Argentinosaurus	1	1	0	0	0	1	1	0	1	3	0	?	0	1	0	0	0	0	?	?	?	?	?	?	?	?	?	?	?	?	?
Argyrosaurus	?	?	?	?	?	?	?	?	?	?	?	?	?	?	?	?	?	?	?	?	?	?	?	?	?	?	?	?	?	?	?
Augustinia	?	0	?	?	?	?	1	1	?	?	?	?	?	?	?	?	?	?	?	?	?	?	?	?	?	?	?	?	?	?	?
Brachiosaurus	0	1	0	0	1	0	0	0	1	2	2	0	0	0	0	0	1	0	0	0	0	1	0	1	1	0	1	0	0	0	0

Camarasaurus	0	1	0	1	0	0	0	0	0	0	0	0	0	0	0	0	1	0	0	1	0	1	0	1	1	0	?	0	0	0	0
Dicraeosaurus	1	0	0	1	0	0	1	1	0	0	0	0	?	?	0	0	0	0	?	0	3	0	1	1	1	0	?	?	?	0	?	0	1	0
Diplodocus	1	0	0	0	1	0	1	1	0	3	3	1	0	0	0	1	0	0	0	0	0	3	1	1	0	0	0	0	1	0	0	0	0	1	0
Epacthosaurus	1	0	?	?	?	?	1	?	?	3	?	?	?	0	0	?	?	?	0	?	0	?	?	?	?	?	?	?	?	?	?
Euhelopus	0	?	0	1	1	0	0	0	0	?	?	1	0	1	0	0	0	1	1	?	0	?	?	0	1	1	0	0	0	0	?	?	?	?
Brazil Series B	1	1	0	0	?	?	1	1	0	3	?	1	?	0	0	?	?	0	1	?	?	?	?	?	?	?	?	?	?	?	?
Jainosaurus	?	?	?	?	?	?	?	?	?	?	?	?	?	?	?	?	?	?	?	?	?	?	?	?	?	?	?	?	?	?	?
Janenschia	?	?	?	?	?	?	?	?	?	?	?	?	?	?	?	?	?	?	?	?	?	?	?	?	?	?	?	?	?	?	?
Lirainosaurus	0	0	?	0	?	1	1	1	0	3	3	1	0	0	0	0	1	0	?	?	?	?	?	?	?	?	?	?	0	?	?	1	1
Magyarosaurus	?	?	?	?	?	?	1	1	0	1	1	?	0	?	?	?	?	?	?	0	?	0	?	?	?	?	?	?	?	?	?
Malawisaurus	?	1	0	0	1	?	1	1	?	?	?	?	?	?	?	?	?	?	?	?	?	?	?	?	?	?	?	?	?	?	?
Nemegtosaurus	?	?	?	?	?	?	?	?	?	?	?	?	?	?	?	?	?	?	?	?	?	?	?	?	?	?	?	?	?	?	?
Neuquensaurus	1	0	0	0	1	1	1	1	3	3	0	1	1	0	0	0	1	1	0	1	0	0	0	0	?	1	1	1	1	?	?	1	1
Opisthocoelicaudia	0	1	1	0	1	0	1	1	1	1	1	0	1	0	0	0	?	1	?	1	?	0	0	1	0	0	?	0	1	?	1	1	2	1
Paralititan	?	?	?	?	?	?	?	?	?	?	?	?	?	?	?	?	?	?	?	?	0	?	?	?	?	?	?	?	?	?	3	?
Phuwiangosaurus	?	?	0	0	1	1	0	0	0	0	?	?	?	?	0	0	1	0	?	0	?	0	?	0	?	?	?	?	?	0	?	?	?	?	?
Quaesitosaurus	?	?	?	?	?	?	?	?	?	?	?	?	?	?	?	?	?	?	?	?	?	?	?	?	?	?	?	?	?	?	?
Rapetosaurus	1	1	0	0	1	0	1	1	0	3	3	1	0	0	1	1	1	0	1	?	0	1	1	0	0	0	1	0	1	0	1	1	0	?	?
Rocasaurus	1	0	?	?	1	1	1	1	?	3	3	?	0	0	1	?	1	0	?	?	?	?	?	?	?	?	?	?	1	?	?	?	?
Saltasaurus	1	0	0	?	1	1	1	1	0	3	3	1	0	0	0	0	1	0	1	0	0	1	0	0	?	?	?	0	1	1	1	?	?	1	0
Titanosaurus	1	0	0	0	0	1	1	1	1	0	1	0	1	0	0	0	0	1	1	0	1	0	0	?	?	?	?	0	?	?	?	?	?	0
Santa Rosa indet.	?	?	?	?	?	?	?	?	?	?	?	?	?	?	?	?	?	?	?	?	?	?	?	?	?	?	?	?	?	?	?
Jabalpur indet.	?	?	?	?	?	?	?	?	?	?	?	?	?	?	?	?	?	?	?	?	?	?	?	?	?	?	?	?	?	?	?
Malagasy Taxon B	?	?	?	?	?	?	?	?	?	?	?	?	?	?	?	?	?	?	?	?	?	?	?	?	?	?	1	?	?	?	?

APPENDIX 2.1. (continued)

```
                              220                 230                 240
Aegyptosaurus     ? ? ? ? 0 ? ? ? ? ? ? ? ? ? ? ? ? ? 0 0 1 ? ? ? ? 0 ? ? ? ? ? ? ? ?
Aeolosaurus       1 0 ? 1 2 0 0 0 ? ? ? 0 0 0 0 1 1 0 0 1 1 2 0 1 1 0 2 ? ? ? ? ? ? ?
Alamosaurus       1 0 0 0 0 1 0 0 1 1 0 1 0 0 ? 1 1 1 0 1 1 0 0 1 0 0 ? ? ? ? ? ? 1 1 1
Ampelosaurus      1 0 ? 0 ? ? ? ? ? ? ? ? ? ? ? ? ? 0 0 1 1 2 1 1 0 ? ? ? ? 1 ? ? ? ?
Andesaurus        0 0 ? 1 1 1 ? ? ? ? ? 0 0 ? ? ? 0 1 0 0 0 0 1 0 1 0 ? ? ? ? ? ? 1 1 1
Antarctosaurus    ? ? ? ? ? ? ? ? ? ? ? ? ? ? ? ? ? ? ? ? ? ? ? ? ? ? ? ? ? ? ? ? ? ?
Apatosaurus       1 0 0 0 2 0 1 1 1 1 0 1 1 1 1 1 1 0 0 0 0 1 0 0 1 1 1 1 0 0 0 0 0
Argentinosaurus   ? ? ? ? ? ? ? ? ? ? ? ? ? ? ? ? ? ? ? ? ? ? ? ? ? ? ? ? ? ? ? ? ? ?
Argyrosaurus      ? ? ? ? ? ? ? ? ? ? ? ? ? ? ? ? ? ? ? ? ? ? ? ? ? ? ? ? ? ? ? ? ? ?
Augustinia        ? ? ? ? ? ? ? ? ? ? ? ? ? ? ? ? ? ? ? ? ? ? ? ? ? ? ? ? ? ? ? ? ? ?
Brachiosaurus     0 0 0 0 2 1 0 0 1 1 0 0 0 0 0 1 1 0 0 0 0 0 1 1 0 ? ? ? 0 ? ? ? ? 0
Camarasaurus      0 0 0 1 1 1 0 0 1 1 0 1 0 0 0 1 1 0 0 0 0 0 0 1 0 0 0 ? ? 0 0 ? 0 1 0
Dicraeosaurus     1 0 0 1 0 0 1 0 1 1 0 1 1 0 ? 1 1 ? 0 0 0 0 0 1 0 0 1 1 ? 1 0 0 0 0 0
Diplodocus        1 1 1 1 0 0 1 1 1 1 0 1 1 1 1 0 1 1 1 0 1 1 0 0 0 1 1 1 1 0 0 0 0 0
Epacthosaurus     ? ? ? ? ? ? ? ? ? ? ? ? ? ? ? ? ? ? ? ? ? ? ? ? ? ? ? ? ? ? ? ? ? ?
Euhelopus         ? ? ? ? ? ? ? ? ? ? ? ? ? ? ? ? ? ? ? ? ? ? ? ? ? ? ? ? 0 1 1 ? ? ?
Brazil Series B   1 0 0 1 0 0 0 0 0 0 0 ? 0 ? 0 1 1 0 0 1 1 0 1 0 1 ? ? ? ? 0 ? ? ? ?
Jainosaurus       ? ? ? ? ? ? ? ? ? ? ? ? ? ? ? ? ? ? ? ? ? ? ? ? ? ? ? ? ? ? ? ? ? ?
Janenschia        ? ? ? ? ? ? ? ? ? ? ? ? ? ? ? ? ? ? ? ? ? ? ? ? ? ? ? ? ? ? ? ? ? ?
Lirainosaurus     1 0 ? 1 1 0 ? 1 1 ? ? 0 ? 0 ? 1 0 1 0 1 1 1 2 1 1 0 0 ? ? ? ? ? ? ?
Magyarosaurus     1 0 ? 0 0 0 0 0 1 1 0 1 0 0 ? 1 0 0 1 0 1 1 1 1 1 0 0 ? ? ? ? ? ? ?
Malawisaurus      1 ? ? ? ? ? ? 0 1 1 1 1 0 ? ? ? ? ? 0 0 0 1 1 1 1 0 ? ? ? 0 1 ? 1 1 1
Nemegtosaurus     ? ? ? ? ? ? ? ? ? ? ? ? ? ? ? ? ? ? ? ? ? ? ? ? ? ? ? ? ? ? ? ? ? ?
Neuquensaurus     1 0 ? 0 0 0 1 1 1 1 1 0 0 0 1 0 0 0 1 1 0 2 0 1 1 2 1 ? ? ? ? 1 1 1
Opisthocoelicaudia 2 0 0 1 0 1 0 0 1 1 0 1 0 0 ? 0 0 1 0 1 2 0 0 1 1 0 1 0 0 ? 1 1 1 1 1
Paralititan       1 ? ? 1 2 1 ? ? ? 1 ? 0 0 ? ? 0 ? ? ? 1 ? ? ? ? 0 ? ? ? ? ? ? ? ? ?
Phuwiangosaurus   0 0 ? 1 2 0 0 0 0 0 0 0 0 0 0 ? 1 0 0 ? ? ? 1 0 1 0 ? ? ? ? 1 1 1
Quaesitosaurus    ? ? ? ? ? ? ? ? ? ? ? ? ? ? ? ? ? ? ? ? ? ? ? ? ? ? ? ? ? ? ? ? ? ?
Rapetosaurus      1 0 0 0 2 0 1 0 1 1 1 0 1 0 ? 1 0 1 0 0 1 0 1 1 1 0 0 0 0 0 1 0 1 1 1
Rocasaurus        ? ? ? ? ? ? ? ? ? ? ? ? ? ? ? ? 1 1 1 ? ? ? 1 1 2 0 ? ? ? ? ? ? ?
Saltasaurus       1 0 ? 0 0 0 1 1 1 1 1 1 1 0 0 1 1 ? 0 1 1 0 2 0 1 1 2 1 ? ? 1 1 1 1 1
Titanosaurus      1 0 0 0 0 0 0 0 1 1 ? 0 0 ? 1 1 0 1 1 1 0 0 1 1 0 ? ? ? 1 0 1 1 1 1
Santa Rosa indet. ? ? ? ? ? ? ? ? ? ? ? ? ? ? ? ? ? ? ? ? ? ? ? ? ? ? ? ? ? ? ? ? ? ?
Jabalpur indet.   ? ? ? ? ? ? ? ? ? ? ? ? ? ? ? ? ? ? ? ? ? ? ? ? ? ? ? ? ? ? ? ? ? ?
Malagasy Taxon B  ? ? ? ? ? ? ? ? ? ? ? ? ? ? ? ? ? 0 0 1 1 2 0 1 1 2 1 ? ? ? ? ? ? ?
```

APPENDIX 2.1. (continued)

```
                        250                 260                 270                 280
Aegyptosaurus     ? ? ? ? 1 1 ? 1 1 ? ? ? ? ? ? ? ? ? 0 0 0 0 0 0 ? 1 0 1 1 1 1 0 1 1
Aeolosaurus       ? ? ? ? ? ? ? ? ? ? ? ? ? ? ? ? ? ? ? ? ? ? ? ? ? ? ? ? 1 1 1 0 1 1
Alamosaurus       0 1 1 0 1 0 0 1 1 2 1 1 1 1 ? ? 0 1 0 0 0 1 0 1 1 1 0 0 0 1 0 0 1 1 0
Ampelosaurus      ? ? 1 0 0 0 2 1 ? ? 1 0 1 0 0 0 1 ? ? 1 1 0 0 0 0 0 1 ? 0 0 1 1 1 1 0 0
Andesaurus        ? ? 1 0 1 ? ? ? ? ? ? ? ? ? ? ? ? ? ? ? ? ? ? ? ? ? ? ? ? ? ? ? ? ?
Antarctosaurus    ? ? ? ? 0 0 1 1 ? ? 1 1 ? ? ? ? ? ? ? ? ? ? ? ? ? ? ? ? ? ? ? ? ? ?
Apatosaurus       0 0 0 0 0 0 1 0 0 ? 1 1 0 0 0 0 0 0 ? 1 1 1 0 0 1 0 0 0 0 1 0 0 0 0 1
Argentinosaurus   ? ? ? ? ? ? ? ? ? ? ? ? ? ? ? ? ? ? ? ? ? ? ? ? ? ? ? ? ? ? ? ? ? ?
Argyrosaurus      ? ? ? ? 1 1 1 0 1 ? 1 1 ? ? ? ? ? ? ? 0 1 2 0 0 1 0 ? 0 0 1 0 1 1 1 1
Augustinia        ? ? ? ? 1 1 1 2 ? ? 1 0 ? ? ? ? ? ? ? ? ? ? ? ? ? ? ? ? ? ? ? ? ? ?
Brachiosaurus     ? ? 1 ? 0 1 0 1 0 1 0 1 0 0 0 0 0 0 0 1 0 0 0 1 0 0 0 1 1 0 0 1 0 1
Camarasaurus      0 0 0 1 0 1 0 0 0 ? 0 1 1 1 0 1 1 0 0 0 0 0 1 0 0 1 1 1 0 1 1 0 1 0 0
Dicraeosaurus     ? ? 0 ? 0 0 0 1 1 2 0 1 0 0 ? 0 ? ? ? 1 0 0 0 0 1 0 ? 0 0 ? 0 0 0 0 ?
Diplodocus        0 0 0 0 0 0 0 0 2 0 1 0 0 ? 0 0 0 0 0 1 0 0 0 0 0 0 0 0 0 1 0 0 1 0 1
Epacthosaurus     ? ? ? ? ? ? ? ? ? ? ? 1 ? 0 ? 1 ? ? ? 1 ? 0 1 ? ? ? ? ? ? ? ? ?
Euhelopus         ? ? ? ? 0 0 ? ? ? ? 1 0 0 0 ? 0 ? ? ? ? ? ? 0 0 ? 0 ? 0 0 ? ? ? ? ?
Brazil Series B   ? ? ? ? ? ? ? ? ? ? ? ? ? ? ? ? ? ? ? ? ? ? ? ? ? ? ? ? ? ? ? ? ? ?
Jainosaurus       ? ? ? ? ? ? ? ? ? ? ? ? ? ? ? ? ? ? ? ? ? ? ? ? ? ? ? ? ? ? ? ? ? ?
Janenschia        ? ? ? ? ? ? ? ? ? ? ? ? ? ? ? ? ? 1 0 1 1 1 0 0 ? 0 0 ? 1 1 1 1 ?
Lirainosaurus     ? ? ? ? 0 0 1 1 ? 0 1 ? 2 1 ? 0 1 1 ? 0 1 0 1 0 0 1 0 0 1 ? ? ? ? ?
Magyarosaurus     ? ? ? ? 0 1 1 1 ? 1 0 1 1 1 0 0 ? ? 0 1 1 1 0 0 1 0 0 0 1 1 0 1 1 0
Malawisaurus      ? ? 1 1 0 1 ? ? ? ? 1 0 1 1 ? 0 ? 1 ? 0 0 1 0 0 0 1 0 0 1 1 0 ? 1 1 1
Nemegtosaurus     ? ? ? ? ? ? ? ? ? ? ? ? ? ? ? ? ? ? ? ? ? ? ? ? ? ? ? ? ? ? ? ? ? ?
Neuquensaurus     ? 1 1 0 1 1 0 1 1 2 1 1 1 1 0 1 0 1 ? 0 1 1 1 1 1 1 0 1 1 1 0 1 1 1 1
Opisthocoelicaudia 0 1 1 0 0 0 0 0 1 0 1 1 1 1 ? 1 0 1 1 1 1 1 0 1 1 1 ? 0 0 ? 1 1 1 1 0
Paralititan       ? ? ? ? ? ? 0 ? ? ? 1 ? ? ? ? ? ? ? 0 1 2 1 0 1 0 ? ? 1 ? ? ? ? ? ?
Phuwiangosaurus   ? 1 1 0 1 1 0 0 0 ? 1 ? ? ? ? ? ? ? ? 0 0 ? 0 0 0 0 ? ? ? 1 0 0 1 1 ?
Quaesitosaurus    ? ? ? ? ? ? ? ? ? ? ? ? ? ? ? ? ? ? ? ? ? ? ? ? ? ? ? ? ? ? ? ? ? ?
Rapetosaurus      1 0 1 1 1 1 0 2 1 ? 1 0 1 0 0 0 1 1 0 1 0 0 1 0 0 0 0 1 0 1 0 0 0 1 0
Rocasaurus        ? ? ? ? ? ? ? ? ? ? ? ? ? ? ? ? ? ? ? ? ? ? ? ? ? ? ? ? ? ? ? ? ? ?
Saltasaurus       ? ? 1 0 0 1 0 0 1 2 1 1 1 1 1 1 0 1 ? 0 1 1 1 1 0 1 ? 0 ? 1 1 0 1 1 1
Titanosaurus      ? 0 1 0 1 0 0 1 1 2 1 0 ? ? ? ? 0 ? ? 0 1 0 0 ? 0 1 ? 0 0 0 1 0 1 1 1
Santa Rosa indet. ? ? ? ? ? ? ? ? ? ? ? ? ? ? ? ? ? ? ? ? ? ? ? ? ? ? ? ? ? ? ? ? ? ?
Jabalpur indet.   ? ? ? ? ? ? ? ? ? ? ? ? ? ? ? ? ? ? ? ? ? ? ? ? ? ? ? ? ? ? ? ? ? ?
Malagasy Taxon B  ? ? ? ? ? ? ? ? ? ? 1 1 1 1 1 ? ? ? ? ? ? ? ? ? ? ? ? ? ? ? ? ? ? ?
```

APPENDIX 2.1. (continued)

	320										330										340										350
Aegyptosaurus	?	?	?	?	?	?	?	?	?	?	?	?	?	?	?	?	1	0	?	1	1	1	0	?	?	?	?	?	?	?	?
Aeolosaurus	?	?	?	?	0	0	1	0	0	0	0	1	1	0	0	0	?	?	?	?	?	?	?	?	?	?	?	?	?	?	?
Alamosaurus	0	1	1	0	0	0	1	?	1	1	1	1	0	0	1	0	?	?	1	1	1	1	0	1	?	?	?	1	?	?	?	0	0	?
Ampelosaurus	?	0	1	?	0	1	2	0	?	?	?	?	?	?	?	?	?	?	?	?	?	0	?	?	?	0	1	1	1	1	0	?
Andesaurus	?	?	?	?	0	0	1	1	1	0	1	1	1	1	0	1	0	?	?	?	?	?	?	?	?	?	?	?	?	?	?
Antarctosaurus	?	?	?	?	?	?	?	?	?	?	?	?	?	?	?	?	?	?	?	?	?	?	?	?	?	?	?	?	?	?	?
Apatosaurus	0	0	0	1	1	0	2	0	0	0	0	1	0	0	0	0	0	0	0	1	1	0	0	1	1	?	0	0	1	0	1	0	0	0
Argentinosaurus	0	1	?	?	?	?	?	?	?	?	?	?	?	?	?	?	?	?	?	?	?	?	?	?	?	?	?	?	?	?	?
Argyrosaurus	?	?	?	?	?	?	?	?	?	?	?	?	?	?	?	?	?	?	?	?	?	?	?	?	?	?	?	?	?	?	?
Augustinia	?	?	?	?	?	?	?	?	?	?	?	?	?	?	?	?	?	?	?	?	?	?	?	?	?	?	?	?	?	?	?
Brachiosaurus	0	0	1	0	1	0	2	2	0	0	1	1	0	0	1	1	0	1	1	1	1	0	0	1	0	0	1	0	1	0	1	0
Camarasaurus	0	1	1	0	1	0	1	0	0	1	1	1	0	0	0	1	0	0	0	1	1	0	0	1	1	1	0	0	1	0	1	0	0	0
Dicraeosaurus	0	0	0	1	1	0	2	0	0	0	0	0	0	0	1	1	0	0	0	1	1	0	0	1	0	1	0	0	1	1	1	0	0	0
Diplodocus	1	1	1	0	0	0	2	0	0	0	0	0	0	0	1	0	1	1	0	1	1	0	1	1	1	1	0	0	1	1	1	0	0	0
Epacthosaurus	?	?	?	?	?	?	?	?	?	?	?	?	?	?	?	?	?	?	?	?	?	?	?	?	?	?	?	?	?	?	?
Euhelopus	0	0	1	0	0	0	1	0	0	0	1	1	0	0	0	1	0	1	1	1	1	0	0	1	1	1	0	0	1	0	1	0	0	?
Brazil Series B	?	?	?	?	?	?	?	?	?	?	?	?	?	?	?	?	?	?	?	?	?	?	?	?	?	?	?	?	?	?	?
Jainosaurus	?	?	?	?	?	?	?	?	?	?	?	?	?	?	?	?	?	?	?	?	?	?	?	?	?	?	?	?	?	?	?
Janenschia	?	?	?	?	?	?	?	?	?	?	?	?	?	?	?	?	?	?	?	?	?	0	1	?	?	?	?	?	?	?	?
Lirainosaurus	?	?	?	?	?	?	?	?	?	?	?	?	1	0	1	?	1	1	1	2	1	?	?	2	1	0	0	0	1	?
Magyarosaurus	?	?	?	?	?	3	?	?	?	?	?	?	?	?	0	0	1	1	0	1	1	0	0	1	0	1	1	0	1	0	1	?
Malawisaurus	?	?	?	?	?	?	0	?	0	0	1	1	1	?	?	?	?	1	?	?	?	?	0	1	1	0	?	0	0	?	0	1	?
Nemegtosaurus	?	?	?	?	?	?	?	?	?	?	?	?	?	?	?	?	?	?	?	?	?	?	?	?	?	?	?	?	?	?	?
Neuquensaurus	0	1	1	?	0	?	1	?	1	1	1	1	1	1	1	0	1	0	1	1	0	1	1	1	0	1	0	0	?	?	?	1	?	?
Opisthocoelicaudia	0	1	?	0	1	2	1	2	1	1	1	1	1	1	0	0	0	1	0	1	2	1	1	0	0	0	0	0	1	1	0	1	1	0	1
Paralititan	?	?	?	?	?	?	?	?	?	?	?	?	?	?	?	?	?	?	?	?	?	?	?	?	?	?	?	?	?	?	?
Phuwiangosaurus	0	0	0	?	0	0	2	0	?	0	1	1	1	0	0	?	0	1	0	0	2	1	1	1	0	1	1	1	1	0	0	1	0	1	?
Quaesitosaurus	?	?	?	?	?	?	?	?	?	?	?	?	?	?	?	?	?	?	?	?	?	?	?	?	?	?	?	?	?	?	?
Rapetosaurus	0	1	1	1	1	1	1	0	1	1	1	1	1	1	1	1	0	1	0	1	0	0	1	1	1	1	?	?	?	1	0	1	0	0	?
Rocasaurus	0	0	1	?	0	0	1	0	1	1	?	1	1	1	0	1	0	?	?	?	?	?	?	?	?	?	?	?	?	?	?
Saltasaurus	0	1	1	?	0	0	1	0	1	1	0	1	1	1	0	0	0	1	1	1	1	0	1	0	0	0	1	1	1	1	1	1	1	?
Titanosaurus	0	1	1	0	?	?	0	2	1	1	1	1	1	1	0	1	0	?	?	?	?	?	?	?	?	?	?	?	?	0	?	?
Santa Rosa indet.	?	?	?	?	?	?	?	?	?	?	?	?	?	?	?	?	?	?	?	?	?	?	?	?	?	?	?	?	?	?	?
Jabalpur indet.	?	?	?	?	?	?	?	?	?	?	?	?	?	?	?	?	?	?	?	?	?	?	?	?	?	?	?	?	?	?	?
Malagasy Taxon B	?	?	?	?	?	?	?	?	?	?	?	?	?	?	?	?	?	?	?	?	?	?	?	?	?	?	?	?	?	?	?

APPENDIX 2.1. (continued)

```
                              290                   300                   310
Aegyptosaurus      1 1 ? 1 1 ? 0 0 ? 2 ? ? ? ? ? ? ? ? ? ? ? ? ? ? ? ? ? ? ? ?
Aeolosaurus        1 0 1 ? 1 ? 0 1 1 ? ? ? ? ? ? ? ? ? 1 ? ? ? ? ? ? ? ? ? ? ?
Alamosaurus        1 2 ? ? 0 ? 0 1 1 ? 1 0 1 2 0 1 1 ? ? 1 ? 0 0 1 ? 1 0 2 ? ? 1 0 1 2 1
Ampelosaurus       0 2 1 1 0 ? 0 1 1 ? ? ? ? ? ? ? ? ? ? ? ? ? ? ? ? ? ? ? ? ? ? 1 ? 2 ?
Andesaurus         ? ? ? ? ? ? ? ? ? ? ? ? ? ? ? ? ? ? ? ? ? ? ? ? ? ? ? ? ? ?
Antarctosaurus     ? ? ? ? ? ? ? ? ? ? ? 0 ? 2 1 1 1 1 ? ? 0 2 1 1 ? 0 ? ? ? ? ? ? ? ?
Apatosaurus        1 0 0 0 0 0 0 0 1 0 0 0 1 0 0 0 1 1 1 0 2 0 1 0 1 1 1 1 0 1 0 0 0 0
Argentinosaurus    ? ? ? ? ? ? ? ? ? ? ? ? ? ? ? ? ? ? ? ? ? ? ? ? ? ? ? ? ? ? ? ? 2 1
Argyrosaurus       1 0 1 0 0 ? 0 0 0 ? ? ? ? ? ? ? ? ? ? ? ? ? ? ? ? ? ? ? ? ? ? ? ? ?
Augustinia         ? ? ? ? ? ? ? ? ? ? ? ? ? ? ? ? ? ? ? ? ? ? ? ? ? ? ? ? ? ?
Brachiosaurus      1 0 0 0 0 2 0 0 0 1 1 0 1 1 0 1 1 1 ? 0 1 ? 1 1 ? 0 0 1 0 0 1 1 1 2 0
Camarasaurus       1 2 0 2 0 2 1 0 1 1 0 0 0 0 0 1 1 1 0 0 2 1 1 0 ? ? 1 1 1 1 0 1 1 1
Dicraeosaurus      ? ? ? ? 0 ? ? ? 0 ? ? ? ? ? ? ? ? ? ? ? ? ? ? ? ? ? ? ? ? 0 0 0 1
Diplodocus         1 0 0 0 0 1 ? 0 0 1 0 ? 0 0 1 1 1 0 1 0 1 1 0 1 0 1 0 0 1 1 0 0 0 2 1
Epacthosaurus      0 ? ? ? 1 ? ? ? ? ? 1 ? ? ? ? ? ? ? ? ? ? ? ? ? ? ? 2 ? ? ? ? ? ? ?
Euhelopus          ? ? ? ? ? ? ? ? ? ? ? ? ? ? ? ? ? ? ? ? ? ? ? ? ? ? ? ? 1 0 1 2 0
Brazil Series B    ? ? ? ? ? ? ? ? ? ? ? ? ? ? ? ? ? ? ? ? ? ? ? ? ? ? ? ? ? ? ? ? ? ?
Jainosaurus        ? ? ? ? ? ? ? ? ? ? ? ? ? ? ? ? ? ? ? ? ? ? ? ? ? ? ? ? ? ?
Janenschia         ? ? ? ? 1 ? 0 ? ? ? ? ? 0 0 ? ? 1 1 0 0 2 0 1 1 1 0 ? ? ? ? ? ? ? ?
Lirainosaurus      0 ? ? ? ? ? ? ? ? ? ? ? ? ? ? ? ? ? ? ? ? ? ? ? ? ? ? ? ? ?
Magyarosaurus      1 2 1 0 1 1 0 1 1 ? ? ? ? ? ? ? 1 ? ? 0 0 0 0 ? ? ? ? ? 1 ? ? ? ?
Malawisaurus       ? ? ? ? ? ? ? ? 0 ? 0 0 2 0 1 ? 1 0 ? ? ? ? ? ? ? ? ? ? ? ? ? ? ?
Nemegtosaurus      ? ? ? ? ? ? ? ? ? ? ? ? ? ? ? ? ? ? ? ? ? ? ? ? ? ? ? ? ? ?
Neuquensaurus      1 1 1 2 0 1 0 1 1 ? ? ? ? ? ? ? ? ? ? ? ? ? ? ? ? ? ? ? 1 1 1 2 1
Opisthocoelicaudia 1 0 1 1 1 0 0 1 1 1 1 1 ? 1 1 1 ? ? 1 ? ? 0 ? ? 1 0 2 ? ? 1 1 1 2 1
Paralititan        ? ? ? ? ? ? ? ? ? ? ? ? ? ? ? ? ? ? ? ? ? ? ? 2 ? ? ? ? ? ? ? ?
Phuwiangosaurus    1 1 0 0 0 ? ? 1 0 ? ? ? ? ? ? ? ? 0 ? ? ? ? ? ? ? ? ? ? ? 0 1 2 1
Quaesitosaurus     ? ? ? ? ? ? ? ? ? ? ? ? ? ? ? ? ? ? ? ? ? ? ? ? ? ? ? ? ? ?
Rapetosaurus       1 2 1 0 0 1 1 1 1 2 0 0 0 1 0 0 0 1 0 0 0 0 0 0 1 1 1 ? ? ? 1 1 1 2 1
Rocasaurus         ? ? ? ? ? ? ? ? ? ? ? ? ? ? ? ? ? ? ? ? ? ? ? ? ? ? ? ? 1 1 2 1
Saltasaurus        0 1 0 2 1 1 0 1 1 ? ? ? ? ? ? ? ? ? ? 0 0 0 0 0 ? ? ? ? 1 1 1 2 0
Titanosaurus       ? ? ? ? ? ? ? ? ? ? ? ? ? ? ? ? ? ? ? ? ? ? ? ? ? ? ? ? 1 1 1 0 0
Santa Rosa indet.  ? ? ? ? ? ? ? ? ? ? ? ? ? ? ? ? ? ? ? ? ? ? ? ? ? ? ? ? ? ?
Jabalpur indet.    ? ? ? ? ? ? ? ? ? ? ? ? ? ? ? ? ? ? ? ? ? ? ? ? ? ? ? ? ? ?
Malagasy Taxon B   ? ? ? ? ? ? ? ? ? ? ? ? ? ? ? ? ? ? ? ? ? ? ? ? ? ? ? ? ? ?
```

APPENDIX 2.1. (continued)

									360				364
Aegyptosaurus	?	?	?	?	?	?	?	?	?	?	?	?	?
Aeolosaurus	?	?	?	?	?	?	?	?	?	?	?	1	0
Alamosaurus	?	?	?	?	?	?	?	?	?	?	?	0	?
Ampelosaurus	?	?	?	?	?	?	?	?	?	?	?	1	1
Andesaurus	?	?	?	?	?	?	?	?	?	?	?	?	?
Antarctosaurus	?	?	?	?	?	?	?	?	?	?	?	?	?
Apatosaurus	0	1	0	0	0	0	0	0	0	0	?	0	0
Argentinosaurus	?	?	?	?	?	?	?	?	?	?	?	?	?
Argyrosaurus	?	?	?	?	?	?	?	?	?	?	?	?	?
Augustinia	?	?	?	?	?	?	?	?	?	?	?	1	?
Brachiosaurus	0	0	?	?	0	1	0	0	?	0	0	?	0	0
Camarasaurus	0	0	0	0	0	1	0	0	0	0	0	0	0
Dicraeosaurus	0	?	?	?	0	0	0	?	0	?	?	?	0	0
Diplodocus	?	1	0	0	0	0	0	0	0	0	0	0	0
Epachthosaurus	?	?	?	?	?	?	?	?	?	?	?	?	?
Euhelopus	0	0	?	1	0	1	0	1	0	?	?	0	0	0
Brazil Series B	?	?	?	?	?	?	?	?	?	?	?	?	?
Jainosaurus	?	?	?	?	?	?	?	?	?	?	?	?	?
Janenschia	1	?	1	0	0	1	0	0	0	0	0	?	?
Lirainosaurus	?	?	?	?	?	?	?	?	?	?	?	1	?
Magyarosaurus	?	?	?	?	?	?	?	?	?	?	?	1	2
Malawisaurus	?	?	?	0	?	?	1	?	?	?	?	1	2
Nemegtosaurus	?	?	?	?	?	?	?	?	?	?	?	?	?
Neuquensaurus	?	?	?	?	?	?	?	?	?	?	?	1	1
Opisthocoelicaudia	1	1	1	?	0	1	0	0	0	?	?	1	0	?
Paralititan	?	?	?	?	?	?	?	?	?	?	?	?	?
Phuwiangosaurus	?	?	?	?	?	?	?	?	?	?	?	?	?
Quaesitosaurus	?	?	?	?	?	?	?	?	?	?	?	?	?
Rapetosaurus	?	0	1	1	1	1	0	0	1	1	1	0	1	2
Rocasaurus	?	?	?	?	?	?	?	?	?	?	?	?	1
Saltasaurus	?	?	?	1	?	0	0	?	?	?	?	?	1	1
Titanosaurus	?	?	?	?	?	?	?	?	?	?	?	0	?
Santa Rosa indet.	?	?	?	?	?	?	?	?	?	?	?	?	?
Jabalpur indet.	?	?	?	?	?	?	?	?	?	?	?	?	?
Malagasy Taxon B	?	?	?	?	?	?	?	?	?	?	?	?	?

APPENDIX 2.2. CHARACTERS ORDERED BY ANATOMICAL REGION

The cladistic codings for the 364 characters (109 skull, 139 axial, 112 appendicular, 2 dermal) used in this analysis are listed below in anatomical order. Approximately 50% of the characters in this work are new and used in a novel manner. Other characters have been used by previous workers as cited. References cited with each character relate to usage of the character in previous phylogenetic analyses and commonly provide a reference for further information or illustrations of the character; they do not necessarily provide a comprehensive list of works that have mentioned or employed the character. In many cases the character definitions, polarities, or distributions have been modified, as indicated. New characters are discussed in detail by Curry (2001).

CRANIAL CHARACTERS

EXTERNAL FEATURES OF THE SKULL

1. Short, deep snout: absent (0); present (1) (modified from Upchurch 1998; Wilson and Sereno 1998).
2. Posterolateral processes of premaxilla and lateral processes of maxilla: without midline contact (0); with midline contact forming marked narial depression, subnarial foramen not visible laterally (1) (Wilson and Sereno 1998).
3. Premaxilla: formed from heavy and thick main body with distinct ascending process (0); formed from transversely narrow main body that is greatly elongated rostrocaudally (1) (Upchurch 1998).
4. Ascending process of premaxilla: directed caudodorsally (0); directed dorsally (1); absent (2) (Gauthier 1986; Wilson and Sereno 1994; Upchurch 1995, 1998).
5. Caudolateral process of premaxilla: present (0); absent (1) (Upchurch 1998).
6. Position of maxillary ascending process: central or anterior to the center of the maxilla (0); posterior to the center of the maxilla (1).
7. Dorsal ascending process of maxilla: contributes less than one-half of its maximum length to the articulation with the contralateral maxilla (0); contributes over one-half of its maximum length to articulation with the contralateral maxilla (1).
8. Maxillary flanges: do not contact each other on midline (0); contact on midline (1).
9. Maxilla, jugal process: maxilla contacts jugal via a blunt posterior process at level of maxillary body (0); maxilla contacts jugal via elongate posterior process that extends posterodorsally from maxillary body (1).
10. Maxilla–nasal contact: present (0); absent (1) (Nowinski 1971).
11. "Shelflike" area lateral to external naris, extending onto rostral end of maxilla: present (0); absent (1) (McIntosh 1990a; Upchurch 1998).
12. Preantorbital fenestra: absent (0); present (1) (Wilson and Sereno 1998).
13. Subnarial foramen on premaxilla–maxilla suture: circular (0); slit-shaped, more than two times longer than broad (1) (Wilson 1998).
14. Subnarial foramen on premaxilla–maxilla suture: faces laterally (0); faces dorsally (1) (Upchurch 1998).
15. Antorbital fenestra, maximum diameter: much shorter than (0), subequal to (1), or much longer than (2) orbital diameter (Wilson 1998).
16. External nares: retracted to level of anterior orbit (0); retracted to position between orbits (1) (Steel 1970; Gauthier 1986; McIntosh 1990a; Upchurch 1995, 1998; Wilson and Sereno 1998).
17. External nares: face laterally or rostrolaterally (0); face dorsally or rostrodorsally (1) (Upchurch 1995, 1998).
18. External nares, maximum diameter: shorter (0); or longer (1) than orbital maximum diameter (McIntosh 1990a; Upchurch 1995, 1998; Wilson and Sereno 1998).
19. Rostral rim of external naris: lies in front of rostral margin of antorbital fenestra (0); lies caudal to rostral margin of antorbital fenestra (1) (McIntosh 1990a; Upchurch 1995, 1998).
20. External naris: divided by contact between maxilla and/or premaxilla and nasals (0); undivided, nasals do not contact maxilla or premaxilla (1) (modified from Gauthier 1986; Wilson and Sereno 1994; Upchurch 1995, 1998).
21. Nasal, medial and lateral processes: medial nasal process substantially longer than lateral nasal process (0); medial and lateral nasal processes equivalent in maximum length (1); lateral nasal process longer than medial nasal process (2).
22. Jugal, contribution to antorbital fenestra: very reduced or absent (0); large, bordering approximately one-third of perimeter (1) (Wilson 1998).

23. Jugal–maxillary process structure: jugal bluntly overlaps posterior border of maxilla (0); jugal has tongue-in-groove articulation with elongate maxillary process (1).
24. Jugal-quadratojugal contact: jugal has broad articulation with quadratojugal (0); jugal has narrow articulation only along dorsal portion of quadratojugal (1).
25. Jugal–ectopterygoid contact: present (0); absent (1) (Wilson and Sereno 1998).
26. Lacrimal: posteriorly inclined (0); vertically oriented (1) (Salgado and Calvo 1997).
27. Lacrimal contribution to external naris: absent (0); present (1) (Nowinski 1971).
28. Postorbital, contact with jugal: postorbital articulates with jugal in simple overlapping contact (0); postorbital articulates with several deep pits on posterior wing of jugal (1).
29. Postorbital contact with laterosphenoid: absent (0); present (1).
30. Prefrontal anterior process: absent (0); present (1) (Wilson and Sereno 1998).
31. Prefrontal, posterior process size: small, not projecting far posterior of frontal nasal suture (0); elongate, approaching parietal (1) (Wilson 1998).
32. Frontals, midline contact (symphysis): sutured (0); fused (1) (Salgado and Calvo 1992; Upchurch 1998; Wilson 1998).
33. Frontal, medial convexity in dorsal view: absent (0); present (1).
34. Frontal contribution to supratemporal fossa: absent (0); present (1) (Wilson and Sereno, 1998).
35. Frontoparietal suture: on same plane as midline suture of each respective bone to its counterpart (0); depressed relative to midline suture of each element's counterpart (1); elevated relative to midline sutures (2).
36. Parietal, anterior inclination with wide caudodorsal exposure of crest: absent (0); present (1) (Salgado and Calvo 1997; Wilson and Sereno 1998; Upchurch 1998).
37. Parietal, elongate lateral process: absent (0); present (1).
38. Parietal, contribution to posttemporal fenestra: present (0); absent (1) (Wilson 1998).
39. Parietal, distance separating supratemporal fenestrae: less than (0); twice (1); or equal to (2) transverse diameter of supratemporal fenestra (modified from Wilson 1998).
40. Postparietal foramen: absent (0); present (1); (Wilson 1998).
41. Supratemporal fenestra, maximum diameter: much longer than (0); subequal to (1); or much less than (2) that of foramen magnum (Wilson 1998).
42. Supratemporal fenestra: faces dorsally or dorsolaterally (0); faces laterally (1); faces anterodorsally (2) (Salgado and Calvo 1992; Upchurch 1998).
43. Supratemporal fenestra, margin: includes squamosal (0); excludes squamosal (1) (Upchurch 1995, 1998).
44. Infratemporal fenestra, rostral margin: lies below midpoint of orbit or more caudally (0); extends as far as, or beyond, rostral margin of orbit (1) (Gauthier 1986; Upchurch 1995, 1998).
45. Temporal bar: parallel (0) or perpendicular (1) to tooth row (Wilson 1998).
46. Quadratojugal, rostral process: straight or curves gently upward toward its rostral tip (0); has "step"-like change of direction at midlength, such that rostral half of process runs rostroventrally (1) (Upchurch 1998).
47. Quadratojugal, angle between rostral and dorsal rami: 90° or less (0); at least 130° (1) (Upchurch 1998).
48. Squamosal–quadratojugal contact: absent (0); present (1) (Gauthier 1986; Wilson and Sereno 1998).
49. Squamosal, general morphology: dorsally convex hook in lateral view (0); shaped like inverted "L" in lateral view (1).
50. Squamosal–postorbital suture: postorbital has overlapping contact with lateral surface of squamosal (0); deep anterodorsal notch in head of squamosal to receive postorbital (1).
51. Quadrate, posterior fossa depth: shallow (0); deeply invaginated (1) (Kurzanov and Bannikov 1983; Wilson and Sereno 1998).
52. Quadrate shaft, long axis: oriented perpendicular to long axis of skull (0); directed caudodorsally (1) (modified from Upchurch 1998).
53. Quadrate head, orientation: directly dorsal to lateral margin of quadrate body (0); extends lateral to lateral margin of quadrate body (1).
54. Quadrate–quadratojugal articulation: via single sutural scar (0); via V-shaped bifid scar (1).

55. Supraoccipital–parietal articulation: supraoccipital has broad contact with caudal border of parietal (0); supraoccipital underlies parietal via two midline, anterior processes (1).

56. Supraoccipital, formation of nuchal crest: formed from single midline prominence (0); results from joining of two lateral prominences, separated by median trough (1).

BRAINCASE AND OCCIPUT

57. Occipital region of skull: anteroposteriorly deep, paroccipital processes oriented posterolaterally (0); flat, paroccipital processes oriented transversely (1) (Wilson 1998).

58. Occipital condyle; caudoventrally directed (0); ventrally directed (1).

59. Crista prootica: rudimentary (0); expanded laterally into "dorsolateral process" (1) (modified from Upchurch 1995; 1998; Wilson 1998).

60. Exit for cranial nerve (CN) I: faces anteriorly (0); faces dorsally (1). Character state assignments are also determined by vertically orientating the supraoccipital (Salgado and Calvo 1997).

61. Laterosphenoid, openings for trigeminal nerve (CN V): single opening between prootic and laterosphenoid (0); double opening, one between prootic and laterosphenoid, one between laterosphenoid and orbitosphenoid (1).

62. Basipterygoid processes: short, approximately twice basal diameter (0); elongate, at least four times basal diameter (1) (Wilson 1998).

63. Basipterygoid processes: perpendicular to (0) or angled approximately 45° to (1) skull roof (Wilson 1998).

64. Basipterygoid processes, angle of divergence: approximately 45° (0); less than 30° (1); over 45° (2) (Wilson 1998).

65. Basipterygoid processes, area between bases: shallowly concave (0); deeply excavated into long narrow pit extending caudally beneath rostral part of braincase (1) (Upchurch 1995, 1998).

66. Basipterygoid angle: shallow U-shaped division (0); V-shaped division (1); deep U-shaped division (2).

67. Basipterygoid processes, orientation of distal ends: faces dorsolaterally (0); faces ventrally (1).

68. Basisphenoid depression: absent (0); present (1) (Wilson 1998).

69. Parasphenoid rostrum: triangular in lateral view with groove along dorsal midline (0); long, slender, strongly laterally compressed process that lacks dorsal groove (1) (McIntosh 1990a; Upchurch 1995, 1998).

70. Basal tubera, anteroposterior depth: approximately half dorsoventral height (0); sheetlike, 20% dorsoventral height (1) (Wilson 1998).

71. Basal tubera, angle of divergence: wide, greater than or equal to 60° (0); narrow, less than 45° (1).

72. Basal tubera, posterior extent: tubera project away from basicranium with considerable relief (0); tubera lie in same plane as basicranium (1).

73. Basal tubera, position of divergence: occurs approximately level with occipital condyle or slightly below (0); occurs well below occipital condyle (1).

74. Paroccipital processes, distal end: slightly expanded with a straight distal margin (0); elongated with a rounded "tongue"-like distal margin (1); elongated at distal tip (2) (modified from Upchurch 1998).

75. Paroccipital processes, transverse projection from basicranium: narrow (ratio of space between occipital condyle and lateral extreme of proximal paroccipital process to occipital condyle height ≤0.5) (0); wide (ratio of space between occipital condyle and lateral extreme of proximal paroccipital process to occipital condyle height >0.8) (1).

PALATE

76. Pterygoid flange, position: posterior to orbit (0); posterior to antorbital fenestra (1); anterior to antorbital fenestra (2) (Wilson 1998).

77. Pterygoid, breadth of main sheet: less than 20% total length of pterygoid (0); approximately 30% total length of pterygoid (1) (McIntosh and Berman 1975; Upchurch 1998:fig.4).

78. Pterygoid, basipterygoid fossa: deeply excavated dorsal fossa with basipterygoid hook present on posterodorsal wing of pterygoid (0); excavation shallow and faces medially (1); shelflike (2) (modified from Upchurch 1998; Wilson and Sereno 1998).

79. Pterygoid: anterior wing of pterygoid contacts opposite element in narrow suture (0); anterior wing of pterygoid contacts opposite element in broad sheet (1).

80. Pterygoid: pterygoids meet at angle of ≤30° in ventral view (0); pterygoids meet at angle of >45° in ventral view (1) (Nowinski 1971).

81. Pterygoid, ectopterygoid process: lies below lacrimal or more posteriorly (0); lies anterior

to lacrimal, under antorbital fenestra (0) (Upchurch 1994, 1998).

82. Pterygoid, ectopterygoid process: robust, projects below level of jaw margin (0); reduced, does not project below margin of upper jaw (1) (Berman and McIntosh 1975; Upchurch 1994, 1998).

83. Palatine, shape of lateral ramus: plate-shaped, long maxillary contact (0); rod-shaped, narrow maxillary contact (1) (Wilson and Sereno 1998).

MANDIBLE AND TEETH

84. Dentary, depth of anterior end of ramus: 150% minimum depth (0); slightly less than that of dentary at midlength (1) (modified from Wilson and Sereno 1998).

85. Dentary tooth rows: teeth distributed one-half of dentary's total length or more (0); teeth restricted to anterior one-third of dentary (1) (modified from Gauthier 1986; Upchurch 1998).

86. Tooth rows, shape of anterior maxillary portions: broadly arched, anterior portion of tooth rows U-shaped (0); rectangular, tooth-bearing portion of jaw perpendicular to jaw rami (1) (Wilson and Sereno 1998).

87. Dentary, relative length of posterior rami: dorsal and ventral rami approximately same length (0); dorsal ramus significantly shorter than ventral (1).

88. Dentary, anteroventral margin: gently rounded (0); sharply projecting triangular "mental" process (1) (Wilson 1998).

89. Dentary, shape of posterodorsal groove: absent (0); present (1) (Wilson 1998).

90. Mandible, angle between long axis of mandibular symphysis and long axis of mandible: approximately 45° (0); approximately 90° (1) (Upchurch 1998; Wilson and Sereno 1998).

91. Surangular depth: less than twice (0) or more than two and one-half times (1) maximum depth of angular (Wilson and Sereno 1998).

92. Surangular, shape: laterally convex (0); almost vertical (1).

93. Coronoid process, elevated relative to surangular: absent (0); present (1) (Salgado and Calvo 1997).

94. Surangular–dentary suture: nearly horizontal with respect to long axis of mandible (0); concave with respect to long axis of mandible (1).

95. Surangular–angular suture: nearly horizontal (0); dorsally convex (1).

96. Surangular, lateral exposure (depth): surangular exposure is significantly greater than that of the angular (0); surangular exposure is less than that of the angular (1).

97. Splenial posterior process: overlaps angular (0); inserts between anterior portions of prearticular and angular (1) (Wilson 1998).

98. Splenial posterodorsal process: present, approaches margin of adductor chamber (0); absent (1) (Wilson 1998).

99. Splenial, participation in medial wall of adductor chamber: present (0); absent (1) (Wilson 1998).

100. Dentary teeth, number: 17 or fewer (0); more than 20 (1) (Wilson and Sereno 1998).

101. Teeth orientation: perpendicular (0) or oriented anteriorly (1) relative to jaw margin (Wilson 1998).

102. Tooth crowns, arrangement: aligned slightly anterolingually, tooth crowns overlapping (0); aligned along jaw axis, crowns not overlapping (1) (Wilson and Sereno 1998).

103. Teeth, shape of wear facets: V-shaped facets (interlocking) (0); high-angled planar facets (1); low-angled planar facets (2) (Wilson and Sereno 1998).

104. Teeth, occlusal pattern: spoonlike, spatulate (0); peglike with high-angle wear facets on one side only (1); peglike with interlocking V-shaped wear facets (2); peglike with low-angle (less than 40°) wear facets (3) (modified from Salgado and Calvo 1997; Upchurch 1998; Wilson and Sereno 1998).

105. Tooth crowns, cross-sectional shape at midcrown: elliptical (0); D-shaped (1); cylindrical (2) (Janensch 1929; Wilson and Sereno, 1998).

106. Teeth, marginal tooth denticles: absent on posterior edge (0); absent on both anterior and posterior edges (1) (McIntosh 1990a; Upchurch 1994, 1998; Wilson and Sereno 1998).

107. Tooth crowns, prominent grooves near mesial and distal margins of labial surface: present (0); absent (1) (Upchurch 1994, 1998).

108. Teeth, SI values (tooth crown length divided by maximum mesiodistal width) for teeth: less than 4.0 (0); greater than 5.0 (1) (Upchurch 1994; 1998).

109. Replacement teeth per alveolus: two or fewer (0); more than four (1).

AXIAL SKELETON

GENERAL PRESACRAL VERTEBRAE

110. Presacral vertebral bone texture: solid (0); spongy, with large, open internal cells (somphospondylus) (1) (Wilson and Sereno 1998).
111. Presacral pneumatic fossae, shape: deep lateral excavations bordered by sharp lip (0); shallow lateral depressions (1).

CERVICAL VERTEBRAE

112. Atlantal intercentrum, occipital facet: rectangular in lateral view, dorsal aspect of intercentrum anteroposterior length subequal to that of ventral aspect (0); expanded anteroventrally in lateral view, dorsal aspect of intercentrum anteroposterior length shorter than that of ventral aspect (1) (Wilson and Sereno 1998).
113. Cervical vertebrae, number: 12 (0); 13 (1); 15 or more (2) (modified from Upchurch 1995, 1998; Wilson and Sereno 1998).
114. Cervical neural arch lamination: well developed, with well-defined pneumatic fossae and laminae (0); rudimentary, diapopohyseal laminae only feebly developed if present (1) (Wilson and Sereno 1998).
115. Cervical parapophysis, dorsal surface excavation: absent (0); present but separated from pneumatic fossa by longitudinal ridge (1) (Upchurch 1998).
116. Anterior cervical centra: height:width ratio of anterior face: approximately 1.25 (0); 1.0 or less (1) (modified from Martin et al. 1999; Upchurch 1998).
117. Anterior cervical centra, anteroposterior length divided by width of posterior face: short (EI ≤ 3.0) (0); elongate (EI > 3.0) (1) (Upchurch 1995, 1998).
118. Anterior cervical centra, midventral keel: absent (0); present (1).
119. Anterior cervical centra, ventral margin: broad and rounded, ventrally convex (0); broadly concave with excavation of ventral centrum (1).
120. Anterior cervical pneumatic fossae: complex, divided by bony septa (0); shallow and simple, divided by single median ridge (1); simple, undivided (2) (modified from Wilson and Sereno 1998; Upchurch 1998).
121. Anterior cervical parapophysis: arises at midcentrum (0); arises in anterior half of centrum (1).
122. Anterior cervicals, neurocentral junction: straight (0); constricted and butterfly-shaped in dorsal view (1).
123. Anterior cervical neural spines: single (0); distally divided into U- or V-shaped trough (bifid) (1); proximally divided by formation of pre- and postspinal coels, and distally fused (2) (modified from McIntosh 1990a; Wilson and Sereno 1998).
124. Anterior cervicals, neural spines: transversely broad, laterally expanded (0); anteroposteriorly expanded and compressed transversely (1).
125. Anterior cervical neural spines, distal morphology: straight (0); posteriorly inclined (1).
126. Anterior cervical prezygapophyses: short, do not extend beyond anterior border of the centrum (0); elongate, extend well beyond anterior articular facet (1).
127. Anterior cervicals, prespinal lamina: present (0); absent (1).
128. Anterior cervicals, postspinal lamina: present (0); absent (1).
129. Infraprezygapophyseal laminae on middle and caudal cervicals: "single" (0); bifurcate toward dorsal ends to form medial and lateral branches (with a triangular hollow in between) (1) (Upchurch 1998).
130. Midcervical centra, anteroposterior length/height of posterior face: >4 (0); 2.5–3.0 (1) (modified from Upchurch 1995, 1998).
131. Middle and posterior cervical centra, articular face shape: height > width (0); height ≤ width (1).
132. Middle and posterior cervical centra, ventral surface shape: with midventral keel (0); without midventral keel (1).
133. Middle and posterior cervical centra, pneumatic fossae: shallow excavations lacking dividing laminae (0); shallow with dividing laminae (1).
134. Middle and posterior cervicals, neurocentral junction: straight (0); constricted and butterfly-shaped (1).
135. Middle and posterior cervicals, neurocentral junction: terminates posterior to anterior articular face (0); extends full length of centrum (1).
136. Middle and posterior cervicals, neural canal shape: dorsoventral and mediolateral dimensions are subequal (0); dorsoventral dimensions are greater than mediolateral dimensions (1).

137. Middle and posterior cervical neural arches, development of laminae and pneumatic fossae: poorly developed shallow fossae bound by illformed laminae (0); deep lateral fossae bound by robust laminae (1) (modified from Upchurch 1998; Wilson and Sereno 1998).

138. Middle and posterior cervicals, pre- and postspinal coels: only prespinal coel present (0); only postspinal coel present (1); both pre- and postspinal coel present (2); both coels absent (3).

139. Middle and posterior cervicals, prespinal lamina: absent (0); present (1).

140. Middle and posterior cervicals, postspinal lamina: absent (0); present (1)

141. Posterior cervical and anterior dorsal neural spines: single (0); bifid (1) (Wilson and Sereno 1998).

142. Posterior cervical and anterior dorsal bifid neural spines, median tubercle: absent (0); present (1) (Wilson and Sereno 1998).

DORSAL VERTEBRAE

143. Dorsal vertebrae: 13 (0); 12 (1); 11 (2); 10 or fewer (3) (modified from Wilson and Sereno 1998).

144. Caudal dorsal centra ratio of centrum length to height of posterior face: <1.0 (0); >3.0 (1) (modified from Upchurch 1998).

145. Anterior dorsal centra, articular face shape: opisthocoelous (0); amphiplatyan (1) (McIntosh 1990a; Upchurch 1998; Wilson and Sereno 1998).

146. Pneumatic fossae in dorsal centra: present (0); absent (1) (Upchurch 1998).

147. Pneumatic fossae in dorsal centra: moderately deep but simple pits (0); deep, ramify extensively within centrum and enter the base of neural arch (1) (Upchurch, 1998).

148. Anterior dorsals, neural canal dimensions: height < width (0); height ≤ width (1).

149. Anterior dorsals, dorsal boundary of neural canal: ventral extension of prespinal fossa (0); interprezygapophyseal lamina (1); bone ventral to interprezygapophyseal lamina and dorsal to neural canal (2).

150. Anterior dorsal neural spines: single (0); bifid (1).

151. Anterior dorsal neural spines: straight, dorsally directed (0); posteriorly inclined (1).

152. Anterior dorsal distal neural spines: straight, equal anteroposterior and transverse dimensions (0); triangular, transversely expanded (1) (modified from Wilson and Sereno 1998).

153. Dorsal neural spines, length measured from prezygapophyses: twice (0) or four times (1) centrum length (modified from Wilson and Sereno 1998).

154. Dorsal transverse processes: directed strongly dorsolaterally (0); directed laterally or slightly upward (1) (Upchurch 1998).

155. Anterior dorsal transverse processes, angle of articulation with neural spine: ≥90° more (0); <70° (1).

156. Anterior dorsals, diapophyses: restricted to position posterior to anterior articular face (0); extend anterior to anterior articular face (1).

157. Anterior dorsal neural arches, infraprezygapophyseal lamina: single (0); divided (1) (modified from Wilson 1998).

158. Anterior dorsal neural arches, infrapostzygapophyseal lamina: single (0); divided (1).

159. Anterior dorsals, interprezygapophyseal lamina: present (0); absent (1).

160. Anterior dorsals, prespinal lamina: absent (0); present, short and restricted to proximal neural spine (1); present, short and restricted to distal neural spine (2); present along entire length of neural spine (3).

161. Anterior dorsals, postspinal lamina: absent (0); present, short and restricted to proximal neural spine (1); present along entire length of neural spine (2).

162. Anterior dorsals, neural arches, hyposphene-hypantra: absent (0); present (1) (modified from McIntosh 1990a; Bonaparte and Coria 1993; Upchurch 1998; Wilson and Sereno 1998).

163. Anterior dorsal neural arches, median infrapostzygapophyseal lamina: absent (0); present (1).

164. Anterior dorsal neural arches, spinodiapophyseal lamina: undivided (0); divided (1).

165. Middle and posterior dorsal centra, ventral concavity: present (0); absent (1).

166. Posterior dorsal centra, articular face shape: amphicoelous (0); opisthocoelous (1) (Salgado et al. 1997; Wilson and Sereno 1998).

167. Middle and posterior dorsal neural spines, height (measured from prezygapophyses): comprise 65% total height of vertebra (0);

comprise ≤50% total height of vertebra (1) (modified from Bonaparte 1986; Upchurch 1998).

168. Middle and posterior dorsal neural spine–transverse process intersection: ≤70° (0); >90° (1).

169. Middle and posterior dorsal neural spines, orientation: vertical (0); posterior, neural spines approach level of diapophyses (1) (Wilson 1998).

170. Middle and posterior dorsal neural arches, hyposphene-hypantrum articulations: present (0); absent (1) (Wilson 1998).

171. Middle and posterior dorsal neural spines, shape: tapering or not flared distally (0); flared distally, with pendant, triangular lateral processes (1) (Upchurch 1995, 1998; Wilson and Sereno 1998).

172. Middle and posterior dorsal neural arches, "infradiapophyseal" pneumatic foramen: absent (0); present (1) (Wilson 1998).

173. Middle and posterior dorsal neural arches, fossa dorsal to neural canal and below postzygapophyses: present (0); absent (1).

174. Posterior dorsal neural spines: rectangular through most of length (0); "petal"-shaped, expanding transversely through 75% of length and then tapering (1) (Wilson 1998).

175. Posterior dorsal vertebrae, transverse processes: lie caudal, or caudodorsal, to parapophysis (0); lie vertically above parapophysis (1) (Upchurch 1998).

176. Posterior dorsal vertebrae, neural spines: wider craniocaudally than transversely (0); compressed craniocaudally (1) (Upchurch 1998).

177. Dorsal neural spines, triangular processes: absent (0); present (1) (Upchurch 1995, 1998).

178. Middle and posterior dorsal neural arches, anterior centroparapophyseal lamina: present (0); absent (1) (Wilson 1998).

179. Middle and posterior dorsal neural arches, prezygoparapophyseal lamina: present (0); absent (1) (Wilson 1998).

180. Middle and posterior dorsal neural arches, posterior centroparapophyseal lamina: absent (0); present (1) (Wilson 1998).

181. Middle and posterior dorsal neural arches spinopostzygapophyseal lamina: divided (0); single (1) (Wilson 1998).

182. Middle and posterior dorsal neural arches, prespinal lamina: absent (0); present, anteroposteriorly expanded and sheetlike (1); present, transversely expanded and triangular (2) (Bonaparte 1986; Calvo and Salgado 1995; Upchurch 1998).

183. Middle and posterior dorsal neural arches, postspinal lamina: absent (0); present, anteroposteriorly expanded and sheetlike (1) (Bonaparte 1986; Calvo and Salgado 1995; Upchurch 1998).

184. Middle and posterior dorsal zygapophyses: prezygapophyses at same level as postzygapophyses (0); postzygapophyses dorsal to prezygapophyses (1). This character is best documented in lateral view.

185. Middle and posterior dorsal prespinal laminae: absent (0); present, short, restricted to proximal one-half of neural spine (1); present, short, restricted to distal one-half of neural spine (2); present, elongate, extending full length of neural spine (3).

186. Middle and posterior dorsal postspinal laminae: absent (0); present, short, restricted to proximal one-half of neural spine (1); present, short, restricted to distal one-half of neural spine (2); present, elongate, extending full length of neural spine (3).

187. Middle and posterior dorsals, anterior centrodiapophyseal lamina: vertical (0); anteriorly inclined (1).

188. Middle and posterior dorsals, intraprezygapophyseal lamina: undivided (0); divided (1).

189. Middle and posterior dorsals, suprapezygapophyseal lamina: present (0); absent (1).

190. Middle and posterior dorsals, fossa between divisions of suprapezygapophyseal lamina: absent (0); present (1).

191. Middle and posterior dorsals, median infrapostzygapophyseal lamina: absent (0); present (1).

192. Middle and posterior dorsals, median infraprezygapophyseal lamina: absent (0); present (1).

193. Middle and posterior dorsal vertebrae, infradiapophyseal lamina: single (0); bifurcated (1) (Salgado et al. 1997).

SACRAL VERTEBRAE

194. Sacral vertebrae, number: 5 (0); 6 (1) (modified from McIntosh 1990a; Salgado et al. 1997;

Upchurch 1995, 1998; Wilson and Sereno 1998).

195. Sacral vertebrae, number supporting acetabulum: two to four (0); five or more (1) (Wilson 1998).
196. Sacral vertebrae, ventral margin: rounded (0); ventrally concave (1).
197. Sacral centra: all opisthocoelous (0); ranging from opisthocoelous to amphiplatyan to procoelous (1); all amphiplatyan (2).
198. Sacral centra, pneumatic fossae or very deep depressions: absent (0); present (1) (Upchurch 1998).
199. Sacral neural spines: approximately twice (0) or more than four times (1) length of centrum (Wilson 1998).
200. Sacral neural spines, prespinal lamina: anteroposteriorly expanded (0); transversely expanded (1).
201. Sacral neural spines, postspinal lamina: present (0); absent (1).
202. Sacral neural spines, prespinal lamina: single proximally (0); divided proximally, arising from prezygapophyses (1).
203. Sacral neural spine: meets transverse process at ≥90° (0); meets transverse process at <70°(1).
204. Sacral ribs, proximal excavation: absent (0); present (1).
205. Sacral ribs, position of apex: high, extending beyond dorsal margin of ilium (0); low, not projecting beyond dorsal margin of ilium (1) (Wilson 1998).

CAUDAL VERTEBRAE

206. Caudal centra, bone texture: solid (0); spongy (somphospondlyus), with large internal cells (1) (Powell 1986; Wilson 1998).
207. Caudal vertebrae: more than 50 (0); 35 or fewer (1) (Berman and McIntosh, 1978; McIntosh 1990a, 1990b; Upchurch 1998).
208. Caudal transverse processes: disappear by caudal 15 (0); only present through caudal 10 (1) (Upchurch 1998; Wilson 1998).
209. First caudal centrum, articular face shape: amphiplatyan (0); procoelous (1); opisthocoelous (2); biconvex (3) (Powell 1986; Wilson 1998).
210. First caudal neural arch, fossa on lateral aspect of neural spine: absent (0); present (1) (Wilson 1998).
211. Anterior caudal centra (excluding the first), articular face shape: amphiplatyan or platycoelous (0); procoelous (1); opisthocoelous (2) (Bonaparte 1986; Powell 1986; Upchurch 1998; Wilson and Sereno 1998).
212. Anterior caudal centra, pneumatic fossae: absent (0); present (1) (modified from Upchurch 1998; Wilson 1998).
213. Anterior caudal centra: maintain same length (0); double in length (1) over first 20 vertebrae (Wilson and Sereno 1998).
214. Anterior caudals, ventral margin of centrum: rounded, convex (0); ventrally concave and marked by a deep midline groove (1).
215. Anterior caudal neural spines: reduced with equal anteroposterior and transverse dimensions (0); transversely widened (1); anteroposteriorly expanded (2).
216. Anterior caudal neural spines: high (neural spine height < centrum height) (0); low (neural spine height ≤ centrum height) (1).
217. Anterior caudal neural arches, spinoprezygapophyseal lamina: absent (0); present and extending onto lateral aspect of neural spine (1) (Wilson 1998).
218. Anterior caudal neural arches, spinoprezygapophyseal lamina–spinopostzygapophyseal lamina contact: absent (0); present, forming a prominent lateral lamina on lateral aspect of neural spine (1).
219. Anterior caudal neural arches, prespinal lamina: absent (0); present (1) (Wilson 1998).
220. Anterior caudal neural arches, postspinal lamina: absent (0); present (1) (Wilson 1998).
221. Anterior caudal neural arches, postspinal fossa: absent (0); present (1) (Wilson 1998).
222. Anterior caudal neural spines, transverse breadth: approximately 50% of (0) or greater than (1) anteroposterior length (Wilson 1998).
223. Anterior caudal transverse processes: triangular, tapering distally (0); "winglike", not tapering distally (1) (Berman and McIntosh 1978; McIntosh 1990a; Calvo and Salgado 1995; Upchurch 1995, 1998).
224. Anterior caudal transverse processes, diapophyseal laminae: absent (0); present (1) (Wilson 1998).
225. Anterior caudal transverse processes, anterior centrodiapophyseal lamina: single (0); bifid (1) (Wilson 1998).

226. Anterior caudal neural arches: anterior margin of neural spine extends beyond anterior margin of centrum (0); anterior margin of neural spine does not extend beyond anterior margin of centrum (1) (Salgado et al. 1997:figs. 3–4).
227. Anterior caudal neural spines, anterior pedicle morphology: straight (0); anteriorly concave and curved (1).
228. Anterior caudal vertebrae, prezygpophyses: face anteriorly (0); face dorsally (1).
229. Middle caudal centra, shape: cylindrical (0); flat ventrally and laterally (1) (Wilson 1998).
230. Middle caudal centra, ventral longitudinal hollow: absent (0); present (1) (Wilson 1998).
231. Middle and posterior caudal centra, anterior articular face shape: flat (0); procoelous (1); opisthocoelous (2) (Bonaparte 1986; Powell 1986; Upchurch 1995, 1998; Wilson and Sereno 1998).
232. Middle caudal neural spine: angled anterodorsally (0); vertical (1); angled posterodorsally (2) (Wilson 1998).
233. Middle and posterior caudal neural spine: tapering distally (0); rectangular (1); low and rounded (2); tapered anteriorly and posteriorly (3) in lateral view (modified from Salgado et al. 1997; Upchurch 1998; Wilson 1998).
234. Midcaudal neural spines: anterodorsal edge of neural spine lies posterior to anterior margin of postzygapophyses (0); anterodorsal edge of neural spine lies anterior to or level with anterior margin of postzygapophyses (1) (Salgado et al. 1997:figs. 3, 4).
235. Neural arches of middle caudals: situated over middle of centrum (0); situated on cranial half of centrum (1) (Calvo and Salgado 1995; Upchurch 1995, 1998).
236. Posterior caudal centra: cylindrical (0); dorsoventrally flattened, breadth at least twice height (1) (Powell 1986, 1992; Wilson et al. 1999; Sanz et al. 1999).
237. Distalmost caudal centra, articular face shape: platycoelous (0); biconvex (1); procoelous (2) (Wilson 1998).
238. Distalmost caudal centra, length-to-height ratio: <4 (0); >5 (1) (Powell 1986; Upchurch 1998; Wilson 1998).
239. Distalmost biconvex caudal centra: 10 or fewer (0); more than 30 (1) (Wilson 1998).

RIBS AND CHEVRONS

240. Cervical ribs, length: much longer than centrum, overlapping as many as three subsequent vertebrae (0); shorter than centrum, no overlap (1) (McIntosh 1990a; Upchurch 1998; Wilson and Sereno 1998; Wedel et al., 2000a, 2000b).
241. Dorsal ribs, proximal pneumatocoels: absent (0); present (1) (Wilson and Sereno 1998).
242. Anterior dorsal ribs, cross-sectional shape: subcircular (0); planklike, anteroposterior breadth more than three times mediolateral breadth (1) (Upchurch 1995, 1998; McIntosh et al. 1996).
243. Middle and posterior caudal chevrons, shape of blade: "forked" with anterior and posterior projections (0); single spine (1) (Upchurch 1998; Wilson and Sereno 1998).
244. Chevrons, proximal "crux" bridging superior margin of hemal canal: present (0); absent (1) (Powell 1992; Calvo and Salgado 1995; Upchurch 1995, 1998; McIntosh 1996; Wilson and Sereno 1998).
245. Chevron, hemal canal length: short, approximately 25% (0), or long, at least 50% (1) of chevron length (Wilson 1998).
246. Chevrons, distribution: disappearing by 50% length of tail (0); persisting throughout at least 80% of tail (1) (Wilson 1998).
247. Posterior chevrons, distal contact: fused (0); unfused (open) (1) (Powell 1992; Upchurch 1995, 1998).
248. Middle and distal chevron distal blades, anteriorly directed process: present (0); absent (1) (Upchurch 1998).
249. Chevrons: proximal division V-shaped (0); proximal division U-shaped (1)

APPENDICULAR SKELETON

PECTORAL GIRDLE

250. Scapular blade, relationships to coracoid: perpendicular to (0) or forming a 45° angle with (1) coracoid (Wilson 1998).
251. Scapular, acromion process: not expanded (0); with rounded expansion on acromial side (1) (modified from Upchurch 1998; Wilson and Sereno 1998).
252. Scapular blade, distal end: flared relative to proximal end (0); same width as proximal end (1); narrower than proximal end (2).

253. Scapular blade and acromion process, angle of intersection: between 50° and 70° (0); >90° (1); <45° (2).
254. Scapular spine: at right angle to long axis of element (0); gently curved, not perpendicular to long axis of element (1).
255. Scapula glenoid contribution vs. coracoid glenoid contribution: 1:1 (0); coracoid has greater contribution (1); scapula has greater contribution (2).
256. Scapular glenoid, orientation: relatively flat (0); strongly beveled medially (1) (Wilson and Sereno 1998).
257. Scapular base, cross-sectional shape: flat or rectangular (0); D-shaped (1) (Wilson and Sereno 1998).
258. Coracoid, proximodistal length: less than (0), approximately twice (1), or subequal to (2) length of scapular articulation of coracoid (Wilson 1998).
259. Coracoid, anteroventral margin shape: rounded (0); rectangular (1) (Wilson 1998).
260. Coracoid, scapular articular surface: anteroposteriorly elongate (0); mediolaterally expanded (1). In *Magyarosaurus*, Malagasy Taxon B, and *Saltasaurus*, the scapular articular surface of the coracoid is mediolaterally expanded.
261. Coracoid, infraglenoid lip: absent (0); present (1) (Wilson 1998).
262. Position of coracoid foramen: deeply inset into coracoid body (0); at margin of coracoid body (1).
263. Sternal plate, shape: oval, lacking distinct concavities (0); crescentic, with strongly concave lateral borders (1) (Salgado et al. 1997; Wilson and Sereno 1998).
264. Maximum length of sternal plate divided by humerus length: 0.65 or less (0); 0.75 or more (1) (McIntosh 1990a; Upchurch 1998).

FORELIMB

265. Humerus, robusticity: gracile (midshaft width:humeral length ≤0.15) (0); robust (midshaft width:humeral length >0.25) (1).
266. Humeral diaphysis, shape of lateral margin: straight (0); concave (1).
267. Humerus, deltopectoral crest: extends less than one-half total length of element (0); extends about one-half total length of element (1); prominent, extends well over one-half total length of element (2) (modified from Upchurch 1998; Wilson and Sereno 1998).
268. Humeral deltopectoral attachment: reduced to a low, rounded crest or ridge (0); prominent, with anterolateral extension (1) (Wilson and Sereno 1998).
269. Humeral deltopectoral crest: relatively narrow throughout length (0); markedly expanded distally (1) (Wilson, 1998).
270. Humerus, head position: level with proximal margin of deltopectoral crest (0); projects above level of proximal margin of deltopectoral crest (1).
271. Humeral proximolateral corner: rounded (0); square (1) (Wilson 1998).
272. Humerus, relative sizes of ulnar and radial condyles: ulnar condyle transversely expanded (0); radial condyle transversely expanded (1) (Powell 1986, McIntosh 1990a).
273. Humeral distal condyles, articular surface shape: restricted to distal portion of humerus (0); expanded onto anterior portion of humeral shaft (1) (Powell 1986; Wilson 1998).
274. Humeral distal end, development of condyles: flat (0); divided (1) (Powell 1986; Wilson 1998).
275. Ulna, robusticity: robust (midshaft width:ulnar length >0.25) (0); gracile (midshaft width:ulnar length ≤0.20) (1).
276. Ulna, length-to-proximal breadth ratio: gracile, maximum width of proximal end is 25% of ulnar length or less (0); stout, maximum width of proximal end is 33% of ulnar length or more (1) (Wilson and Sereno 1998).
277. Craniomedial process of proximal ulna, shape of articular surface: flat (0); strongly concave (1) (Upchurch 1995:fig. 14, 1998).
278. Ulnar proximal condylar processes: subequal (0); unequal, anterior arm longer (1) (modified from Wilson and Sereno 1998).
279. Ulnar olecranon process: prominent, projecting above proximal articulation (0); rudimentary, level with proximal articulation (1) (McIntosh 1990a; Wilson and Sereno 1998).
280. Ulna, distal articular surface: triangular with anteromedial apex (0); circular (1).
281. Radius: robust (midshaft width:radial length >0.25) (0); gracile (midshaft width:radial length ≤0.15) (1) (modified from McIntosh 1990a; Upchurch 1995, 1998).

282. Radius: proximal and distal ends equally expanded (0); proximal end more transversely expanded (1); distal end more transversely expanded (2).
283. Radius, well-defined interosseous ridge: absent (0); present (1).
284. Radius, proximal articular surface: rounded/subrectangular, with subequal transverse and anteroposterior dimensions (0); triangular, with expanded anteroposterior dimensions (1); transversely expanded (2).
285. Radius, proximal end: slender, maximum width of proximal end 25% of radius length or less (0); robust, maximum width of proximal end 33% of radius length or more (1) (McIntosh 1990a; Upchurch 1995, 1998).
286. Radius, distal articular surface: oval, with subequal anteroposterior and transverse dimensions (0); transversely expanded, elliptical (1); anteroposteriorly expanded (2).
287. Radial distal condyle: subrectangular, flattened posteriorly and articulating in front of ulna (0); circular (1) (Wilson and Sereno 1998).
288. Radius, distal breadth: slightly larger than (0) or approximately twice (1) midshaft breadth (Wilson 1998).
289. Distal radius, transverse axis orientation: perpendicular to (0) or beveled approximately 20° proximolaterally (1) relative to long axis of shaft (Curry Rogers and Forster 2001).
290. Humerus-to-femur ratio: <0.6 (0); 0.7–0.8 (1); ≥0.9 (2) (modified from McIntosh 1990a; Upchurch 1994, 1995, 1998; Wilson and Sereno 1998).

MANUS

291. Longest metacarpal-to-radius ratio: ≤0.3 (0); ≥0.45 (1) (modified from Salgado et al. 1997; Wilson and Sereno. 1998).
292. Metacarpal I: shorter than metacarpals II and III (0); longer than metacarpal III and subequal to or longer than metacarpal II (1) (Upchurch 1998).
293. Metacarpal I: shorter than (0) or longer than (1) metacarpal IV (Wilson and Sereno 1998).
294. Metacarpal I, proximal surface shape: broad quadrangle (0); oval with posterior process (1); triangular (2).
295. Metacarpal I, proximal and distal ends: slightly expanded relative to midshaft (0); widely flared relative to midshaft (1).
296. Metacarpal I, divided trochlea on distal condyle: present (0); absent (1) (modified from Wilson and Sereno 1998).
297. Metacarpal I distal condyle, transverse axis orientation: beveled approximately 20° proximodistally (0) or flat (1) with respect to axis of shaft (modified from Wilson and Sereno 1998).
298. Metacarpal II: shaft oval/rounded in cross section (0); shaft triangular in cross section (1).
299. Metacarpal II, distal end shape: oval (0); subrectangular (1).
300. Metacarpal III: longest metacarpal (0); shorter than II or IV (1).
301. Metacarpal III, shaft cross section: triangular (0); oval to circular (1).
302. Metacarpal III, proximal end: triangular, with posterior apex (0); triangular, with medially concave edge (1); triangular, with posterior base (2).
303. Metacarpal IV, shape in anterior view: proximal and distal ends flared (0); proximal and distal ends not flared (1).
304. Metacarpal IV, proximal end: anteroposteriorly elongate oval (0); triangular (1).
305. Metacarpal IV, distal surface: rounded (0); triangular (1).
306. Metacarpal V: gracile (ratio midshaft width:metacarpal length <0.10) (0); robust (ratio midshaft width:metacarpal length ≥0.10) (1).
307. Metacarpal V: proximal and distal ends subequally expanded (0); proximal end more expanded (1).
308. Phalangeal number: 2-3-4-3-2 or more (0); reduced, 2-2-2-2-2 or fewer (1); ossified phalanges absent (2) (Wilson and Sereno 1998).
309. Manual phalanx I.1: rectangular (0); wedge-shaped (1) (Wilson 1998).
310. Manual nonungual phalanges: broader transversely than long proximodistally (0); longer proximodistally than broad transversely (1) (Wilson and Sereno, 1998).

PELVIC GIRDLE

311. Pelvis, shape: narrow, ilia longer anteroposteriorly than distance separating preacetabular processes (0); broad, distance between preacetabular processes exceeds anteroposterior length of ilia (1) (Wilson and Sereno 1998).

312. Ilium, shape of preacetabular process in lateral view: triangular and tapering to a point (0); broad, with rounded anterior tip (1) (Calvo and Salgado 1995; Upchurch 1998).

313. Ilium, position of highest point of iliac blade: central or posterior to acetabulum (0); centered over pubic peduncle (1) (Salgado et al. 1997: fig.1).

314. Ilium, pubic peduncle contribution to acetabulum: shares acetabular border equally with ischial peduncle (0); comprises less than one-half acetabular border (1); comprises over one-half acetabular border (2).

315. Ilium, pubic peduncle: projects at acute angle relative to sacral axis (0); projects at right angle relative to sacral axis (1) (Salgado et al. 1997: figs. 8, 9).

316. Ilium, ischial peduncle size: low, rounded (0); large, prominent (1) (McIntosh 1990a; Wilson and Sereno 1998).

317. Ilium, preacetabular process (when resting on pubic and ischial peduncles): projects anterolateraly (0) or perpendicularly (1) to body axis (Salgado et al., 1997).

318. Ilium, preacetabular process, shape (when resting on pubic and ischial peduncles): pointed, arching ventrally (0); semicircular, anterodorsally oriented (1) (modified from Salgado et al. 1997; Wilson and Sereno 1998).

319. Pubis, ambiens process development: small, confluent with (0), or prominent, projecting anteriorly from (1), anterior margin of pubis (Wilson and Sereno 1998).

320. Pubis, relationship of acetabular and ischial surfaces: perpendicular to one another (0); meet at obtuse angle (1).

321. Pubis, ratio of length of acetabular surface to length of ischial surface: ischial surface longer (0); acetabular surface longer (1); two surfaces subequal (2).

322. Pubis, expansion of proximal and distal ends: unexpanded (0); only proximal end widened relative to shaft (1); distal end widened relative to midshaft (2); both proximal and distal ends expanded (3).

323. Puboischial contact: approximately one-third total length (0), one-half total length (1), or entire length of pubis (2) (Wilson and Sereno 1998).

324. Ischium:pubis length ratio: ≥0.90 (0); <0.75 (1) (Calvo and Salgado 1995; Upchurch 1998).

325. Ischial blade: equal to or longer than (0) or much shorter than (1) pubic blade (Wilson and Sereno 1998).

326. Ischial blade, emargination distal to pubic peduncle: present (0); absent (1) (Salgado et al. 1997).

327. Ischial distal shaft: triangular, depth of ischial shaft increases medially (0); bladelike, medial and lateral depths subequal (1) (Wilson and Sereno 1998).

328. Ischial distal shafts, cross-sectional shape of articulated ischia: V-shaped, forming an angle of nearly 50° with each other (0); flat, nearly parallel (1) (McIntosh 1990a; Upchurch 1998; Wilson and Sereno 1998).

329. Ischium, width across ischial shaft at midlength divided by length of ischium: <0.15 (0); >0.20 (1) (Jacobs et al. 1993; Upchurch 1998).

330. Ischium, distal end: elongated and narrow (0); short, reduced, and broad, does not extend beyond body (1).

331. Ischium, length of pubic peduncle: dorsoventrally short (distance from upper corner of ischium's pubic blade to posterior border of iliac peduncle is subequal to or longer than pubic articular surface) (0); dorsoventrally extended (distance from upper corner of ischium's pubic blade to posterior border iliac peduncle is shorter than pubic articular surface) (1) (Salgado et al. 1997: fig. 5).

332. Ischium, iliac peduncle: distinctive and well separated from body of ischium (0); low and rounded, not separated from ischial body (1).

HINDLIMB

333. Femur, robusticity:robust, midshaft width:femoral length >0.20 (0); gracile, midshaft width:femoral length ≤0.10 (1).

334. Femur, location of head: dorsal to greater trochanter (0); level with greater trochanter (1).

335. Femoral shaft, lateral margin: straight (0); proximal one-third deflected medially (1) (Wilson and Sereno 1998; Wilson and Carrano 1999).

336. Femoral midshaft transverse diameter: subequal to (0), 125%–150% of (1), or at least 185% of (2) anteroposterior diameter (Wilson 1998).

337. Femoral distal condyles, relative transverse breadth: subequal to one another (0); medial condyle much broader than lateral (1) (Wilson 1998).

338. Femoral distal condyles, transverse axis orientation: perpendicular or slightly beveled dorsolaterally (0), or beveled dorsomedially approximately 10° (1), relative to femoral shaft (Wilson and Carrano 1999).

339. Femoral distal condyles, articular surface: restricted to distal portion of femur (0); expanded onto anterior portion of femoral shaft (1) (Wilson and Carrano 1999).

340. Tibia, proximal articular condyle shape: mediolaterally narrow, long axis anteroposterior (0); expanded transversely, condyle subcircular (1); subequal expansions of anterior and posterior dimensions (2) (Wilson and Sereno 1998).

341. Tibia, cnemial crest: projects anteriorly (0); projects laterally (1) (Wilson and Sereno 1998).

342. Tibia, shape of distal end: oval (0); triangular (1).

343. Tibia, distal end: transversely expanded and anteroposteriorly compressed (0); transverse and anteroposterior dimensions subequal (1) (Salgado et al. 1997).

344. Tibia, distal breadth: approximately 125% of (0), more than 200% of (1), or subequal to (2) midshaft breadth.

345. Fibula, robusticity: robust, midshaft width:fibular length >0.25 (0); gracile, midshaft width:fibular length ≤0.15 (1).

346. Fibula, proximal articular surface shape: flat to slightly convex (0); steeply angled (1).

347. Fibula, lateral trochanter: absent (0); present (1) (Wilson and Sereno 1998).

348. Fibula, muscle scar on lateral surface at midlength: oval (0); elongate ridge running subparallel to long axis of shaft (1) (Powell 1992; Upchurch 1998).

349. Fibular distal condyle, relative width: subequal to width of shaft (0); expanded transversely, greater than midshaft width (1) (Wilson 1998).

350. Astragalar posterior fossa, shape: divided by vertical crest (0); undivided (1) (Wilson and Sereno 1998).

351. Astragalus, transverse length: 50% more than (0) or subequal to (1) proximodistal height (Wilson 1998)

352. Calcaneum: present (0); ossified calcaneum absent (1) (McIntosh, 1990a; Upchurch, 1998).

PES

353. Metatarsal I, length: shortest metatarsal (0); metatarsal V shorter than metatarsal I (1).

354. Metatarsal I, anterior extension on proximal condyle: absent (0); present (1).

355. Metatarsal I distal condyle, transverse axis orientation: angled dorsomedially to (0) or perpendicular to (1) axis of shaft (Wilson 1998).

356. Metatarsal I distal condyle, posterolateral projection: present (0); absent (1) (Wilson 1998).

357. Metatarsal I and V proximal condyles: subequal to (0) or smaller than (1) those of metatarsals II and IV (Wilson and Sereno 1998).

358. Metatarsal III: proximal ends and distal ends expanded and broad (0); proximal and distal ends not expanded (1).

359. Metatarsals III and IV, minimum transverse shaft diameters: less than 65% of (0) or subequal to (1) those of metatarsals I or II (Wilson and Sereno 1998).

360. Metatarsal V, shaft in cross section: circular (0); triangular (1).

361. Metatarsal V, proximal articular surface: single, flat surface (0); surface divided into two portions by sharp angle (1).

362. Pedal digit I ungual, length relative to pedal digit II ungual: 25% larger than (0) or subequal to (1) that of digit II (Wilson and Sereno 1998).

DERMAL

363. Osteoderms: absent (0); present (1) (Depéret 1896a, 1896b; Wilson and Sereno 1998).

364. Osteoderm morphology: absent (0); rounded and platelike with ventral keels (1); triangular, with disklike external surface and flat ventral surface, lacking ventral keel (2) (Powell 1986; Csiki 1999).

APPENDIX 2.3. LIST OF SYNAPOMORPHIES FOR THE TOTAL-EVIDENCE PHYLOGENETIC ANALYSIS WITH NO TAXA REMOVED.

This appendix lists character support for each of the nodes documented in figures 2.10 and 2.110. The distribution of apomorphies is based on the Accelerated Transformation option in PAUP* (Swofford 1999). Character numbers listed below correspond to characters numbered in the character–taxon matrix (appendix 2.1) and in the character list (appendix 2.2).

STRICT CONSENSUS TREE

TITANOSAURIFORMES
12, absence of preantorbital fenestra; 36, presence of parietal crest; 83, platelike contact between maxilla and platine; 94, surangular-angular contact dorsally concave; 113, presence of 13 cervical vertebrae; 123, single anterior cervical neural spines; 246, chevron blades curve backward and down.

No unambiguous support for Somphospondylii.

ADAM'S CONSENSUS TREE

TITANOSAURIFORMES
12, absence of preantorbital fenestra; 34, frontal contributes to boundary of supratemporal fenestra; 36, parietal with wide exposure of posterior crest; 83, palatine with plate-shaped maxillary contact; 94, concave surangular–dentary with respect to long axis of mandible; 123, undivided anterior cervical neural spines; 125, anterior neural spines slightly posteriorly inclined; 137, poorly developed middle and posterior cervical neural arch pneumatic fossae; 150, single anterior dorsal neural spines; 151, anterior dorsal neural spines posteriorly inclined; 173, middle and posterior dorsal neural arches lack fossa between neural canal and prezygapophyses; 222, anterior caudal neural spines transversely thin; 226, anterior border of neural spine extends anterior to centrum in proximal caudals; 235, mid–caudal neural arches anteriorly positioned; 248, chevron blades curve backward and downward; 335, proximal one-third of femoral shaft deflected medially.

NODE A: TITANOSAURIA
Unambiguous support not calculated in Adam's consensus.

NODE B: LITHOSTROTIA
226, anterior caudal neural spine extends beyond anterior margin of centrum; 276, stout ulna; 285, proximal end of radius is robust; 289, distal end of radius is beveled proximolaterally.

NODE C: UNNAMED
268, prominent humeral deltopectoral attachment; 281, presence of robust radius.

NODE D: UNNAMED
165, absence of ventral concavity in middle and posterior dorsals; 174, posterior dorsal neural spine expands transversely for 75% of length, then tapers.

NODE E: RAPETOSAURUS CLADE
161, postspinal lamina restricted to proximal neural spine in anterior dorsals; 169, middle and posterior dorsal neural spines posteriorly directed; 171, middle and posterior dorsal neural spines distally flared; 175, transverse processes of posterior dorsals lie vertically above parapophysis; 181, middle and posterodorsal neural arches with divided spinodiaphyseal lamina; 257, scapular basis flat or rectangular; 280, ulnar distal articular surface triangular with anteromedial apex; 282; distal end of radius transversely expanded; 321, pubis acetabular surface longer than ischial articular surface.

NODE F: UNNAMED
111, shallow presacral pneumatic fossa; 163, presence of median infrapostzygapophyseal lamina in anterior dorsals; 233, low and rounded midcaudal neural spines; 277, craniomedial process of ulna flat proximally; 281, presence of gracile radius; 284, radius, proximal articular surface rounded with subequal anteroposterior and transverse dimensions; 364, osteoderms triangular and lack ventral keel.

NODE G: SALTASAURINAE
1, absence of short, deep snout; 16, external nares retracted to level of orbits; 64, basipterygoid processes diverge at < 30°; 70, basal tubera sheetlike; 108, SI vales for teeth exceed 5.0.

NODE H: UNNAMED
144, posterior dorsal centra length:height ratios exceed 1.0; 218, presence of prominent lateral lamina in anterior caudal neural arches.

NODE I: OPISTHOCOELICAUDIINAE
No unambiguous support calculated in Adams consensus tree.

NODE J: SALTASAURIDAE
No unambiguous support calculated in Adams consensus tree.

APPENDIX 2.4. LIST OF SYNAPOMORPHIES FOR THE TOTAL-EVIDENCE PHYLOGENETIC ANALYSIS WITH SOME TAXA REMOVED

This appendix lists character support for each of the nodes documented in figures 2.12 and 2.13. Taxa less than 15% completely coded have been removed from the analysis. The distribution of apomorphies is based on the Accelerated Transformation option in PAUP. Character numbers listed below correspond to characters* numbered in the character–taxon matrix (appendix 2.1) and in the character list (appendix 2.2). (*) Indicates synapomorphies also present in the strict consensus tree of the reduced data set.

NODE A: SOMPHOSPONDYLII

*5, presence of caudolateral process of premaxilla; *8, maxillary flanges do not contact one another on midline; *14, subnarial foramen faces laterally; 84, dentary, depth of anterior end of ramus slightly less than at midlength; *101, teeth oriented perpendicular relative to jaw margin; *110, spongy presacral bone texture; 114, rudimentary cervical lamination; 117, elongate anterior cervicals; 147, pleurocoels in dorsal centra deep but simple pits; 154, dorsal transverse processes directed dorsolaterally; *160, prespinal lamina present along entire length of anterior dorsal neural spine; *162, absence of hyposphene-hypantra in anterior dorsals; 187, anterior centrodiapophyseal lamina in middle and posterior dorsals anteriorly inclined; *256, scapular glenoid beveled medially; 257, scapular base flat or rectangular; 320, pubis, acetabular and ischial surfaces meet at obtuse angle.

NODE B: UNNAMED

182, presence of prespinal lamina in middle and posterior dorsal neural arches; 183, presence of anteroposteriorly expanded postspinal lamina in middle and posterior dorsals; 193, single infradiapophyseal lamina in midposterior dorsals; 230, presence of midventral hollow in middle caudals; 259, coracoid with rectangular anteroventral margin.

NODE C: SALTASAURIDAE

149, dorsal boundary of neural canal in anterior dorsals either interprezygapophyseal lamina or a ventral extension of the prespinal fossa; 222, anterior caudal neural spines transversely narrow; 262, coracoid foramen deeply inset into coracoid body; 341, cnemial crest curves anteriorly.

NODE D: LITHOSTROTIA

111, shallow, lateral presacral pneumatic fossae; 204, presence of proximal excavation in sacral ribs; 251, acromial edge of scapula lacks expansion; 304, proximal end of metacarpal IV anteroposteriorly elongate and oval; 337, subequal breadth of femoral distal condyles.

NODE E: SALTASAURINAE

236, posterior caudal centra dorsoventrally flattened; 237, distalmost caudal centra procoelous.

NODE F: UNNAMED

123, anterior cervical neural spines proximally divided and distally fused; 238, distalmost caudal centra extremely elongate.

NODE G: RAPETOSAURUS CLADE

167, middle and posterior dorsal neural spines comprise less than 50% of total vertebral height; 169, middle and posterior dorsal neural spines posteriorly inclined; 171, middle and posterior dorsal neural spins distally flared with triangular processes; 175, posterior dorsal transverse processes vertical to parapophysis; 177, presence of triangular lateral process on dorsal neural spines; 233, rectangular midposterior caudal neural spines.

NODE H: UNNAMED

230, midcaudal centra lack midline, ventral hollow.

NODE I: UNNAMED

120, anterior cervical pleurocoels shallow and simple but divided by single median ridge; 154, dorsal transverse processes directed dorsolaterally (reversal); 155, anterior dorsal transverse processes meet neural spine at an angle of at least 90°; 221, postspinal fossa in anterior caudal neural spines; 266, straight lateral margin of humerus; 341, cnemial crest curves laterally.

NODE J: UNNAMED

*1, absence of short, deep snout; *16, external nares retracted to between orbits; *17, external nares face dorsally or rostrodorsally; 90, angle between mandibular symphysis and long axis ~90°; *108, teeth SI that exceeds 5.0.

NODE K: OPISTHOCOELICAUDIINAE

No unambiguous support for this clade.

THREE

Phylogenetic and Taxic Perspectives on Sauropod Diversity

Paul Upchurch and Paul M. Barrett

THE SAUROPODA REPRESENT ONE of the most diverse and geographically widespread dinosaurian radiations. After their origin during the Late Triassic, sauropods increased rapidly in diversity, and acquired a nearly global distribution by the Middle Jurassic. Diversity appears to peak in the Late Jurassic, at which point sauropods were very abundant and dominated many terrestrial environments. Although sauropod diversity apparently declined very rapidly in the Early Cretaceous, several lineages remained as a significant component in many faunas, and the titanosaurs underwent a major radiation during the mid- or Late Cretaceous.

These changes in diversity took place against a backdrop of profound geological and biotic events. The gradual fragmentation of Pangaea affected physical environments by altering sea levels and climatic regimes, which in turn could have had an impact on sauropod diversity. In addition, evolutionary interactions with predators, competing large-bodied herbivores, and plants may also have affected sauropod diversity. Before we can attempt to tease apart the effects of these physical and biological factors, the basic pattern of sauropod diversity must be established accurately.

No summary of sauropod evolutionary history and palaeobiology is complete without a consideration of diversity (e.g., how the number of sauropod taxa fluctuated through time). Yet this aspect of sauropod history has been relatively neglected. Most studies have been based on direct observation of numbers of sauropod species and/or genera in the fossil record. Although this approach has the potential to provide an accurate measure of diversity, it can be seriously affected by non-random sampling biases. In principle, such problems can be partially corrected when the phylogenetic relationships of taxa are taken into account (Smith 1994), but to date only one study (Weishampel and Jianu 2000) has applied this method to sauropod diversity. This neglect is not surprising given the confused nature of much of sauropod species- and genus-level systematics and the lack (until recently) of detailed cladistic analyses for this group. Fortunately, several recent studies (Upchurch 1995, 1998, 1999; Wilson and Sereno 1998; Curry Rogers and Forster 2001; Wilson 2002; Upchurch et al.,

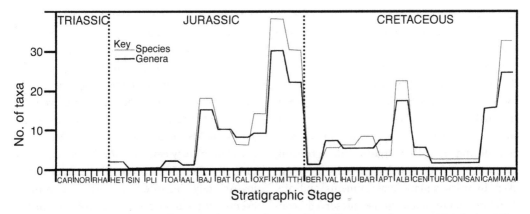

FIGURE 3.1. Taxic diversity estimates for sauropod genera and species obtained by Hunt et al. (1994). AAL, Aalenian; ALB, Albian; APT, Aptian; BAJ, Bajocian; BAR, Barremian; BAT, Bathonian; BER, Berriasian; CAL, Callovian; CAM, Campanian; CAR, Carnian; CEN, Cenomanian; CON, Coniacian; HAU, Hauterivian; HET, Hettangian; KIM, Kimmeridgian; MAA, Maastrichtian; NOR, Norian; OXF, Oxfordian; PLI, Pliensbachian; RHA, Rhaetian; SAN, Santonian; SIN, Sinemurian; TTH, Tithonian; TOA, Toarcian; TUR, Turonian; VAL, Valanginian.

2004) have gone a long way toward removing the obstacles to detailed analysis of sauropod diversification patterns. Thus, the main aims of the current chapter are (1) to present and compare alternative estimates of sauropod lineage diversity and (2) to evaluate the implications of these diversity patterns for our understanding of sauropod evolutionary history and sampling biases. The reader is also referred to the following companion chapter by Barrett and Upchurch, which examines the relative diversity of different sauropodomorph clades and considers the possible relevance of morphological characters related to feeding strategy.

PREVIOUS WORK

Few studies have examined sauropod diversity in detail. This topic has, however, been treated in a superficial fashion by several studies that have attempted either an "overview" of dinosaur evolutionary history (Bakker 1977, 1978; Sereno 1997, 1999) or an assessment of the factors controlling dinosaur diversity (Horner 1983; Weishampel and Horner 1987; Haubold 1990; Barrett and Willis 2001). All of these studies, as well as the detailed treatment of sauropod diversity presented by Hunt et al. (1994), are based on the taxic approach to diversity estimation, in which diversity is measured by counting the number of taxa present in the fossil record at each point in time. The resulting "consensus" is that sauropods gradually increased in diversity throughout the Jurassic, reaching a peak in the Kimmeridgian–Tithonian. At the Jurassic–Cretaceous boundary, however, sauropods seem to undergo a major extinction event: although many of the familial lineages survive into the Early Cretaceous, the overall level of genus or species diversity is drastically reduced (Bakker 1978). Sauropods maintained a lowered level of diversity until the end of the Cretaceous, at which point the number of genera increased as a result of the titanosaur radiation. In many of these studies, fluctuations in the numbers of sauropod genera or species are accepted as genuine changes in diversity, even though it is possible that such variations actually reflect changes in the quality of the fossil record.

The larger data sets employed by Hunt et al. (1994) and Barrett and Willis (2001) enabled them to detect a somewhat more complex pattern of peaks and troughs in sauropod diversity (fig. 3.1). In particular, these studies noted the presence of peaks in diversity during the Middle Jurassic (Bajocian), Late Jurassic (Kimmeridgian), early Early Cretaceous (Valanginian–Barremian), late Early Cretaceous (Albian), and Late Cretaceous

(Campanian–Maastrichtian). The results of these analyses agree with those for the Dinosauria as a whole obtained by Haubold (1990). These fluctuations in diversity appear to correlate with changes in sea level (Haubold 1990; Hunt et al., 1994) as determined by Haq et al. (1987). Haubold (1990) and Hunt et al. (1994) proposed a non-biotic explanation for this correlation: during periods of marine transgression, the remains of terrestrial organisms are more likely to reach aquatic environments and are, therefore, more likely to be preserved; conversely, regressions are predicted to produce a decrease in preservation. Thus, peaks and troughs in apparent diversity could reflect a taphonomic effect caused by the impact of sea level on preservation potential for terrestrial taxa. This view receives some additional support from Hunt and coworkers' (1994) observation that the peaks in sauropod diversity are positively correlated with the number of sauropod-bearing formations for each point in time. If this "taphonomic bias" hypothesis is correct, then important features of sauropod diversity, such as the Jurassic–Cretaceous boundary extinction, could be artifacts rather than real evolutionary events.

Other authors have noted the same correlation between diversity and sea level but have developed biological mechanisms to explain this phenomenon (Bakker 1977; Horner 1983; Weishampel and Horner 1987). For example, during periods of marine transgression, lowland areas are likely to become constricted in size and potentially separated from each other. This promotes allopatric speciation and may therefore result in a peak in diversity. Conversely, marine regression removes the barriers between separate lowland areas, allowing the mixing of previously isolated biotas and potentially resulting in extinction events. Haubold (1990:101) considered these mechanisms but rejected them on the basis that "the 'habitat bottlenecks' (Weishampel and Horner, 1987) in transgressive phases, or extinctions due to regressions (Bakker, 1977), reduce the fossil record regionally, but do not diminish the diversity noticeably." To this, it might be added that the impact of area fragmentation may be more complex than often assumed. Although the formation of geographic barriers can promote increased diversity through allopatric speciation, it may also result in extinction events as the size of certain habitats decreases. Nevertheless, if biological factors are responsible for the correlation between sea level and sauropod diversity, then fluctuations in the latter can be legitimately regarded as real evolutionary events rather than artifacts caused by uneven preservation rates.

Lockley et al. (1994) presented the only attempt to use a large data set of sauropod trackways to assess changes in diversity and abundance. Sauropods appear to be most abundant in the Late Jurassic and Early Cretaceous. Narrow-gauge trackways (i.e., non-titanosaur tracks) are most common in the Jurassic, whereas after the Kimmeridgian the wide-gauge trackways of titanosaurs dominate Cretaceous biomes (Wilson and Carrano 1999). Although the trackway data are in broad agreement with diversity estimates based on body fossils, the usefulness of the former is severely limited by our inability to identify the track-makers to lower taxonomic levels. Nevertheless, trackways can make some important contributions to our understanding of sauropod diversity, such as the recent discovery of wide-gauge titanosaur trackways in the Middle Jurassic of England, which extend the temporal range of this clade back by approximately 12 Ma (Day et al. 2002, 2004).

Sereno (1997, 1999) presented cladograms for the Dinosauria that were calibrated against time and used these to assess changes in lineage diversity. Unfortunately, Sauropoda was represented by only a small data set (9 and 11 lineages in Sereno [1997] and Sereno [1999], respectively), so the details of their diversity are poorly resolved. Nevertheless, Sereno noted that sauropod diversity peaked in the Late Jurassic and agreed with Upchurch (1995) that this suggested a Middle Jurassic radiation for the major neosauropod clades.

Weishampel and Jianu (2000) presented the first detailed analysis of sauropodomorph diversity that utilized both observed stratigraphic ranges and phylogenetic relationships. The sauropodomorph cladogram employed by Weishampel and Jianu (2000) was a composite of the trees produced by Galton (1990; prosauropods) and Upchurch (1995, 1998; sauropods). This composite tree was enlarged further by (1) replacing genera by monophyletic, but polytomous, clusters of species and (2) adding those taxa not considered by Galton (1990) or Upchurch (1995, 1998), through identification of their probable relationships on the basis of synapomorphies (i.e., the taxa were added directly to the tree, rather than their relationships being assessed through incorporation in a data matrix and application of cladistic analysis). Thus, the Weishampel and Jianu (2000) sauropodomorph cladogram is virtually comprehensive in terms of its species sampling, but this is achieved at the cost of poorer resolution and assumptions regarding relationships (see below). The main focus of Weishampel and Jianu's study was a comparison of the lineage diversity of Sauropodomorpha and Ornithischia to investigate possible evolutionary interactions between these groups and with plants. As a result, Weishampel and Jianu (2000) did not discuss changes in sauropod diversity in detail, though some important points emerge from their study. First, Weishampel and Jianu's diversity curves for Sauropoda show very little correspondence between the taxic and the phylogenetic diversity estimates (hereafter referred to as TDEs and PDEs, respectively). This could be interpreted as indicating that sauropods have a particularly poor fossil record. Second, the PDEs indicate the presence of a much higher peak in diversity during the Middle Jurassic (approximately 75% of that observed for the Late Jurassic) than had been estimated previously. These results, and other aspects of the Weishampel and Jianu study, are compared with our analyses under "Results and Discussion", below.

METHODS AND MATERIALS

TAXIC VERSUS PHYLOGENETIC DIVERSITY ESTIMATES

There are many approaches to the estimation of lineage diversity through time (Smith 1994), though they generally fall into one of two broad categories—*taxic* (or "probabilistic") and *phylogenetic*. Taxic methods assess diversity by counting the number of taxa observed at each point in time. Ideally, these taxa should be species, but it is often necessary to use higher taxonomic ranks as a proxy for species level diversity. The main reason for this is that the fossil record is inevitably incomplete: as a result, the representation of lower-level taxa, especially species, will be particularly poor. Sampling higher-level taxa should help to compensate for such sampling biases (Foote 1996). This is particularly noticeable when dealing with "Lazarus" taxa (i.e., groups that appear, apparently go extinct, and then reappear later). Thus, the use of higher taxa, rather than species, allows gaps in stratigraphic ranges to be identified as sampling errors, but this benefit is achieved at the cost of poorer pattern resolution (Smith 1994).

A variety of more sophisticated probabilistic approaches has been developed to ameliorate sampling errors. For example, Alroy (2000) introduced the "appearance event ordination" method, which, like many other similar approaches, uses a large database to infer confidence limits on the stratigraphic ranges of taxa. Such techniques, however, cannot deal with "singletons" (i.e., taxa known from only a single point in time). Given that we are dealing with a relatively small data set for sauropod genera (chapter 4, appendix 4.2), and ~70% of these taxa are only known from single occurrences (or several specimens from a single horizon), we have not attempted to apply this technique here.

There are obvious difficulties with taxic approaches, especially when large data sets are not available. For example, such approaches cannot identify gaps in the fossil record that

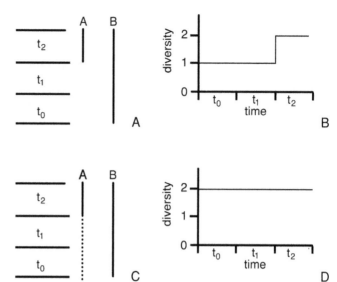

FIGURE 3.2. A schematic diagram illustrating the differences between the taxic and the phylogenetic approaches to diversity estimation. (A) Stratigraphic ranges for the taxa A and B. (B) Taxic diversity estimate through time. (C) Stratigraphic ranges and inferred ghost range for taxa A and B. (D) Diversity estimate based on both observed and inferred ranges.

occur prior to a taxon's first stratigraphic appearance (Smith 1994). This and other concerns prompted the development of the PDE (Fisher 1982; Paul 1982; Cripps 1991; Norell 1992; Norell and Novacek 1992; Smith 1994), in which both observed stratigraphic ranges and inferred "ghost" ranges are used to determine diversity by counting the number of lineages that pass through each designated time period (fig. 3.2). A *ghost range* is defined as the minimal portion of a lineage's stratigraphic range that is missing but that must have existed because of the stratigraphic range of its sister-taxon. Thus, consider two taxa, or evolutionary lineages, A and B (fig. 3.2). A and B are sister-taxa according to a cladistic analysis, and neither can be the ancestor of the other because each possess autapomorphies. A and B have different stratigraphic ranges: in particular, the first appearance of A at time t_2 is later than the first appearance of B at time t_0. Since A and B must have descended from a common ancestor that existed at or prior to the first appearance of B, there must be a lineage linking this ancestor to the first appearance of A. Thus, in figure 3.2, a ghost range is inferred for A, extending from time t_0 to time t_2. The PDE can then be determined by counting the total number of lineages (observed and inferred) for each designated time period. This simple example also illustrates how the TDEs and PDEs may differ even though based on the same stratigraphic range data. In figure 3.2, the TDE suggests that only one taxon (B) is present during time t_1, so that the appearance of A in t_2 would be interpreted as an increase in diversity. The PDE, in contrast, suggests that diversity has remained constant during t_0–t_2.

The apparent "superiority" of PDEs over TDEs has led several authors to suggest that the former technique should replace the latter (Novacek and Norell 1982; Norell 1992; Smith 1994; Rieppel 1997). Some recent studies, however, have demonstrated a number of serious problems with the PDE approach (Foote 1996; Wagner 2000a, 2000b; Wagner and Sidor 2000).

First, practical constraints mean that many cladistic analyses do not sample all of the taxa within a particular clade. This could create errors in PDEs in at least two ways. First, the taxa included or excluded from the analysis may not be randomly distributed in the stratigraphic record. The earliest taxa may be particularly fragmentary and workers may therefore choose not to include them. Alternatively, workers may focus on early "basal" forms because of an a priori belief that these are particularly important in determining phylogenetic relationships. Any systematic bias regarding the temporal distributions of taxa could lead to incorrect inferences of diversity change through time. Second, omission of taxa from a cladistic analysis could increase the probability of errors in the topology of the most parsimonious trees: this in turn could produce systematic errors in PDEs (see below).

Second, Foote (1996) and Wagner (2000a) have noted that PDEs introduce an asymmetrical "correction" into diversity estimates: that is, they potentially extend origination times backward but do not offer a corresponding correction that extends extinction times forward. Sampling theory suggests that corrections should be made nearly equally to both origination and extinction times (Foote 1996:5), and it therefore seems probable that PDEs introduce a systematic bias into diversity pattern reconstructions.

Third, Wagner (2000a) used simulations to study the impact of various factors (e.g., speciation mode, speciation rate, extinction rate, and errors in tree topology) on the accuracy of PDEs. This approach demonstrated that PDEs tend to overestimate the amount of range extension, with this error typically peaking in the first time interval (Wagner 2000a: 76–77). This problem becomes more severe as phylogenetic trees decrease in their topological accuracy.

Although the above three points raise some serious concerns regarding the accuracy of PDEs, there are ways in which errors can be minimized. First, it seems that PDEs (and measures of fossil record quality based on phylogenetic topologies) are most accurate when the operational taxonomic units (OTUs) have certain characteristics. For example, greater accuracy is achieved when the OTUs are (1) stem-based taxa, (2) species (or at least the lowest feasible taxonomic level), and (3) treated as ancestors in the phylogeny if they lack autapomorphies (Wagner 2000a, 2000b; Wagner and Sidor 2000). Second, Wagner (2000a) noted that PDEs should be "conservative" when it comes to the detection of mass extinction. This is because the PDE approach produces a backward "smearing" of origination times that tends to diminish the significance of mass extinction events. Thus, if a major decrease in diversity is suggested by a PDE method, it is legitimate to regard this as evidence for a genuine evolutionary event (i.e., a mass extinction) rather than an artifact caused by a patchy fossil record.

In short, both the TDE and the PDE have advantages and disadvantages. We therefore employ both approaches in the pluralistic spirit advocated by Foote (1996).

PRESERVATION/SAMPLING BIASES

One of the fundamental difficulties associated with the reconstruction of diversity patterns concerns the interpretation of "apparent" diversity curves. Whichever method is applied, we ultimately obtain a graph that suggests fluctuations in diversity through time, but we can never be sure whether these apparent changes in diversity represent "real" evolutionary events or differences in preservation or sampling rates. There are many reasons why sampling rates might be uneven through time (and indeed between different clades). These include geological factors (such as differences in the volumes of fossiliferous sediments from different time periods); biological traits (organisms have different preservation potentials depending on their size, structure, dispersal strategies, and so on), and "human" factors (such as the ease or difficulty associated with the logistics of collection from particular rock units).

The impact of sampling biases might be evaluated by determining whether our opportunities to observe/collect specimens have remained approximately evenly distributed across time. For example, one could plot the volume of sedimentary rock for a particular facies, or the number of fossiliferous formations, against time. This approach was adopted by Hunt et al. (1994), who noted that their TDEs for sauropods seemed to correlate closely with the numbers of sauropod-bearing formations from the Late Triassic to the Late Cretaceous (see above). This result suggests that the apparent fluctuations in sauropod diversity are artifacts caused by different preservation rates. If we wish to identify true evolutionary patterns, therefore, we must demonstrate that apparent diversity is not purely controlled by the number of opportunities to observe (NOOs).

Although assessment of preservation rates is highly desirable, there exist a number of problems with such an approach.

First, there will be occasions when NOOs will correlate with genuine rather than artifactual changes in diversity. For example, as a clade radiates, in terms of lineage diversity and/or geographic distribution, we might expect its members to appear in increasing numbers of formations or rock units. Similarly, as a group declines toward its extinction, it would be expected to be found in decreasing numbers of formations. Thus, because Hunt et al. (1994) designated their "opportunities to observe" as numbers of sauropod-bearing formations, we cannot really be sure what the correlations between this and the sauropod TDEs mean. We suggest that this problem can be circumvented, or at least ameliorated, by defining opportunities to observe in a broader way. Thus, in this study, we consider any dinosaur-bearing formation as an opportunity to observe, since any rock unit capable of preserving large terrestrial vertebrate material is presumably capable of preserving a sauropod. Thus, if we see, at a particular point in time, that there are a large number of dinosaur-bearing formations but a relatively low sauropod diversity, it seems legitimate to infer that the latter is a genuine evolutionary signal rather than an artifact of preservation.

Second, estimating NOOs is somewhat arbitrary. What, for example, do the units of measurement actually mean? Is a formation from the Late Jurassic of North America equivalent to one from the Middle Jurassic of China? Should a rock unit that has produced one fragmentary and indeterminate dinosaur specimen be given the same weight as one that has produced multiple skeletons of many different diagnosable taxa? These ambiguities may not cause too much distortion in our estimate of NOOs provided the various sources of error are randomly distributed: but it is conceivable that this is not the case. For example, the numbers of formations may be exaggerated in areas where sedimentary analysis began relatively recently, whereas in more thoroughly understood areas, a plethora of formations may be reduced through careful correlation and synonymization. Similarly, time periods of particular interest (e.g., the Kimmeridgian–Tithonian boundary) may receive a disproportionate amount of attention, resulting in inflation of NOOs.

Third, determining what is, or is not, an opportunity to observe members of a particular clade is somewhat ambiguous because the decision depends on a complex combination of factors, including the paleoecology, physiology, biogeography, and evolutionary history of members of a clade. Wagner (2000b:351–352) discussed opportunities to observe with regard to lower-level taxa, such as species. Suppose, for example, that species A and B occur in a particular facies type (X) in a geographic region (Y) at time t_o. A and B can be assumed to have the same preservation potential because they are similar in terms of their size, structure, and so on. We should expect to find the same species at later points in time provided there is a continuation of the existence of facies X in area Y from t_o to the subsequent time periods concerned. If these conditions are fulfilled, but we do not find species A (despite intense sampling and the continued presence of B) during later time periods, we can conclude that this is probably the result of genuine extinction rather than failure of preservation or sampling. Such constraints, however, do not apply when we are considering the history of a clade of taxa rather than a single species. This is because changes in the behavior, physiology, or structure of organisms through their evolutionary history may allow them to cross biogeographic barriers, invade new facies types, or change their preservation potentials.

Despite the difficulties outlined above, we feel that it is preferable to consider some approximate estimate for the quality of the fossil record, rather than none at all, when evaluating diversity curves. We have, therefore, plotted the number of dinosaur-bearing formations against time (see below) based on the data provided by Weishampel et al. (2004).

STRATIGRAPHIC RANGES AND TIMESCALES

Both the TDE and the PDE methods require information on the stratigraphic ranges of taxa. Previous studies have adopted subtly different approaches in terms of how stratigraphic ranges have been treated.

The first issue concerns whether the age of a taxon should be represented in terms of its presence during a particular stratigraphic unit (e.g., assigning the taxon to one or more of the Standard European Stages, such as the Kimmeridgian) or by applying an absolute age based on radiometric dating of the relevant strata. Hunt et al. (1994) and Barrett and Willis (2001) used stratigraphic stages, whereas Weishampel and Jianu (2000) employed absolute dates. The latter method has one major advantage: when diversity curves are plotted against absolute time (or at least a linear scale of relative time), then the slopes of the curves give a direct estimate of the tempo of radiation and extinction events. This will not necessarily be the case when the "timescale" is a sequence of stratigraphic units because different Standard European Stages are of different absolute lengths (e.g., the Kimmeridgian and Albian Stages last for 3.4 and 13.3 Ma, respectively [Gradstein et al. 1995]). The use of absolute ages, however, is an ideal that is difficult to achieve at present. Outside of Europe and North America, the absolute ages of many stratigraphic units are not well constrained by radiometric dates. Most studies that report the discovery of new sauropod taxa provide information only on stratigraphic position, rather than absolute age. In many cases, therefore, the absolute ages for sauropod species, utilized by Weishampel and Jianu (2000), are unlikely to be based on direct radiometric dating: rather, many would have to be based on biostratigraphic correlations between the deposits in which the sauropod was found and other deposits whose absolute age has been determined radiometrically. The danger with the use of an absolute timescale is that it can give a false sense of precision that, currently, is not obtainable. Here, we have used a nonlinear stage-based scale for the x-axes in figures 3.1, 3.4, and 3.5 but have converted standard European Stages to a linear absolute timescale based on Gradstein et al. (1995) in figure 3.6.

The second issue, concerning dating, involves the treatment of uncertainty or poor stratigraphic constraint. Hunt et al. (1994) noted that the ages of many sauropod taxa are not well constrained. Thus, *Barapasaurus*, from the Kota Formation of India, is dated as "Hettangian–Pliensbachian", reflecting uncertainty in the dating of the deposits rather than the presence of a long-lived genus. Hunt et al. (1994) noted that such uncertainty could distort diversity estimates. For example, if the true age of *Barapasaurus* were Hettangian, then counting this taxon as present during the Pliensbachian would contribute to an overestimate of diversity. To compensate for this, Hunt et al. (1994) down-weighted the contribution of taxa that occur in more than one Stage. Thus, *Barapasaurus* would contribute one-third of a sauropod genus in each case to the diversity estimates for the Hettangian, Sinemurian, and Pliensbachian. We sympathize with this attempt to correct for factors that might overweight diversity estimates, but we do not employ Hunt and coworkers' strategy for three important reasons.

First, the proportional down-weighting employed by Hunt et al. (1994) will also distort diversity estimates, albeit in a subtly different way compared to a simple equally weighted approach. Suppose that, a particular genus is dated as coming from one of two consecutive stages, X and Y. Hunt and coworkers' down-weighting would yield 0.5 of a genus in each of X and Y. Suppose that, in reality, the true age for the genus is restricted to X alone. The strategy of Hunt et al. does reduce the overestimate of diversity for Y, but also underestimates the diversity for X. The simpler equally weighted strategy would count one genus present in X and one in Y. This doubles the error for Y but at least there is no underestimate for X. It seems, therefore, that both down- and equal-weighted schemes are equally problematic.

Second, the Hunt et al. down-weighting strategy seems somewhat arbitrary because it does not take into account the absolute duration of stages. Consider genus A from the Kimmeridgian–Tithonian and genus B from the Albian. In absolute terms, the age of genus A is better constrained than that for B, since the total duration of the Kimmeridgian + Tithonian is ~9 Ma, whereas that for the Albian is 13.3 Ma (Gradstein et al. 1995). Yet because genus A cannot be constrained to a single stage, its contribution to the diversity estimate is halved, whereas that of B remains at its full weight. If unequal weighting is to be applied in a rigorous fashion, therefore, it must be applied in conjunction with an absolute, or linear relative, timescale.

Third, down-weighting is partially incompatible with the application of a phylogenetic approach to diversity estimation. For example, suppose the age of genus A cannot be constrained beyond the fact that it comes from one of two consecutive stages, X and Y (where X precedes Y). Suppose also that the age of genus B, the sister-taxon to A, is known to be constrained to stage X. Under these conditions, there will be occasions when the uncertainty in the age of A does not affect the diversity estimate and, therefore, down-weighting would be inappropriate. For example, if the true age of A is Y, then the phylogenetic diversity approach infers the presence of a ghost range extending back through stage X. In this case, down-weighting the presence of A in X is inappropriate since the PDE indicates that A, or a lineage leading to A, was present during X. Down-weighting would only be appropriate if the true age of A is X, because the possible presence of A in Y now results in an overestimate of diversity for Y. Since we do not know what the true age of A is, we cannot be sure whether any form of down-weighting is appropriate when obtaining PDEs. In short, one of the advantages of the PDE is that it automatically partially compensates for uncertainties in the ages of taxa, and therefore no additional down-weighting is required.

In summary, we do not believe that down-weighting strategies offer any real benefits, and in some cases such approaches will be more misleading than a simple equally weighted strategy. We therefore apply the latter in the following analyses.

TAXONOMIC LEVEL OF ANALYSIS

The two largest previous studies of sauropod diversity utilized genera and species (Hunt et al., 1994) or just species (Weishampel and Jianu 2000) as their taxonomic units of analysis. We prefer to analyze sauropod diversity at the generic level, and it is therefore worth considering the costs and benefits of the alternative approaches.

The most important advantage associated with diversity estimates at the species level is that a more detailed picture of diversity trends may be obtained. However, as Weishampel and Jianu (2000) noted, many sauropod genera are monospecific (~85% according to the taxonomic revision of Upchurch et al. [2004]), and we might not expect there to be major differences between species- and genus-level diversity curves. There are also several disadvantages associated with any attempt to utilize sauropod species at the current time. First, sauropod species-level taxonomy is often rather problematic. For example, a recent revision of the Middle Jurassic English genus *Cetiosaurus* (Upchurch and Martin 2002, 2003) found that only 1 of 13 published species could be confirmed as genuinely belonging to that genus. Although taxonomic and nomenclatural problems also exist at the generic level (Upchurch et al. 2004), in general sauropod genera are considerably more "stable" than the species they contain. Second, all major analyses of sauropod phylogeny (Calvo and Salgado 1995; Upchurch 1995, 1998, 1999; Salgado et al. 1997; Wilson and Sereno 1998; Curry Rogers and Forster 2001; Wilson 2002; Upchurch et al. 2004) have used the genus as the standard OTU. If we wish to analyze sauropod diversity using information from cladograms, therefore, we are somewhat constrained to consider genera. Weishampel and Jianu (2000) attempted to circumvent this point by using the generic-level cladogram of Upchurch (1998), with each multispecies genus replaced by a "star burst" (i.e., polytomous) cluster of species. Furthermore, those taxa not considered by Upchurch (1998) were added to the sauropod cladogram via determination of possible synapomorphies, rather than direct cladistic analysis (Weishampel and Jianu 2000: 131). As a result, Weishampel and Jianu (2000) produced a virtually complete phylogenetic tree for sauropod species, although the topology inevitably contained many polytomies. This increased the sample size in the phylogeny utilized by Weishampel and Jianu, but this benefit may have been outweighed by the potential errors introduced into the analysis. The first potential source of error concerns the relationships of those taxa that were added to the Upchurch (1998) cladogram on the basis of synapomorphies. Many such taxa contain unique combinations of character states that mean that they cannot be unequivocally "slotted" into a particular position: indeed, more recent enlarged analyses (Upchurch et al. 2004) indicate that consideration of additional taxa has some important effects on the relationships between the original 26 sauropod genera considered by Upchurch (1998). Thus, it seems likely that the topology utilized by Weishampel and Jianu (2000) cannot be fully justified in terms of cladistic analysis. The second problem concerns the presence of multiple polytomies in the Weishampel and Jianu tree. The precise effect of such polytomies on diversity estimation will vary depending on the ages and number of species involved. As an example, however, consider the hypothetical situation in figure 3.3. Suppose we observe the fossil record of genus A that is known from five species (A_1, A_2, etc.). These species are each known from a separate time period, such that A_1 occurs during t_1, A_2 occurs during t_2 and so on (fig. 3.3). When the relationships among the five species are unknown, the total amount of inferred ghost range equals 10 time units (fig. 3.3A): this is the same result that would be achieved if the fully resolved species-level cladogram happened to be maximally incongruent with the stratigraphic distributions of the five species (figs 3.3B, C). Figures 3.3D–F show the required ghost ranges (4 time units in total) when the fully resolved species-level cladogram happens to be maximally congruent with the stratigraphic ranges of the species. This simple example illustrates one of the intrinsic biases introduced into the PDE method by polytomies: the ghost ranges will be overestimated because all missing lineages will have to extend back as far as the oldest known taxon within the polytomy. In other words polytomies are undesirable because they introduce a systematic bias toward inflated PDEs, rather than a random error that might be at least partially cancelled out in large enough data sets. Such errors will be most severe when the stratigraphic range of a supposed species cluster is particularly extensive, as occurs with problematic taxa such as *Cetiosaurus* and *Titanosaurus*. These conclusions are consistent with the results of the simulation studies carried out by Wagner (2000a) and Wagner and Sidor (2000), in which errors in the estimation of tree topology are shown to result in a systematic overestimation of diversity early in a clade's history.

Here, therefore, we prefer to avoid polytomies (and therefore sauropod species) via the use of a generic-level analysis of diversity. Later in this chapter, we examine the similarities and differences between the PDEs obtained by Weishampel and Jianu's analysis of species and those for genera obtained by ourselves.

DATA

Both TDEs and PDEs require information on the stratigraphic ranges of sauropod genera. The taxonomy and stratigraphic ranges of sauropod

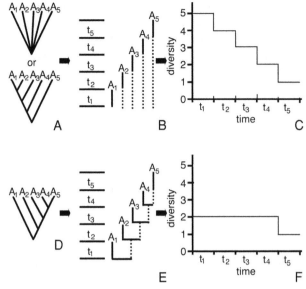

FIGURE 3.3. A schematic diagram showing the effect of polytomies on phylogenetic diversity estimation. (A) Phylogenetic relationships of the five taxa A_1–A_5. Top cladogram shows no resolution; bottom cladogram is resolved so that it is maximally incongruent with the stratigraphic ranges of the taxa. (B) Stratigraphic ranges of taxa A_1–A_5 and ghost ranges required by the relationships in (A). (C) Diversity through time based on the observed and inferred ranges shown in (B). (D) Cladogram for taxa A_1–A_5 that is maximally congruent with the stratigraphic ranges of the taxa shown in (E). (E) Stratigraphic ranges of taxa A_1–A_5 and ghost ranges required by the relationships in (D). (F) Diversity through time based on the observed and inferred ranges shown in (E). In the diagrams showing observed and ghost ranges, solid lines indicate known stratigraphic range, while dotted lines indicate ghost ranges.

genera have been reviewed and revised recently by Upchurch et al. (2004). These authors concluded that there are currently approximately 98 valid, or potentially valid, sauropod genera. These taxa, and their stratigraphic ranges, form the basis for our TDEs and are summarized in chapter 4 (appendix 4.2).

Upchurch et al. (2004) also provide the most detailed cladogram of sauropod genera currently available. The data matrix comprises 309 characters for 41 ingroup sauropod genera and 6 dinosaurian and dinosauromorph outgroups. The Heuristic search in PAUP 4.0 (Swofford 2000) produced 1,056 most parsimonious trees (MPTs), which were then subjected to reduced consensus analysis (Wilkinson 1994). The latter indicated that a single fully resolved topology can be obtained if five taxa (*Andesaurus*, *Argentinosaurus*, *Lapparentosaurus*, *Nigersaurus*, and 'Texan *Pleurocoelus*') are deleted *a posteriori* from the 1,056 original MPTs. The resulting reduced consensus tree therefore contains 36 sauropod genera and is shown, plotted against stratigraphic Stages, in figure 3.4. Our PDE values were calculated directly from the inferred ghost ranges and observed stratigraphic ranges in figure. 3.4, with conversion to an absolute timescale using Gradstein et al. (1995).

RESULTS AND DISCUSSION

The results of our diversity analyses are shown in figures 3.5 and 3.6. Figure 3.5 presents a comparison between the phylogenetic estimates of sauropod diversity produced in the current chapter (UB) and Weishampel and Jianu (2000) (WJ). Figure 3.6 compares the TDE and PDE curves produced by our data alone and

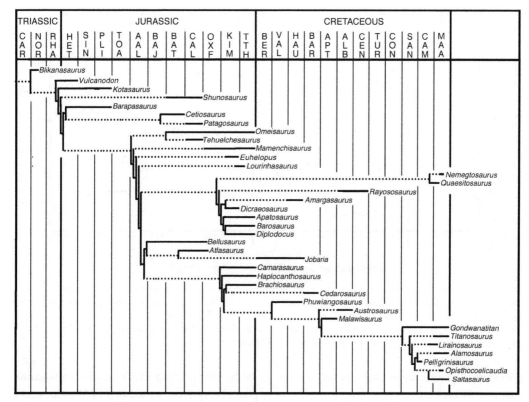

FIGURE 3.4. A time-calibrated version of the reduced consensus cladogram produced by Upchurch et al. (2004) (see text for details). Stage name abbreviations are listed in the legend to figure 3.1. Solid lines show the observed stratigraphic range for each sauropod genus; dotted lines indicate inferred ghost ranges.

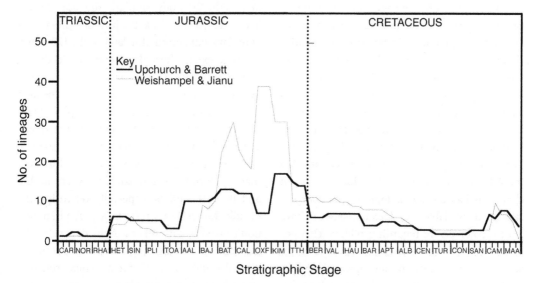

FIGURE 3.5. Comparison of the phylogenetic diversity estimates obtained by Weishampel and Jianu (2000) and in the current chapter. Stage name abbreviations are listed in the legend to figure 3.1.

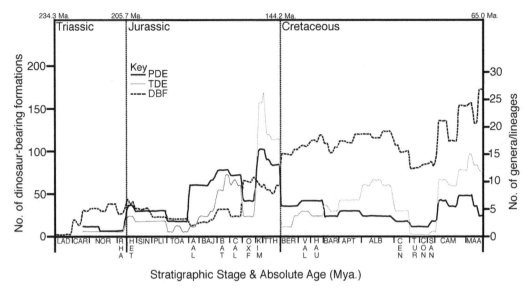

FIGURE 3.6. Comparison of the taxic and phylogenetic diversity estimates for sauropod genera, and the number of dinosaur-bearing formations, obtained in the current chapter (see text for details). Stage name abbreviations are listed in the legend to figure 3.1. DBF, number of dinosaur-bearing formations (i.e., number of opportunities to observe); PDE, phylogenetic diversity estimate; TDE, taxic diversity estimate. Numbers of dinosaur-bearing formations were calculated using the data of Weishampel et al. (2004). Absolute ages are based on Gradstein et al. (1995).

also shows the number of dinosaur-bearing formations through time.

COMPARISON WITH WEISHAMPEL AND JIANU (2000)

The PDEs derived by WJ and UB are shown in figure 3.5. The absolute diversity estimates often differ between the two studies, presumably because of differences in taxon sampling (see above). In general, however, the two curves are in close agreement in terms of the relative changes in diversity through time. This agreement is important since it indicates that the potential pitfalls associated with the WJ and UB analytical approaches have not led to significantly different diversity patterns. Thus, the utilization of sauropod species, and the introduction of polytomies, has not adversely affected the WJ study. Similarly, the fact that the current chapter is based on a cladogram that does not sample all available taxa does not seem to have led to major errors regarding relative changes in diversity, though clearly the absolute diversity levels have been underestimated. Future studies may gain further insights into the impact of differences in phylogenetic hypotheses by constructing PDEs based on the sauropod cladograms presented by Pisani et al. (2002) and Wilson (2002).

There are, however, some subtle differences between the WJ and the UB PDE curves that deserve discussion. Most notably, the Middle Jurassic increase in sauropod diversity commences somewhat earlier in the UB PDE curve than in the WJ study. Similarly, the rapid decline in diversity, toward or at the Jurassic–Cretaceous boundary, happens a little later in the UB curve. To some extent these disagreements in timing may reflect the different timescales employed by the two studies: conversion of the absolute dates presented by WJ into stratigraphic stages (or even parts of stages) may have introduced a small distortion into the direct comparison between the PDE values. We suspect, however, that much of the discrepancy between the UB and the WJ studies, especially regarding the Middle Jurassic, represents a real phenomenon caused by differences in the data sets. The new cladogram (fig. 3.4), used here, contains several Middle Jurassic forms whose relationships have important effects on the number and extent of ghost lineages. These crucial taxa, such as *Bellusaurus,* have not previously been incorporated into a formal cladistic analysis, and hence their precise relationships would not have been available to WJ. These differences mean that the WJ study generates fewer early Middle Jurassic ghost ranges than does the UB analysis.

SAUROPOD DIVERSITY PATTERNS

The taxic and phylogenetic diversity curves in figure 3.6 agree strongly on many aspects of sauropod evolutionary history. In particular, both TDEs and PDEs indicate that (1) diversity increased throughout the Jurassic, reaching a peak in the Kimmeridgian–Tithonian; (2) the Jurassic increase in diversity was punctuated by two short-term decreases, one in the Toarcian and one in the Oxfordian; (3) there was a dramatic decrease in diversity at the Jurassic–Cretaceous boundary; and (4) diversity reached another peak during the Campanian–Maastrichtian. There are, however, discrepancies between the PDE and the TDE values at various points in time, and these potentially represent important sampling biases or problems with taxon sampling.

PATTERNS AND ARTIFACTS

The earliest sauropod body fossils currently occur in the Late Triassic. These include *Blikanasaurus* from the Carnian or early Norian of South Africa (Galton and van Heerden 1985), which was regarded as a prosauropod until the recent cladistic analysis by Galton and Upchurch (2004), and *Isanosaurus* from the late Norian or Rhaetian of Thailand (Buffetaut et al. 2000). This provides confirmation of the suggestion that sauropods diverged from their prosauropod sister-group at the beginning of the Carnian at the latest (Upchurch 1995). The fossil record apparently indicates that sauropod diversity was relatively low during the Late Triassic, compared to that for prosauropods, and this conclusion is not currently contradicted by ghost range data (see also Barrett and Upchurch, chapter 4).

Sauropod diversity started to increase noticeably during the Early Jurassic, though there may have been a slight decline in the Toarcian. To some extent, apparent diversity, especially that inferred from the TDEs, seems to be affected by preservation rates during the Late Triassic and Early Jurassic. PDEs, TDEs, and NOOs all gradually decline from a peak at the Triassic–Jurassic boundary to a low point in the Toarcian. Nevertheless, it is interesting to note that NOOs in the Early Jurassic are generally no higher than those for the Late Triassic, yet there is a clear increase in sauropod diversity in the former time period.

The most serious discrepancy between the diversity curves occurs in the Aalenian to lower Bathonian interval, with PDEs as much as five times higher than the TDEs. This observation should be treated with some caution because the large number of ghost lineages extending back to the Aalenian have been created partly by the age and relationships of *Bellusaurus* (fig. 3.4). The stratigraphic age of this genus is not well constrained: the 'Wucaiwan' or 'Shishigou' Formation, in which *Bellusaurus* is found, has been suggested to be somewhere within the Aalenian–Callovian (see chapter 4, appendix 4.2). Furthermore, the cladistic analysis by Upchurch et al. (2004) places *Bellusaurus* in a clade along with *Atlasaurus* and *Jobaria*, representing an unexpected and potentially controversial clustering of taxa that is not strongly supported by numerous unequivocal synapomorphies. Alteration of either the age or the relationships of *Bellusaurus*, therefore, could have a profound impact on the mid-Jurassic PDE values. These concerns reflect the more general point made by Foote (1996) and Wagner (2000a), that PDEs may overestimate diversity, especially toward the beginning of a clade's history. Notwithstanding these points, however, several lines of reasoning suggest that the Middle Jurassic increase in sauropod diversity may reflect a real signal. First, although the TDEs lag behind the PDEs, they also reach a similar peak in the Middle Jurassic (albeit not until the Bathonian–Callovian). Second, NOOs during the Aalenian–Bathonian are similar to those for the Norian and Hettangian, even though the latter stages are associated with much lower PDE values. Finally, other lines of evidence suggest that the major neosauropod lineages (brachiosaurs, titanosaurs, diplodocoids) had already radiated by the Bathonian. These include the presence of putative diplodocoid remains and titanosaur trackways in the Bathonian of the United Kingdom (Upchurch and Martin 2003; Day et al. 2002, 2004) and the presence of a vicariance pattern consistent with Pangaean fragmentation (Upchurch et al. 2002). Sauropod diversity may have been considerably higher during the early Middle Jurassic than has been proposed previously. For example, the TDE values obtained by Hunt et al. (1994) and the UB analysis suggest that Middle Jurassic sauropod diversity was only approximately 40% of that for the Late Jurassic, whereas the WJ and UB PDE values indicate that this relative diversity measure is closer to 75%. Thus, the significant increase in sauropod diversity, which includes the important early radiation of neosauropod lineages, may have commenced somewhat earlier and been considerably more important than previously recognized from taxic estimates of diversity. The cause of the discrepancy between PDEs and TDEs is not entirely clear, though it should be noted that the Aalenian represents one of the poorest parts of the dinosaur fossil record as judged from NOOs (fig. 3.6), and therefore sampling error may be at least partly responsible.

Both PDEs and TDEs indicate a dramatic drop in sauropod diversity during the Oxfordian. This apparent extinction is less marked in the PDEs, but this may represent an artifact caused by backward "smearing" of origination times (see above). In the Jurassic, NOOs reach their highest level during the Oxfordian, higher even than in the Kimmeridgian and Tithonian. Thus, the Oxfordian drop in diversity potentially represents a genuine extinction event rather than an artifact. We hypothesize that the Oxfordian may represent an important transition in sauropod evolution. During the Middle Jurassic, cetiosaurids and basal eusauropods are relatively diverse, whereas neosauropods may have been represented by only a relatively small number of lineages. The Oxfordian extinction may have been particularly serious for the cetiosaurid and basal eusauropod taxa, allowing neosauropods to radiate into several vacant niches during the Kimmeridgian and Tithonian. At present, however, the fossil record for sauropods is too sparse to allow direct and detailed testing of this possibility.

The Late Jurassic has long been regarded as the pinnacle of sauropod diversity. This view is fully supported by the TDE and PDE values for the Kimmeridgian and Tithonian. During these stages, there are at least 27 valid or potentially valid genera. This peak in diversity could have been partly caused by a particularly rich fossil record at this time, especially the well-preserved and intensively studied Morrison and Tendaguru faunas. However, NOOs are actually lower for these stages than they are for the Oxfordian and Berriasian (fig. 3.6). The UB PDE values for the Kimmeridgian–Tithonian are approximately 67%–75% those obtained from TDEs, indicating that the cladogram in figure 3.4 has somewhat under sampled the diversity present in the Late Jurassic. It will be important to add these unsampled taxa to future cladistic analyses, to see whether their incorporation alters the shape of the Jurassic PDE curve.

Both TDEs and PDEs indicate a dramatic decrease in sauropod diversity across the Jurassic–Cretaceous boundary: from the Tithonian to the Berriasian, TDEs drop from 18 to 2, while the corresponding PDEs fall from 14 to 6. This represents a decrease in diversity of 57%–89%, returning sauropod diversity to the levels seen briefly during the Oxfordian. This extinction event is even more marked when the diversity levels for the Kimmeridgian are compared with those of the Berriasian—a decrease of 65–93% over 5.5 Ma. Two factors suggest that this diversity decrease represents a genuine mass extinction. First, NOOs actually increase dramatically (from an average of ∼60 to ∼100) at the Jurassic–Cretaceous boundary. Second, as Wagner (2000a) has noted, PDEs are conservative with respect to mass extinctions (see above). It is interesting to note that much of this extinction affected generic-level diversity but left many of the higher taxa with lineages that survive into the Early Cretaceous. Thus, although diplodocids and all non-neosauropods die out at the end of the Jurassic, many other distinct clades (nemegtosaurids, rebbachisaurids, dicraeosaurids, brachiosaurids, and titanosaurs) are represented by body fossils and/or ghost ranges in the Berriasian.

There is a significant discrepancy between TDE and PDE values in the mid-Cretaceous (Albian), with the former being up to 2.75 times higher than the latter. This could indicate that the PDEs are artificially low because the cladistic analysis did not sample sufficient taxa from the mid-Cretaceous. The cladistic analysis by Upchurch et al. (2004) originally contained 41 sauropod genera, though 5 were deleted a posteriori to obtain the reduced consensus tree shown in figure 3.4. Four of these five genera (*Andesaurus, Argentinosaurus, Nigersaurus,* and 'Texan *Pleurocoelus*') are of mid-Cretaceous age (see chapter 4, appendix 4.2). Assignment of these taxa to their "probable" relationships (i.e., *Andesaurus, Argentinosaurus,* and 'Texan *Pleurocoelus*' as basal titanosaurs that are closer to other titanosaurs than to *Phuwiangosaurus,* with *Nigersaurus* as a rebbachisaurid) would not introduce major new ghost ranges back into the early part of the Cretaceous. For example, *Nigersaurus* is the earliest of the three known rebbachisaurid genera (see chapter 4, appendix 4.2). These four additional taxa, therefore, are likely to increase the PDE values for the mid-Cretaceous (i.e., Aptian–Cenomanian) without increasing the values for the earliest Cretaceous (Berriasian–Barremian). This would tend to bring the PDE and TDE curves into closer agreement, though it would not be enough by itself to produce a major mid-Cretaceous peak in PDEs. There is also evidence that the disagreement between the UB TDEs and the UB PDEs for the mid-Cretaceous is not solely caused by undersampling of taxa in the cladistic analysis. For example, the "comprehensive" WJ PDEs show the same pattern as the UB PDEs: that is, a gradual decline in diversity from the Valanginian to the Coniacian, with no Albian peak (fig. 3.5). This "true" diversity pattern may have been modified to give an Albian peak in TDEs by the presence of a sampling bias: NOOs are higher during the Aptian–Albian than they are in the Early Cretaceous and Cenomanian–Santonian. Given the evidence currently available, it is possible that sauropod diversity remained fairly constant, or declined only slightly, from the Berriasian to the Albian and then decreased more rapidly during the Cenomanian.

The presence of a depressed preservation rate in the Turonian and Coniacian is obvious from the fact that no diagnosable sauropod genera are known from these stages. There are a number of fragmentary and indeterminate sauropods from these stages, such as material from the Turonian of Uzbekistan and Kazakhstan (Weishampel et al. 2004) and the Coniacian of Argentina (Bonaparte 1996), but even this poor-quality material is scarce. Prior to the application of phylogenetic diversity estimation, it might have been predicted that the Turonian–Coniacian diversity trough would be at least partially eliminated by the discovery of numerous ghost ranges extending back from the major titanosaur radiation of the Late Cretaceous. Both the WJ and the UB PDE curves (fig. 3.5), however, show that this is not the case. Clearly a few sauropod lineages existed during the Turonian–Coniacian, but there is currently no evidence to suggest that sauropods were diverse at this time. This view is supported by the NOOs for the Cretaceous (fig. 3.6). The number of dinosaur-bearing formations decreases in the early Late Cretaceous, especially during the Turonian, Coniacian, and early Santonian. This pattern may explain part of the dip in apparent sauropod diversity between the Albian and the Campanian peaks. Yet it should also be noted that NOOs during the early Late Cretaceous are substantially higher than those for portions of the Jurassic that have produced numerous sauropod genera. Thus, the scarcity of sauropod material from the Turonian and Coniacian certainly represents some form of sampling bias, but this appears to have been exacerbated by a genuine rarity of sauropods.

The TDE and PDE curves agree that a major radiation of sauropods (essentially titanosaurs) occurred during the Santonian–Maastrichtian. The diversity levels achieved at this time rival those observed in the Middle Jurassic, though not approaching the peak seen in the Late Jurassic. The TDEs exceed the PDEs by approximately 1.5–3 times, reflecting the latter's underestimation of diversity due to taxon sampling problems. At least part of the Campanian–Maastrichtian pattern can be interpreted as resulting from variation in preservation rates. Figure 3.6 suggests a close correspondence between fluctuations in NOOs and PDE or TDE values (e.g., peaks in the early Campanian and late Campanian–early Maastrichtian). However, relative to the rest of the Cretaceous, NOOs are only moderately elevated (except for the extreme peak in the latest Maastrichtian), suggesting that the Campanian peak in sauropod diversity also partly reflects a genuine radiation of new taxa.

The diversity curves in figure 3.6 also indicate that sauropods went into decline prior to the final K–T extinction, rather than disappearing "suddenly" at the latter boundary. Whether this decline represents a real phenomenon, or is an artifact caused by the current uncertainty surrounding the stratigraphic ages of many sauropods, is not entirely clear. Nevertheless, the most detailed evidence currently available suggests that, as with several other dinosaur groups, sauropods were undergoing a gradual decrease in diversity immediately prior to the final "mass" extinction event.

PROCESSES AFFECTING SAUROPOD DIVERSITY

Most studies of sauropod, or even dinosaur, diversity have attempted to explain the observed fluctuations in terms of one or more processes. These explanations fall into two main categories, so that peaks and troughs in apparent diversity are regarded as (1) artifacts produced by geological processes or (2) real patterns caused by the impact of physical and biological processes on evolutionary history.

The principal artifact-based explanation argues that peaks and troughs in diversity reflect the impact of sea level change on the preservation potential of terrestrial taxa (Haubold 1990; Hunt et al., 1994). The apparent close correlation between peaks in diversity and sea level high-stands, and between troughs in diversity and sea level low-stands, is impressive. Furthermore, there is a plausible causal link between sea level and observed taxic diversity in the fossil record, since the preservation potential of terrestrial organisms could be higher when shallow seas spread into lowland areas. However, this "taphonomic bias" explanation for fluctuations in sauropod apparent diversity was proposed at a time when phylogenetic diversity estimates were not available: PDE values provide an important opportunity to test this hypothesis rigorously because they can suggest which diversity peaks and troughs are real and which are probably artifacts.

Haubold (1990) and Hunt et al. (1994) identified five major peaks in taxic diversity that correlate with sea level high-stands: (1) Bajocian, (2) Kimmeridgian, (3) Valanginian–Barremian, (4) Albian, and (5) Campanian–Maastrichtian. Four of these peaks (Nos. 1–3 and 5) are seen in the WJ and UB PDEs (figs. 3.5 and 3.6), suggesting that they represent real increases in diversity and cannot be explained simply as artifacts of preservation. There are also four notable troughs in diversity that correlate with sea level low-stands: (1) Toarcian–Aalenian, (2) Oxfordian, (3) Berriasian, and (4) Turonian–Coniacian. Two of these (Nos. 2 and 4) appear to be genuine low points in diversity, although the Turonian–Coniacian trough must also represent some form of sampling bias. Thus, of the nine major points of comparison between sauropod diversity and sea level, only one peak (Albian) and two troughs (Toarcian–Aalenian and Berriasian) provide support for the "taphonomic bias" hypothesis. The remaining peaks and troughs in diversity, although still correlated with sea level, seem to represent real evolutionary phenomena. This result casts some doubt on the validity of the taphonomic bias hypothesis, at least in terms of its being the main control on observed taxic diversity. Furthermore, the graph of dinosaur-bearing formations through time (fig. 3.6) also suggests that there is not a simple relationship between sea level and preservation rate. For example, NOOs do not peak in the Bajocian, Kimmeridgian–Tithonian, or Valanginian–Barremian relative to adjacent stages. This tends to undermine the mechanism proposed by Haubold (1990) and Hunt et al. (1994), which supposedly links apparent diversity directly to preservation rates. This is a complex issue, however, since "number of dinosaur-bearing formations" is not the only measure by which the quality of the fossil record can be measured. Certain facies may preserve a small quantity of high-quality and therefore diagnosable material, while others may produce large quantities of indeterminate fragments. This may explain why, despite the lower NOO values, the Late Jurassic has produced considerably more valid sauropod species than the apparently well-represented Turonian–Coniacian.

The above analysis suggests that other processes, apart from uneven sampling rates, may have been responsible for the correlation between sea level and diversity. Several studies have proposed possible mechanisms by which sea level changes could affect the evolution of clades in a manner that would modify diversity. In particular, increases in sea level could promote the separation and isolation of terrestrial areas, resulting in allopatric speciation. Similarly, decreases in sea level could cause previously isolated areas to come into contact, allowing biotic mixing and perhaps extinction (Bakker 1977, 1978; Weishampel and Horner 1987). Thus, sea level fluctuations can be plausibly linked to real, as well as artifactual, changes in diversity. Our analyses, however, disrupt some of the correlation between real peaks and troughs in diversity and highs and lows in sea level. For example, the Toarcian–Aalenian sea level low-stand seems to coincide with a major increase in PDE values, while the Albian increase in sea level is not currently correlated with a corresponding increase in PDE values. Furthermore, we may wish to question the validity of the correlation between overall dinosaur diversity and sea level. Although the sauropod diversity curves match sea level fluctuations reasonably well, this may not be the case for theropods and ornithischians. So far, phylogenetic diversity estimation has not been applied to theropods, and the PDEs for Ornithischia obtained by Weishampel and Jianu (2000:fig. 5.8) show major peaks and troughs in diversity that do not match those for sauropods.

While major extrinsic events, such as sea level fluctuations, may have played some role in controlling diversity levels, it is also conceivable that the main constraints on sauropod diversity were factors relating to the particular ecology and evolutionary potential of this clade. Much of sauropod diversity seems to be related to modifications to the skull, neck, and appendicular skeleton, which would have had a direct or indirect effect on the feeding mechanisms the animals could employ. Upchurch and Barrett (2000) noted that, when the number of possible feeding strategies is plotted against time, there is a positive correlation with diversity. This issue is explored in detail by Barrett and Upchurch (chapter 4).

CONCLUSION

The taxic and phylogenetic diversity estimation methods are sometimes portrayed as rival techniques. While each approach has its advantages and disadvantages, the current chapter demonstrates the valuable insights that can be obtained when both methods are applied to the same data set. Sauropod diversity fluctuated considerably during their evolution, with major radiations occurring during the Middle and Late Jurassic. These were followed by a dramatic extinction at the Jurassic–Cretaceous boundary, then a gradual decline throughout most of the Cretaceous, and, finally, a brief but important radiation of titanosaurs just prior to the K–T mass extinction. Although much of this pattern has been suspected from previous studies, the current analysis indicates that the Middle Jurassic radiation commenced earlier, and was more important, than has been recognized hitherto. The timing and magnitude of this Middle Jurassic radiation have been obscured by a severe sampling bias affecting the Aalenian and early Bajocian.

The step from "pattern identification" to "process inference," regarding the detection and interpretation of diversity, is inevitably difficult: not only are we potentially dealing with "one–off" historical events, but also it seems likely that several processes interacted in a complex manner. There appears to be some correlation between sea level and TDE or PDE values, suggesting an impact on some aspect of evolution, such as speciation and extinction rates. It is also clear that the sauropod fossil record has been affected by sampling biases, most notably in the early part of the Middle Jurassic and the Turonian–Coniacian.

The results of the current chapter should be treated with some caution since they depend so heavily on the accuracy of stratigraphic dating, taxonomic identifications, phylogenetic reconstruction, and the cladistic topology. All of these components inevitably contain errors, and future work could result in some major shifts in the magnitude and timing of the diversity peaks and troughs. Nevertheless, it is interesting to note that the taxic and phylogenetic diversity estimates obtained by Hunt et al. (1994), by Weishampel and Jianu (2000), and in the current analysis generally agree quite closely. Although this agreement must partly reflect the overlapping and therefore non-independent nature of the data sets, it seems that a consensus on sauropod diversity history is beginning to emerge.

ACKNOWLEDGMENTS

P.U.'s research is funded by a Natural Environment Research Council Fellowship (No. GT59906ES). P.U. also wishes to thank the Department of Earth Sciences, Cambridge University, for supporting this work. We wish to thank Dan Chure, Kristi Curry Rogers, and Jeff Wilson for inviting us to contribute to this book. Jack McIntosh has provided encouragement and advice on many occasions, and has always been supportive and interested in our work on sauropods—Thanks, Jack.

LITERATURE CITED

Alroy, J. 2000. New methods for quantifying macroevolutionary patterns and processes. Paleobiology 26: 707–733.

Bakker, R.T. 1977. Tetrapod mass extinctions—Model of the regulation of speciation rate and immigration by cycles of topographic diversity. In: Hallam, A. (ed.). Patterns of Evolution. Elsevier, Amsterdam. Pp. 439–468.

———. 1978. Dinosaur feeding behaviour and the origin of flowering plants. Nature 274: 661–663.

Barrett, P.M., and Willis, K.J. 2001. Did dinosaurs invent flowers? Dinosaur–angiosperm coevolution revisited. Biol. Rev. 76(3): 411–447.

Bonaparte, J.F. 1996. Dinosaurios de America del Sur. Museo Argentino de Ciencias Naturales, Buenos Aries. 174 pp.

Buffetaut, E., Suteethorn, V., Cuny, G., Tong, H., Le Loeuff, J., Khansubha, S., and Jongautchariyakul, S. 2000. The earliest known sauropod dinosaur. Nature 407: 72–74.

Calvo, J.O., and Salgado, L. 1995. *Rebbachisaurus tessonei* sp. nov. A new Sauropoda from the Albian–Cenomanian of Argentina; new evidence on the origin of the Diplodocidae. GAIA 11: 13–33.

Cripps, A.P. 1991. A cladistic analysis of the cornutes (stem chordates). Zool. J. Linn. Soc. London 102: 333–366.

Curry Rogers, K.A., and Forster, C.A. 2001. The last of the dinosaur titans: a new sauropod from Madagascar. Nature 412: 530–534.

Day, J.J., Upchurch, P., Norman, D.B., Gale, A.S., and Powell, H.P. 2002. Sauropod trackways: evolution and behavior. Science 296: 1659.

Day, J.J., Norman, D.B., Gale, A.S., Upchurch, P., and Powell, H.P. 2004. A Middle Jurassic dinosaur trackway site from Oxfordshire, UK. Palaeontology 47: 319–348.

Fisher, D.C. 1982. Phylogenetic and macroevolutionary patterns in the Xiphosurida. North Am. Paleontol. Conv. Proc. 3: 175–180.

Foote, M. 1996. Perspective: evolutionary patterns in the fossil record. Evolution 50: 1–11.

Galton, P.M. 1990. Basal Sauropodomorpha—Prosauropoda. In: Weishampel, D.B., Osmólska, H., and Dodson, P. (eds.). The Dinosauria, 1st ed. University of California Press, Berkeley. pp.320–344.

Galton, P.M., and Upchurch, P. 2004. Basal Sauropodomorpha—Prosauropoda. In: Weishampel, D.B., Osmólska H., and Dodson, P. (eds.). The Dinosauria, 2nd ed. University of California Press, Berkeley. Pp. 232–258.

Galton, P.M., and van Heerden, J. 1985. Partial hindlimb of *Blikanasaurus cromptoni* n. gen. and n. sp., representing a new family of prosauropod dinosaurs from the Upper Triassic of South Africa. Géobios 18: 509–516.

Gradstein, F.M., Agterberg, F.P., Ogg, J.G., Hardenbol, J., van Veen, P., Thierry, J., and Huang, Z. 1995. A Triassic, Jurassic and Cretaceous time scale. SEPM (Soc. Sediment. Geol.) Special Publ. 54: 95–126.

Haq, B.U., Hardenbol, J., and Vail, P.R. 1987. Chronology of fluctuating sea levels from the Triassic. Science 235: 1156–1167.

Haubold, H. 1990. Dinosaurs and fluctuating sea levels during the Mesozoic. Hist. Biol. 4: 75–106.

Horner, J.R. 1983. Cranial osteology and morphology of the type specimen of *Maiasaura peeblesorum* (Ornithischia: Hadrosauridae), with discussion of its phylogenetic position. J. Vertebr. Paleontol. 3: 29–38.

Hunt, A.P., Lockley, M.G., Lucas, S.G., and Meyer, C.A., 1994. The global sauropod fossil record. GAIA 10: 261–279.

Lockley, M.G., Meyer, C.A., Hunt, A.P., and Lucas, S.G. 1994. The distribution of sauropod tracks and trackmakers. GAIA 10: 233–248.

Norell, M.A. 1992. Taxic origin and temporal diversity: the effect of phylogeny In: Novacek, M.J., and Wheeler, Q.D. (eds.). Extinction and Phylogeny. Columbia University Press, New York. Pp. 89–118.

Norell. M.A., and Novacek, M.J. 1992. Congruence between suprapositional phylogenetic patterns: comparing cladistic patterns with the fossil record. Cladistics 8: 319–337.

Novacek, M.J., and Norell, M.A. 1982. Fossils, phylogeny and taxonomic rates of evolution. Syst. Zool. 31: 266–275.

Paul, C.R.C. 1982. The adequacy of the fossil record. In: Joysey, K.A., and Friday, A.E. (eds.). Problems of Phylogenetic Reconstruction. Academic Press, London. Pp. 75–117.

Pisani, D., Yates, A.M., Langer, M.C., and Benton, M.J. 2002. A genus-level supertree of the Dinosauria. Proc. Roy. Soc. of London Ser. B 269: 915–921.

Rieppel, O. 1997. Falsificationist versus verificationist approaches to history. J. Vertebr. Paleontol. 17(3-suppl): 71A.

Salgado, L., Coria, R.A., and Calvo, J.O. 1997. Evolution of titanosaurid sauropods. I. Phylogenetic analysis based on the postcranial evidence. Ameghiniana 34(1): 3–32.

Sereno, P.C. 1997. The origin and evolution of dinosaurs. Annu. Rev. Earth Planet. Sci. 25: 435–489.

Sereno, P.C. 1999. The evolution of dinosaurs. Science 284: 2137–2147.

Smith, A.B. 1994. Systematics and the Fossil Record. Blackwell Scientific, Oxford, 223 pp.

Swofford, D.L. 2000. PAUP: phylogenetic analysis using parsimony, version 4.0b4a. MacMillan, London.

Upchurch, P. 1995. The evolutionary history of sauropod dinosaurs. Philos. Trans. Roy. Soc. London Ser. B 349: 365–390.

———. 1998. The phylogenetic relationships of sauropod dinosaurs. Zool. J. Linn. Soc. London 124: 43–103.

———. 1999. The phylogenetic relationships of the Nemegtosauridae (Saurischia, Sauropoda). J. Vertebr. Paleontol. 19(1): 106–125.

Upchurch, P., and Barrett, P.M. 2000. The evolution of sauropod feeding mechanisms. In: Sues, H.-D. (ed.). The Evolution of Herbivory in Terrestrial Vertebrates, Perspectives from the Fossil Record. Cambridge University Press, Cambridge. Pp. 79–122.

Upchurch, P., and Martin, J. 2002, The Rutland *Cetiosaurus*: the anatomy and relationships of a Middle Jurassic British sauropod dinosaur. Palaeontology 45(6): 1049–1074.

———. 2003. The anatomy and taxonomy of *Cetiosaurus* (Saurischia, Sauropoda) from the Middle Jurassic of England. J. Vertebr. Paleontol. 23(1): 208–231.

Upchurch, P., Hunn, C.A., and Norman, D.B. 2002. An analysis of dinosaurian biogeography: evidence for the existence of vicariance and dispersal patterns caused by geological events. Proc. Roy. Soc. London Ser. B 269: 613–622.

Upchurch, P., Barrett, P.M., and Dodson, P., 2004. Sauropoda; In: Weishampel, D.B., Osmólska, H., and Dodson, P., (eds.). The Dinosauria, 2nd Ed. University of California Press, Berkeley. Pp. 259–322.

Wagner, P.J. 2000a. The quality of the fossil record and the accuracy of phylogenetic inferences about sampling and diversity. Syst. Biol. 49: 65–86.

———. 2000b. Phylogenetic analyses and the fossil record: tests and inferences, hypotheses and models. Paleobiology 26: 341–371.

Wagner, P. J., and Sidor, C. A., 2000. Age rank/clade rank metrics-sampling, taxonomy, and the meaning of "stratigraphic consistency." Syst. Biol. 49: 463–479.

Weishampel, D. B., and Horner, J. R., 1987. Dinosaurs, habitat bottlenecks, and the St. Mary River Formation. In: Currie, P. J., and Koster, E. H., (eds.). Fourth Symposium on Mesozoic Terrestrial Ecosystems, Short Papers. Royal Tyrell Museum, Drumheller. Pp. 224–229.

Weishampel, D. B., and Jianu. C.-M., 2000. Plant-eaters and ghost lineages: dinosaurian herbivory revisited. In: Sues, H.-D., (ed.), The Evolution of Herbivory in Terrestrial Vertebrates. Perspectives from the Fossil Record. Cambridge University Press, Cambridge. Pp. 123–143.

Weishampel, D. B., Barrett, P. M., Coria, R. A., Le Loeuff, J., Xu, X., Zhao, X.-J., Sahni, A., Gomani, E. M. P. and Noto, C. R. 2004. Dinosaur distribution. In: Weishampel, D. B., Osmólska, H., and Dodson, P. (eds.). The Dinosauria, 2nd ed. University of California Press, Berkeley. Pp. 517–606.

Wilkinson, M. 1994. Common cladistic information and its consensus representation: Reduced Adams and cladistic consensus trees and profiles. Syst. Biol. 43: 343–368.

Wilson, J. A. 2002. Sauropod dinosaur phylogeny: critique and cladistic analysis. Zool. J. Linn. Soc. London 136: 217–276.

Wilson, J. A., and Carrano. M. T., 1999. Titanosaurs and the origin of 'wide gauge' trackways: a biomechanical and systematic perspective on sauropod locomotion. Paleobiology 25: 252–267.

Wilson, J. A., and Sereno, P. C., 1998. Early evolution and higher-level phylogeny of the sauropod dinosaurs. Mem. Soc. Vertebr. Paleontol. 5: 1–68.

FOUR

Sauropodomorph Diversity through Time

PALEOECOLOGICAL AND MACROEVOLUTIONARY IMPLICATIONS

Paul M. Barrett and Paul Upchurch

SAUROPODOMORPH DINOSAURS were incredibly successful animals, by any standard. They attained high levels of alpha-taxonomic diversity, with more than 100 valid genera known currently (Galton 1990; McIntosh 1990; Galton and Upchurch 2004; Upchurch et al. 2004) and were the dominant animals, in terms of numerical abundance and biomass, in many Middle and Late Mesozoic terrestrial biomes (e.g., Young 1951; Dodson et al. 1980; Russell et al. 1980; Galton 1986; Dong 1992).

Phylogenetic evidence, and the known stratigraphic distributions of these animals, suggests that the Sauropodomorpha originated in the early Late Triassic, though there are also recent reports of Middle Triassic prosauropod material (Flynn et al. 1999). Sauropodomorph monophyly is well established (Charig et al. 1965; Gauthier 1986; Benton 1990; Sereno 1997, 1999; Galton and Upchurch 2004), and early in its history the clade split into two major lineages: the Prosauropoda and Sauropoda. Although some authors regard the prosauropods as paraphyletic to the Sauropoda (Gauthier 1986; Benton 1990; Yates 2003), the current consensus is that both groups are monophyletic (Galton 1990; Gauffre 1993; Upchurch 1995, 1998; Sereno 1997, 1999; Wilson and Sereno 1998; Benton et al. 2000; Pisani et al. 2002; Galton and Upchurch 2004; Upchurch et al. 2004). Prosauropods attained a global distribution during the Late Triassic and Early Jurassic, before disappearing close to the Early–Middle Jurassic boundary (Galton 1990). Sauropods first appeared in the Late Triassic (Buffetaut et al. 2000) and achieved a global distribution by the Middle Jurassic, persisting thereafter on almost all of the major continental landmasses until the latest Cretaceous (McIntosh 1990; Weishampel 1990).

One of the reasons for this success may be the early adoption of facultative or obligate herbivory in the Sauropodomorpha. In analyses on living amniotes, Moore and Brooks (1995, 1996) demonstrated an evolutionary association between the origin of herbivory and amniote diversification by showing that clades of extant herbivorous amniotes (principally mammals) are, on average, 16.8 times more speciose than their respective carnivorous sister-groups. Comparisons between taxic diversity estimates for herbivorous and carnivorous dinosaurs throughout the Mesozoic have yielded similar results, with herbivores being between 3 and 10 times more diverse than carnivores at any one time (Barrett 1998). Such analyses, and the apparent rapid diversification of amniotes after the origin of herbivory in this clade during the Late Carboniferous, provide support for the suggestion that herbivory can be regarded as a "major adaptive zone" or a "key innovation" (e.g., Moore and Brooks 1995, 1996; Hotton et al. 1997; Reisz and Sues 2000). Adaptive zones represent particular discrete ways of life that, although they may be difficult to enter evolutionarily, allow the diversification of a group once the zone has been entered, whereas key innovations represent the acquisition of a particular character state or states that promote the radiation of a clade (e.g., Brooks and McLennan 1991).

Prosauropod feeding mechanisms appear to have been relatively uniform, with little variation in either the craniodental or the postcranial characters involved in the collection and processing of plant food (Galton 1985, 1986; Barrett 1998). Jaw actions were simple orthal movements; there was little or no tooth–tooth contact; oral processing of food was probably negligible; and fermentative digestion in elongate digestive tracts and/or the mechanical action of gastric mills seem to have been the dominant methods for breaking down plant food (Galton 1986; Farlow 1987; Norman and Weishampel 1991; Barrett 2000). Some prosauropods may have been omnivorous, while others were facultative or obligate herbivores (Barrett 2000). The long necks, large body size, and bipedal gaits of these animals further suggest that some were capable of browsing up to 5 m above ground level (Barrett 1998).

Until recently, sauropod feeding was also viewed as simple and stereotyped, with little variation in the feeding apparatus between taxa (e.g., Bakker 1971; Coombs 1975; Dodson 1990). However, detailed reconsideration of sauropod anatomy, phylogeny, and paleoecology has revealed a surprising diversity of form and function in the feeding mechanisms of these animals

(Fiorillo 1991, 1998; Barrett and Upchurch 1994, 1995; Calvo 1994; Christiansen 1999, 2000; Upchurch and Barrett 2000). All sauropods appear to have been obligate herbivores, and they experimented with a wide range of tooth types, occlusal patterns, and jaw actions; in addition, many taxa had specialised craniodental structures for food gathering, such as the presence of tooth combs. Sauropod postcranial adaptations were likewise variable, including differences in body size, absolute and relative neck length, neck function (lateral sweeping, dorsoventral flexion, or alternation between these modes), and limb proportions (related to specialization for high or low browsing) (Barrett and Upchurch 1994, 1995; Martin et al. 1998; Stevens and Parrish 1999; Upchurch and Barrett 2000). Use of gastric mills seems to have been important in some taxa, and fermentative digestion was also likely to have been important (Farlow 1987). Sauropod feeding mechanisms have a strong phylogenetic component, and different clades can be characterized in part by specific characteristics, or combinations of various characteristics, of the feeding apparatus (Upchurch and Barrett 2000; *contra* Sereno and Wilson 2001).

Several studies have noted various broad-scale changes in sauropodomorph diversity through time (Hunt et al. 1994; Barrett 1998; Weishampel and Jianu 2000; Barrett and Willis 2001; Upchurch and Barrett, chapter 3). Here we consider a number of potential macroevolutionary mechanisms that may account for these patterns, including various potential clade–clade interactions (coevolution and competition). We explore the possibility that the observed changes may be at least partially accounted for by fluctuations in the frequencies of different sauropodomorph feeding mechanisms through time.

MATERIALS AND METHODS

There are various techniques for estimating diversity through time (Smith 1994). Taxic diversity estimates (TDEs) for Prosauropoda and

Sauropoda were obtained by plotting stratigraphic range data for all known valid genera, as listed in the recent reviews by Galton and Upchurch (2004) and Upchurch et al. (2004), and summing the number of valid taxa for each European standard stage (see appendixes 4.1 and 4.2). These data are presented as a series of simple histograms. Genera were used, rather than species, as the majority of sauropodomorph taxa are monospecific (approximately 85% of sauropod genera and 75% of prosauropod genera are monotypic [Galton and Upchurch 2004; Upchurch et al. 2004] and there is, therefore, little difference between the species-level and the genus-level diversity curves. Moreover, the designation of fossil species is beset by a number of taxonomic problems—sexual dimorphism, individual variation, ontogenetic differences, and ecological differences being paramount among them—whereas differences between fossil genera are more substantial, with larger discontinuities in morphology between genera than between species. As a result, genera tend to be more stable than species and may be more amenable to diversity analysis, though it should be acknowledged that this represents only an approximation of species-level patterns (Raup and Boyajian 1988; King 1990).

TDEs have several advantages over phylogenetic diversity estimates (PDEs) in an analysis of this kind. TDEs reflect "real" empirical data and are not as dependent on the assumption of a correct phylogeny for any particular clade as PDEs. Moreover, PDEs often include fewer taxa for operational reasons: unstable taxa (e.g., those with large amounts of missing data or character conflict) are often pruned from analyses a posteriori to increase resolution, and computational requirements can limit the number of taxa included in an analysis. Conversely, TDEs can include all valid genera, even if these taxa are too poorly known to be included in a cladogram: poorly known taxa can be referred to clades on the basis of synapomorphies, even if their position cannot be fully resolved in the phylogeny. Finally, although PDEs can extract more information from the fossil record on diversity than TDEs (on the condition that as many taxa as possible have been included in the latter), we have shown elsewhere that the PDEs and TDEs for sauropods are generally in close agreement (Upchurch and Barrett, chapter 3). We believe, therefore, that it is possible to use TDEs in the current chapter without the ensuing diversity patterns being distorted by major preservational biases that are also not expressed in the PDEs. Nevertheless, it must be remembered that there are a number of problems inherent to the TDE approach (dependency on accurate taxonomy and stratigraphic ranges, problems with preservational biases, etc.) (see Upchurch and Barrett, this chapter 3).

Sauropod taxa were classified according to feeding mechanism, tooth morphology, and systematic position to explore changes in the frequencies of these categories through time. Sauropod feeding mechanisms are extremely varied and 11 functional complexes were recognized in this study. Many of these functional complexes are defined in Barrett and Upchurch (1995) and Upchurch and Barrett (2000), though additional research has resulted in the identification of several novel feeding mechanisms.

1. Cetiosaurid type: Broad-crowned (BC) teeth; interdigitating occlusion (mesial and distal wear facets); orthal jaw action; cropping of vegetation; intermediate FL/HL (forelimb length:hindlimb length) ratio (ca. 0.84); short (12 or 13 cervical vertebrae), laterally inflexible, dorsoventrally flexible neck; low- to medium-level browser (up to 5 m).

2. *Shunosaurus* type: BC teeth; interdigitating occlusion (mesial and distal wear facets, occasional apical wear); orthal jaw action; cropping of vegetation; very low FL/HL ratio (0.64); short (12 cervical vertebrae), laterally and dorsoventrally flexible neck; low- to medium-level browser (up to 4.5 m).

3. Higher eusauropod type: BC teeth; interdigitating occlusion (mesial and distal

wear facets); orthal jaw action; cropping of vegetation; intermediate Fl/HL ratios (0.85–0.88); elongate (up to 17 cervical vertebrae), dorsoventrally flexible, laterally inflexible necks; high browsers (up to 11 m).

4. Nemegtosaurid type: Narrow-crowned (NC) teeth; precision-shearing occlusion (mainly apical wear facets, occasional mesial and distal wear); ?propalinal jaw action; "nipping" of vegetation; postcrania unknown at present.

5. *Nigersaurus* type: NC teeth; ? precision-shearing occlusion (apical wear facets); transversely oriented tooth rows; postcranial material undescribed (Sereno et al. 1999; see Sereno and Wilson, chapter 5).

6. Dicraeosaurid type: NC teeth; ?propalinal jaw action; precision-shearing occlusion (mainly apical wear facets, occasional mesial and distal wear); "nipping" of vegetation; very low FL/HL ratio (0.65–0.70); short (11 or 12 cervical vertebrae), laterally flexible, dorsoventrally inflexible necks; specialist low browser (up to 4 m).

7. *Apatosaurus* type: NC teeth; propalinal jaw action; occlusion (apical wear facets); branch-stripping/nipping of vegetation; low FL/HL ratio (ca. 0.70); elongate (15 cervical vertebrae), laterally and dorsoventrally flexible neck; tripodal ability; low and medium browsing (up to 8 m) (Connely and Hawley 1998; Stevens and Parrish 1999; P.M.B. and P.U., pers. obs.).

8. *Diplodocus* type: NC teeth; propalinal jaw action; lack of occlusion; branch-stripping; low FL/HL ratio (ca. 0.70); elongate (15 or 16 cervical vertebrae), laterally and dorsoventrally flexible neck; tripodal ability; low, medium, and high browsing (up to 12 m).

9. *Camarasaurus* type: BC teeth; interdigitating occlusion (mesial and distal wear facets); propalinal jaw action; cropping of vegetation; intermediate FL/HL ratio (0.79–0.87); short (12 cervical vertebrae), laterally inflexible and dorsoventrally flexible neck; medium to high browsing (up to 9 m).

10. Brachiosaurid type: BC teeth (but approaching the NC condition); precision-shearing occlusion (apical wear facets with occasional mesial and distal wear facets); orthal jaw action; cropping and "nipping" of vegetation; high FL/HL ratio (1.1); long (12 or 13 extremely elongate cervicals), laterally inflexible, dorsoventrally flexible neck; high browsing (up to 14 m).

11. Titanosaur type (additional postcranial data from Curry Rogers and Forster 2001; see Curry Rogers, chapter 2): NC teeth; precision-shearing occlusion (apical wear facets); orthal jaw action; "nipping" of vegetation; intermediate FL/HL ratios (0.75–0.83); 13 to 16 cervical vertebrae present, suggesting some variation in neck function within the group. Titanosaur anatomy is very poorly known at present: it is likely that new discoveries will significantly alter our interpretations of the feeding mechanisms utilised by this clade.

Taxa were assigned to these categories on the basis of the characters listed above or on phylogenetic grounds, following mapping of these characters on to the sauropod cladogram (cf. Upchurch and Barrett 2000; phylogeny after Upchurch et al. 2004). In cases where character optimization was ambiguous (i.e., where possession of one of two different feeding systems was equally likely), taxa were not assigned a feeding mechanism and were excluded from the feeding related parts of the analysis.

BC teeth are defined as those with slenderness indices (SIs) of ≤4.0, whereas NC teeth have a SI ≥4.0. SI is calculated by dividing the apicobasal length of a tooth crown by its mesiodistal width (Upchurch 1998). In addition, BC teeth have a "spatulate" (D-shaped) cross section; NC teeth, in contrast, have a cylindrical or subcylindrical cross section. This division reflects a fundamental difference in the feeding mechanism: BC genera generally crop vegetation, whereas those taxa with NC teeth rake or "nip" vegetation (though there is great variety in the exact mechanism; see above). Where tooth morphology is unknown, as a result of poor preservation, genera were assigned to one of these categories on the basis of their phylogenetic position (see appendix 4.2).

Prosauropoda was not subdivided due to the apparent uniformity of the feeding apparatus throughout the group.

A full list of the genera included in the TDEs, along with information on the stratigraphic range, systematic position and feeding mechanism of each genus, is included in appendices 4.1 and 4.2.

RESULTS: SAUROPODOMORPH DIVERSITY THROUGH TIME

SAUROPODOMORPH DIVERSITY PATTERNS

The earliest prosauropod (*Saturnalia*) appears in the middle Carnian (Langer et al. 1999). Thereafter, the group radiates rapidly and seven genera are present during the Norian (fig. 4.1A). The Rhaetian witnesses a rapid drop in the TDE, to three genera, but following this trough prosauropod diversity apparently increases suddenly again and reaches its acme (a peak composed of nine genera) in the early Sinemurian. During the remainder of the Early Jurassic, prosauropod taxic diversity declines in a stepwise manner to four genera in the late Pliensbachian. Prosauropods apparently became extinct close to the Early/Middle Jurassic boundary: their remains have not been identified in any deposits younger than Pliensbachian (Weishampel et al. 2004). If the TDE reflects the "real" diversity of these animals, it suggests that prosauropod decline was relatively gradual until the late Pliensbachian, at which time all remaining taxa went extinct (fig. 4.1A).

Potential problems with this TDE include uncertainties in the dating of several key genera. Although the dates for many European and South American forms are apparently well constrained, those from southern Africa, North America and China can often be referred only to broad stratigraphic intervals. This problem is particularly acute in the Early Jurassic. For example, the age of the North American East Coast prosauropod fauna (Galton 1976) is poorly constrained by radiometric and biostratigraphical data and is currently considered to extend over the duration of the Hettangian–Pliensbachian. Similarly, many other Early Jurassic prosauropods have ranges that extend over two or more stratigraphic stages. This imprecision may have artificially "smeared" the tail end of the TDE, obscuring its true shape. Furthermore, it is unclear, at present, whether the Rhaetian reduction in prosauropod diversity (fig. 4.1A) represents a "real" evolutionary event or a preservational bias: fewer prosauropod-bearing formations are known from the Rhaetian than from either the Norian or the lowermost Jurassic (see Upchurch and Barrett, chapter 3). A PDE, which could help to test these differing hypotheses, has yet to be calculated for prosauropods. For the time being, we cautiously accept the results of the TDE but are mindful that it contains a number of potential pitfalls.

The sauropod TDE and its potential shortcomings are discussed in detail in chapter 3 so are not dwelt on here. In outline, sauropods first appear in the late Carnian (*Blikanasaurus* [Galton and Upchurch 2004; Upchurch et al. 2004]) and are rare in Late Triassic and Early Jurassic deposits, though they exhibit a slight increase in diversity during this interval (fig 4.1B). Sauropod diversification accelerates in the Middle Jurassic, reaching a peak of 12 genera in the middle Callovian before falling off in the Oxfordian. A second peak in the TDE occurs during the Kimmeridgian and Tithonian, representing the acme of sauropod diversity, with 26 valid genera. An apparent extinction event at the Jurassic/Cretaceous boundary results in a massive drop in diversity (though the PDE suggests that this drop is not as marked as that suggested by the TDE; see chapter 3), which gradually

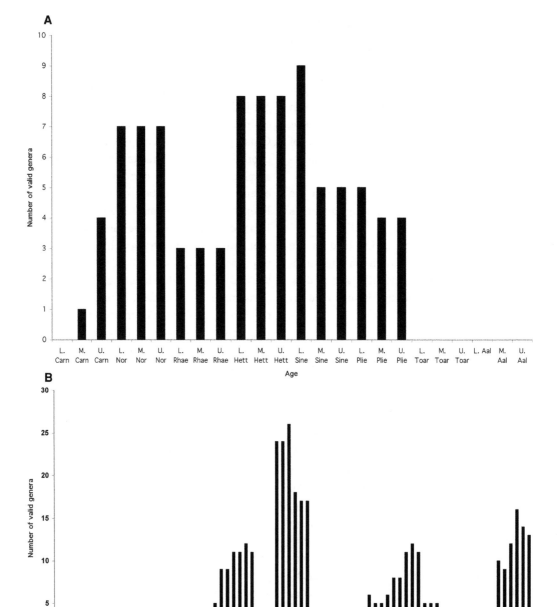

FIGURE 4.1. Sauropodomorph taxic diversity estimates (TDEs). (A) TDE for the Prosauropoda (based on data from Galton and Upchurch [2004] and Weishampel et al. [2004]; see appendix 4.1). $N = 20$. (B) TDE for the Sauropoda (based on data from Upchurch et al. [2004] and Weishampel et al. [2004]; see appendix 4.2). $N = 98$. (C) Plot of the relative abundances (expressed as proportions of the total number of sauropodomorph genera) of prosauropods and sauropods in the Late Triassic and Early Jurassic. See text for description and further details. Aal, Aalenian; Alb, Albian; Apt, Aptian; Baj, Bajocian; Barr, Barremian; Bath, Bathonian; Berr, Berriasian; Call, Callovian; Camp, Campanian; Carn, Carnian; Ceno, Cenomanian; Con, Coniacian; Hauv, Hauterivian; Hett, Hettangian; Kimm, Kimmeridgian; L, lower; M, middle; Maas, Maastrichtian; Nor, Norian; Oxf, Oxfordian; Plie, Pliensbachian; Rhae, Rhaetian; Sant, Santonian; Sine, Sinemurian; Tith, Tithonian; Toar, Toarcian; Turo, Turonian; U, upper; Val, Valanginian.

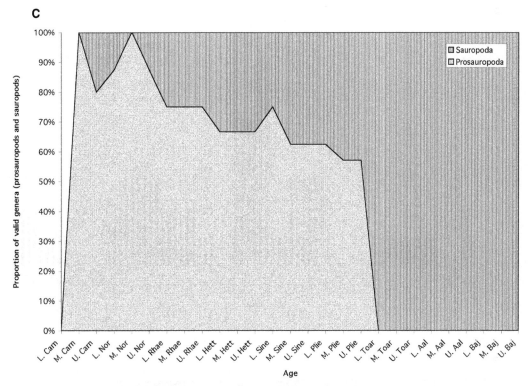

FIGURE 4.1. (Continued)

recovers through the Early Cretaceous to reach a third peak (of 12 genera) in the Albian. This is followed by another apparent crash in diversity in the early Late Cretaceous (also reflected in the PDE; see chapter 3) before a final radiation in the Campanian and Maastrichtian that resulted in a peak of 16 genera in the lower Maastrichtian (fig. 4.1B).

There was considerable temporal overlap between prosauropods and sauropods, extending from the late Carnian to the end of the Pliensbachian. This is not surprising, as phylogenies that recovered a monophyletic Prosauropoda predict that sauropod origin must be at least as far back in time as the first appearance of prosauropods.

FEEDING MECHANISM DIVERSITY

Lack of direct information on the feeding mechanisms of various taxa (e.g., basal sauropods, *Bellusaurus*) results in the zero frequency values for sections of the histogram in the Late Triassic, early Middle Jurassic (Toarcian–middle Bathonian), and basal Cretaceous (Berriasian–middle Valanginian) (fig. 4.2A), though the TDE and PDE clearly demonstrate the continuity of several sauropod lineages through each of these intervals (Upchurch and Barrett, chapter 3). Absence of valid genera from the Turonian and Coniacian stages accounts for the zero value of the frequency histogram at this point. These deficits have been partially compensated for by extrapolating the stratigraphic ranges of the various feeding mechanisms so that they are continuously present from their earliest to their latest known occurrences. For example, the presence of titanosaurs in the Turonian can be inferred from their presence in the Cenomanian and Campanian, and their feeding mechanism can be added to the frequency curve shown in figure 4.2A. Future work on sauropod phylogeny (with the inclusion of more taxa) and the resulting PDEs for component sauropod clades (e.g., Titanosauria,

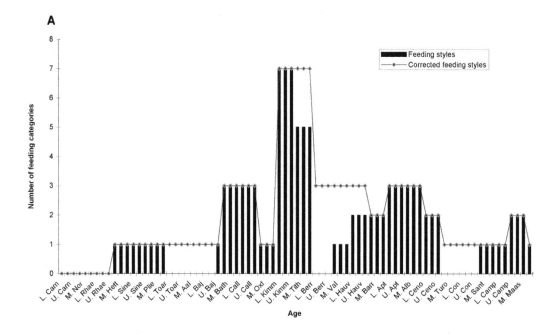
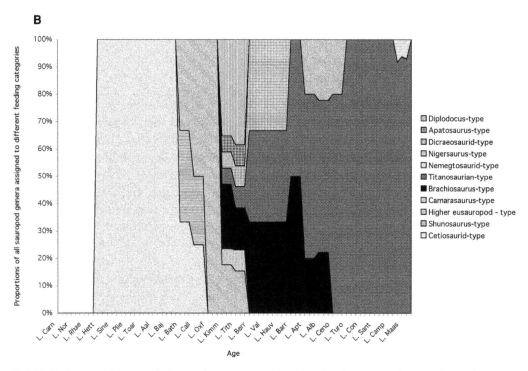

FIGURE 4.2. Diversity of sauropod feeding mechanisms through time (based on data given in the text and appendix 4.2). (A) Frequency of different sauropod feeding mechanisms through time, with a partial correction to account for missing data (see text for details). (B) Plot of the relative abundances (as proportions of the total number of sauropod genera present) of the various sauropod feeding mechanisms through time. Total number of sauropod genera represented = 61 (this reflects the lack of information, anatomical or phylogenetic, on the feeding mechanisms of many taxa). Abbreviations as in the legend to figure 4.1.

Cetiosauridae) may also help to close these temporal gaps.

In general terms, the frequency histogram for different sauropod feeding mechanisms through time is similar in shape to the sauropod TDE (fig. 4.2A). There is a strong, statistically significant, positive correlation between sauropod taxic diversity and the number of feeding mechanisms at any one point in time ($r^2 = 0.759$, $p < 0.0001$). Feeding mechanism diversity is low in the Early Jurassic, being represented by only a single category (*Cetiosaurus* type), though this paucity may simply reflect our ignorance of early sauropod craniodental adaptations. The Middle Jurassic witnesses an increase in feeding mechanism diversity, to three categories (*Cetiosaurus* type, *Shunosaurus* type, and higher eusauropod type), but this is followed by a sharp decrease in diversity during the Oxfordian. Both the TDE and the PDE indicate that the Oxfordian may have been a genuine low point in sauropod diversity (Upchurch and Barrett, chapter 3), so this corresponding drop in the number of different feeding mechanisms at this time is not surprising. Feeding diversity reached its maximum in the Late Jurassic, at which time up to seven different mechanisms are present (higher eusauropod type, *Camarasaurus* type, brachiosaurid type, titanosaur type, *Diplodocus* type, *Apatosaurus* type, and dicraeosaurid type). The basal Cretaceous is marked by an apparent crash in diversity that is exacerbated by the current lack of information on the feeding mechanism of the only valid Berriasian genus (*Jobaria*). Nevertheless, both the sauropod PDE (Upchurch and Barrett, chapter 3) and the extrapolation of feeding mechanism stratigraphic ranges indicates that several sauropod lineages were present at this time and that a minimum of three feeding categories was probably in place (brachiosaurid type, titanosaur type, and dicraeosaurid type). The extrapolated frequency curve suggests that the same level of feeding mechanism diversity was maintained throughout much of the Early Cretaceous, though the actual number of feeding categories observed on empirical evidence generally remains below this level until the Aptian–Albian. The extinction of dicraeosaurids at the end of the Hauterivian causes a drop in predicted diversity during the Barremian, but the appearance of the *Nigersaurus*-type feeding mechanism in the Aptian restores feeding diversity levels to three categories.

Late Cretaceous feeding diversity does not track the sauropod TDE. Cenomanian–Santonian sauropod faunas decrease in feeding mechanism diversity, with the loss of the brachiosaurid-type and then the *Nigersaurus*-type categories. A small increase in feeding diversity occurs in the late Campanian–middle Maastrichtian, reflecting the appearance of the nemegtosaurid-type feeding mechanism. It is possible that the latter had an extensive history prior to its first appearance, as a long ghost lineage is inferred if Nemegtosauridae occupies a basal position within Diplodocoidea (Upchurch 1999; Upchurch et al. 2004). Only the titanosaur-type feeding mechanism continues through to the end of the Cretaceous. Although the sauropod TDE is high in the Late Cretaceous, almost all of this diversity is confined to a single clade (Titanosauria), accounting for the low diversity of feeding categories at this time. It should be noted that alternative phylogenies place Nemegtosauridae within Titanosauria (Salgado and Calvo 1997; Wilson and Sereno 1998; Curry Rogers and Forster 2001; Wilson 2002), but this does not currently affect the number of sauropod feeding categories in the Late Cretaceous. Existing data on nemegtosaurid and titanosaur feeding mechanisms warrant separation between these categories for the time being, though it is recognized that new material from either clade may significantly alter this conclusion.

The relative abundances of each feeding category also vary through time (fig. 4.2B). The *Shunosaurus*-type, *Apatosaurus*-type, and *Camarasaurus*-type feeding categories are all restricted in stratigraphic range and are currently limited to a single genus, but each of these categories may gain more members when more is known about the feeding mechanisms in various basal sauropod, diplodocoid, and basal macronarian lineages. Cetiosaurid-type feeding mechanisms are the most abundant in the Early and early Middle Jurassic, but these are "replaced" by *Shunosaurus*-type and "higher" eusauropod-type feeding mechanisms in the late Middle Jurassic and early Late Jurassic. (Diplodocoid ['*Cetiosaurus*' *glymptonensis*] and basal titanosauriform (*Lapparentosaurus*) lineages were also present at this time, but their feeding mechanisms cannot be assessed with confidence on the basis of the criteria outlined above and, so, do not appear in the graph). During the Late Jurassic, taxa with *Diplodocus*-type and brachiosaurid-type feeding mechanisms are the most abundant: the former disappear in the basal Cretaceous, however, whereas the latter continue to be moderately important until the end of the Albian. Sauropods with titanosaur-type and dicraeosaurid-type mechanisms are also important in the Early Cretaceous, though it should be remembered that all feeding categories in this interval are based on a very small sample of valid genera. Titanosaur-type feeding mechanisms increase in relative diversity through the latter part of the Early Cretaceous, becoming the dominant category in the Albian. *Nigersaurus*-type sauropods also make a moderate contribution to late Early and early Late Cretaceous feeding diversity. Most of the Late Cretaceous is dominated by titanosaur-type feeding mechanisms, though there is a minor contribution from nemegtosaurid-type genera in the late Campanian–middle Maastrichtian.

OCCURRENCE OF TAXA WITH EITHER NARROW-CROWNED OR BROAD-CROWNED TEETH

During the Late Triassic to middle Bathonian the sauropod TDE is composed exclusively of taxa with spatulate, BC teeth (basal sauropods, cetiosaurids, and other non-neosauropod eusauropod lineages) (fig. 4.3A). The earliest probable NC form appears in the late Bathonian (the diplodocoid '*Cetiosaurus*' *glymptonensis*). Thereafter, NC forms rapidly increase in both relative and absolute diversity, with the appearance of several diplodocoid and titanosaur genera in the late Middle Jurassic and Late Jurassic. In the same interval, BC sauropods (non-neosauropod sauropods, basal macronarians, basal titanosauriforms) decline in relative diversity, though their absolute diversity actually increases slightly, and by the Kimmeridgian, NC and BC sauropods each account for approximately 50% of the total sauropod diversity (fig. 4.3A, B). BC and NC sauropods decline in diversity from the Tithonian to the Barremian (reflecting the overall TDE), but the BC sauropods suffer a greater loss of genera (e.g., the extinction of all non-neosauropod eusauropods) than the NC forms, so that in relative terms NC sauropods increase in importance during this period (apart from an isolated peak in BC diversity in the Barremian, at which time NC sauropods appear to be rare). The increase in the TDE during the Aptian–Albian is composed mainly of NC taxa (rebbachisaurids, titanosaurs), which account for approximately 75% of the sauropod fauna. The last BC sauropods (all of which are basal titanosauriforms) disappear by the end of the Early Cretaceous, and the Late Cretaceous peak in the TDE is composed solely of NC taxa (fig. 4.3A, B).

Thus, sauropod faunas are initially composed of BC genera; there is an extended period of overlap between NC and BC genera during the late Middle Jurassic–Early Cretaceous; BC forms gradually decline in importance in the Early Cretaceous; and this trend culminates in Late Cretaceous faunas that contain only NC genera.

COMPARISON BETWEEN NARROW-CROWNED CLADES (DIPLODOCOIDEA AND TITANOSAURIA)

The radiation of diplodocoids begins in the late Bathonian, but the group remains poorly represented in the Callovian–Oxfordian interval (fig. 4.4A). The Kimmeridgian heralds a dramatic radiation of diplodocoid taxa, including diplodocids (e.g., *Diplodocus*, *Tornieria*), dicraeosaurids (*Dicraeosaurus*), and a variety of

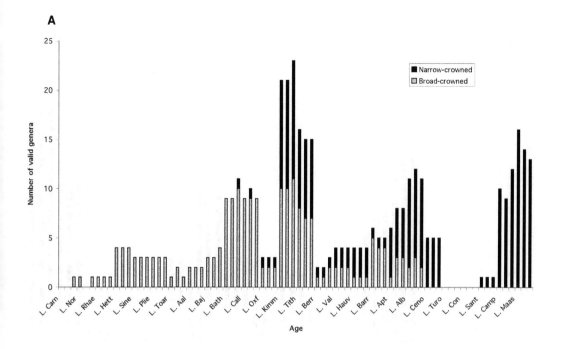

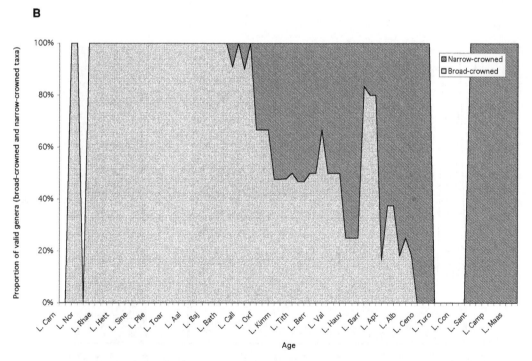

FIGURE 4.3. Comparison of the diversities of sauropods with broad-crowned (BC) versus narrow-crowned (NC) teeth (raw data given in appendix 4.2). (A) Abundances of BC and NC sauropods through time. Total number of genera included = 92 (reflecting lack of information, both anatomical and phylogenetic, on tooth morphology in a number of genera). (B) Plot of the relative abundances (expressed as proportions of the total number of sauropod genera) of sauropod taxa with either BC or NC teeth through time. Abbreviations as in the legend to figure 4.1.

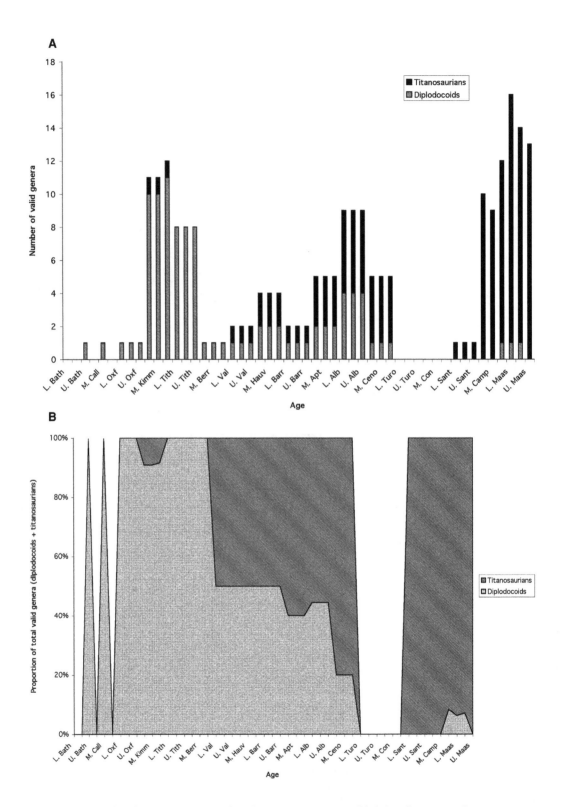

FIGURE 4.4. Diplodocoid and titanosaur TDEs through time. (A) Frequencies of diplodocoid ($N = 19$) and titanosaur ($N = 36$) genera (see appendix 4.2). (B) Plot of the relative abundances of the two clades (expressed as proportions of the total number of diplodocoid + titanosaur genera) through time. Abbreviations as in the legend to figure 4.1.

diplodocoids of uncertain systematic position (e.g., *Amphicoelias*), with up to 11 valid genera present at this time. This represents the peak of diplodocoid diversity: thereafter, diversity declines steeply in the basal Cretaceous. The late Early Cretaceous sees a moderate recovery, with the appearance of rebbachisaurids, but only four valid diplodocoid genera are present in the Albian interval. Diplodocoids are apparently absent in the early Late Cretaceous, though the appearance of the Nemegtosauridae in the late Campanian indicates the presence of an extensive ghost lineage (Upchurch 1999; chapter 3; but see comments above relating to the possible systematic position of nemegtosaurids).

The earliest confirmed titanosaur is *Janenschia* from the Kimmeridgian of Tanzania. Titanosaur diversity apparently remains very low through the Late Jurassic and basal Cretaceous: few valid genera are known at this time, though phylogeny predicts the continuity of one or two lineages through this interval (Upchurch and Barrett, chapter 3). The group has a modest radiation in the late Early Cretaceous, reaching a maximum of five genera in the Albian, but dwindles rapidly during the early Late Cretaceous. An extensive, and apparently rapid, radiation occurs in the Campanian and Maastrichtian that is composed almost exclusively of derived titanosaurs (fig. 4.4A).

Comparison of the TDEs for each of these clades demonstrates that they coexisted for most of their respective histories but that diplodocoids were more diverse than titanosaurs (in both relative and absolute terms) during the Late Jurassic, whereas the converse is true in the Late Cretaceous. Only in the late Early Cretaceous do the two clades closely approach each other in diversity (fig. 4.4B).

DISCUSSION

Fluctuations in clade diversity are often assumed to have some macroevolutionary significance, and many studies have attempted to identify abiotic or biotic factors that may influence or control such phenomena (e.g., Valentine 1985; Jablonski et al. 1996). Possible controlling factors include large-scale environmental and tectonic perturbations (e.g., climate change, continental fragmentation, changes in atmospheric composition) or modifications to community structure and dynamics (e.g., appearance of new predators, loss of prey species, competitive exclusion). Sauropodomorph diversity patterns have been largely exempt from such consideration, apart from examination of the role of prosauropods in the initial stages of the dinosaur radiation during the Late Triassic (e.g., Benton 1983, 1986). Nevertheless, several recent studies have attempted to rectify this deficit (Haubold 1990; Hunt et al. 1994; Weishampel and Jianu 2000; Barrett and Willis 2001; Upchurch and Barrett, chapter 3).

Haubold (1990) and Hunt et al. (1994) have proposed that the fluctuations in sauropod TDEs are not a direct product of evolutionary or environmental processes, but nothing more than taphonomic artifacts under the control of changes in global sea level that affect the preservation potential of sauropod remains. It has also been suggested that sea level had a more direct influence on dinosaur diversity in terms of the consequent geographical separation or confluence of terrestrial habitats, either by promoting allopatric speciation or as a result of biotic mixing (e.g., Bakker, 1977; Weishampel and Horner 1987). However, many of the suggested correlations between sauropod diversity patterns and sea level changes collapse when PDEs are considered (Upchurch and Barrett, chapter 3). Here we evaluate the possible roles of several biological processes that may have influenced sauropod diversity through time.

FEEDING MECHANISMS AND DIVERSITY

The strong positive correlation between sauropod taxic diversity and the number of different feeding mechanisms present at any one time confirms previous suggestions that feeding ecology played a central role in sauropod evolutionary history (figs. 4.1B, 4.2A, B) (Barrett 1998; Upchurch and Barrett 2000). The strong linkage between the Late Jurassic radiation of sauropods and the diversification of feeding mechanisms, and the coincident decline in the sauropod TDE and in feeding mechanism diversity during the Early Cretaceous, makes it tempting to speculate that there was a direct causal link between these various phenomena.

These observations are also concordant with the suggestion that niche partitioning, on the basis of differences in feeding ecology, was important in faunas that contained a variety of sympatric sauropod taxa (Fiorillo 1991, 1998; Barrett and Upchurch 1995; Upchurch and Barrett 2000). The peaks in the sauropod TDE and in feeding mechanism diversity coincide in the Late Jurassic at a time when the majority of well-known dinosaur faunas, such as those of the Morrison Formation (USA), the Tendaguru Beds (Tanzania), and the Upper Shaximiao Formation (China), contain a melange of sauropods from a number of clades, which possess a number of different feeding mechanisms. The decline in the TDE and the reduction in feeding mechanism diversity in the Early Cretaceous are concurrent with a change in sauropod faunas at this time: with rare exceptions these faunas now contain only one or two sauropod genera and a correspondingly low diversity of feeding ecologies (Weishampel et al. 2004).

The correlation between generic-level diversity and feeding mechanism variety does not continue into the Late Cretaceous, however (Figs. 4.1B, 4.2A, B). During this period there is a significant radiation of titanosaurs, culminating in a maximum of 15 genera, which is supplemented by the appearance of the Nemegtosauridae. This represents the second highest peak in sauropod diversity: however, it appears that only two feeding mechanisms underpin this radiation. The co-occurrence of several titanosaur genera in South American, Indian, and Madagascan faunas (Curry Rogers and Forster 2001; Weishampel et al. 2004) suggests that niche partitioning has occurred in these instances despite the paucity of different feeding mechanisms. This may simply reflect our ignorance of the variation in the titanosaur feeding apparatus: lack of information on both the skull and the postcrania of many genera may be obscuring a diversity of feeding mechanisms within the group. For example, complete cervical series are unknown in many titanosaurs, so basic comparative data on neck flexibility and length are wanting. Moreover, the dearth of partially complete crania confounds the identification of any variation in the feeding apparatus. Alternatively, a number of other possible mechanisms may permit niche partitioning among sympatric titanosaur genera, including differences in adult body mass (which would affect browse height, assimilation rates, mobility, population sizes, etc.) and differing dietary preferences.

COMPETITIVE REPLACEMENTS: PREVIOUS WORK

Although the role of competitive replacement in macroevolution has been the subject of intense debate, there is currently little consensus on its frequency or on its importance (relative to other biological and abiological processes) in controlling clade diversity over geological time scales (e.g., Benton 1996; Sepkoski 1996). A recent attempt to quantify the prevalence of competitive replacements in the evolutionary history of tetrapods recognized a number of "candidate competitive replacements" (CCRs) and compared the number of familial originations that could have been the result of a competitive replacement with the total number of origination events through time (Benton 1996). CCRs were identified where two tetrapod families displayed (1) an overlap in geographical and stratigraphic ranges and (2) a similar mode of life (in terms of body mass, diet, and habitat). Similar criteria for recognizing potential competitive replacements were also advocated by Sepkoski (1996).

Of the CCRs proposed by Benton (1996), three involve sauropodomorph clades: replacement of Melanorosauridae by the Vulcanodontidae and Massospondylidae in the Hettangian; replacement of Camarasauridae by the Cetiosauridae in the Oxfordian; and replacement of Cetiosauridae by the Titanosauridae in the Kimmeridgian. However, the identifications of these CCRs are founded on a database (Benton 1993) that contains a large number of taxonomic and systematic errors. According to Benton (1993), data on sauropodomorph families were taken from Weishampel (1990), whose taxonomy was based in turn on the compilations of Galton (1990; on prosauropods) and McIntosh (1990; on sauropods). Subsequent work on sauropodomorph taxonomy and systematics has resulted in a large number of changes to the classifications proposed by Galton and McIntosh (e.g., Upchurch 1995, 1998; Wilson and Sereno 1998; Wilson 2002; Galton and Upchurch 2004; Upchurch et al. 2004) that impact directly on Benton's (1996) proposed CCRs. Each CCR is dealt with in turn, below.

MELANOROSAURIDAE VS. VULCANODONTIDAE/ MASSOSPONDYLIDAE

'Vulcanodontidae' is a paraphyletic grouping of basal sauropods from the Early Jurassic (Upchurch 1995). As such, its geographical and stratigraphic ranges cannot be compared to those of a monophyletic clade, as these attributes of a paraphyletic taxon are dependent on a subjective choice on which taxa to include within it. Melanorosauridae is currently considered to be a monophyletic clade, but it does not include the Chinese form *Lufengosaurus* (Galton 1990; Galton and Upchurch 2004; *contra* Benton 1993). As a result, this clade ranges in time from the ?late Carnian to the Rhaetian but does not extend into the Early Jurassic as suggested by Benton (1993).

Therefore, melanorosaurids could not have interacted with the Early Jurassic genus *Massospondylus* (the only member of the monotypic Massospondylidae) or with any of the basal sauropods included in the 'Vulcanodontidae' (*sensu* Benton 1993). The validity of this CCR is, therefore, highly doubtful.

CAMARASAURIDAE VS. CETIOSAURIDAE

The Camarasauridae and Cetiosauridae of McIntosh (1990) and Benton (1993) have been demonstrated to be artificial groupings of sauropods with diverse taxonomic affiliations (Upchurch 1995, 1998; Wilson and Sereno 1998; Wilson 2002; Upchurch et al. 2004). For example, *Euhelopus* and *Opisthocoelicaudia* were once considered to be "camarasaurids" but have since been realised to be only distantly related to *Camarasaurus*: *Euhelopus* is either a non-neosauropod or a titanosauriform, whereas *Opisthocoelicaudia* is a titanosaur (Upchurch 1995; Wilson 2002). Similarly, the 'cetiosaurid' *Haplocanthosaurus* is currently considered to be a basal macronarian or diplodocoid. As both groups were so poorly defined and characterized, any comparison between their geographical and their stratigraphic ranges becomes somewhat meaningless. Furthermore, as these groups (*sensu* McIntosh 1990; Benton 1993) contain animals with different feeding mechanisms, they cannot be regarded as ecologically homogeneous and so mode of life comparisons between them are also undermined.

Current classifications, which reflect the topology of sauropod phylogenetic trees, restrict membership of the Camarasauridae to the Late Jurassic North American genus *Camarasaurus*. Use of the term Cetiosauridae has also been limited, so that it refers only to a monophyletic clade that includes the Early and Middle Jurassic genera *Barapasaurus* (India), *Cetiosaurus* (UK), and *Patagosaurus* (Argentina) (Upchurch et al. 2004). As the spatial and temporal distributions of these families did not coincide, it is unlikely that any competitive interactions could have occurred between them and this CCR is rejected.

CETIOSAURIDAE VS. TITANOSAURIDAE

The more restrictive definition of Cetiosauridae (see above) removes the temporal and spatial overlap between this clade and the Titanosauridae. The last cetiosaurid (*Patagosaurus*: Argentina) becomes extinct at the end of the Callovian, well before the first appearance of the first titanosaur (*Janenschia*: East Africa). Consequently, the conditions for a potential competitive replacement event are not met.

There has been some debate regarding the effects of taxonomic revision on estimates of clade diversity (e.g., Patterson and Smith 1987; Sepkoski and Kendrick 1993; Robeck et al. 2000). The foregoing examples from sauropods demonstrate that revision of a group and removal of para- and polyphyletic taxa can indeed have a major impact on the interpretation of the diversity patterns obtained from such studies. Examination of sauropod diversity within a rigorous phylogenetic framework has fatally undermined several perceived, but erroneous, patterns that were based on more "traditional" noncladistic classifications. Although assessment of the above-mentioned CCRs suggests that competitive exclusion was not an important process in sauropodomorph evolutionary history (see also Benton 1996), several patterns revealed by the prosauropod and sauropod TDEs are suggestive of possible competitive replacement events (see below).

COMPETITION BETWEEN PROSAUROPODS AND SAUROPODS?

A growing body of evidence indicates that there was considerable overlap between the stratigraphic and geographical ranges of prosauropods and sauropods during the Late Triassic and Early Jurassic (fig. 4.1). In the Late Triassic, remains of the sauropod *Isanosaurus* (late Norian–Rhaetian, Thailand [Buffetaut et al. 2000]) were recovered from the same unit as indeterminate prosauropod material (Buffetaut et al. 1995); similarly, the sauropod *Blikanasaurus* (?late Carnian–?early Norian, South Africa [Upchurch et al. 2004]) was a member of the diverse Lower Elliot Formation dinosaur fauna, which also included a minimum of two prosauropod genera (*Euskelosaurus* and *Melanorosaurus* [Weishampel et al. 2004]). The association between prosauropods and sauropods continues in the Early Jurassic, particularly in China, where the prosauropods of the Lower Lufeng Formation and adjacent penecontemporaneous faunas (Hettangian–Pliensbachian [Young 1951; Dong 1992; Luo and Wu 1994; Barrett 2000]) lived alongside a variety of early sauropods (Dong et al. 1983; He et al. 1998; Barrett 1999, 2000; P.M.B. and P.U., pers. obs.). In southern Africa, the prosauropod *Massospondylus* was a contemporary of the sauropod *Vulcanodon*, though remains of both animals have yet to be recovered from the same locality (Weishampel et al. 2004). Furthermore, there is overlap between the modes of life of the various members of these two clades. Sauropods were obligate herbivores, while many prosauropods were at least facultative herbivores (Farlow 1987; Barrett 2000; Upchurch and Barrett 2000) and members of both clades are found in the same deposits, indicating that they shared the same habitats (see above). There is also significant overlap in body size: prosauropods range from approximately 1.5 m (*Saturnalia* [Langer et al. 1999]) to 10 m (*Riojasaurus* [Bonaparte 1972]) in length, whereas the earliest sauropods occupied the size range from 5 m (*Blikanasaurus* [Galton 1990]) to 14 m (*Gongxianosaurus* [He et al. 1998]).

Prosauropods were the dominant vertebrates of Late Triassic and Early Jurassic terrestrial faunas in terms of biomass, and it has been estimated that in some faunas they accounted for up to 95% of the standing crop (Galton 1985). In contrast, sauropods were rare during this time interval and the majority of genera are represented only by fragmentary remains. However, comparison of prosauropod and sauropod TDEs demonstrates that during the Early Jurassic the number of prosauropod genera steadily declined while sauropod diversity remained roughly constant (fig. 4.1). In terms of relative diversity, prosauropods gradually decreased in importance from the Norian (when they made up close to 100% of the sauropodomorph fauna) to the point immediately before their final extinction at the Pliensbachian–Toarcian boundary (at which time they constituted approximately 60% of sauropodomorph taxa). Conversely, sauropods increased in relative diversity during the same interval, from close to 0% to approximately 40% of the sauropodomorph fauna. Such a pattern is strongly reminiscent of the classic "double-wedge" (cf. Benton 1983) that is the sine qua non of competitive replacements (fig. 4.1C).

The evidence presented above demonstrates that prosauropods and sauropods fulfill all of the criteria proposed by Benton (1996) for the recognition of a CCR: therefore, it is possible that the early stages of the sauropod radiation may have driven prosauropod tabefaction (see also Barrett 1998). The small number of taxa involved makes it difficult to assign this CCR to either a "full competitive replacement" or an "incumbent replacement" model (cf. Benton 1996). Nevertheless, both of these models assume that possession of a "key innovation" gives the "successful" incoming group a competitive advantage over the "unsuccessful" outgoing group. In this case, convergence in mode of life does suggest that prosauropods and sauropods may have been competing directly for the same resources over an extended time scale. Although the modes of life of these animals apparently overlapped considerably, the development of a systematic occlusion (extensive tooth–tooth contact) may have been a "key innovation" that could have given sauropods a competitive advantage over prosauropods in the Early Jurassic. No prosauropod is known to have employed tooth–tooth contact in the collection or mechanical breakdown of plant food: although such a mechanism had been proposed for *Yunnanosaurus* (Galton 1985), reexamination of the pertinent specimens demonstrated that this was not the case (Salgado and Calvo 1997; Wilson and Sereno 1998; Barrett 1999, 2000; P.M.B. and P.U., pers. obs.). In contrast, one unnamed basal sauropod from the Lower Lufeng Formation of China possesses teeth that have well-developed high-angled mesial and distal wear facets (Barrett 1999, 2000); *Gongxianosaurus* also appears to display tooth–tooth wear (He et al. 1998). Occlusion would have permitted a higher degree of oral processing of plant food and/or may have opened up food sources that were unavailable to animals that lacked it. Unfortunately, the craniodental material of other basal sauropods is not sufficiently well preserved to confirm the presence or absence of occlusion. However, other basal sauropods did acquire novel features in the feeding apparatus that are absent from prosauropods, including the development of a lateral plate and various modifications to tooth crown morphology (Barrett 1999; Upchurch and Barrett 2000): these indicate that sauropod feeding mechanisms were somewhat different from those of prosauropods.

Several authors have suggested that prosauropods may have been paraphyletic with respect to sauropods (Gauthier 1986; Benton 1990; Yates 2003), but this does not significantly affect the proposed prosauropod/sauropod interactions outlined above. As prosauropod feeding mechanisms are so conservative, they can all be accommodated within a single feeding category: consequently, the only correction that would need to be applied to the discussion above would be to regard prosauropods as a paraphyletic grade, rather than a monophyletic clade.

Although differences in the feeding apparatus may have provided sauropods with an evolutionary advantage, these suggestions must be regarded as tentative: the small number of taxa involved precludes rigorous statistical testing and our knowledge of Late Triassic and Early Jurassic dinosaurs is poor. Future work on basal sauropodomorph phylogeny and functional morphology, discovery of new material, and more fine-grained comparisons of prosauropod and sauropod TDEs and PDEs (by component monophyletic lineages) will permit more rigorous evaluation of this hypothesis. Finally, it must also be remembered that other factors could also have influenced sauropodomorph diversity during this time interval (see below).

COMPETITION BETWEEN SAUROPOD GRADES AND CLADES?

BROAD- AND NARROW-CROWNED CLADES

The BC and NC sauropod groupings (see above) are artificial constructs in a phylogenetic sense: BC clades are paraphyletic with respect to NC clades, and the NC group has diphyletic origins. Nevertheless, comparisons between them may be justified, as each group can be defined as a distinct ecomorphotype or grade (on the basis of general resemblance in tooth morphology). Plotting the frequency of sauropods with BC teeth relative to that of NC forms reveals an apparent negative correlation between the diversity of these two groups, forming another "double-wedge" pattern that may be indicative of a competitive replacement (fig. 4.3B). However, while the stratigraphic ranges of the groups exhibit considerable overlap (as do various mode of life traits, such as body size), comparison of their geographical ranges suggests that competition was unlikely, at least on the basis of current data.

Although BC and NC sauropods coexisted in several late Middle and Late Jurassic faunas from Europe, North America, and Africa, there were no NC forms in Asia or South America during this interval (appendix 4.2 [Weishampel et al. 2004]). In the Early Cretaceous most NC forms are found in Africa and South America: both are areas in which BC taxa were either rare or absent. Examples of Early Cretaceous faunas that contain both NC and BC taxa (such as the Wessex Formation [Barremian] of the Isle of Wight) are rare: most faunas of this age do not contain both BC and NC taxa (appendix 4.2) (Weishampel et al. 2004). Consequently, it is unlikely that there was significant competition for resources between these two groups in the Early Cretaceous due to their limited spatial overlap; competition may have been more important in Late Jurassic faunas but would have been limited to several distinct regions.

Furthermore, patterns arising from assessing the relative diversity of taxa are misleading in this case. Although BC sauropods apparently lose ground to NC genera in the Late Jurassic in plots of relative diversity, this change in proportions coincides with a substantial increase in the total number of sauropod taxa (due largely to the rapid radiation of NC forms, particularly diplodocoids). Thus, while BC sauropods decline in relative terms in the Late Jurassic, the actual number of BC taxa remains high, and is equal to the number of NC taxa, until the Early Cretaceous (compare fig. 4.3A and B).

These independent lines of evidence suggest that factors other than competitive replacement may need to be sought to account for the decline in BC sauropod diversity during the Early Cretaceous. Furthermore, it might be unrealistic to expect competition between BC and NC sauropods, the apparent "double-wedge" notwithstanding: the varied feeding mechanisms of the component taxa in each grouping probably helped to minimize competition, allowing niche partitioning in diverse sauropod faunas, such as those of the Late Jurassic Morrison Formation and Tendaguru Beds (Fiorillo 1991, 1998; Barrett and Upchurch 1995; Upchurch and Barrett 2000).

DIPLODOCOIDS AND TITANOSAURS

Plotting the frequencies of diplodocoid and titanosaur taxa against each other through time reveals an apparent negative correlation between the relative diversities of these two clades (fig. 4.4B). However, this pattern cannot be regarded as strong evidence for competitive interactions between these two groups, for several reasons. During the Late Jurassic diplodocoids peak in diversity and are found in Europe (*Dinheirosaurus*), Africa (*Tornieria*, *Dicraeosaurus*), and North America (*Diplodocus*, *Barosaurus*, and many others). In contrast, titanosaurs are extremely rare at this time and are apparently confined to Africa (*Janenschia*): consequently, there would have been little interaction between these two clades. This period is followed by a significant drop in diplodocoid diversity at the Jurassic/Cretaceous boundary, with the extinction of diplodocids and several related genera of uncertain systematic position (e.g., *Amphicoelias*): however, the extremely limited interactions between diplodocoids and titanosaurs before this event suggests that competition between these clades was not an important factor in diplodocoid tabefaction. Although titanosaurs increase in relative abundance during the basal Cretaceous, this pattern results from a combination of extinctions among diplodocoid lineages and the small sample size for valid sauropod genera during this interval.

During the Berriasian–Barremian there are only one or two genera representing each of these clades at any one time: indeed, for much of the Early Cretaceous a single genus represents the diplodocoid clade (*Losillasaurus*, from Spain). The exact age of *Losillasaurus* is uncertain and it is currently unknown whether it pertains to either the Late Jurassic or the Early Cretaceous (Casanovas et al. 2001): hence, more precise dating of this single genus could result in a major change to the relative abundances of diplodocoids and titanosaurs during this interval. Moreover, geographical overlap between Diplodocoidea and Titanosauria is restricted during the basal Cretaceous: no faunas of this age contain valid genera from both clades (see appendix 4.2). Several faunas apparently contain identifiable elements that provide evidence for some spatial overlap between these groups (Weishampel et al. 2004), but many of these identifications may be suspect, as they are based on very fragmentary material, such as isolated teeth or vertebral centra. As a result, the evidence for competition between these clades during this time interval is weak.

The TDE appears to indicate that diplodocoids were absent for much of the Late Cretaceous, but PDEs demonstrate that at least one lineage (Nemegtosauridae) was present throughout this interval (Upchurch and Barrett, chapter 3). This suggests that competition between titanosaurs and nemegtosaurids was possible, as the two clades overlapped geographically and stratigraphically. Moreover, mode of life comparisons are in accord with this suggestion: recent work has demonstrated extensive homoplasy in the skull morphology of titanosaurs and diplodocoids (Upchurch 1999; Coria and Chiappe 2001) so it is possible that they were competing for the same food resources in their shared habitats. Nevertheless, the extremely limited diversity of nemegtosaurids, coupled with their very restricted geographical distribution, suggests that they would have been very vulnerable to extinction and that a large number of potential mechanisms could have been responsible for their final disappearance. Moreover, as so few nemegtosaurid genera are known it is currently impossible to demonstrate the presence of a "double-wedge" in diversity. Finally, the long stratigraphic overlap between titanosaurs and nemegtosaurids (which potentially extends back into the latest Jurassic; see chapter 3) suggests that competition between these clades may have been negligible. As a result, it is extremely difficult to decide whether competition, or some other factor, was the driving force behind the final extinction of the Nemegtosauridae.

The case for a competitive replacement is stronger in the late Early Cretaceous and early Late Cretaceous, at a time when titanosaurs and rebbachisaurid diplodocoids coexist in several African and South American faunas (see appendix 4.2). Diversity of both groups in this interval is relatively low, with a maximum of five titanosaur and four diplodocoid genera present. Titanosaurs are consistently more diverse than diplodocoids at this time, but the TDEs for each group display coincident rises in diversity through the Aptian and Albian stages (fig. 4.4A). In the Cenomanian, both groups decrease in diversity, but diplodocoids suffer a much greater loss of genera, with rebbachisaurids becoming extinct at the end of this stage. The geographical and stratigraphic overlap of the two clades suggests that competition with titanosaurs may have been a factor in rebbachisaurid decline. Modes of life may also have been similar due to the observed convergence in skull form between titanosaurs and diplodocoids (see above). However, more information is needed on the feeding mechanisms (both cranial and postcranial features) of rebbachisaurids and titanosaurs to confirm or reject this proposition. Additional data from PDEs may also help to clarify the patterns of titanosaur and diplodocoid radiation and tabefaction, particularly for the poorly known basal Cretaceous and early Late Cretaceous intervals.

POTENTIAL INTERACTIONS BETWEEN SAUROPODOMORPHS AND ORNITHISCHIANS

Galton (1973, 1976) suggested that the decline of prosauropods in the Early Jurassic may have been linked to the appearance of ornithischian dinosaurs with superior food processing and locomotor abilities. As mentioned above, prosauropod feeding systems were simple and stereotyped, and did not permit extensive chewing of plant material (Galton 1985, 1986; Crompton and Attridge 1986; Norman and Weishampel 1991). In contrast, the development of occlusion and the inferred presence of extensive fleshy cheeks in ornithischians were viewed as key adaptations that greatly increased the efficiency of mastication (Galton 1973, 1976).

A brief survey of prosauropod and ornithischian distribution patterns demonstrates that the two groups did coexist in several Early Jurassic faunas (Weishampel et al. 2004), including those of the Kayenta Formation in the western United States (containing the prosauropod *Massospondylus* and at least two ornithischian taxa—*Scutellosaurus* and an undescribed heterodontosaurid ornithopod) and the Upper Elliot and Clarens Formations of South Africa (which include *Massospondylus*, a variety of heterodontosaurids, and the basal ornithischian *Lesothosaurus*). Moreover, ornithischian diversity was increasing in the Early Jurassic, with the radiation of heterodontosaurids, and basal thyreophorans, at a time when prosauropod diversity was declining (Weishampel and Jianu 2000; Barrett and Willis 2001). Nevertheless, several lines of evidence suggest that competition with ornithischians was not an important factor in prosauropod extinction.

First, the geographical overlap between prosauropods and ornithischians in the Early Jurassic was limited. Ornithischians are not present (or are only represented by rare and fragmentary material) in many faunas that contain abundant prosauropods, such as those of the Lower Lufeng Formation of China, the McCoy Brook Formation of Canada, and the Portland Formation of the eastern United States (Weishampel et al. 2004). Conversely, prosauropods are absent from several faunas that yield Early Jurassic ornithischians (Weishampel et al. 2004). Second, ornithischians were rare components of Early Jurassic

faunas and did not account for a significant proportion of the biomass (Galton 1985), even in those faunas that contain a diversity of taxa. For example, although three species of heterodontosaurid are known from the Upper Elliot and Clarens Formations of South Africa, each species is known from only a handful of specimens (Weishampel and Witmer 1990), whereas it has been estimated that the contemporaneous prosauropod *Massospondylus* is represented by at least 80 partial skeletons (Galton 1990). Third, most Early Jurassic ornithischians were small, with maximum body lengths ranging between 1 and 2 m and maximum browse heights of 0.5 to 1 m (Barrett 1998). In contrast, Early Jurassic prosauropods ranged from 2.5 to 7 m in length, with maximum browse heights of 3 to 5 m (Barrett 1998; P.M.B. and P.U., pers. obs.). Consequently, Early Jurassic prosauropods and ornithischians do not fulfill Benton's (1996) criteria for the recognition of a CCR. The differences in body size and maximum browse height indicate that these animals foraged at different levels: it is likely that the food requirements of the two groups also differed significantly. Moreover, the low biomass of ornithischians and their patchy geographical distribution suggest that their interactions with prosauropods and contemporary vegetation would have been limited in extent.

Although the geographical and stratigraphic overlaps between sauropod and ornithischian taxa are more extensive and numerous than those between the latter and prosauropods (see listings in Weishampel et al. 2004), there is no evidence for the competitive replacement of any sauropod clade by ornithischians, or vice versa. Substantial differences in the modes of life of these two groups probably precluded competition in most cases. Many sauropod taxa possessed adaptations for high browsing (up to a potential 14 m above ground level) that would have enabled them to gather vegetation that was inaccessible to ornithischians (with maximum browse heights of between 1 and 5 m [e.g., Bakker 1978]). Order of magnitude variance in the modal body masses of the two groups, coupled with fundamental differences between the masticatory apparatuses of the various ornithischian and sauropod clades, may also have resulted in radically different food plant preferences or foraging strategies that would minimize competition for resources. Sauropods that were specialised for low browsing (e.g., dicraeosaurids) may have competed with large ornithischians (e.g., iguanodontian ornithopods), but problems with regional sampling biases (see below) in combination with the low diversity of low-browsing sauropod clades make it difficult to test this hypothesis. In particular, it is difficult to determine whether the coincident diversity changes that are necessary to satisfy the CCR criteria occurred (i.e., did low-browsing sauropods become less diverse as ornithischians increased in diversity in the same region?).

The global TDEs and PDEs of the two groups are broadly similar for much of the Mesozoic, with peaks in diversity during the Late Jurassic, late Early Cretaceous, and latest Cretaceous and times of lower diversity in the Middle Jurassic, basal Cretaceous, and early Late Cretaceous (see Weishampel and Jianu 2000; Barrett and Willis 2001; Upchurch and Barrett, chapter 3). Consequently, there are no periods in which global ornithischian diversity increases coincide with significant decreases in sauropod diversity. For example, the extinction of many sauropod lineages at the Jurassic/Cretaceous boundary cannot be linked to the perceived increase in ornithischian abundance (cf. Bakker 1978), for three reasons. (1) Ornithischian taxa were less diverse and abundant than sauropods in the Late Jurassic (e.g., Bakker 1978), so competition between the groups was probably not intense at this time. (2) Both sauropod and ornithischian lineages suffered extinction at the Jurassic/Cretaceous boundary, resulting in approximately equal diversities for the two clades in basal Cretaceous faunas (Barrett and Willis 2001)—ornithischian diversity did not markedly increase relative to that of sauropods during this period. (3) The major radiations of Cretaceous ornithischians occurred in the late Early Cretaceous (iguanodontian ornithopods, nodosaurid ankylosaurs) and Late Cretaceous (hadrosaurian ornithopods, ceratopians, ankylosaurid ankylosaurs) and, on the basis of current data, cannot be implicated in sauropod decline in the basal Cretaceous.

Nevertheless, it is interesting to note that faunas dominated by sauropods, such as those from the Late Jurassic of North America (Dodson et al. 1980), China (e.g., Dong 1992), and East Africa (Russell et al. 1980) and from the Late Cretaceous of South America (Bonaparte 1996) and Madagascar (Curry Rogers and Forster 2001), contain relatively few ornithischians. Conversely, ornithischian-dominated faunas, such as those of Early Cretaceous Europe (e.g., Martill and Naish 2001) and Late Cretaceous North America (e.g., Russell 1989) do not yield abundant sauropod remains. However, strong sampling biases currently prevent analysis of the fine-grained changes in regional dinosaur faunas through time that would permit the identification (or robust refutation) of CCRs. For example, our knowledge of North American basal Cretaceous faunas is extremely limited: consequently, there are simply not enough data to determine whether the change from the sauropod communities of Morrison times to the ornithischian communities of the late Early Cretaceous was gradual and incremental (and thus potentially a series of competitive replacements) or whether the sauropod decline was abrupt and not linked to the diversity or abundance patterns of any ornithischian group. Similarly, dinosaur faunas from the pre-Late Cretaceous of South America and Madagascar are rare, so it is not possible to determine the sequence of events that led to the dominance of titanosaurs in these regions, at least on the basis of current data.

As the criteria for sauropod/ornithischian CCRs are not met (see above), reasons other than competition need to be sought to explain the Cretaceous decline of sauropods in some regions (principally in Laurasia) and their persistence/diversification in others. One possibility is that regional extinction of specific sauropod clades was responsible for creating the observed regional diversity and abundance patterns, though evaluating the evidence for (and the causes of) such extinctions is beyond the scope of this chapter.

SAUROPODOMORPH–PLANT INTERACTIONS

Observations on extant ecosystems have demonstrated that large herbivores have a considerable impact on the growth, structure, and diversity of vegetation: consequently, we might expect that extinct herbivores had similar effects on coeval floras (e.g., Weishampel 1984; Wing and Tiffney 1987; Weishampel and Norman 1989; Tiffney 1992; Barrett and Willis 2001). With the exception of a few small prosauropods, almost all sauropodomorphs had body masses in excess of 1,000 kg and can be considered to be "megaherbivores" (*sensu* Owen-Smith 1988). It is likely that populations of such large herbivores would have had a significant impact on Mesozoic floras, through the agencies of browsing and other forms of habitat disturbance (such as trampling). Conversely, it would also be expected that large-scale floral replacements, or the evolution of novel plant defences (chemical or physical), would have affected the diversity and/or distributions of contemporary herbivores.

During the early Late Triassic (Carnian–early Norian), Gondwanan lowland floras were dominated by the seed fern *Dicroidium* (with smaller numbers of conifers and other plants), whereas floras in Europe and North America lacked seed ferns and were composed largely of conifers, ferns, and cycadophytes (Benton 1983). Investigation of contemporaneous animal communities suggested that the southern "*Dicroidium* flora" was associated with a vertebrate herbivore fauna dominated by rhynchosaurs (in terms of both diversity and biomass), with rare dinosaurs, while the northern "conifer flora" coexisted with a diverse, abundant dinosaur fauna consisting mainly of prosauropods. Benton (1983) proposed that extinction of the "*Dicroidium* flora"/rhynchosaur fauna in the late Norian permitted a global extension of the northern conifer flora that promoted the rise to dominance, and further diversification, of dinosaurs (largely prosauropods) on the southern continents. However, examination of the prosauropod TDE, and consideration of prosauropod distribution and palaeoecology, indicates that this hypothesis is flawed in two respects. First, Gondwanan prosauropods are not rare in the Norian: indeed, this is the time at which they reach their peak diversity, with abundant material of several African and South American genera (e.g., Galton 1990; Bonaparte 1996). Second, the late Norian extinction of the "*Dicroidium* flora" did not correlate with an increase in Gondwanan prosauropod diversity, but with an abrupt decline, from a maximum of six taxa in the Norian to two taxa in the Early Jurassic (appendix 4.1). Thus, if the record is read literally, this might suggest that the floral replacement drove the tabefaction of Gondwanan prosauropods, rather than their radiation (*contra* Benton 1983). However, although the drop in prosauropod diversity coincides with the disappearance of the "*Dicroidium* flora," the large number of prosauropod specimens from the Early Jurassic strata of southern Africa (Cooper 1981; Gow et al. 1990; Galton 1990) indicates that prosauropod biomass remained high and that they continued to be the dominant terrestrial vertebrates (though it should be noted that remains of prosauropods are extremely rare in other Early Jurassic Gondwanan localities, although this may be due to poor sampling). These contradictory observations make it difficult to evaluate the significance of this floral replacement with respect to the contemporary fauna: more detailed study of floral and faunal distributions, and of potential plant–animal interactions, is needed. Nevertheless, extinction of the "*Dicroidium* flora" does not seem to have promoted the radiation of Gondwanan prosauropods.

There were no major floral replacement events in the Jurassic and no major plant clades originated during this period (Niklas et al. 1985;

Wing et al. 1992). Moreover, the diversity of vascular plants remained relatively constant throughout this interval, with very low species turnover rates (Niklas et al. 1985; Niklas 1997). Consequently, it is unlikely that the major changes in sauropod diversity that occurred during the Middle and Late Jurassic resulted from any major restructuring of contemporary plant communities. The late Early Cretaceous peak in the sauropod TDE and PDE coincides with the initial stages of the angiosperm radiation, but the low biomass and restricted geographical distribution of angiosperms at this time suggest that sauropod–angiosperm interactions would have been extremely limited (Barrett and Willis 2001). There is limited circumstantial evidence for the coevolution of titanosaurs and angiosperms in the latest Cretaceous: the radiation of titanosaurs coincides with the angiosperms rise to ecological dominance (Barrett and Willis 2001). However, the support for this hypothesis is weak and direct evidence for specific titanosaur–angiosperm interactions (coprolites, enterolites, and faunal/floral associations) is lacking: more work is needed on titanosaur feeding mechanisms, Late Cretaceous angiosperm physiognomy, and Late Cretaceous floristics.

CONCLUSIONS

It is difficult to infer the various processes that controlled the observed patterns of sauropodomorph diversity. Rigorous analysis is hampered by small sample sizes, significant biases in the record, and the likelihood that several processes operate concurrently, thereby obfuscating the various signals generated by each process independently. Nevertheless, combination of diversity data (from both TDEs and PDEs) with information on organismic distributions, paleoecology, and modes of life allows some tentative conclusions to be drawn.

Current data do not support the validity of many of the CCRs that have been proposed to account for the extinction of various sauropod clades. One potential CCR, the replacement of the rebbachisaurid clade by titanosaurs in the early Late Cretaceous, is only weakly supported at present. Mode of life differences and disparities in geographical distribution suggest that competition between sauropodomorphs and ornithischians was restricted or unimportant, at least in macroevolutionary terms: evidence for the supposed CCRs between these clades is extremely weak. There are stronger indications that competition from sauropods may have been a significant factor in prosauropod decline during the Early Jurassic, as both of these clades fulfil the requirements for a CCR in this interval. However, this suggestion is made tentatively, given the limited nature of the evidence. It is difficult to quantify the role of competition in sauropodomorph evolution at present: if the prosauropod/sauropod CCR is supported by additional data, it is arguably an important process; if not, competition would appear to be of local paleoecological significance only. Further analysis of interactions among sauropodomorph clades is possible (e.g., detailed clade-by-clade comparisons between different sauropod groups) but is beyond the scope of this chapter. Continued testing of all of these CCRs is necessary in the light of mutable phylogenetic relationships and new discoveries.

Comparisons of sauropodomorph diversity with information on Mesozoic floras reveal one possible incidence of coevolution, between titanosaurs and angiosperms, in the Late Cretaceous. However, the evidence for coevolution is circumstantial and weak: much more information is needed to either confirm or refute this hypothesis. Examination of the potential interactions between prosauropods and contemporary floras in the Late Triassic does not support existing hypotheses of faunal succession, but neither does it provide a clear unambiguous macroevolutionary signal. The decline of the "*Dicroidium* flora" may have driven the tabefaction of Gondwanan prosauropods (not their radiation [*contra* Benton 1983]), but the apparent abundance of these animals (albeit in paucispecific faunas) after the floral replacement is currently paradoxical.

The strong correlation between diversity and feeding ecology demonstrates that herbivory and the various adaptive complexes that arose to deal with a diet of plants were central to the radiation of sauropodomorphs in the Middle and Late Jurassic. Similarly, the acquisition of novel feeding adaptations may have given basal sauropods a competitive advantage over sympatric prosauropods. Finally, these analyses are consistent with the suggestion that the evolution of varied feeding mechanisms allowed niche partitioning, which, in turn, would have promoted sauropod diversification. Nevertheless, knowledge of sauropod feeding mechanisms cannot currently account for the high diversity of titanosaurs in the Late Cretaceous.

Much functional morphological, phylogenetic, and paleoecological work remains to be done on sauropodomorphs, but preliminary studies such as this one demonstrate that they have huge potential for contributing to more general macroevolutionary and paleoecological debates.

ACKNOWLEDGMENTS

We dedicate this contribution to Professor. J. S. McIntosh in recognition of the many important contributions he has made to the study of sauropods. Jack has always been ready with a smile, a sympathetic ear, and bucket-loads of good advice, even when he clearly thought that we were talking complete nonsense. We are grateful to Dr. Chris Brochu for providing useful comments on a draft of this chapter. Many thanks go to Drs. Kristina Curry Rogers and Jeff Wilson for their invitation to contribute to this book. S. M. Feerick provided discussion on the statistical testability (or otherwise) of the various ideas presented herein. P.U. is supported by a NERC Research Fellowship (NERC/GT/ 59906/ES).

LITERATURE CITED

Bakker, R. T. 1971. Ecology of the brontosaurs. Nature 229: 172–174.

———. 1977. Tetrapod mass extinctions—Model of the regulation of speciation rate and immigration by cycles of topographic diversity. In: Hallam, A. (ed.). Patterns of Evolution. Elsevier, Amsterdam. Pp. 439–468.

———. 1978. Dinosaur feeding behaviour and the origin of flowering plants. Nature 274: 661–663.

Barrett, P.M. 1998. Herbivory in the Non-avian Dinosauria. Unpublished Ph.D. dissertation, University of Cambridge, Cambridge. 308 pp.

———. 1999. A sauropod dinosaur from the Lower Lufeng Formation (Lower Jurassic) of Yunnan Province, People's Republic of China. J. Vertebr. Paleontol. 19: 785–787.

———. 2000. Prosauropods and iguanas: speculation on the diets of extinct reptiles; In: Sues, H.-D. (ed.). The Evolution of Herbivory in Terrestrial Vertebrates. Perspectives from the Fossil Record. Cambridge University Press, Cambridge. Pp. 42–78.

Barrett, P. M. and Upchurch, P. 1994. Feeding mechanisms in *Diplodocus*. GAIA 10: 195–203.

———. 1995. Sauropod feeding mechanisms: their bearing on palaeoecology. In: Sun, A.-L., and Wang, Y.-Q. (eds.). Sixth Symposium on Mesozoic Terrestrial Ecosystems and Biota, Short Papers. China Ocean Press, Beijing. Pp. 107–110.

Barrett, P.M., and Willis, K.J. 2001. Did dinosaurs invent flowers? Dinosaur–angiosperm coevolution revisited. Biol. Rev. 76: 411–447.

Benton, M.J. 1983. Dinosaur success in the Triassic: a non-competitive ecological model. Q. Rev. Biol. 58: 29–55.

———. 1986. The Late Triassic tetrapod extinction events. In: Padian, K. (ed.). The Beginning of the Age of Dinosaurs: Faunal Change Across the Triassic–Jurassic Boundary. Cambridge University Press, Cambridge. Pp. 303–320.

———. 1990. Origin and interrelationships of dinosaurs. In: Weishampel, D. B., Dodson, P., and Osmólska, H. (eds.). The Dinosauria, 1st ed. University of California Press, Berkeley. Pp. 11–30.

———. 1993. Reptilia. In: Benton, M.J. (ed.). The Fossil Record 2. Chapman and Hall, London. Pp. 681–715.

———. 1996. On the nonprevalence of competitive replacement in the evolution of tetrapods. In: Jablonski, D., Erwin, D. H., and Lipps, J. H. (eds.). Evolutionary Paleobiology. University of Chicago Press, Chicago. Pp. 185–210.

Benton, M.J., Juul, L., Storrs, G.W., and Galton, P.M. 2000. Anatomy and systematics of the prosauropod dinosaur *Thecodontosaurus antiquus* from the Upper Triassic of southwest England. J. Vertebr. Paleontol. 20: 77–108.

Bonaparte, J. F. 1996. Dinosaurios de America del Sur. Museo Argentino de Ciencias Naturales, Buenos Aires. 174 pp.

Brooks, D. R., and McLennan, D. A. 1991. Phylogeny, ecology, and behaviour: a research programme in comparative biology. University of Chicago Press, Chicago. 434 pp.

Buffetaut, E., Martin, V., Sattayarak, N., and Suteethorn, V. 1995. The oldest known dinosaur from southeast Asia: a prosauropod from the Nam Phong Formation (Late Triassic) of northeastern Thailand. Geol. Mag. 132: 739–742.

Buffetaut, E., Suteethorn, V., Cuny, G., Tong, H., Le Loeuff, J., Khansubha, S., and Jongautchariyakul, S. 2000. The earliest known sauropod dinosaur. Nature 407: 72–74.

Calvo, J. O. 1994. Jaw mechanics in sauropod dinosaurs. GAIA 10: 183–194.

Casanovas, M. L., Santafé, J. V., and Sanz, J. L. 2001. *Losillasaurus giganteus*, un nuevo saurópodo del tránsito Jurásico–Cretácico de la cuenca de 'Los Serranos' (Valencia, España). Paleontol. Evol. 32–33:99–122.

Charig, A. J., Attridge, J., and Crompton, A. W. 1965. On the origin of the sauropods and the classification of the Saurischia. Proc. Linn. Soc. London 176: 197–221.

Christiansen, P. 1999. On the head size of sauropodomorph dinosaurs: implications for ecology and physiology. Hist. Biol. 13: 269–297.

———. 2000. Feeding mechanisms of the sauropod dinosaurs *Brachiosaurus*, *Camarasaurus*, *Diplodocus* and *Dicraeosaurus*. Hist. Biol. 14: 137–152.

Connely, M. V., and Hawley, R. 1998. A proposed reconstruction of the jaw musculature and other soft cranial tissues of *Apatosaurus*. J. Vertebr. Paleontol. 18: 35A.

Coombs, W. P., Jr. 1975. Sauropod habits and habitats. Palaeogeogr. Palaeoclimatol. Palaeoecol. 17: 1–33.

Cooper, M. R. 1981. The prosauropod dinosaur *Massospondylus carinatus* Owen from Zimbabwe: its biology, mode of life and phylogenetic significance. Occas. Papers Natl. Mus. Monuments Rhodesia Ser. B Nat. Sci. 6: 689–840.

Coria, R. A., and Chiappe, L. M. 2001. Tooth replacement in a sauropod premaxilla from the Upper Cretaceous of Patagonia, Argentina. Ameghiniana 38: 463–466.

Crompton, A. W., and Attridge, J. 1986. Masticatory apparatus of the larger herbivores during Late Triassic and Early Jurassic times. In: Padian, K. (ed.). The Beginning of the Age of the Dinosaurs. Cambridge University Press, Cambridge. Pp. 223–236.

Curry Rogers, K., and Forster, C. A. 2001. The last of the dinosaur titans: A new sauropod from Madagascar. Nature 412: 530–534.

Dodson, P. 1990. Sauropod paleoecology. In: Weishampel, D. B., Dodson, P., and Osmólska, H. (eds.). The Dinosauria, 1st ed. University of California Press, Berkeley. Pp. 402–407.

Dodson, P., Behrensmeyer, A. K., Bakker, R. T., and McIntosh, J. S. 1980. Taphonomy and paleoecology of the Upper Jurassic Morrison Formation. Paleobiology 6: 208–232.

Dong, Z.-M. 1992. Dinosaurian Faunas of China. Springer-Verlag, New York. 188 pp.

Dong, Z.-M., Zhou, S., and Zhang, Y. 1983. [The dinosaurian remains from Sichuan Basin, China]. Palaeontol. Sinica Ser. C 23: 139–145. (In Chinese with English summary.)

Farlow, J. O. 1987. Speculations about the diet and digestive physiology of herbivorous dinosaurs. Paleobiology 13: 60–72.

Fiorillo, A. R. 1991. Dental microwear on the teeth of *Camarasaurus* and *Diplodocus*: implications for sauropod palaeoecology. In: Kielan-Jaworowska, Z., Heintz, N., and Nakrem, H. A. (eds.). Fifth Symposium on Mesozoic Terrestrial Ecosystems and Biota, Extended Abstracts. Contributions from the Palaeontological Museum, University of Oslo, 364. Pp. 23–24.

———. 1998. Dental microwear patterns of the sauropod dinosaurs *Camarasaurus* and *Diplodocus*: evidence for resource partitioning in the Late Jurassic of North America. Hist. Biol. 13: 1–16.

Flynn, J. J., Parrish, J. M., Rakotosamimanana, B., Simpson, W. F., Whatley, R. L., and Wyss, A. R. 1999. A Triassic fauna from Madagascar, including early dinosaurs. Science 286: 763–765.

Galton, P. M. 1973. The cheeks of ornithischian dinosaurs. Lethaia 6: 67–89.

———. 1976. Prosauropod dinosaurs of North America. Postilla 169: 1–98.

———. 1985. Diet of prosauropods from the Late Triassic and Early Jurassic. Lethaia 18: 105–123.

———. 1986. Herbivorous adaptations of Late Triassic and Early Jurassic dinosaurs. In: Padian, K. (ed.). The Beginning of the Age of the Dinosaurs. Cambridge University Press, Cambridge. Pp. 203–221.

———. 1990. Basal Sauropodomorpha—Prosauropoda. In: Weishampel, D. B., Dodson, P., and Osmólska, H., (eds.). The Dinosauria, 1st ed. University of California Press, Berkeley. Pp. 320–344.

Galton, P. M., and Upchurch, P. 2004. Basal Sauropodomorpha—Prosauropoda. *In:* Weishampel, D. B., Dodson, P., and Osmólska, H. (eds.).

The Dinosauria 2nd ed. University of California Press, Berkeley. Pp. 232–258.

Gauffre, F.-X. 1993. The prosauropod dinosaur *Azendohsaurus laarousii* from the Upper Triassic of Morocco. Palaeontology 36: 897–908.

Gauthier, J. A. 1986. Saurischian monophyly and the origin of birds. Mem. Calif. Acad. Sci. 8: 1–55.

Gow, C. E., Kitching, J. W., and Raath, M. A. 1990. Skulls of the prosauropod dinosaur *Massospondylus carinatus* Owen in the collections of the Bernard Price Institute for Palaeontological Research. Palaeontol. Afr. 27: 45–58.

Haubold, H. 1990. Dinosaurs and fluctuating sea levels during the Mesozoic. Hist. Biol. 4: 75–106.

He, X.-L., Wang, C.-S., Liu, S.-Z., Zhou, F.-Y., Liu, T.-Q., Cai, K.-J., and Dai, B. 1998. A new species of sauropod from the Early Jurassic of Gongxian County, Sichuan. Acta Geol. Sichuan 18: 1–7 (In Chinese, with English abstract.)

Hotton, N., Olson, E. C., and Beerbower, R. 1997. Amniote origins and the discovery of herbivory. In: Sumida, S. S., and Martin, K. L. M. (eds.). Amniote Origins: Completing the Transition to Land. Academic Press, San Diego. Pp. 207–264.

Hunt, A. P., Lockley, M. G., Lucas, S. G., and Meyer, C. A. 1994. The global sauropod fossil record. GAIA 10: 261–279.

Jablonski, D., Erwin, D. H., and Lipps, J. H. (eds.). 1996. Evolutionary Paleobiology. University Chicago Press, Chicago. 484 pp.

King, G. M. 1990. Dicynodonts and the end-Permian event. Palaeontol. Afr. 27: 31–39.

Langer, M. C., Abdala, F., Richter, M., and Benton, M. J. 1999. A sauropodomorph dinosaur from the Upper Triassic (Carnian) of southern Brazil. C. R. Acad. Sci. Paris Sci. Terre Planettes 329: 511–517.

Luo, Z.-X., and Wu, X.-C., 1994. The small tetrapods of the Lower Lufeng Formation, Yunnan, China. In: Fraser, N. C., and Sues, H.-D. (eds.). In the Shadow of the Dinosaurs. Cambridge University Press, Cambridge. Pp. 251–270.

Martill, D. M., and Naish, D. (eds.). 2001. Dinosaurs of the Isle of Wight. Palaeontological Association, London. 433 pp.

Martin, J., Martin-Rolland, V., and Frey, E. 1998. Not cranes or masts, but beams: the biomechanics of sauropod necks. Oryctos 1: 113–120.

McIntosh, J. S. 1990. Sauropoda. In: Weishampel, D. B., Dodson, P., and Osmólska, H. (eds.). The Dinosauria, 1st ed. University of California Press, Berkeley. Pp. 345–401.

Moore, B. R., and Brooks, D. R. 1995. A comparative analysis of herbivory and amniote diversification in recent terrestrial ecosystems. J. Vertebr. Paleontol. 15: 45A.

———. 1996. A comparative analysis of herbivory and amniote diversification in recent terrestrial ecosystems. Spec. Publ. Paleontol. Soc. 8: 280.

Niklas, K. J. 1997. The Evolutionary Biology of Plants. University of Chicago Press, Chicago. 449 pp.

Niklas, K. J., Tiffney, B. H., and Knoll, A. H. 1985. Patterns in vascular land plant diversification: an analysis at the species level. In: Valentine, J. W. (ed.). Phanerozoic Diversity Patterns: Profiles in Macroevolution. Princeton University Press and American Association for the Advancement of Science, Princeton, NJ, and San Francisco. Pp. 97–128.

Norman, D. B., and Weishampel, D. B. 1991. Feeding mechanisms in some small herbivorous dinosaurs: processes and patterns. In: Rayner, J. M. V., and Wootton, R. J. (eds.). Biomechanics in Evolution. Cambridge University Press, Cambridge. Pp. 161–181.

Owen-Smith, R. N. 1988. Megaherbivores: The Influence of Very Large Body Size on Ecology. Cambridge University Press, Cambridge. 369 pp.

Patterson, C., and Smith, A. B. 1987. Is periodicity of mass extinctions a taxonomic artifact? Nature 330: 248–251.

Pisani, D., Yates, A. M., Langer, M. C., and Benton, M. J. 2002. A genus-level supertree of the Dinosauria. Proc. Roy. Soc. London Ser. B 269: 915–921.

Raup, D. M., and Boyajian, G. E. 1988. Patterns of generic extinction in the fossil record. Paleobiology 14: 109–123.

Reisz, R. R., and Sues, H.-D. 2000. Herbivory in late Palaeozoic and Triassic terrestrial vertebrates. In: Sues, H.-D. (ed.). The Evolution of Herbivory in Terrestrial Vertebrates. Perspectives from the Fossil Record. Cambridge University Press, Cambridge. Pp. 9–41.

Robeck, H. E., Maley, C. C., and Donoghue, M. J. 2000. Taxonomy and temporal diversity patterns. Paleobiology 26: 171–187.

Russell, D. A. 1989. An odyssey in time: the dinosaurs of North America. University of Toronto Press/National Museum of Natural Sciences, Toronto. 240 pp.

Russell, D. A., Béland, P., and McIntosh, J. S. 1980. Paleoecology of the dinosaurs of Tendaguru (Tanzania). Mem. Soc. Geol. France 139: 169–175.

Salgado, L., and Calvo, J. O. 1997. Evolution of titanosaurid sauropods. II. The cranial evidence. Ameghiniana 34: 33–48.

Sepkoski, J. J., Jr. 1996. Competition in macroevolution: The double wedge revisited. In: Jablonski, D., Erwin, D. H., and Lipps, J. H. (eds.).

Evolutionary Paleobiology. University of Chicago Press, Chicago. Pp. 211–255.

Sepkoski, J. J., Jr. and Kendrick, D. C. 1993. Numerical experiments with model monophyletic and paraphyletic taxa. Paleobiology 19: 168–184.

Sereno, P. C. 1997. The origin and evolution of dinosaurs. Annu. Rev. Earth Planet. Sci. 25: 435–489.

———. 1999. The evolution of dinosaurs. Science 284: 2137–2147.

Sereno, P. C., and Wilson, J. A. 2001. A sauropod tooth battery: Structure and function. J. Vertebr. Paleontol. 21: 100A–101A.

Sereno, P. C., Beck, A. L., Dutheil, D. B., Larsson, H. C. E., Lyon, G. H., Moussa, B., Sadleir, R. W., Sidor, C. A., Varricchio, D. J., Wilson, G. P., and Wilson, J. A. 1999. Cretaceous sauropods from the Sahara and the uneven rate of skeletal evolution among dinosaurs. Science 286: 1342–1347.

Smith, A. B. 1994. Systematics and the Fossil Record. Blackwell Scientific, Oxford. 223 pp.

Stevens, K. A., and Parrish, J. M. 1999. Neck posture and feeding habits of two Jurassic sauropod dinosaurs. Science 284: 798–800.

Tiffney, B. H. 1992. The role of vertebrate herbivory in the evolution of land plants. Palaeobotanist 41: 87–97.

Upchurch, P. 1995. The evolutionary history of sauropod dinosaurs. Philos. Trans. Roy. Soc. London Ser. B 349: 365–390.

———. 1998. The phylogenetic relationships of sauropod dinosaurs. Zool. J. Linn. Soc. London 124: 43–103.

———. 1999. The phylogenetic relationships of the Nemegtosauridae (Saurischia, Sauropoda). J. Vertebr. Paleontol. 19: 106–125.

Upchurch, P., and Barrett, P. M. 2000. The evolution of sauropod feeding mechanisms. In: Sues, H.-D. (ed.). Evolution of Herbivory in Terrestrial Vertebrates: Perspectives from the Fossil Record. Cambridge University Press, Cambridge. Pp. 79–122.

Upchurch, P., Barrett, P. M., and Dodson, P. 2004. Sauropoda. In: Weishampel, D. B., Dodson, P., and Osmólska, H. (eds.). The Dinosauria, 2nd ed. University of California Press, Berkeley. Pp. 259–322.

Valentine, J. W. (ed.). 1985. Phanerozoic Diversity Patterns: Profiles in Macroevolution. Princeton University Press and American Association for the Advancement of Science, Princeton, NJ, and San Francisco. 441 pp.

Weishampel, D. B. 1984. Interactions between Mesozoic plants and vertebrates: fructifications and seed predation. Neues Jahrb. Geol. Paleontol. Abhandlungen 167: 224–250.

———. 1990. Dinosaurian distribution. In: Weishampel, D. B., Dodson, P., and Osmólska, H. (eds.). The Dinosauria, 1st ed. University of California Press, Berkeley. Pp. 63–139.

Weishampel, D. B. and Horner, J. R. 1987. Dinosaurs, habitat bottlenecks, and the St. Mary River Formation. In: Currie, P. J., and Koster, E. H. (eds.). Fourth Symposium on Mesozoic Terrestrial Ecosystems, Short Papers. Royal Tyrell Museum, Drumheller. Pp. 224–229.

Weishampel, D. B. and Jianu, C.-M. 2000. Plant-eaters and ghost lineages: dinosaurian herbivory revisited. In: Sues, H.-D. (ed.). Evolution of Herbivory in Terrestrial Vertebrates: Perspectives from the Fossil Record. Cambridge University Press, Cambridge. Pp. 123–143.

Weishampel, D. B., and Norman, D. B. 1989. Vertebrate herbivory in the Mesozoic; jaws, plants and evolutionary metrics. Spec. Papers Geol. Soc. Am. 238: 87–100.

Weishampel, D. B., and Witmer. L. M. 1990. Heterodontosauridae. In: Weishampel, D. B., Dodson, P., and Osmólska, H. (eds.). The Dinosauria, 1st ed. University of California Press, Berkeley. Pp. 486–497.

Weishampel, D. B., Barrett, P. M., Coria, R. A., Le Loeuff, J., Xu, X., Zhao, X.-J., Sahni, A. Gomani, E. M. P., and Noto, C.R. 2004. Dinosaur distribution. In: Weishampel, D. B., Dodson, P., and Osmólska, H. (eds.). The Dinosauria, 2nd ed. University of California Press, Berkeley. Pp. 517–606.

Wilson, J. A. 2002. Sauropod dinosaur phylogeny: critique and cladistic analysis. Zool. J. Linn. Soc. London 136: 217–276.

Wilson, J. A. and Sereno, P. C. 1998. Early evolution and higher-level phylogeny of the sauropod dinosaurs. Mem. Soc. Vertebr. Paleontol. 5: 1–68.

Wing, S. L., and Tiffney, B. H. 1987. The reciprocal interaction of angiosperm evolution and tetrapod herbivory. Rev. Palaeobot. Palynol. 50: 179–210.

Wing, S. L., Sues, H.-D., Tiffney, B. H., Stucky, R. K., Weishampel, D. B., Spicer, R. A., Jablonski, D., Badgley, C. E., Wilson, M. V. H., and Kovach, W. L. 1992. Mesozoic and Early Cenozoic terrestrial ecosystems. In: Behrensmeyer, A. K., Damuth, J. D., DiMichele, W. A., Potts, R., Sues, H.-D., and Wing, S. L. (eds.). Terrestrial Ecosystems through Time: Evolutionary Paleoecology of Plants and Animals. University of Chicago Press, Chicago. Pp. 327–416.

Yates, A. M. 2003. A new species of the primitive dinosaur *Thecodontosaurus* (Saurischia: Sauropodomorpha) and its implications for the systematics of early dinosaurs. J. Syst. Palaeontol. 1: 1–42.

Young, C. C. 1951. The Lufeng saurischian fauna. Palaeontol. Sinica Ser. C 13: 1–96.

APPENDIX 4.1. PROSAUROPOD GENERA INCLUDED IN THE TAXIC DIVERSITY ESTIMATES

After Galton and Upchurch (2004), Weishampel et al. (2004), and Barrett and Upchurch (pers. obs.).

GENUS	STRATIGRAPHIC RANGE
Ammosaurus	Hettangian–Pliensbachian
Anchisaurus	Hettangian–Pliensbachian
Azendohsaurus	late Carnian
Camelotia	Rhaetian
Coloradisaurus	Norian
Euskelosaurus	late Carnian–early Norian
"Gyposaurus" sinensis	Hettangian–early Sinemurian
Jingshanosaurus	Hettangian–early Sinemurian
Lessemsaurus	Norian
Lufengosaurus	Hettangian–early Sinemurian
Massospondylus	Hettangian–Pliensbachian
Melanorosaurus	late Carnian–Pliensbachian
Mussaurus	Norian
Plateosaurus	late Norian
Riojasaurus	Norian
Saturnalia	middle Carnian
Sellosaurus	middle Norian
Thecodontosaurus	late Carnian–Rhaetian
Yimenosaurus	Hettangian–early Sinemurian
Yunnanosaurus	Sinemurian–early Pliensbachian

APPENDIX 4.2. SAUROPOD GENERA INCLUDED IN THE TAXIC DIVERSITY ESTIMATES

After Upchurch et al. (2004), Weishampel et al. (2004), and Barrett and Upchurch (pers. obs). A question mark (?) denotes genera for which tooth morphology and feeding mechanism could not be predicted due to unknown phylogenetic position; these genera are excluded from morphology counts. Abbreviations for tooth type: NC, narrow-crowned (cylindrical or subcylindrical; SI ≥ 4.0); BC, broad-crowned (spatulate; SI ≤ 4.0); inferred, tooth morphology predicted on the basis of phylogenetic position (actual teeth unknown). Abbreviations for feeding mechanisms: APATO, Apatosaurus type; BRACH, brachiosaurid type; CAMARA, Camarasaurus type; CETIO, cetiosaurid type; DICRA, dicraeosaurid type; DIPLO, Diplodocus type; HIGHEU, "higher eusauropod" type; NEMEGT, nemegtosaurid type; NIGER, Nigersaurus type; SHUNO, Shunosaurus type; TITAN, titanosaurian type.

GENUS	STRATIGRAPHIC RANGE	CLASSIFICATION	TOOTH TYPE	MECHANISM
Abrosaurus	Bathonian–Callovian	Basal macronarian	BC	?
Aegyptosaurus	Cenomanian	Basal titanosaurian	Inferred NC	TITAN
Aeolosaurus	Campanian	Titanosauroidea	Inferred NC	TITAN
Agustinia	Aptian	Titanosauria incertae sedis	Inferred NC	TITAN
Alamosaurus	Maastrichtian	Titanosauroidea	NC	TITAN
Amargasaurus	Hauterivian	Diplodocoidea: Dicraeosauridae	Inferred NC	DICRAE
Ampelosaurus	early Maastrichtian	Titanosauroidea	NC	TITAN
Amphicoelias	Kimmeridgian–Tithonian	Diplodocoidea incertae sedis	Inferred NC	?
Amygdalodon	Bajocian	Eusauropoda incertae sedis	BC	?
Andesaurus	Albian–Cenomanian	Basal titanosaurian	Inferred NC	TITAN
Antarctosaurus	Campanian–Maastrichtian	Titanosauroidea incertae sedis	NC	TITAN
Apatosaurus	Kimmeridgian–Tithonian	Diplodocoidea: Diplodocidae	NC	APATO
"Apatosaurus" minimus	Kimmeridgian–Tithonian	Sauropoda incertae sedis	?	?
Aragosaurus	early Barremian	Eusauropoda incertae sedis	BC	?
Argentinosaurus	Albian–Cenomanian	Basal titanosaurian	Inferred NC	TITAN
Argyrosaurus	Campanian–Maastrichtian	Titanosauria incertae sedis	Inferred NC	TITAN
Astrodon	middle–late Aptian	Basal titanosauriform	BC	?
Atlasaurus	Bathonian–Callovian	Basal macronarian	BC	?
Austrosaurus	Albian	Basal titanosaurian	Inferred NC	TITAN
Barapasaurus	Hettangian–Pliensbachian	Cetiosauridae	BC	CETIO
Barosaurus	Kimmeridgian–Tithonian	Diplodocoidea: Diplodocidae	Inferred NC	DIPLO
Bellusaurus	Aalenian–Callovian	Basal macronarian	BC	?
Blikanasaurus	?late Carnian or ?early Norian	Basal sauropod	Inferred BC	?

APPENDIX 4.2. (continued)

GENUS	STRATIGRAPHIC RANGE	CLASSIFICATION	TOOTH TYPE	MECHANISM
Brachiosaurus	Kimmeridgian–Tithonian	Brachiosauridae	BC	BRACH
"*Brachiosaurus*" *atalaiensis*	Kimmeridgian	Brachiosauridae	Inferred BC	BRACH
"*Brachiosaurus*" *nougaredi*	Albian	Brachiosauridae	Inferred BC	BRACH
Camarasaurus	Kimmeridgian–Tithonian	Basal camarasauromorph	BC	CAMARA
Cardiodon	late Bathonian	Sauropoda *incertae sedis*	BC	?
Cedarosaurus	Barremian	Brachiosauridae	Inferred BC	BRACH
Cetiosauriscus	middle Callovian	Diplodocoidea *incertae sedis*	Inferred NC	?
Cetiosaurus	late Bajocian–Bathonian	Cetiosauridae	Inferred BC	CETIO
"*Cetiosaurus*" *glymptonensis*	late Bathonian	Diplodocoidea *incertae sedis*	Inferred NC	?
"*Cetiosaurus*" *humerocristatus*	Kimmeridgian	Brachiosauridae	Inferred BC	BRACH
Chubutisaurus	Albian	Basal titanosaurian	Inferred NC	TITAN
Datousaurus	Bathonian–Callovian	Sauropoda *incertae sedis*	BC	?
Dicraeosaurus	Kimmeridgian	Diplodocoidea: Dicraeosauridae	NC	DICRA
Dinheirosaurus	late Kimmeridgian	Diplodocoidea *incertae sedis*	Inferred NC	?
Diplodocus	Kimmeridgian–Tithonian	Diplodocoidea: Diplodocidae	NC	DIPLO
Dyslocosaurus	Kimmeridgian–Tithonian (or Maas)	Diplodocoidea: Diplodocidae	Inferred NC	DIPLO
Dystrophaeus	Callovian–Oxfordian	Sauropoda *incertae sedis*	?	?
Dystylosaurus	Kimmeridgian–Tithonian	Brachiosauridae	Inferred BC	BRACH
Epachthosaurus	Campanian–Maastrichtian	Basal titanosaurian	Inferred NC	TITAN
Euhelopus	Kimmeridgian	Non-neosauropod eusauropod	BC	HIGHEU
Gondwanatitan	Santonian–Maastrichtian	Titanosauroidea	Inferred NC	TITAN
Gongxianosaurus	Hettangian–Toarcian	Basal sauropod	BC	?
Haplocanthosaurus	Kimmeridgian–Tithonian	Basal camarasauromorph	Inferred BC	?
Huabeisaurus	?Campanian	Basal titanosaurian	NC	TITAN
Hudiesaurus	Kimmeridgian–Tithonian	Eusauropoda *incertae sedis*	BC	?
Isanosaurus	late Norian–Rhaetian	Basal sauropod	Inferred BC	?
Jainosaurus	Maastrichtian	Titanosauroidea	Inferred NC	TITAN
Janenschia	Kimmeridgian	Basal titanosaurian	Inferred NC	TITAN
Jobaria	Berriasian–Hauterivian	Basal macronarian	BC	?
Klamelisaurus	Aalenian–Callovian	Sauropoda *incertae sedis*	BC	?
Kotasaurus	Hettangian–Pliensbachian	Basal sauropod	BC	?
Laplatasaurus	Campanian–Maastrichtian	Titanosauroidea	Inferred NC	TITAN

Genus	Age	Classification		
Lapparentosaurus	Bathonian	Titanosauriformes *incertae sedis*	BC	?
Lirainosaurus	late Campanian–early Maastrichtian	Titanosauroidea	NC	TITAN
Losillasaurus	Late Jurassic–Early Cretaceous	Diplodocoidea *incertae sedis*	Inferred NC	?
Lourinhasaurus	late Kimmeridgian–early Tithonian	Non-neosauropod eusauropod	Inferred BC	?
Magyarosaurus	late Maastrichtian	Titanosauroidea	Inferred NC	TITAN
Malawisaurus	Aptian	Titanosauroidea	NC	TITAN
Mamenchisaurus	Oxfordian–Tithonian	Non-neosauropod eusauropod	BC	HIGHEU
"*Morosaurus*" *agilis*	Kimmeridgian–Tithonian	Sauropoda *incertae sedis*	?	?
Nemegtosaurus	middle Maastrichtian	Diplodocoidea: Nemegtosauridae	NC	NEMEGT
Neuquensaurus	Campanian–early Maastrichtian	Titanosauroidea	Inferred NC	TITAN
Nigersaurus	Aptian–Albian	Diplodocoidea: Rebbachisauridae	NC	NIGER
Ohmdenosaurus	Middle Toarcian	Basal sauropod	Inferred BC	?
Omeisaurus	Bathonian–Tithonian	Non-neosauropod eusauropod	BC	HIGHEU
Opisthocoelicaudia	middle Maastrichtian	Titanosauroidea	Inferred NC	TITAN
Oplosaurus	Barremian	Sauropoda *incertae sedis*	BC	?
Ornithopsis	Barremian	Titanosauriformes *incertae sedis*	Inferred BC	?
Paralititan	Cenomanian	Titanosauroidea	Inferred NC	TITAN
Patagosaurus	Callovian	Cetiosauridae	BC	CETIO
Pellegrinisaurus	Early Campanian	Titanosauroidea	Inferred NC	TITAN
"*Pelorosaurus*" *conybearei*	late Berriasian–Valanginian	Titanosauriformes *incertae sedis*	Inferred BC	?
Phuwiangosaurus	Valanginian–Hauterivian	Basal titanosaurian	NC	TITAN
Pleurocoelus	middle–late Aptian	Titanosauriformes *incertae sedis*	BC	?
"*Pleurocoelus*" Texas	Middle Albian	Titanosauriformes *incertae sedis*	BC	?
Quaesitosaurus	late Campanian–early Maastrichtian	Diplodocoidea: Nemegtosauridae	Inferred NC	NEMEGT
Rapetosaurus	Maastrichtian	Titanosauroidea	NC	TITAN
Rayososaurus	Albian–Cenomanian	Diplodocoidea: Rebbachisauridae	NC	NIGER
Rebbachisaurus	Albian	Diplodocoidea: Rebbachisauridae	Inferred NC	?
Rhoetosaurus	Bajocian	Sauropoda *incertae sedis*	?	?
Rocasaurus	Campanian–Maastrichtian	Titanosauroidea	Inferred NC	TITAN
Saltasaurus	late Campanian–Maastrichtian	Titanosauroidea	Inferred NC	TITAN
Sauroposeidon	Aptian–Albian	Brachiosauridae	Inferred BC	BRACH
Seismosaurus	Kimmeridgian–Tithonian	Diplodocoidea: Diplodocidae	Inferred NC	DIPLO

APPENDIX 4.2. (continued)

GENUS	STRATIGRAPHIC RANGE	CLASSIFICATION	TOOTH TYPE	MECHANISM
Shunosaurus	Bathonian–Callovian	Non-neosauropod eusauropod	BC	SHUNO
Supersaurus	Kimmeridgian–Tithonian	Diplodocoidea: Diplodocidae	Inferred NC	DIPLO
Tangvayosaurus	Aptian–Albian	Basal titanosaurian	Inferred NC	TITAN
Tehuelchesaurus	Callovian	Non-neosauropod eusauropod	Inferred BC	HIGHEU
Tendaguria	Kimmeridgian	Sauropoda *incertae sedis*	?	?
Titanosaurus	Maastrichtian	Titanosauroidea	NC	TITAN
"*Titanosaurus*" *colberti*	Maastrichtian	Titanosauroidea	Inferred NC	TITAN
Tornieria	Kimmeridgian	Diplodocoidea: Diplodocidae	NC	DIPLO
Venenosaurus	Barremian	Basal titanosaurian	Inferred NC	TITAN
Volkheimeria	Callovian	Sauropoda *incertae sedis*	?	?
Vulcanodon	Hettangian	Basal sauropod	Inferred BC	?

FIVE

Structure and Evolution of a Sauropod Tooth Battery

Paul C. Sereno and Jeffrey A. Wilson

DURING THE JURASSIC, SAUROPOD dinosaurs rose to predominance among vertebrate herbivores, in terms of both species diversity and biomass (e.g., Romer 1966; McIntosh 1990). Their perceived decline on northern landmasses during the Cretaceous has been linked to the evolution of tooth batteries in ornithischian herbivores (e.g., Lull and Wright 1942; Ostrom 1961; Bakker 1978; Lucas and Hunt 1989). On southern landmasses, in contrast, sauropod diversity increased during the Cretaceous (Weishampel 1990; Hunt et al. 1994), and a newly discovered southern sauropod, the rebbachisaurid *Nigersaurus taqueti*, is now known to have evolved a complex tooth battery (Sereno et al. 1999).

Rebbachisaurids are a poorly known sauropod clade, reported thus far only from Cretaceous rocks in South America (Calvo and Salgado 1995; Bonaparte 1996; Apesteguía et al. 2001; Lamanna et al. 2001), Africa (Lavocat 1954; Taquet 1976; Sereno et al. 1999), and Europe (Dalla Vecchia 1998; Pereda-Suberbiola et al. 2003). Many of these are fragmentary finds, leaving much of the skeletal anatomy of this group in question, especially the skull. In this chapter, we describe the tooth-bearing bones and dental battery of *Nigersaurus taqueti* and provide an initial cranial reconstruction. We outline the feeding specializations common to diplodocoids and how these were modified within rebbachisaurids.

Institutional Abbreviation: MNN, Musée National du Niger, Niamey.

SYSTEMATIC PALEONTOLOGY

SAUROPODA MARSH 1878

EUSAUROPODA UPCHURCH 1995

NEOSAUROPODA BONAPARTE 1986

DIPLODOCOIDEA MARSH 1884

REBBACHISAURIDAE BONAPARTE 1997

NIGERSAURUS SERENO ET AL. 1999

TYPE SPECIES: *Nigersaurus taqueti* Sereno et al. 1999.

AGE: Early Cretaceous (Aptian–Albian).

DISCUSSION: *Nigersaurus taqueti* is the most common sauropod and one of the most common species recovered in the rich vertebrate fauna described from Gadoufaoua, Niger Republic

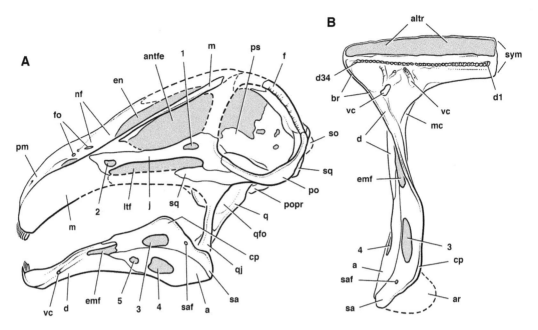

FIGURE 5.1. Preliminary reconstruction of the skull of *Nigersaurus taqueti* based on MNN GDF512. (A) Skull in left lateral view. (B) Left lower jaw in dorsal view with tooth batteries removed. 1–5, accessory cranial fenestrae; **a**, angular; **altr**, alveolar trough; **antfe**, antorbital fenestra; **ar**, articular; **br**, buccal ridge; **cp**, coronoid process; **d**, dentary; **d1–d34**, replacement foramina for dentary tooth positions 1–34; **emf**, external mandibular fenestra; **en**, external naris; **f**, frontal, **fo**, foramen; **j**, jugal; **ltf**, laterotemporal fenestra; **m**, maxilla; **mc**, Meckel's canal; **nf**, narial fossa; **pm**, premaxilla; **po**, postorbital; **popr**, paraoccipital process; **ps**, parasphenoid; **q**, quadrate; **qfo**, quadrate fossa; **qj**, quadratojugal; **sa**, surangular; **saf**, surangular foramen, **so**, supraoccipital; **sq**, squamosal; **sym**, symphyseal surface; **vc**, vascular canal.

(Taquet 1976; Sereno et al. 1998, 1999, 2001, 2003; Larsson and Gado 2000). Nevertheless, because the skull and skeleton are delicately constructed and highly pneumatic, there are no complete skulls and only a few partially articulated skeletons. *Nigersaurus taqueti* was named and identified as a rebbachisaurid by Sereno et al. (1999). An earlier report from the same beds of a dicraeosaurid allied with titanosaurs (Taquet 1976:53) very likely pertains to the same species.

NIGERSAURUS TAQUETI SERENO ET AL. 1999 (FIGS. 5.1, 5.3–5.8)

HOLOTYPE: MNN GDF512, partial disarticulated skull and partially articulated neck preserved in close association on 1 m² of sandstone outcrop. Sereno et al. (1999:1346) also list a "scapula, forelimbs, and hind limbs" as part of the holotype. These and other bones, which we regard as referable to this species, were found at some distance from the skull and neck and cannot be reliably associated with the holotypic specimen. The partial skeleton described by Taquet (1976) also was discovered in the vicinity of the holotype and may pertain to the same species.

LOCALITY AND HORIZON: Gadoufaoua (16° 27′N, 9°8′′E), eastern edge of the Ténéré Desert, Niger Republic; Elrhaz Formation.

REFERRED MATERIAL: MNN GDF513, worn crown. Additional skeletal and dental material is described elsewhere.

REVISED DIAGNOSIS: Rebbachisaurid sauropod characterized by five accessory fenestrae in the jugal, surangular, and angular; tooth number increased to 20 and to 34 in the maxilla and dentary, respectively; tooth replacement increased to as many as 10–12 in a single column; premaxilla and dentary lacking alveolar septa; maxilla with oval (vertically elongated) replacement foramina; extension of the dentary tooth row lateral to the sagittal plane of the lower jaw; subcircular mandibular symphysis; crowns with prominent mesial (medial) and distal (lateral) ridges; scapula with prominent rugosity on the medial aspect of the proximal end of the blade.

DESCRIPTION

SKULL AND DENTITION

The skull and neck of the holotypic specimen of *Nigersaurus taqueti* were found in close association. Most of the dorsal skull roof is preserved (fig. 5.1A). The braincase is intact, with the proximal end of the stapes in place in the fenestra ovalis. The quadrate is the only palatal bone preserved. All of these bones, with the exception of the frontal and braincase, are composed of thin laminae or narrow struts and are extremely delicate. Five unique accessory fenestrae are present, two in the jugal and three in the surangular and angular (figs. 5.1–5.5). These accessory openings are bordered by bone that tapers gradually in width to a paper-thin edge that has a smooth margin. It is highly unlikely, therefore, that they represent lesions or some other kind of bone pathology, like the healed openings reported in aged individuals of other dinosaurian species (e.g., Brochu 2003).

The snout is proportionately much shorter and the dental arcade is less prognathous than in diplodocids or dicraeosaurids (figs. 5.1, 5.2). As mentioned above, the external naris is not retracted above the orbit in *Nigersaurus*. A statement to the contrary—that the external naris is positioned as in diplodocids—was made on the basis of the maxilla before the premaxilla was exposed (Sereno et al. 1999:1344). Although the nasal and thus the posterior margin of the external naris are not known, the anterior portion of the border is far anterior to that in any known diplodocid. The external naris is large, laterally facing, separated in the midline from its opposite by a vertical premaxilla–nasal septum, and surrounded by a more pronounced narial fossa. As in other diplodocoids, the jaw articulation and laterotemporal fenestra are shifted anteriorly under the orbit (figs. 5.1, 5.2). The supratemporal fenestra is very reduced or absent altogether, in contrast to those in other diplodocoids (Holland 1924) or other sauropods.

The transversely expanded form of the alveolar ramus of the dentary is unique among dinosaurs (fig. 5.1B). No other dinosaur has a tooth row that extends lateral to the longitudinal plane of the lower jaw. The maximal width across the anterior end of the paired dentaries slightly exceeds the length of the entire lower jaw. Despite the gaping alveolar trough that housed hundreds of slender teeth packed together as a tooth battery, the dentary and other bones of the lower jaw in *Nigersaurus* are as lightly built as those in *Diplodocus* and much more slender than those in *Camarasaurus* (figs. 5.1, 5.2). The description below is limited to the tooth-bearing bones and the teeth.

PREMAXILLA

The premaxilla is a slender bone, the ventral third of which houses a battery of teeth aligned in four columns (fig. 5.3, table 5.1). The dorsal two-thirds of the premaxilla extends as a thin lamina appressed to its opposite in the midline. In lateral view, these two parts of the premaxilla meet at an angle of approximately 30°, greater than that in *Diplodocus* (figs. 5.1A, 5.2B). Four replacement foramina and a connecting groove for the dental lamina are visible on the posterior aspect of the bone. Unlike those in other diplodocoids, the alveolar margin and space within the premaxilla housing replacement teeth are not divided by bony septa into discrete alveoli. Rather, they are developed as an open trough, the arched anterior wall of which is thin and extends more than 1 cm farther ventrally than the posterior wall (figs. 5.3, 5.4).

Although the active, or functioning, teeth have fallen from the trough, a battery of replacement teeth is visible in computed tomography

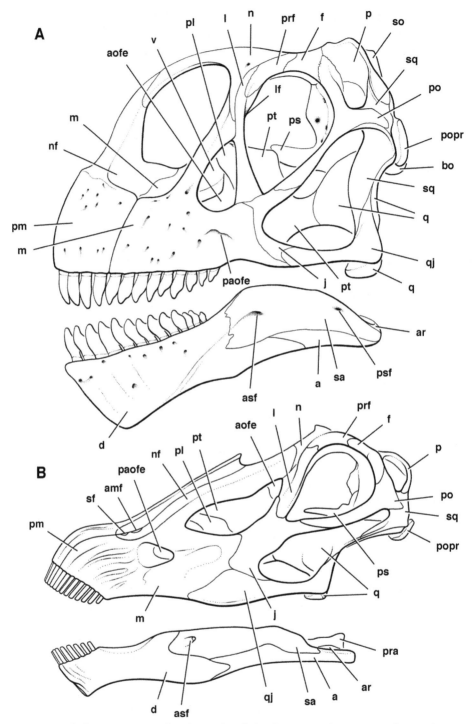

FIGURE 5.2. Skull reconstructions in lateral view of (A) the basal neosauropod *Camarasaurus lentus* and (B) the diplodocid *Diplodocus longus* (from Wilson and Sereno 1998). **a**, angular; **amf**, anterior maxillary foramen; **aofe**, antorbital fenestra; **ar**, articular; **asf**, anterior surangular foramen; **bo**, basioccipital; **d**, dentary; **f**, frontal; **j**, jugal; **l**, lacrimal; **lf**, lacrimal foramen; **m**, maxilla; **n**, nasal; **nf**, narial fossa; **p**, parietal; **paofe**, preantorbital fenestra; **pl**, palatine; **pm**, premaxilla; **po**, postorbital; **popr**, paraoccipital processes; **pra**, prearticular; **prf**, prefrontal; **ps**, parasphenoid; **psf**, posterior surangular foramen; **pt**, pterygoid; **q**, quadrate; **qj**, quadratojugal; **sa**, surangular; **sf**, surangular foramen; **sq**, squamosal; **so**, supraoccipital; **v**, vomer.

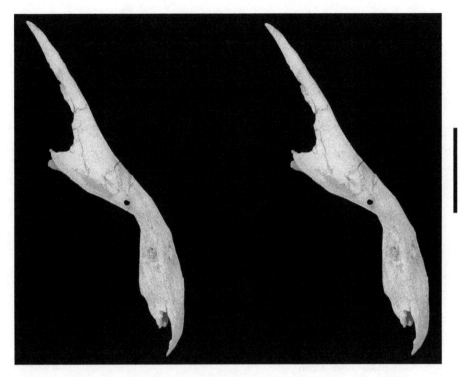
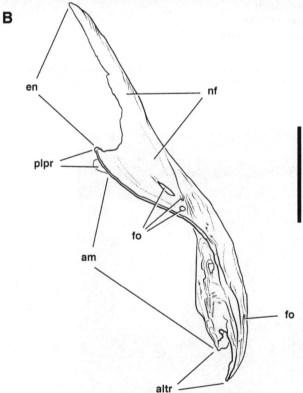

FIGURE 5.3. (A) Stereopair and (B) matching drawing of the right premaxilla of *Nigersaurus taqueti* (MNN GDF512) in lateral view. **am**, articular surface for the maxilla; **altr**, alveolar trough; **en**, external naris; **fo**, foramen; **nf**, narial fossa; **plpr**, palatal processes. Scale bars equal 5 cm.

TABLE 5.1
Measurements of the Tooth-Bearing Bones and Teeth of Nigersaurus taqueti

Premaxilla	
Maximum length	209
Maximum width of maxillary articular surface, midheight	18
Alveolar trough, anteroposterior width	10
Posterodorsal ramus, length	77
Maxilla	
Alveolar ramus	
Preserved width	94
Depth (to ventral rim of antorbital fenestra)	73
Replacement foramen, height	11
Replacement foramen, width	3
Dentary	
Alveolar ramus, width	115
Alveolar trough, anteroposterior width of medial end	12
Symphysis	
Dorsoventral height	21
Anteroposterior depth	23
Teeth	
Maxillary tooth	
Crown height	(25)
Basal crown width	4
Dentary tooth	
Crown height	(20)
Basal crown width	3

NOTE: Measurements are for the holotypic skull of *Nigersaurus taqueti* (MNN GDF512). Parentheses indicate estimated measurement. Measurements are in millimeters.

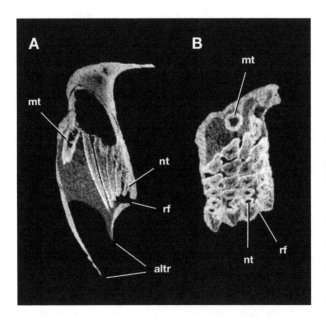

FIGURE 5.4. High-resolution computed tomographic sections through the right premaxilla of *Nigersaurus taqueti* (MNN GDF512) in (A) longitudinal and (B) cross-sectional views. The packing pattern of the tooth battery and the asymmetrical distribution of enamel on individual crowns are visible. **altr**, alveolar trough; **mt**, mature tooth; **nt**, new tooth; **rf**, replacement foramen.

(CT) scans in the upper half of the alveolar ramus (fig. 5.4B). The premaxillary teeth are self-supporting and in mutual contact along the lengths of their crowns, which presumably erupted and wore as a unit. The teeth originate just anterior to the replacement foramina. As the teeth increase in length, their roots extend into the upper half of the alveolar ramus before migrating ventrally toward the alveolar trough (fig. 5.4A).

MAXILLA

The ventral one-fifth of the maxilla houses a battery of teeth arranged in columns internal to each oval replacement foramen (Sereno et al. 1999:fig.2D; fig. 5.5, table 5.1). The dorsal four-fifths of the bone is developed proximally as a thin plate and distally as a narrow and delicate strut that divides the external naris and antorbital fenestra (fig. 5.1). As in the premaxilla, the alveolar margin is developed as an undivided trough, as seen in posterior view, with the labial (anterior) wall extending approximately 1 cm farther ventrally than the lingual (posterior) wall. Unlike the premaxilla and dentary, bony septa are present within the body of the alveolar ramus. These septa join the posterior and anterior walls, enclosing crypts for each replacement series that are approximately twice as deep labiolingually (anteroposteriorly) as mesiodistally (transversely).

Although all worn maxillary teeth have fallen from the alveolar trough, columns of replacement teeth are visible in the two most mesial (medial) tooth positions (Fig. 5.5). As in the premaxilla and dentary, the teeth originate just labially (anteriorly) to each replacement foramen, extend with growth dorsally and ventrally within the alveolar ramus, and then migrate in series toward the alveolar trough.

DENTARY

The dentary has a very unusual structure in *Nigersaurus taqueti* (figs. 5.1, 5.6, table 5.1). Other rebbachisaurids may eventually be shown to share some or all of these apomorphies. Several aspects of the dentary, however, currently have no parallel among reptiles.

In dorsal view, the T-shaped bone is divided into a mandibular ramus, oriented anteroposteriorly, and an alveolar ramus, oriented transversely (figs. 5.1B, 5.6). The posterior end of the mandibular ramus, as in other dinosaurs, is oriented in a parasagittal plane. It is divided into a slender dorsal process, which contacts the surangular, and a broader, tongue-shaped ventral process. The anterior end of the mandibular ramus, unlike that in any other sauropod, is dorsoventrally compressed and flares in transverse width before joining the alveolar ramus (fig. 5.1B). As a consequence, this portion of the mandibular ramus appears unusually slender in lateral view (fig. 5.1A). In dorsal view, this portion of the mandibular ramus is subtriangular with a concave dorsal surface. Laterally it is bounded by a sharp, upturned edge that joins the lateral extremity of the alveolar ramus. A large oval foramen opens in the center of this portion of the mandibular ramus. This foramen lies dorsal to a sizable vascular canal that passes anteriorly within the dentary toward the alveolar trough. Shallow grooves also exit the foramen and pass anteriorly toward the row of replacement foramina (Sereno et al. 1999:fig. 2C; fig. 5.6).

The breadth of the alveolar ramus exceeds the length of the mandibular ramus, a remarkable proportion that results in lower jaws that are as broad as they are long (figs. 5.1B, 5.6, table 5.1). The extraordinary width of the anterior end of the lower jaws is the result of the lateral extension of the alveolar ramus, approximately 30% of which is positioned lateral to the sagittal plane of the mandibular ramus. The medial portion of the alveolar ramus is subcylindrical. The mandibular symphysis, as a result, is subcircular (table 5.1). The symphysial surface is smooth and gently concave, suggesting that a small amount of flexion about the sagittal plane may have been possible.

The alveolar trough is nearly straight in dorsal view (fig. 5.1B). It narrows in width toward its lateral end. Other changes toward the lateral side of the trough include a decrease in the size of the replacement foramina and a decrease in

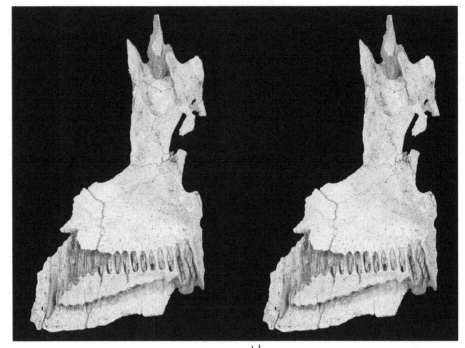
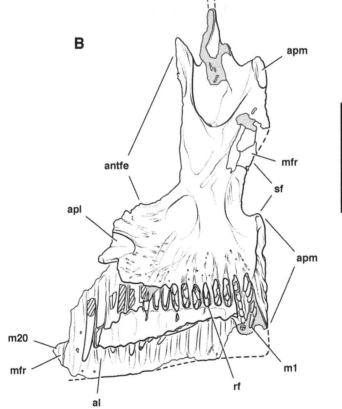

FIGURE 5.5. (A) Stereopair and (B) matching drawing of the left maxilla of *Nigersaurus taqueti* (MNN GDF512) in posterior view with two replacement tooth columns exposed by erosion. **al**, alveolus; **antfe**, antorbital fenestra; **apm**, articular surface for the premaxilla; **apl**, articular surface for the palatine; **m1– m20**, maxillary tooth positions 1–20; **mfr**, missing fragment; **rf**, replacement foramen; **sf**, subnarial foramen. Scale bars equal 5 cm.

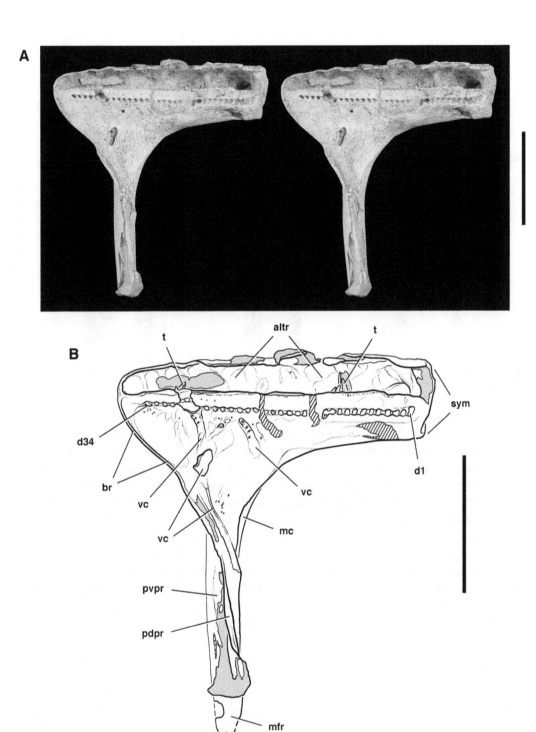

FIGURE 5.6. (A) Stereopair and (B) matching drawing of the left dentary of *Nigersaurus taqueti* (MNN GDF512) in dorsal view. **altr**, alveolar trough; **br**, buccal ridge; **d1–d34**, replacement foramina for dentary tooth positions 1–34; **mc**, Meckel's canal; **mfr**, missing fragment; **pdpr**, posterodorsal process; **pvpr**, posteroventral process; **sym**, symphysis; **t**, tooth; **vc**, vascular canal. Scale bars equal 5 cm.

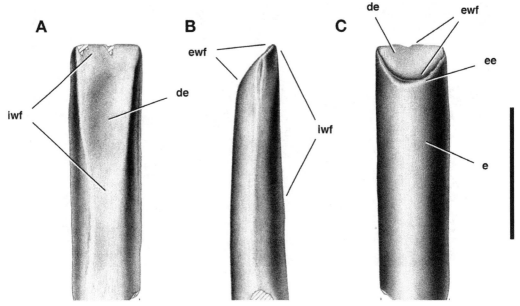

FIGURE 5.7. Isolated tooth of *Nigersaurus taqueti* (MNN GDF513) in (A) lingual (internal), (B) mesial or distal (medial or lateral), and (C) labial (external) views. **de**, dentine; **e**, enamel; **ee**, enamel edge; **ewf**, external wear facet; **iwf**, internal wear facet. Scale bar equals 1 cm.

the distance between the replacement foramina and the edge of the alveolar trough. This gradual decrease in the width of the trough and the size and position of the replacement foramina suggests that the replacement rate and/or tooth size decreased toward the lateral end of the tooth row. Several loose teeth are preserved in the bottom of the alveolar trough and confirm the gradual decrease in tooth size laterally along the battery. There are 34 replacement foramina and, thus, 34 columns of teeth, which is 20 more than are typically present in the dentary of the diplodocid *Diplodocus* (Holland 1924; fig. 5.2B). As there are no septa at any depth within the alveolar trough, dentary teeth would have erupted and migrated toward the functioning wear surface as a self-supporting dental battery. No other sauropod has a dentary with an alveolar ramus of similar form lacking septa.

TEETH

The upper and lower teeth are virtually identical in form (Sereno et al. 1999: fig. 2C; fig. 5.7). These similarities are based on the premaxillary, maxillary, and dentary teeth in situ in MNN GDF513. Upper and lower teeth differ only in size; dentary teeth are smaller, perhaps by as much as 20%–30%. This is difficult to estimate more accurately because the available dentary teeth, a few of which remain within the alveolar trough, are of unknown maturity. A similar size differential, with upper teeth larger and longer than the lowers, has been observed in *Diplodocus* (Holland 1924:389) and the titanosaur *Nemegtosaurus* (Wilson 2005: 305).

The numbers of premaxillary, maxillary, and dentary tooth positions are 4, at least 24, and 34, respectively. Maxillary and dentary tooth counts are approximately double those in other diplodocoids such as *Diplodocus* (Holland, 1924; fig. 5.2B). Although the maxillary tooth row is incomplete laterally, there are probably only a few missing tooth positions, because the length of the combined premaxillary and maxillary tooth rows in MNN GDF513 is only slightly less than that for the dentary. The higher number of tooth positions in the dentary (34 versus approximately 30 for the premaxilla plus maxilla) is consistent with the smaller size of the dentary teeth.

Crown shape and structure are uniform among upper and lower teeth. Unworn crowns are gently curved along their length and taper gradually to a narrow, rounded tip, which shows subtle wrinkles on its thickly enameled labial (external) surface. Teeth in place in the premaxilla and maxilla and near their natural position in the dentary confirm that the crowns are convex labially (anteriorly) in both the upper and the lower jaws.

At midlength, the crown has a trapezoidal cross section, with smooth enamel and a longitudinal groove on each side that accommodates the edges of adjacent crowns (Sereno et al. 1999:fig. 2D; fig. 5.7). Toward its tip, the crown has an oval cross section. The enamel is approximately eight times thicker on the labial (external), as opposed to the lingual (internal), side of the crown in both upper and lower teeth. Whereas other diplodocids and some titanosaurians have similar-shaped, narrow, cylindrical crowns, markedly asymmetrical enamel has not been reported previously among sauropods and is absent in other diplodocoids (*Dicraeosaurus* [Janensch 1935–1936:pl. 12, fig. 16], *Diplodocus* [Holland 1924:fig. 6]).

TOOTH WEAR

Although wear facets are not present in teeth preserved in the holotypic jaw bones of *Nigersaurus taqueti*, worn teeth referable to this species have been recovered from many sites in the Gadoufaoua beds and show a distinctive pattern of wear. There is little reason to doubt the reference of these teeth to *Nigersaurus taqueti*, despite the presence of an unnamed titanosaurian in the same horizons; the crowns have narrow cylindrical proportions and highly asymmetrical enamel.

Unlike the teeth in nearly all other sauropods, those in *Nigersaurus taqueti* have a pair of wear facets located on opposite sides of the crown. The first—a labial (external) facet—is typical of dicraeosaurids and diplodocids; it cuts the crown at a high angle and appears to be the product of nonocclusal abrasion (Holland 1924; Upchurch and Barrett 2000; fig. 5.7B, C). The elevated, rounded rim of thick enamel suggests that the wear facet was produced by ingested plant matter rather than the harder enameled edge of an opposing crown. The second—a lingual (internal) facet—resembles the internal facet on some narrow-crowned titanosaurians; it cuts the crown at a low angle and appears to be the product of occlusal abrasion (fig. 5.7B, C).

TOOTH BATTERY

STRUCTURE

Archosaur teeth, primitively, are anchored by alveoli that are separated from adjacent teeth by bony septa. They erupt *en echelon* along the jaw rather than in unison as a single unit. In most dinosaurian herbivores, wear facets develop from tooth-to-tooth occlusion at several points along the tooth row rather than developing as a continuous wear surface.

A *tooth battery*, by contrast, is here defined as a tooth composite composed of self-supporting teeth that erupt and wear in unison. The teeth are supported by adjacent teeth and erupt as a unit with wear surfaces that are continuous from one tooth to the next. In *Nigersaurus* part of the dentition fulfills these criteria—the premaxilla and dentary tooth rows. These bones do not have alveoli, intervening septa, or even well-developed grooves to guide erupting tooth columns. Rather, the teeth are packed into an open alveolar trough.

Although all functioning (worn) teeth have fallen away from the jaws of the holotypic specimen, the tooth battery is intact within the body of the premaxilla (fig. 5.3), as visualized in cross and longitudinal sections with high-resolution CT (fig. 5.4). Embryonic teeth migrate into the alveolar trough via replacement foramina. As they grow in length and diameter, they migrate deeper into the body of the premaxilla before passing ventrally out of the alveolar trough (fig. 5.4A). There appear to be as many as 10 teeth at a single tooth position, from embryonic tooth to wearing crown. The four tooth columns in the premaxilla are arranged *en echelon* so that the widest portion of a given crown contacts the more tapered portion of adjacent crowns, with the prominent edge of enamel lodged in a groove on the side of adjacent crowns (fig. 5.4B).

Less is known about the structure of the tooth battery of the dentary, although it was probably very similar. The crowns have the same shape, structure, and orientation. The position of the replacement foramina near the margin of the alveolar trough indicates that embryonic teeth began their trajectory closer to the open end of the trough and grew deep into the alveolar trough before emerging at the functional end of the tooth battery. The most distal (lateral) tooth columns in the dental battery of the dentary must have angled anterolaterally as shown by the extension of the trough lateral to the lateral most replacement foramen (fig. 5.1B).

The body of the maxilla has septa separating columns of teeth (fig. 5.5). Closer to the ventral end of the maxilla, the septa give way to an open trough with a groove for each tooth column on anterior and posterior walls. As in the premaxilla, there appear to have been at least 10 teeth to a column medially and fewer laterally; an incomplete sequence of 8 teeth is visible in the first column (fig. 5.5). The maxillary teeth, presumably, emerged from the alveolar trough as a self-supported tooth battery.

FUNCTION

Exactly how the tooth batteries of *Nigersaurus* functioned to produce the wear facets evident on isolated teeth remains unsolved, despite knowledge of the general structure of the tooth batteries. What we can outline at this point are several observations that provide some insight into the mystery.

We know that teeth in lower and upper tooth rows do not match one-to-one given their differing sizes and numbers. We also know that both upper and lower tooth batteries were constructed in a similar manner, with the convex, thickly enameled crown surface facing labially (anteriorly). Many mammals, such as rodents, have analogous anteriorly positioned, self-sharpening lower and upper incisors with lingual (internal) wear facets (Taylor and Butcher 1951). As in *Nigersaurus*, both lower and upper incisors are externally convex, with thickened enamel on their labial (external) sides and wear facets facing lingually (internally). This is the closest analogy to the structure and orientation of the individual teeth in *Nigersaurus*. Mammalian incisors like these, however, are much more robust and are sharpened in a manner very different from that in any dinosaur. The chisel-shaped edge is maintained by breakage (preferential chipping of the softer dentine) and by tooth-to-tooth abrasion. Unlike *Nigersaurus*, the facets on lower and upper incisors are not symmetrical (the uppers typically have higher-angle, stepped facets), and the shearing breakage that keeps the leading edge sharp rarely produces facets that are uniformly planar.

We strongly suspect that the low-angle, lingual (internal) wear facet was produced by tooth-to-tooth abrasion, because the facet is extremely flat and cuts smoothly across the external margin of enamel. The high-angle labial (external) wear facet, in contrast, appears to have been produced by tooth-to-plant abrasion, because the facet is concave, with a rounded, polished rim of enamel along its trailing labial (external) edge (fig. 5.7).

Because all of the isolated teeth from Niger, England (Naish and Martill 2001), and Brazil (Kellner 1996) show a similar pattern of wear—with the low-angle facet first to appear—it is likely that both lower and upper teeth are represented and wear in a similar manner. Yet it is not clear how this is functionally possible. Hadrosaurid and ceratopsian tooth batteries have opposing, mirrored patterns of tooth wear: the enamel is thickest on opposing sides of the crown (medially in maxillary teeth, laterally in dentary teeth); the crowns curve in opposite directions (laterally in maxillary teeth, medially in dentary teeth); and the wear facets occur in opposing orientations (medially or ventromedially in maxillary teeth, laterally or dorsolaterally in dentary teeth). Although an isolated ornithischian tooth from a tooth battery may be difficult to assign to either the upper or the lower tooth row (when upper and lower crowns have a similar shape and ornamentation), these teeth have mirrored positions when they are found in place. One possible explanation for the uniform pattern of wear in *Nigersaurus* is that the lingual (internal), low-angle facet is produced by the lower crowns passing lingually (internally) to the upper crowns (the usual tetrapod condition) but that the lower crowns are worn away and eventually obliterated in the process. If this were true, the entire sample of isolated teeth of *Nigersaurus* and related taxa found now on several continents would include only premaxillary and maxillary teeth from the upper tooth batteries. If not, and if we are correct that the low-angle facet is produced by tooth-to-tooth occlusion, there must have existed an occlusal mechanism unlike any described to date among tetrapods—a mechanism capable of producing low-angle, lingual (internal) wear facets on both lower and upper crowns.

Another conundrum involves the cutting edge of the crown. In ornithischians with tooth batteries and in mammalian herbivores, thickened enamel is always positioned along the cutting edge. This is not the case in *Nigersaurus*. Not only are the lower and upper teeth oriented with their curvature and thickened enamel on the same (external) side, but the leading wedge-shaped edge of the crown is formed entirely of dentine. It is difficult to understand how this edge, which is perfectly straight, is maintained in the course of wear without the protection of enamel (fig. 5.7). Even if one envisages adjacent crowns, it is hard to understand how the soft leading edge of dentine would not be concave from wear rather than straight.

A final observation suggests that the tooth battery of *Nigersaurus* is quite different from that in ornithischians. To produce an elongate, low-angle wear facet on the lingual (internal) side of the crown (fig. 5.7A), most of this side of the crown must have been exposed. The succeeding replacement crown, in other words,

must have been positioned at a good distance from the cutting edge of the functional crown. Only one crown in each tooth column could have been functional. In ornithischian dentitions, higher-angle wear facets allow more than a single crown in the same tooth column to participate in the active, cutting surface of the dental battery.

In the future, we plan to digitally define and prototype the intact portion of the tooth battery within the premaxilla in the hope that it will shed further light on how the battery functioned during mastication.

DISCUSSION

COMPARISONS

SKULL

In general proportions, the skull of *Nigersaurus taqueti* bears little resemblance to that of diplodocids (figs. 5.1, 5.2B). It has a more abbreviate, less prognathous snout; the depth of the cranium is approximately 90% of its length (as measured from the snout to the quadrate condyle). This cranial proportion is even more abbreviate than that of *Camarasaurus* (fig. 5.2A) and *Jobaria* (Sereno et al. 1999), which have a cranial depth between 50% and 60% of its length. The external naris in *Nigersaurus* is large and parasagittal in position as in *Camarasaurus* and *Jobaria*. In diplodocids, in contrast, the external naris is smaller, dorsally facing, and retracted to a position anterodorsal to the orbit (the condition in dicraeosaurids is as yet unknown). The laterotemporal fenestra in *Nigersaurus* is proportionately elongate and extends anteriorly as far as the antorbital fenestra and external nares, farther than in any sauropod described to date. There is no development of a preantorbital fenestra as occurs in other neosauropods (fig. 5.2B). The dermal bones of the skull roof in *Nigersaurus* are remarkably slender and delicate compared to those of other sauropods including diplodocids. The posterodorsal ramus of the maxilla, which separates the external naris and antorbital fossa, is reduced to an extremely delicate, strap-shaped lamina 1 mm thick and a few millimeters wide (fig. 5.1).

The lower jaw is easy to distinguish from that in diplodocids (figs. 5.1, 5.2B). The coronoid process on the surangular is prominent and deep in lateral view, more closely resembling that in the titanosaurian *Rapetosaurus* (Curry Rogers and Forster 2004) than the low profile jaws of diplodocids (fig. 5.2B). In *Nigersaurus* the teeth are restricted to the transverse portion of the anterior end of the skull, and at least the lower tooth row extends lateral to the parasagittal plane of the lower jaw. In both of these attributes, *Nigersaurus* is unique among dinosaurs. Diplodocids show an incipient condition in these regards; all but the lateral extremities of the tooth rows are positioned along the anterior, transverse margin of the skull, and the dentary tooth row flares just beyond the parasagittal plane of the posterior portion of the lower jaw.

A dentary from Upper Cretaceous rocks in South America referred to *Antarctosaurus wichmannianus* (Huene 1929:69, pl. 29, fig. 5) is similar to that in *Nigersaurus taqueti* in the extreme breadth of the transverse portion of the ramus and the concomitant increase in tooth count. There are at least 24 teeth in the dentary, with the majority (approximately 18) located in the broad, transverse portion of the ramus. The dentary of this South American form, however, is not as derived as that in *Nigersaurus*. The teeth are set in sockets rather than an undivided alveolar trough, the tooth row is L-shaped rather than straight and restricted to the transverse portion of the ramus, and the symphysial surface is narrow rather than circular. Despite these similarities, the phylogenetic affinity of *Antarctosaurus* is not yet resolved. There is a possibility that this taxon, which was found together with titanosaur cranial and postcranial remains, may have acquired these features convergently with *Nigersaurus*.

TEETH AND TOOTH WEAR

Teeth that closely match those of *Nigersaurus taqueti* in form, structure, and wear have been described recently from slightly older Barremian-age beds on the Isle of Wight (Naish and Martill 2001:pl. 36) and from the Upper Cretaceous Bauru Group in Brazil (Kellner 1996:fig. 7). The crowns are narrow and sub-cylindrical and, at midlength, have a trapezoidal cross section. The size, form, and angle of the pair of wear facets are exactly like those described above for *Nigersaurus taqueti*. As noted by Kellner (1996:619), the low-angle lingual (internal) facet is the first to appear and is always more elongate than the labial (external) facet, as preserved in progressively worn crowns. The enamel may have an asymmetrical distribution on the crown, but this needs firsthand verification. Kellner provisionally regarded these teeth as titanosaurian, because of the predominance of titanosaurian postcranial bones from the same beds. It is very probable, however, that these Brazilian teeth belong to a rebbachisaurid diplodocoid that lived on South America in the Late Cretaceous. (e.g., *Limaysaurus*; Salgado et al. 2004).

In other diplodocoids, the teeth are larger relative to the jaw bones and anchored in individual alveoli (fig. 2B; *Dicraeosaurus* [Janensch 1935–1936:figs. 107, 111]; fig. 2B). The crowns have a circular cross section, symmetrical enamel, and a single low-angle, labial (external) wear facet (Holland 1924). Despite some variation, isolated teeth of *Dicraeosaurus* show the same external facet (Janensch 1935–1936:pl. 12, figs. 23, 25). This facet is characterized by a rounded lip of enamel along its trailing (external) edge (fig. 5.7; ewf) and scratches that course across the dentine from the internal to the external sides (Fiorillo 1991; Calvo 1994b; Upchurch and Barrett 2000). How this wear facet formed has remained a mystery ever since Holland (1924) summarized early speculation, which included scraping cycad trunks (Holland), procuring fish hiding in stream beds (Tornier), and munching on freshwater bivalves (Sternfeld). Because Holland (1924:fig. 4) depicted "*Diplodocus* seizing a mussel," that hypothesis gained the upper hand, although Holland maintained no personal preference. More recent proposals include stripping leaves from branches (Dodson 1990; Barrett and Upchurch 1994) and raking

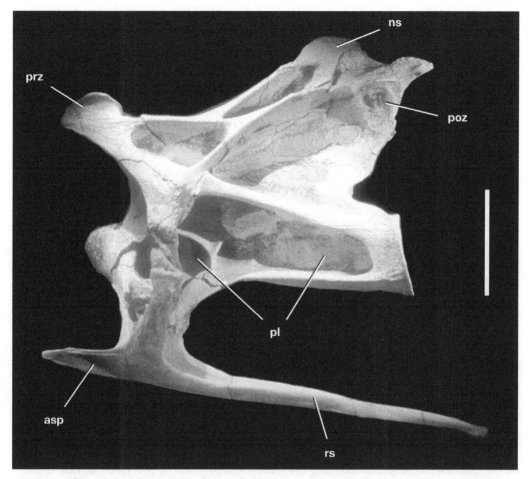

FIGURE 5.8. Fifth cervical vertebra and co-ossified rib of *Nigersaurus taqueti* (MNN GDF512) in lateral view. **asp**, anterior spine; **ns**, neural spine; **pl**, pleurocoel; **poz**, postzygapophysis; **prz**, prezygapophysis; **rs**, rib shaft. Scale bar equals 5 cm.

bark (Bakker 1986). *Nigersaurus* adds a new wrinkle. Here we see the same kind of external facet in teeth set into a tooth battery. These batteries, in turn, are housed in a skull with a very different shape and even more delicate construction than in diplodocids (fig. 5.2).

Some titanosaurians have narrow-crowned teeth that bear a general resemblance to those of *Nigersaurus taqueti*, but these teeth have more robust proportions (e.g., Kellner and Mader 1997:fig.2; *Rapetosaurus* [Curry Rogers and Forster 2004:fig. 32]). The internal (lingual) facet—the only one present—cuts the crown at a low angle. An external (labial) facet has never been described outside Diplodocoidea. *Nemegtosaurus*, which has internal V-shaped and apical wear facets (Nowinski 1971; Wilson 2005), has continued to confuse discussions of tooth form and masticatory style among diplodocoids (e.g., Upchurch and Barrett 2000). *Nemegtosaurus* and the conspecific, or closely allied, *Quaesitosaurus* from the Late Cretaceous of Asia are better understood as titanosaurians rather than diplodocoids (Calvo 1994a; Wilson and Sereno 1998; Wilson 2002, 2005).

EVOLUTION OF A SAUROPOD TOOTH BATTERY

Nigersaurus taqueti must be placed within the context of diplodocoid phylogeny to better understand how its novel tooth battery evolved (fig. 5.9). Diplodocoids include diplodocids, dicraeosaurids, and rebbachisaurids. Diplodocids are best known from the Late Jurassic of North America and include *Diplodocus*, *Apatosaurus*,

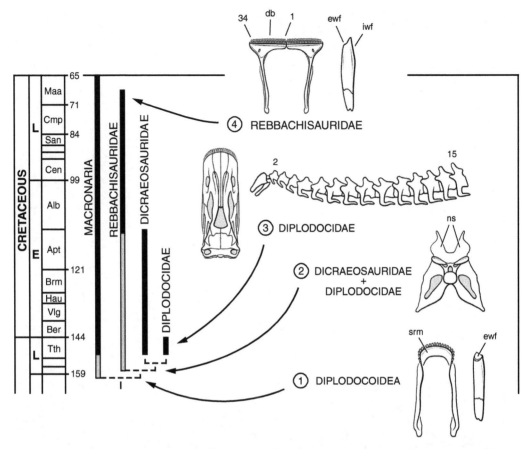

FIGURE 5.9. Phylogram showing the relationships and recorded durations of the three principal diplodocoid clades (Rebbachisauridae, Dicraeosauridae, Diplodocidae) and some of their feeding specializations. Node 1 (Diplodocoidea): slender cylindrical crowns, low-angle external (labial) wear facet, teeth restricted anteriorly in subrectangular muzzle. Node 2 (Dicraeosauidae + Diplodocidae): bifid cervical neural spines. Node 3 (Diplodocidae): nares retracted, 15 cervical vertebrae. Node 4 (Rebbachisauridae): low-angle, internal wear facet, increase in tooth number, asymmetrical enamel, dental batteries in some taxa. 1–34, dentary teeth 1 through 34; 2–15, cervical vertebrae 2–15; **db**, dental battery; **en**, external naris; **ewf**, external wear facet; **iwf**, internal wear facet; **ns**, neural spine; **srm**, subrectangular muzzle.

Barosaurus, and others (Hatcher 1901; Holland 1924; Berman and McIntosh 1978; Ostrom and McIntosh 1999). Dicraeosaurids include *Dicraeosaurus* (Janensch 1935–1936) from the Late Jurassic of Africa and the long-spined *Amargasaurus* from the Early Cretaceous of South America. Rebbachisaurids include *Limaysaurus* (Calvo and Salgado 1995; Salgado et al. 2004) from the Early Cretaceous of South America and *Nigersaurus* (Sereno et al. 1999) and *Rebbachisaurus* (Lavocat 1954) from the mid- and Late Cretaceous of Africa. Rebbachisaurid teeth and other fragmentary remains indicate that the group was also present in Europe and persisted to the end of the Cretaceous at least in South America. As a sister taxon to dicraeosaurids and diplodocids, rebbachisaurids must have diverged from other diplodocoids before the end of the Late Jurassic, although no trace of the group has yet been found from this period (fig. 5.9).

Although the teeth of *Nigersaurus* are particularly slender, all diplodocoids are characterized by proportionately narrow, subcylindrical crowns that are only weakly expanded and flattened (fig. 5.9, node 1). The rate of tooth replacement, in addition, appears to have been accelerated across the group, although this is most apparent in *Nigersaurus*. In this regard, we assume that the number of teeth in a single column is related to the replacement rate. *Diplodocus*, for example, has as many as six teeth in a single column in the maxilla (Marsh 1884:pl. 4, fig. 3). Worn diplodocoid teeth have a characteristic external, low-angle, wear facet on both upper and lower teeth (fig. 5.7 ewf) that must have formed from tooth-to-plant abrasion (*contra* Calvo 1994b; Barrett and Upchurch 1994; Upchurch and Barrett 2000). Most diplodocoid teeth are positioned in a transverse row along the squared anterior margin of the snout (fig. 5.9, node 1).

Nigersaurus has further developed a number of these features (fig. 5.9, node 2). Tooth size decreases relative to the size of the jaw bones, and tooth number increases to 30 or more in lower and upper tooth rows. In *Nigersaurus* the teeth are restricted to the anterior margin of the snout and extend lateral to the sagittal plane of the lower jaw, features unknown elsewhere among dinosaurs. More teeth are present in a single column, and the rate of replacement, presumably, has increased as well. Most of the teeth in the jaws are pressed so close to one another that there are no intervening septa and, for the first time among sauropods, batteries of self-supporting teeth erupt as a single unit. Although still a functional conundrum, an additional low-angle wear facet appears on the lingual (internal) side of the active crowns in both lower and upper tooth batteries.

Dicraeosaurids and diplodocids evolved other features that likely impacted food procurement but are not present in *Nigersaurus* or other rebbachisaurids. These include a reduction in the number of teeth, a projecting "chin" on the dentary, a relative lengthening of the snout, retraction of the external nares, and an increase in the length and number of cervical vertebrae (fig. 5.9, nodes 3, 4). In dicraeosaurids and diplodocids, in addition, the resting curvature of the neck arches anteroventrally (Janensch 1929; Stevens and Parrish 1999). This is not the case in *Nigersaurus*, as exemplified by the fifth cervical vertebra (fig. 5.8).

COMPARISON TO ORNITHISCHIAN TOOTH BATTERIES

Tooth batteries evolved twice among ornithischian dinosaurs from a hypothetical common ancestor with a simple dentition characterized by a suite of ornithischian features related to herbivory—a predatary, cheek embayments on the dentary and maxilla, asymmetrical enamel in dentary and maxillary crowns, and wear facets from tooth-to-tooth occlusion on the buccal (lateral) and lingual (medial) sides of dentary and maxillary teeth, respectively. Evolution of tooth-supported batteries occurred first among ornithopods, with initial changes toward a dentition occurring before the close of the Jurassic, and later among ceratopsian dinosaurs, where all structural changes occurred during the Late Cretaceous (Sereno 1997, 2000).

The fully developed tooth batteries in hadrosaurids and ceratopsids are structurally very similar, evolved in comparable stages, and involved an increase in body size of approximately an order of magnitude (Sereno 1997: fig. 7; 2000: fig. 25.9). Independent but similar structural changes include a relative decrease in tooth size and increase in tooth columns and replacement rate, the loss of alveolar septa, restriction of the enamel to opposing sides of the crown in maxillary versus dentary teeth, an increase in the prominence of a ridge on the enameled side of the crown, adjustment of the crown shape for efficient packing, an increase in the volume of supporting bone in the maxilla and dentary, a reduction of postdentary elements in the lower jaw, and the development of a coronoid process with an expanded process for muscular attachment.

The circumstances surrounding the evolution of tooth batteries in rebbachisaurid sauropods bear a few similarities to and many striking differences from those in ornithischians. Similarities include the reduction of tooth size, increase in number of tooth columns, increase in replacement rate (or at least the number of teeth per column), loss of alveolar septa between tooth columns, thickened enamel on one side of the crown and near-loss of enamel on the other, and adjustment of crown shape for efficient packing. These similarities, thus, are confined to the size, shape, number, and rate of replacement of the teeth and the asymmetrical distribution of enamel.

Fundamental differences begin with timing. Available rebbachisaurid fossils suggest that a tooth battery among sauropods had evolved sometime during the Early Cretaceous; the teeth from the Isle of Wight (Naish and Martill 2001) look very similar to those of *Nigersaurus* and are Barremian (ca. 125 Ma; middle Early Cretaceous) in age. This postdates the establishment of tooth-supported dentitions in ornithopods (Late Jurassic) but predates the appearance of ceratopsian tooth batteries (Late Cretaceous). Another basic difference involves body size. There was no increase in body size among sauropods concomitant with the development of tooth batteries. Although *Rebbachisaurus* ranks among the largest of sauropods, *Nigersaurus* ranks among the smallest, with a body length of approximately 15 m.

The orientations of the tooth batteries are diametrically opposed. The tooth batteries have an anteroposterior, rather than a transverse, orientation in ornithischians and rebbachisaurids, respectively. The ornithischian tooth battery is located posteriorly within the jaws and used for food processing; cropping is a function of an expanded, toothless bill. The sauropod dental battery, in contrast, is located anteriorly and may have been used primarily in cropping.

The presence or absence of gastroliths as an accessory means to break down plant matter may be correlated with the aforementioned fundamental functional differences. The absence of gastroliths among ornithischians with advanced dentitions (euornithopods, neoceratopsians) including dental batteries suggests that more efficient oral processing of plant matter has replaced gut-processing by gastroliths (Sereno 1997:473). In sauropods, in contrast, the group in which dental batteries evolved (diplodocoids) has the greatest proven incidence of gastroliths, suggesting that their derived dental features did not function primarily in the breakdown of plant matter.

Gastroliths are present in rebbachisaurids (Calvo 1994b), dicraeosaurids (Janensch 1929), and diplodocids (Cannon 1906; Brown 1941; Gillette 1990). The absence of gastroliths among nondiplodocoid sauropods is based on many articulated skeletons of *Shunosaurus*, *Camarasaurus*, *Jobaria*, and *Opisthocoelicaudia*. *Cedarosaurus*, a macronarian sauropod of uncertain affinity, is thus far the only nondiplodocoid sauropod with gastroliths (Sanders et al. 2001).

The pattern of wear in the rebbachisaurid tooth battery is completely different from that in ornithischians, which uses thickened enamel and tooth-to-tooth occlusion to form a self-sharpening cutting margin. In rebbachisaurids, only one of a pair of wear facets is formed by tooth-to-tooth occlusion, the sharp leading edge of worn crowns is formed in dentine rather than enamel, the thickened enamel is located on the same (labial) side of lower and upper crowns, and lower and upper crowns apparently have identical wear patterns.

The locus of the most rapid replacement and wear is different in ornithischian and rebbachisaurid tooth batteries. In ornithischians, the crown size, the number of teeth in a column, and the distance of the replacement foramen from the alveolar margin are all greatest in the middle of the tooth battery. In *Nigersaurus*, in contrast, the crown size, the number of teeth in a column, and, to a lesser degree, the distance of the replacement foramen from the alveolar margin are all greatest toward the midline.

Finally, the tooth-bearing bones are constructed differently in ornithischians and sauropods with dental batteries. In ornithischians with tooth batteries, the dentary, in particular, is robustly constructed, with a thick and prominent coronoid process for attachment of substantial adductor musculature. Postdentary elements are greatly reduced in size. This pattern of change in the lower jaw closely mirrors changes that occurred earlier in the evolution of the mammalian masticatory apparatus (Allin 1975). In *Nigersaurus*, in contrast, the tooth-bearing elements are constructed of thin laminae, the dentary does not gain in relative length in the lower jaw, the coronoid process is developed as a thin plate of bone, and the supratemporal fenestra (the usual origin of adductor musculature) is closed by approximation of surrounding bones.

CONCLUSIONS

Early in their evolution, sauropods adopted tooth-to-tooth occlusion and, in consequence, evolved lower and upper tooth rows of equivalent length, characteristic patterns of wear facets, a more substantial coronoid process, and a robust mandibular symphysis (McIntosh 1990; Calvo 1994a; Wilson and Sereno 1998; Upchurch and Barrett 2000). Among sauropods, it is now apparent that diplodocoids evolved complex dentitions during the Cretaceous, as exemplified by the dental batteries of a recently named African rebbachisaurid, *Nigersaurus taqueti* (Taquet 1976; Sereno et al. 1999).

The dental battery on each side of the upper and lower jaws is composed of more than 30 columns of teeth that are packed into a tight self-supporting unit in the premaxilla and dentary. Individual teeth have slender rod-shaped crowns characterized by highly asymmetrical enamel. Dentary teeth are somewhat smaller than but otherwise similar to premaxillary and maxillary teeth. The crowns in both lower and upper jaws have thickened enamel on their convex labial (external) side. Wear produces two stereotypical facets, the first appearing as a low-angle, lingual (internal) facet produced by tooth-to-tooth occlusion and the second as a high-angle, labial (external) facet produced by tooth-to-plant abrasion. Both facets are well developed on crowns with significant wear, resulting in a straight, sharp apical wedge of dentine where the facets intersect. How either of these facets was produced remains a significant, and largely unanswered, question.

Unlike the parasagittal dental batteries of ornithischians, the dental battery in *Nigersaurus* is oriented transversely and may have been used for cropping rather than prolonged oral processing. Although the rebbachisaurid dental battery is preserved only in *Nigersaurus*, isolated teeth from Lower Cretaceous horizons on the Isle of Wight and rocks of Late Cretaceous age in Brazil suggest that related forms with potentially a similar degree of dental complexity were present on other continents. It is highly unlikely, however, that rebbachisaurids with dental batteries ever achieved the taxonomic diversity of ornithischians with dental batteries (hadrosaurids, ceratopsids) in Late Cretaceous faunas of North America and Asia.

Dental batteries evolved three times independently within Dinosauria—in euornithopod and neoceratopsian ornithischians and in rebbachisaurid sauropods. Fundamental functional differences coupled with their diachronous appearance suggest that dinosaurian dental batteries did not evolve in response to a single environmental cue, such as the rise of angiosperms during the mid-Cretaceous (Sereno 1997: Sereno 1999; Barrett and Willis 2001).

ACKNOWLEDGMENTS

We thank members of the 1997 and 2000 expeditions for discovering the material, C. Abraczinskas for drawing from specimens and executing final drafts of reconstructions, E. Dong, T. Keillor, and R. Masek for preparing fossils, and B. Gado (Institut de Recherche en Science Humaine), and I. Kouada (Ministère de L'Enseignement Supérieur de la Recherche et de la Technologie) for granting permission to conduct fieldwork. This research was funded by The David and Lucile Packard Foundation, the National Geographic Society, the Comer Science and Education Foundation, and Nathan Myhrvold.

LITERATURE CITED

Allin, E. F. 1975. Evolution of the mammalian middle ear. J. Morphol. 147: 403–438.

Apesteguía, S., Valais, S. D., Gonzales, J. A., Gallina, P. A., and Agnolin, F. L. 2001. The tetrapod fauna of 'La Buitrera,' new locality from the basal Late Cretaceous of North Patagonia, Argentina. J. Vertebr. Paleontol. 21: 29A.

Bakker, R. T. 1978. Dinosaur feeding behavior and the origin of flowering plants. Nature 274: 661–663.

———. 1986. The Dinosaur Heresies. Bath Press, Avon.

Barrett, P. M., and Upchurch, P. 1994. Feeding mechanisms of *Diplodocus*. GAIA 10: 195–204.

Barrett, P. M., and Willis, K. J. 2001. Did dinosaurs invent flowers? Dinosaur-angiosperm coevolution revisited. Biol. Rev. 76: 411–447.

Berman, D. S., and McIntosh, J. S. 1978. Skull and relationships of the Upper Jurassic sauropod *Apatosaurus* (Reptilia: Saurischia). Bull. Carnegie Mus. Nat. Hist. Pittsburgh 8: 1–35.

Bonaparte, J. F. 1996. Dinosaurios de America del Sur. Impreso en Artes Gráficas Sagitario Iturri, Buenos Aires.

———. 1997. *Rayososaurus agrioensis* Bonaparte 1995. Ameghiniana 34: 116.

Brochu, C. A. 2003. Osteology of *Tyrannosaurus rex*: insights from a nearly complete skeleton and high-resolution computed tomographic analysis of the skull. J. Vertebr. Paleontol. Suppl. 22: 1–138.

Brown, B. 1941. The last dinosaurs. Nat. Hist. 48: 290–295.

Calvo, J. O. 1994a. Jaw mechanics in sauropod dinosaurs. GAIA 10: 183–193.

———. 1994b. Feeding Mechanisms in Some Sauropod Dinosaurs. Master's thesis. University of Illinois at Chicago, Chicago.

Calvo, J. O., and Salgado, L. 1995. *Rebbachisaurus tessonei* sp. nov. a new Sauropoda from the Albian-Cenomanian of Argentina; new evidence on the origin of Diplodocidae. GAIA 11: 13–33.

Cannon, G. L. 1906. Sauropodan gastroliths. Science 24: 116.

Curry Rogers, K., and Forster, C. A. 2004. The skull of Rapetosaurus krausei (Sauropoda: Titanosauria) from the Late Cretaceous of Madagascar. J. Vertebr. Paleontol. 24: 121–144.

Dalla Vecchia, F. M. 1998. Remains of Sauropoda (Reptilia, Saurischia) in the Lower Cretaceous (upper Hauterivianl/lower Baremian) limestones of SW Istria (Croatia). Geol. Croatia 51: 105–134.

Dodson, P. 1990. Sauropod paleoecology. In: Weishampel, D. B., Dodson, P., and Osmólska, H. (eds.). The Dinosauria. University of California Press. Pp. 402–407.

Fiorillo, A. R. 1991. Dental microwear on the teeth of *Camarasaurus* and *Diplodocus*: implications for sauropod paleoecology. In: Kielan-Jaworowska, Z., Heintz, N., and Nakren, N. A. (eds.). Fifth Symposium on Mesozoic Ecosystems and Biota, Extended Abstracts. Contributions of the Paleontological Museum, University of Oslo, Oslo. Pp. 23–24.

Gillette, D. 1990. Gastoliths of a saurpod dinosaur from New Mexico. J. Vertebr. Paleontol. 10: 24A.

Hatcher, J. B. 1901. *Diplodocus* (Marsh): its osteology, taxonomy, and probable habits, with a restoration of the skeleton. 1: 1–63.

Holland, W. J. 1924. The skull of *Diplodocus*. Mem. Carnegie Mus. 9: 379–403.

Huene, F. von. 1929. Los Saurisquios y Ornithisquios de Cretacéo Argentino. Ann. Mus. La Plata 3: 1–196.

Hunt, A. P., Lockley, M. G., Lucas, S. G., and Meyer, C. A. 1994. The global sauropod record. In: Lockley, M. G., dos Santos, V. F., Meyer, C. A., and Hunt, A. (eds.). Aspects of Sauropod Paleobiology. GAIA 10: 261–279.

Janensch, W. 1929. Die Wirbelsäule der Gattung *Dicraeosaurus*. Palaeontographica 2: 39–133.

———.1935-1936. Die Schädel der Sauropoden *Brachiosaurus*, *Barosaurus* und *Dicraeosaurus* aus den Tendaguruschichten Deutsch-Ostrafrikas. Palaeontographica 2(Suppl. 7): 147–298.

Kellner, A.W.A. 1996. Remarks on Brazilian dinosaurs. Mem. Queensland Mus. 39: 611–626.

Kellner, A.W.A. and Mader, B.J. 1997. Archosaur teeth from the Cretaceous of Morocco. Journal of Paleontology 71: 525–527.

Lamanna, M.C., Martinez, R.D., Luna, M., Casal, G., Dodson, P., and Smith, J.B. 2001. Sauropod faunal transition through the Cretaceous Chubut Group of central Patagonia. Journal of Vertebrate Paleontology 21: 71A.

Larsson, H.C.E., and Gado, B. 2000. A new Early Cretaceous crocodyliform from Niger. Neues Jahrb. Geol. Palaontol. Abhandlungen 217: 131–141.

Lavocat, R. 1954. Sure les dinosauriens du Continental Intercalaire des kem-Kem de la Daoura. Comptes rendus de la Dix–Neuviéme Session, Congrès Géologique International, Alger, ASGA, fasc. 21. Pp. 65–68.

Lucas, S. G., and Hunt, A. P. 1989. *Alamosaurus* and the sauropod hiatus in the Cretaceous of the North American Western Interior. In: Farlow, J. O. (eds.). Paleobiology of the Dinosaurs. Geological Society of America, Special Papers. Pp. 75–85.

Lull, R.S., and Wright, N.E. 1942. Hadrosaurian dinosaurs of North America. Geol. Soc. Am. Spec. Papers 40: 1–242.

Marsh, O.C. 1884. Principal characters of American Jurassic dinosaurs. Part VII. On the Diplodocidae, a new family of the Sauropoda, American Journal of Science (series 3) 27: 161–168.

McIntosh, J. S. 1990. Sauropoda. In: Weishampel, D. B., Dodson, P., and Osmólska, H. (eds.). The Dinosauria. University of California Press. Pp. 345–401.

Naish, D., and Martill, D. M. 2001. Saurischian dinosaurs I: Sauropods. In: Martill, D. M., and Naish, D. (eds.). Dinosaurs of the Isle of Wight. Palaeontological Association, London. Pp. 185–211.

Nowinski, A. 1971. *Nemegtosaurus mongoliensis* n. gen., s. sp, (Sauropoda) from the Uppermost Cretaceous of Mongolia. Palaeontologia Polonica 25: 57–81.

Ostrom, J. H. 1961. Cranial morphology of the hadrosaurian dinosaurs of North America. Bull. Am. Mus. Nat. Hist. 122: 33–186.

Ostrom, J. H., and McIntosh, J. S. 1999. Marsh's Dinosaurs: The Collections from Como Bluff. Yale University Press, New Haven, CT, London.

Pereda-Suberbiola, X., Torcida, F., Izquierdo, L. A., Huerta, P., Montero, D., and Perez, G. 2003. First rebbachisaurid dinosaur (Sauropoda, Diplodocoidea) from the early Cretaceous of Spain: paleobiological implications. Bull. Soc. Geol. France 174: 471–479.

Romer, A. S. 1966. Vertebrate Paleontology. University of Chicago Press, Chicago.

Salgado, L., A., Garrido, S. Cocca and J.R. Cocca. 2004. Lower Cretaceous rebbachisaurids from Cerro Aguada del León (Lohan Cura Formation). Neuquén Province, northwestern Patagonia, Argentina. Journal of Vertebrate Paleontology 24: 903–912.

Sanders, F., Manley, K., and Carpenter, K. 2001. Gastroliths from the Lower Cretaceous sauropod *Cedarosaurus weiskpfae*. In: Tanke, D. H., and Carpenter, K. (eds.). Mesozoic Vertebrate Life. Indiana University Press, Bloomington. Pp. 166–180.

Sereno, P. C. 1997. The origin and evolution of dinosaurs. Annu. Rev. Earth Planet. Sci. 25: 435–489.

———. 2000. The fossil record, systematics and evolution of pachycephalosaurs and ceratopsians from Asia. In: Benton, M., Kurochkin, E., Shishkin, M., and Unwin, D. (eds.). The Age of Dinosaurs in Russia and Mongolia. Cambridge University Press, Cambridge. Pp. 480–516.

Sereno, P. C., Beck, A. L., Dutheil, D. B., Gado, B., Larsson, H. C. E., Lyon, G. H., Marcot, J. D., Rauhut, O. W. M., Sadlier, R. W., Sidor, C. A., Varricchio, D. J., Wilson, G. P., Wilson, J. A. 1998. A long-snouted predatory dinosaur from Africa and the evolution of spinosaurids. Science 282: 1298–1302.

Sereno, P. C. Beck, A. L., Dutheil, D. B., Larsson, H. C. E., Lyon, G. H., Moussa, B., Sadlier, R. W., Sidor, C. A., Varricchio, D. J., Wilson, G. P., and Wilson, J. A. 1999. Cretaceous sauropods from the Sahara and the uneven rate of skeletal evolution among dinosaurs. Science 286: 1342–1347.

Stevens, K. A., and Parrish, J. M. 1999. Neck posture and feeding habits of two Jurassic sauropod dinosaurs. Science 284: 798–800.

Taquet, P. 1976. Géologie et paléontologie du gisement de Gadoufaoua (Aptian du Niger). Cahiers Paleontol. 1976: 1–191.

Taylor, A.C., and Butcher, E.C. 1951. The regulation of eruption rate in the incisor teeth of the white rat. Journal of Experimental Zoology 117: 165–188.

Upchurch, P., and Barrett, P.M. 2000. The evolution of sauropod feeding mechanisms. In: Sues, H.-D. (eds.). Evolution of Herbivory in Terrestrial Vertebrates: Perspectives from the Fossil Record. Cambridge University Press, Cambridge. Pp. 79–122.

Weishampel, D.B. 1990. Dinosaur distributions. In: D.B. Weishampel, P. Dodson, and H. Osmólska (eds.). The Dinosauria. University of California Press, Berkeley. Pp. 63–139.

Wilson, J. A. 2003. Sauropod dinosaur phylogeny: critique and cladistic analysis. Zool. J. Linn. Soc. 136: 217–276.

Wilson, J. A., and Sereno, P. C. 1998. Higher-level phylogeny of sauropod dinosaurs. J. Vertebr. Paleontol. 18 (Suppl.): 1–68.

Wilson, J. A. 2005. Redescription of the Mongolian sauropod *Nemegtosaurus mongoliensis* Nowinski (Dinosauria: Saurischia) and comments on Late Cretaceous sauropod diversity. Journal of Systematic Palaeontology 3: 283–318.

SIX

Digital Reconstructions of Sauropod Dinosaurs and Implications for Feeding

Kent A. Stevens and J. Michael Parrish

IN RECENT YEARS, SAUROPODS HAVE been interpreted primarily as quadrupedal herbivores, with sympatric taxa differentiated in their feeding behavior presumably according to their dentition and feeding height in a quadrupedal stance (e.g., Fiorillo 1998; Upchurch and Barrett 2000). In order to generate detailed hypotheses concerning sauropod paleoecology, it is essential to start with as accurate a reconstruction of their body plans as can be afforded from their fossils. An accurate rendering of the life posture of a sauropod is necessary in order to determine the feeding envelope for each taxon in its conventional quadrupedal stance. We review here the body plan of several major sauropod groups, emphasizing the use of whole-body reconstructions to determine the approximate head height when the animal was standing quadrupedally, supporting its weight symmetrically from left to right, and holding the axial skeleton in an undeflected state. The undeflected state is termed the "neutral pose," defined geometrically, and analyzed on the basis of osteological determinants in extant vertebrates as a guide to their reconstruction for sauropods. Neutral position head height is one key point for analyzing variation in feeding behaviors across sauropods, another being variation in dentition; both are set against the backdrop of available fodder.

What was the relationship between a sauropod's preferred feeding height and the height at which its head was held when the neck was undeflected? A bridging assumption is necessary to relate these two parameters for an extinct species. The habitual feeding posture of a terrestrial herbivore can relate to the neutral position of its neck in three ways: (1) the herbivore can deflect its neck ventrally relative to the neutral position for browsing, or "browse by ventriflexion" (BV); it can raise its neck relative to the neutral position, or "browse by dorsiflexion" (BD); or it can feed at or near the neutral position of the neck, or "browse neutrally"(BN). As browsing behavior is not directly preserved in the fossil record, it is necessary to consider the phylogenetic and functional distribution of these three feeding models among extant tetrapods.

The form of browsing (BV, BD, or BN) is, in principle, independent of the neutral pose head height (relative to shoulder height) of a given herbivore. Many extant low browsers and grazers, such as Thomson's gazelle (e.g., Leuthold 1977:table 2), deer, and horses, engage in BV feeding primarily, but not exclusively. In a neutral pose, their heads are held in a high position, presumably for vigilance during periods of inactivity (Walther 1969). Muscular effort is expended to lower the head to feed, increasing the tension on the epaxial nuchal ligaments. One argument posed against low browsing in sauropods, even in diplodocids, is that it would leave them vulnerable to attack (Paul 2000).

While the high head heights that gazelles and other fleet-footed herbivores maintain when inactive allow them to detect and flee approaching predators, flight from predation was not a practical option for sauropods, the speeds of which are generally estimated as being much slower than those of their most likely predators (e.g., Alexander 1989; Thulborn 1990). Today, large BN to BV megaherbivores respond to the presence of carnivores by charging (e.g., rhinos, hippos, elephants [Owen-Smith 1988]) or indifference (e.g., elephants, hippos [Owen-Smith 1988]).

The giraffe is of particular importance to this chapter, as it has been cited as an extant model for those sauropods, including brachiosaurids, euhelopids, and camarasaurids, that are sometimes reconstructed as giraffe-like (e.g., Paul 1987:figs. 16, 17; Currie 1987:figs. 2, 3; Christian and Heinrich 1998), effortlessly feeding while in a cervical neutral position (BN browsing) with the head held high above the shoulders. We review the osteological basis for the elevation of the giraffe neck, then examine the validity of proposing such a posture for any sauropod. It should first be noted that, perhaps surprisingly, giraffes frequently browse by ventriflexion, with the head at or below shoulder height (Leuthold and Leuthold 1972; Pellew 1984; Young and Isbell 1991; Woolnough and du Toit 2001). The elongate neck of the giraffe is not a simple consequence of vertical niche partitioning (Simmons and Scheepers 1996). Pincher (1949) proposes that predation provided the selection pressure for limb elongation in giraffes, and that neck elongation secondarily provided the ability to drink and reach low fodder.

There are many other modern examples of herbivores with neutral head height taller than the shoulders that primarily engage in browsing by ventriflexion, including many cervids, rhinoceratids, and equids (e.g., Leuthold 1972; Owen-Smith 1988). There are also modern herbivores that have the head situated much lower in neutral position, again relative to shoulder height, which engage in BN, such as some large bovids and other cervids such as the greater and lesser kudu (e.g., Leuthold 1972; Owen-Smith 1988). We are, however, unable to find examples of extant herbivores with heads well below shoulder height in neutral position that predominantly feed quadrupedally in a BD position.

If vigilance were not a factor for sauropods, then it might be inferred that the neutral pose would be related closely to their preferred feeding heights. Indeed, as we show, the neutral pose for some sauropods places the head very low to the ground. A sauropod with shorter forelimbs than hindlimbs and a steadily descending neck that brings the head to near-ground level certainly appears well suited for BN (and BV) browsing, without compromise to vigilance, and less adapted to BD. Large extant grazing herbivores, such as bison, that hold their head closer to feeding height (e.g., Leuthold 1977:table 2) are perhaps better analogues for those sauropods with low head height in neutral position.

Osteologically based reconstruction of neutral head height provides some refinement on the question of sauropod feeding behavior, when combined with consideration of whether a given taxon browsed by ventriflexion, by dorsiflexion, or near the cervical neutral position. We suggest that careful analysis of the osteologically defined neutral pose along the axial skeleton is an indicator of the mean feeding height of the sauropods, particularly as they do not, in any of the sauropod taxa we have examined, place the necks in a giraffelike high slope with heads held high, a position consistent with a predominant state of vigilance. As reviewed below, neutral pose reconstructions suggest that most sauropods would have their necks held horizontally or subhorizontally when not actively feeding or otherwise raising their heads.

THE NEUTRAL POSE AS A BASIS FOR ESTIMATING FEEDING HEIGHT

In describing the sauropod body plan, it is useful to start by establishing the height of the acetabular axis above ground level. Fortunately, maximum hindlimb length can often be reconstructed with some confidence, with the primary unknowns being the precise amount of cartilage separating the limb elements and the degree of flexure at the knee (e.g., Paul 1987; Bonnan 2001). With the femoral heads inserted into their associated acetabula, the acetabular axis constitutes a pivot point for the whole-body reconstruction—a fulcrum about which the axial skeleton tilts depending on a given reconstruction of the trunk. The literature presents a range of interpretations regarding the arch to the span of vertebral column that supported the trunk. The degree of flexure of the arch is important here, for the greater the curvature, the lower the resulting head height. A given reconstruction of the dorsal vertebrae, with associated ribcage, forms an armature on which to place the pectoral girdles and forelimbs. In creating a skeletal reconstruction to quantitatively estimate mean feeding height, it is important to determine potential sources of variability in the appendicular and axial skeleton. These are reviewed below.

For modern quadrupedal herbivores, the characteristic posture associated with a standing pose is the starting point for exploring the range of feeding movements achieved by movements of the head and neck. A skeletal reconstruction of an extinct form, such as a sauropod, must similarly be posed in a neutral position as a basis for examining its characteristic feeding envelope. As considered here, the feeding envelope for a given taxon can be visualized as the extremes of head reach allowed by the flexibility of its neck (e.g., Martin 1987:fig. 3; Stevens and Parrish 1999). Note that the envelope, thus defined, reflects only variation in head position allowed by neck mobility, not the contributions of trunk and forelimb movements that undoubtedly broadened this envelope.

THE NEUTRAL POSE OF THE APPENDICULAR SKELETON

The nature of the articulations between the elements of the sauropod appendicular skeleton cannot be inferred entirely from osteology, although the hindlimb, because of the solid joints between sacrum and pelvis, is better constrained than the forelimb and pectoral girdles, which have no osteological connection to the axial column.

Whereas hindlimb length determines the height of the caudal end of the dorsal vertebral column, forelimb length only defines the height of the glenoid; the inferred height and slope of the anterior part of the trunk depend on the position and orientation of the scapulae on the ribcage and the degree to which the dorsal vertebral column is arched.

In extant quadrupeds, pectoral girdles do not have bony articulations with the trunk but, instead, are suspended from the thorax by musculature and soft tissues. While the presence of subtle depressions in the dorsal ribs in some sauropods such as *Apatosaurus* and *Diplodocus* provide some indication of the position and the alignment of the scapulocoracoid (Parrish and Stevens 2002b), determining its precise location requires considering the scapulocoracoids within the context of the overall function of the girdles, ribcage, and forelimbs.

Changing the relative height and inclination of the scapulocoracoid on the trunk can result in significant differences in the height of the cranial end of the thorax, with a corresponding change in head height (Stevens and Parrish 2005a). Early sauropod mounts differed in the orientation of the pectoral girdles relative to the ground, and thus, indirectly, in the angle of the scapula relative to the cranial part of the dorsal vertebral column. For instance, Gilmore (1936:pl. 34) placed the scapulae of *Apatosaurus* in a subhorizontal orientation, nearly parallel to the anterior dorsal column, whereas Osborn and Mook (1921) placed the scapulae of *Camarasaurus* in a far more vertical orientation. The effect of shoulder girdle placement on the overall reconstruction can be appreciated by comparing the illustrations of the *Apatosaurus* in Gilmore (1936:pl. 34) versus McIntosh et al. (1997) or those of the *Diplodocus* mounted at the Carnegie Museum of Natural History (e.g., McGinnis 1982:68–69) versus that at the Senckenberg Museum in Frankfurt (e.g., Beasley 1907). The effect of variation in shoulder girdle placement is also apparent below in digital reconstructions of *Brachiosaurus brancai*, in combination with variation in dorsal vertebral column curvature.

A range of interpretations has also been proposed for the articulation of the elements of the sauropod forelimbs, ranging from a vertical, columnar arrangement (e.g., Christian et al. 1999) to partial flexure of the limbs (e.g., Janensch 1950b), resulting in a range of heights and locations for the humeral head. Most recent investigators (e.g., Bonnan 2001; Christian et al. 1999; Wilhite 2003) predict minimal flexion of the elbow during standing, in keeping with the columnar forelimb posture of extant graviportal animals, but some reconstructions (e.g., the Humboldt Museum mount of *Brachiosaurus* [Christian et al. 1999:fig. 1]) depict a more reptilian sprawling pose (e.g., Christian et al. 1999).

THE NEUTRAL POSE IN THE AXIAL SKELETON

The neutral pose of the axial skeleton is strongly constrained. The presacral vertebral column in sauropods is characterized by opisthocoelous central articulations, a character shared with many large herbivorous modern quadrupeds such as rhino, giraffe, horse, and camel. Opisthocoely, fortunately, provides particularly strong osteological clues to the state of neutral position.

Presacral intervertebral articulation involves synovial joints between centra and between left and right pairs of zygapophyses. Dorsoventral and mediolateral angular deflection results in gliding contact between the zygapophyses that limit axial rotation (Stevens and Wills 2001). The postzygapophyses are displaced posteriorly relative to their associated prezygapophyses during dorsiflexion and anteriorly during ventriflexion. The zygapophyseal facets are superimposed and centered in a state of neutral deflection.

The state of null deflection is also apparent at the central synovial capsule. The condyle, inserted deeply within cotyle, is surrounded by a broad circumferential ligamentous capsule. Within the capsule, the articular facets are closely spaced in modern vertebrates with opisthocoelous centra. In a rhino, for instance, the separation between cotyle and condyle is only a few millimeters, and in the giraffe, whose cervical vertebrae are also strongly opisthocoelous, the intervertebral gaps are only slightly wider, again, of the order of millimeters (B. Curtice, pers. comm. 2001, pers. obs.). Note that this close intervertebral separation may not be apparent from cursory observation. The capsule is substantially wider than the intervertebral separation, of course, to accommodate the displacement undertaken by the cotyle during deflection. Articulated sauropod cervical vertebral series are likewise very closely spaced, in all instances we have examined.

In modern vertebrates, dorsiflexion separation between the cotylar and the condylar margins of the capsule increases ventrally (and diminishes dorsally), placing the surrounding ligaments in progressively greater tension ventrally. Conversely, in ventriflexion, the separation and ligament tension increases dorsally (and diminishes ventrally). In a state of neutral deflection the margins of the associated cotyle and condyle are parallel; the gap is uniform around the perimeter of the synovial capsule, thus providing a second osteological indictor of neutral deflection. When undeflected, the posterior edge of the cotyle is parallel to the attachment scar surrounding the condyle, corresponding to the state of minimal stress on the synovial capsule.

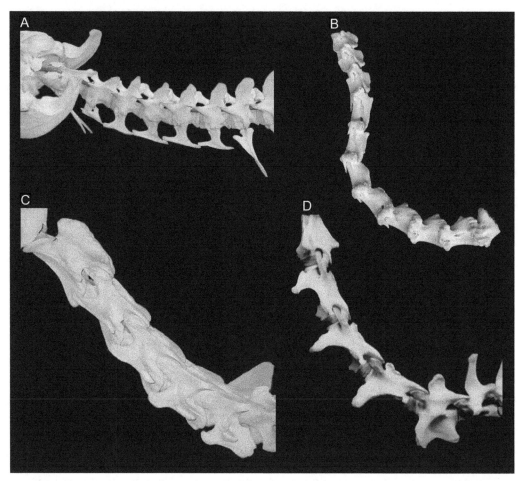

FIGURE 6.1. The cervical vertebral columns of articulated skeletons of (A) crocodilian (*Gavialis gangeticus*), (B) turkey (*Meleagris gallopavo*), (C) horse (*Equus przevalskii*), and (D) camel (*Camelus dromedarius*), each mounted in the undeflected, or "neutral," pose wherein apposed pre- and postzygapophyses are aligned and centered, and simultaneously, the margins of cotyle and condyle at each intervertebral joint are parallel. In each case, the neck in this undeflected state is naturally curved in the manner characteristic of that animal.

When the central articulation is undeflected according to the above criterion, the associated pre- and postzygapophyses are also undeflected (i.e., superimposed and centered). That is, the two criteria are satisfied simultaneously (fig. 6.1) in all extant vertebrates that we have observed. Their redundancy is particularly useful in reconstructing the neutral pose for vertebrae that are missing their zygapophyses (see the *Brachiosaurus brancai* reconstruction below). When successive vertebrae are placed in neutral position, joint by joint, the vertebral column forms the intrinsic curvature characteristic of the given extant animal.

When the vertebrae of extant mammals are placed in neutral pose, they replicate their habitual, characteristic posture (figs. 6.1, 6.2). For instance, in neutral pose, the neck of the camel exhibits its familiar catenary shape, and the sharp change in angulation observed at the base of the neck of the giraffe is clearly visible (fig. 6.2). Likewise, the sigmoid curves in the necks of theropod dinosaurs, including extant birds (fig. 6.1A), are associated with an undeflected neck and derive from the geometry of the vertebrae. Therefore, the neutral pose of the cervical vertebral column, and the cranial end of the dorsal vertebral column,

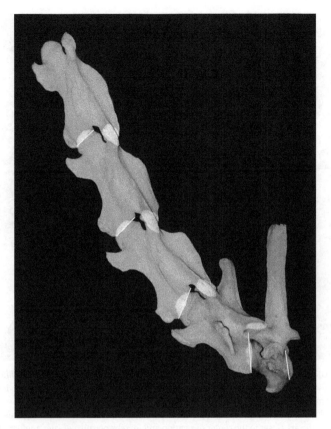

FIGURE 6.2. The steeply ascending neck characteristic of the giraffe arises in the undeflected state, not by bending of the neck. Photographs of vertebrae C4–T1 of an adult giraffe are placed in neutral position. Transparency is utilized to show the insertion of each central condyle within the cotyle and the overlap and alignment of the zygapophyses. The characteristic neck elevation derives from wedge- or keystone-shaped centra, especially apparent in the sixth and seventh cervical vertebrae. (Individual photographs courtesy of Brian Curtice.)

can be shown to be consistent with the preferred head height in a variety of extant quadrupedal herbivores, and thus we consider it a robust predictor of preferred head position in sauropods as well.

The cervical and anterior dorsal vertebrae of sauropods are strongly opisthocoelous, with prominent condyles associated with deep cotyles (figs. 6.3, 6.4). Articulation of axial elements in sauropods is geometrically similar to that in many extant vertebrates, allowing inference of their neutral or undeflected state to be made with some confidence.

Given the commonality in morphology of the articular facets of sauropod cervical vertebrae and those of a wide range of extant vertebrates, these criteria are presumed to hold for the cervical vertebrae of sauropods as well, permitting the reconstruction of their undeflected state and, hence, their intrinsic curvature.

RECONSTRUCTING SAUROPOD CERVICAL COLUMNS

METHOD

The neutral pose of a vertebral column can be reconstructed by creating a composite of lateral view illustrations or photographs, all depicted at the same scale, placed such that each successive pair of vertebrae is in neutral deflection. This method has long been used in traditional axial skeleton reconstructions (e.g., Osborn and Mook 1921:fig. 28). To introduce the technique, consider a pair of platycoelous centra. The apposed central facets are parallel to one another when the intervertebral joint is undeflected, hence their margins also appear parallel when viewed laterally. This neutral alignment is readily reconstructed graphically by two-dimensional rotation, translation of images of the two vertebrae (e.g., using layers in Photoshop)

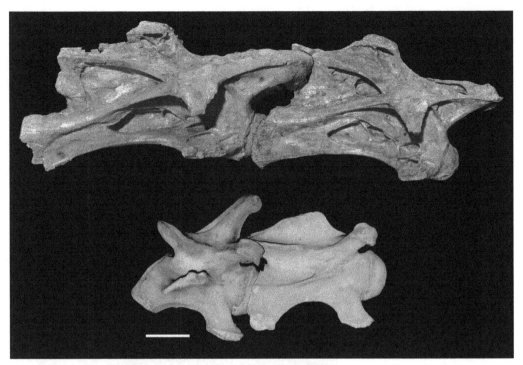

FIGURE 6.3. Intervertebral articulation in the neutral pose for the sauropod *Brachiosaurus brancai* (Top, cervical vertebrae C4 and C5 of the Humboldt Museum specimen SI) and for a giraffe (Bottom, cervical vertebra 7 and first thoracic vertebra), shown at the same scale. Note the similarity in articulation geometry, both being strongly opisthocoelous with condyles of circular profile making the center of rotation for dorsoventral flexion unambiguous. The relatively larger zygapophyses of the giraffe are closer to the center of rotation and thereby permit a greater angular range of motion. Scale bar equals 10 cm. (*Brachiosaurus* and giraffe photographs courtesy of Christopher McGowan and Brian Curtice, respectively.)

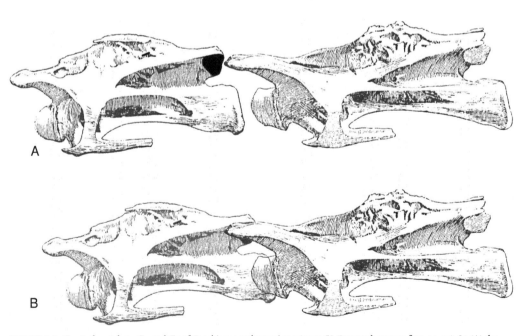

FIGURE 6.4. Cervical vertebrae C4 and C5 of *Brachiosaurus brancai* specimen SI (Janensch 1950a: figs. 34, 37). In (A) the left postzygapophysis of C4 is shaded. In (B) the two vertebrae are composited in neutral position such that the pre- and postzygapophyses are centered and the margins of the cotyle and condyle at the central articulation are parallel.

until the gap between centra is uniform and the zygapophyses are centered (i.e., both criteria of neutral position are met). When an entire vertebral series is composed from individual illustrations by this means, an intrinsic curve is often revealed, not due to flexion, but from centra that are sometimes subtly keystone- or wedge-shaped.

In sauropods, the articulations of the centra throughout the cervical vertebral column are strongly opisthocoelous. Fortunately, since the curvature of the sauropod neck is critical to understanding their paleobiology, opisthocoely helps, rather than hinders, the reconstruction of the neutral state of deflection. With condyle inserted into cotyle and placed in an undeflected state, the postzygapophyses are centered above prezygapophyses, and the posterior margin of the cotyle is parallel to the attachment scar of the synovial capsule surrounding the central condyle. In figure 6.4, line drawings of two cervical vertebrae of *Brachiosaurus brancai* have been composited into neutral position according to these criteria. Note that transparency reveals the insertion of the condyle within cotyle and the centering of the zygapophyses.

A second method is conceptually similar but performed three-dimensionally. Neutral pose can be determined by direct manipulation, as part of a process of exploring the range of motion in the axial skeleton. Although a pair of vertebrae, or even a complete neck, of an extant animal can be articulated and manipulated, sauropod vertebrae are not only unwieldy, but usually too distorted to allow proper rearticulation and manipulation. An alternative is to create digital representations of the vertebrae, which can then be articulated and posed virtually in three dimensions. A parametric skeletal modeling approach, DinoMorph (Stevens 2002), has been developed and used to estimate the feeding envelopes of two diplodocids (e.g., Stevens and Parrish 1999). For the present study, dimensionally- accurate digital skeletons of three sauropods are posed in neutral position to estimate mean feeding heights. In each case published dimensional data provide the basis for reconstructing the overall body plan, with the morphology of individual elements ranging from detailed to schematic. A hybrid approach is also used for the reconstruction of the cervical vertebral series: two-dimensional reconstructions based on original artwork provide an estimate of the intrinsic curvature of the undeflected cervical vertebral column, which is then used to pose the neck of the digital three-dimensional model (Stevens and Parrish 2005a).

RECONSTRUCTING NECK CURVATURE

DIPLODOCIDS

Apatosaurus and *Diplodocus* were initially reconstructed with necks that were quite straight and extended from an arched back so that the necks descended gently from the shoulders (e.g., Holland 1906:fig. 2). Usually the necks are depicted in a state of mild dorsiflexion at the base, which raises the heads to about shoulder height. Recent renditions of these two diplodocids often provide their necks with a more pronounced sigmoid curve, dorsiflexed caudally and ventriflexed cranially (e.g., McIntosh et al. 1997:figs. 20.11, 20.12; Wilson and Sereno 1998:foldout 1). Similarly, in the original skeletal reconstruction of the diplodocid *Dicraeosaurus* the neck was shown in a pronounced sigmoid curve, abruptly dorsiflexed at the base and more gradually deflected downward cranially (Janensch 1929:pl. 16). Unfortunately, it is difficult to determine from drawings to what extent the curvature was intended to depict active flexion versus a shape intrinsic to the neck.

The vertebrae in the original descriptions of the above diplodocids were rendered with excellent dimensional accuracy as detailed line drawings, and the original material was, in most cases, sufficiently undistorted to permit reconstructions of their neutral position by graphical compositing (Gilmore 1936:pl. 24, 25; Hatcher 1901:pl. 4, 6; Janensch 1929:pl. 1). These images were digitally scanned, then composited (fig. 6.5) to create reconstructions of the axial columns in neutral pose. The composites reveal that these diplodocids had remarkably straight

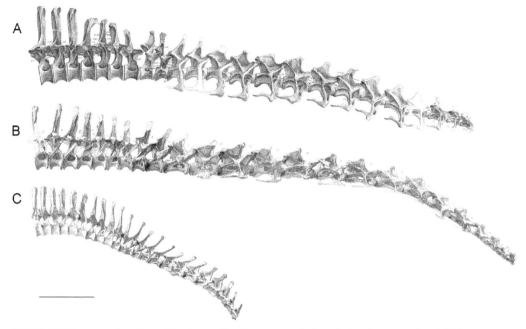

FIGURE 6.5. The presacral vertebrae of the diplodocids (A) *Apatosaurus louisae*, (B) *Diplodocus carnegii*, and (C) *Dicraeosaurus hansemanni*, composited from the original figures, placing each successive pair in neutral deflection (from Stevens and Parrish 2005b). *Apatosaurus* composite mirrored left-for-right for uniformity; all vertebral columns at the same scale. Scale bar equals 1 m.

cervicodorsal transitions; their necks were straight extensions of their backs. The cervical columns of *Apatosaurus* and *Dicraeosaurus* show a gentle degree of ventral curvature in the neutral pose, which brings the head to a position well adapted to low browsing. The familiar sigmoid curve attributed to the neck of *Dicraeosaurus* (Janensch 1929:pl. 16) was certainly within the limits achievable by dorsiflexion at the base and ventriflexion more cranially, but this pose was not reflected in the osteology of the neck. But presuming that the osteologically determined neutral pose was also the habitual posture for *Dicraeosaurus*, it usually held its head close to ground level, in common with other diplodocids, but at a steeper angle due to its relatively shorter neck.

BRACHIOSAURUS AND OTHER SAUROPODS OFTEN DEPICTED AS GIRAFFE-LIKE

The original reconstruction of *Brachiosaurus* (Janensch 1950b:pl. 6–8) shows remarkable similarity to the modern giraffe, in part because of its tall limbs and ascending dorsal column but, especially, as a result of the neck that rises steeply at its base. The osteological basis for such a posture is reviewed here. This giraffe-like posture is even more dramatic in some reconstructions of *Euhelopus* and *Camarasaurus*, which are also considered.

Most reconstructions of *Brachiosaurus* provide the sauropod with a steeply upturned neck and depict the cervicodorsal vertebrae as wedge-shaped, with centra longer ventrally than dorsally, much as those at the base of the neck in the giraffe. Figure 6.6 shows in detail the cervicodorsal region as originally reconstructed, plus three subsequent depictions, all of which can be compared with a detailed figure of the original fossil material in this critical region.

Accurate line drawings of the available vertebrae of *Brachiosaurus brancai* were provided by Janensch (1950a:figs. 14–49). The lateral views are amenable to composition into a reconstructed column, for both specimen SI

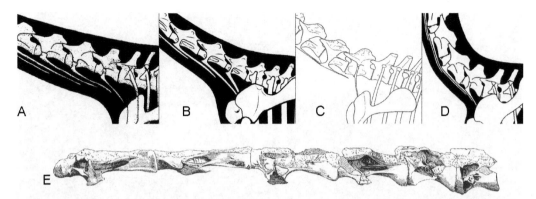

FIGURE 6.6. Details from four depictions of *Brachiosaurus brancai*: (A) Czerkas and Czerkas (1991:132); (B) Wilson and Sereno (1998:foldout 1); (C) Janensch (1950b:pl. 8); (D) McIntosh et al. (1997 fig. 20.16). Note the differing degrees to which the cervicodorsal region is depicted as upcurved between the posterior of D2 and the posterior of C10. This region, in (A) is curved merely 5°, while the same region is curved 68° in (D) largely due to illustrating the centra as if they were distinctly wedge- or keystone-shaped (longer ventrally than dorsally). In (A) approximately 2° of the curvature is due to keystoning, the rest presumably resulting from dorsiflexion. In (D) approximately 48° is accumulated due to the shape of the centra, especially C12 and C13, and approximately 20° reflects dorsiflexion, which further contributes to the near-vertical posture favored by that illustrator. Compare with original illustration (E; from Janensch 1950a:fig. 49) of cervicals C10–C13 plus first two dorsals, where no wedge shape is apparent.

FIGURE 6.7. Neutral pose reconstructions of *Brachiosaurus brancai* specimens SII (Top, C3–D2) and SI (Bottom, C2–C7) from individual line drawings (Janensch 1950a:figs. 14–49). SI is mirrored left-for-right to facilitate comparison with SII (from Stevens and Parrish in 2005b). The slight ventral curvature in neutral position appears intrinsic to *Brachiosaurus* and is consistent with head-down feeding. Note that vertebrae C10–D2 of SII were collinear in situ, and while the neural arches were ablated, the centra show no keystoning or other osteological evidence to support a giraffe-neck interpretation (see also fig. 6.6). In fact, the partially exposed ventral surface of the central condyle of D2 is evidence of some dorsiflexion between D1 and D2. Scale bar equals 1 m.

(vertebrae C2 through C7) and specimen SII (C3 through D2). Cranially, both exhibit the gradual ventral curvature in *Dicraeosaurus* (fig. 6.5). In specimen SII, although the neutral arches were not preserved caudal to C9, the central articulations provide clear evidence for the neutral deflection between subsequent pairs of vertebrae from the midneck to the second dorsal (Fig. 6.7). The result is a remarkably straight neck at the base, quite contrary to most restorations of this taxon (but see Czerkas and Czerkas 1991:132). Properly restoring the neck of *Brachiosaurus* as extending straight from the shoulders, however, does not change its undisputed role as a high browser; see figure 6.8.

Euhelopus zdanskyi (Wiman 1929:fig. 3) is another sauropod traditionally depicted as giraffelike, with the life pose originally drawn having the same upturned neck as found in situ (fig. 6.9). The dorsiflexion at the base of the neck (primarily between C16 and C17 and between C17 and D1), however, reflects a "death pose" resulting from shrinkage of the nuchal ligaments. This conclusion is drawn from observing that the degree of dorsiflexion as measured between successive centra equals the angle of dorsiflexion at the zygapophyses

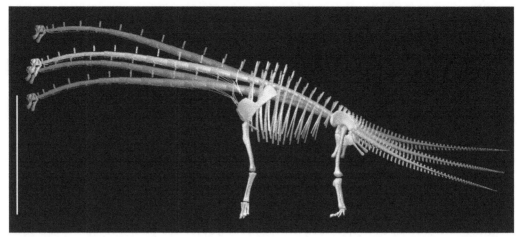

FIGURE 6.8. A DinoMorph model of *Brachiosaurus brancai* shows the effect on head height of pectoral girdle placement and dorsal column curvature. The neck curvature is held constant in the neutral pose derived from digital composites of the original steel engravings (fig. 6.7). The postcervical skeleton is based on specimen SII (Janensch 1950a, 1950b, 1961). Four alternative head heights result from combinations of high versus low arch to the dorsal column and high versus low placement of the pectoral girdles on the ribcages. The high arch and low pectoral girdle placement case is shown in full contrast, while the others are depicted at lower contrast. The highest head height is associated with the combination of low arch and low girdles. The lowest head height derives from the combination of high arch and high girdles, which brings the head only to shoulder height. Vertical scale bar equals 5 m.

FIGURE 6.9. The original life reconstruction of *Euhelopus zdanskyi* (A) was depicted with a giraffe like neck, ascending with about 38° of slope (from Wiman 1929: fig. 3). The sharp curvature at the base of the neck replicated that of the original specimen as found in situ (B; from Wiman 1929:pl. 3). In (C) (from Paul 2000: appendix A), the slope is increased to about 65°. With removal of the "death pose" dorsiflexion that is localized to the base of the neck (D; from Stevens and Parrish 2005b), the neck of *Euhelopus* emerged from the shoulder as a straight extension of the dorsals, sloping more or less downward depending on the arch of the dorsal column and the relative height of the (unknown) forelimb. The resultant low head height was similar to that of the diplodocoids and consistent with low browsing.

(Stevens and Parrish 2005b). Removing this degree of dorsiflexion restores the neutral pose and *Euhelopus* is revealed to be a low browser, not at all giraffelike.

Similarly, the juvenile *Camarasaurus* at the Carnegie Museum (CMNH11338 [Gilmore 1925:fig. 14]) also exhibits the extreme curvature of a death pose. The neck was preserved with the swanlike curvature that is now popularly portrayed in life reconstructions of this taxon. Examination of the original specimen reveals that many of the postzygapophyses are completely disarticulated from their associated prezygapophyses (Parrish and Stevens 1998). Even if this condition could have been tolerated in life, such an extreme of dorsiflexion certainly did not constitute the neutral pose for the neck. Furthermore, quantitatively, unlike in *Euhelopus*, the angular deflection at the zygapophyses was generally greater than that at the centra, suggesting that the neck in neutral pose was at least partially ventriflexed, in common with our observation of other sauropod taxa. In summary, we have yet to find any sauropod with evidence of osteological adaptations for an upraised neck in the undeflected, neutral pose. Dorsiflexion could undoubtedly greatly increase the head height for feeding many meters above the ground, but the popular rendition of sauropods such as *Brachiosaurus*, *Euhelopus*, and *Camarasaurus* as giraffe-like is unwarranted.

Finally, we note that while the neutral pose represents the approximate center of a tetrapod's potential feeding envelope, the predominant feeding position is not necessarily the neutral position. As noted, many open habitat ungulates assume close to the neutral position while inactive yet frequently engage in BV feeding, often far from the neutral position. There are well-defined limits, fortunately, to how far the neck may deviate from the neutral position. Based on behavioral observations and manipulation of living and preserved necks of extant birds and mammals, we are confident that the feeding positions of sauropods in life would be limited by mechanical soft tissue constraints that might be violated taphonomically. In estimating the range of deflection that might have been achieved about the osteologically defined neutral position, we turn to extant vertebrates.

ESTIMATING NECK FLEXIBILITY IN SAUROPODS

In seeking modern analogues on which to base estimates of sauropod neck flexibility, potential candidates include long-necked vertebrates such as the ratite, giraffe, and camel. All have considerable mediolateral flexibility. A giraffe, for instance, can reach insects biting the base of its own neck. Some mammals achieve greater lateral angular deflection per vertebral joint than avians, having only 7, compared to 11 to 25, cervical vertebrae over which to distribute the curvature. Dorsoventrally, however, these long necks vary considerably in flexibility, with the giraffe showing little flexibility except at the cervical–dorsal transition, whereas the camel can touch the back of its head to its shoulder (Gauthier-Pilters and Dagg 1981:fig. 29).

With such a wide range of observed flexibility, there might appear to be little ground for postulating any particular range of neck flexibility in the sauropods. Specifically, it has been suggested (Sereno et al. 1999) that estimating sauropod neck mobility based on the observed intervertebral limitations of avian cervical vertebrae (Stevens and Parrish 1999; fig. 6.10) might be too conservative in light of the camel's remarkable flexibility. In order to test the validity of the assumptions made regarding the neutral pose and estimates of the range of mobility in sauropods, we performed several comparisons between the observed ranges of movement in extant long-necked mammals and the amount of mobility produced by manually manipulating skeletonized and disarticulated specimens of the cervical columns of individuals of the same species (Parrish and Stevens 2002a). Independent studies involving the manipulation of extant ostrich are also consistent with our estimates regarding the angular constraints imposed by zygapophyseal synovial

FIGURE 6.10. DinoMorph model of *Apatosaurus louisae* showing the extremes of dorsoventral motion and lateral flexibility estimated by Stevens and Parrish (1999). The three-dimensional skeletal model is based on Gilmore (1936) and data from Philip R. Platt (pers. comm. 2003). The neck of *Apatosaurus* was found to be capable of a substantial range of motion (see also fig. 6.15), with more lateral and dorsal flexibility than the longer-necked *Diplodocus*, but it shared with *Diplodocus* the theoretical ability to lower its head below mean ground level, a potential adaptation for feeding on subaqueous plants from shore.

capsules (Wedel and Sanders 1999). In avians, these synovial capsules limit displacement between associated pre- and postzygapophyses such that they remain overlapping, with a safety margin, throughout the range of motion. The remarkable dorsal flexibility of the camel neck is likewise achieved without zygapophyseal disarticulation (fig. 6.11). The elongate zygapophyseal facets provide sufficient travel to permit this extraordinary flexibility.

Neck flexibility in some extant vertebrates is delimited by osteological stops that prevent excessive displacement. In the domestic turkey, for instance, dorsiflexion causes the postzygapophyses to translate caudally until they contact the ascending base of the dorsal spine of the next vertebra, often nesting into a matching depression just posterior to the associated prezygapophyses medial to the paradiopophyseal lamina. The zygapophyseal pair is still articulated, that is, overlapping, at this limit. This mechanism operates unilaterally when dorsiflexion is combined with lateral deflection; for example, adding left lateral deflection causes the left postzygapophysis to eventually make contact, and at this limit both the left and the right zygapophyseal pairs preserve a safe degree of overlap. The osteological stop is present along most of the neck of the domestic turkey and the rhino, but in the horse, for example, it is apparent only at the base of the neck. The extreme dorsal flexibility of the camel is likewise limited by bony contact that prevents the neck from disarticulation (fig. 6.12), despite the external appearance of disarticulation in the behaving animal.

Although a camel's death pose is achievable in life (Gauthier-Pilters and Dagg 1981:fig. 60 vs. 33), it bears reminding that the camel neck permits such flexibility by design and it is osteologically prevented from disarticulation in dorsiflexion. The flexion of the camel neck, while considerable at its limits, is delimited by the same biomechanical means as in avians and, by

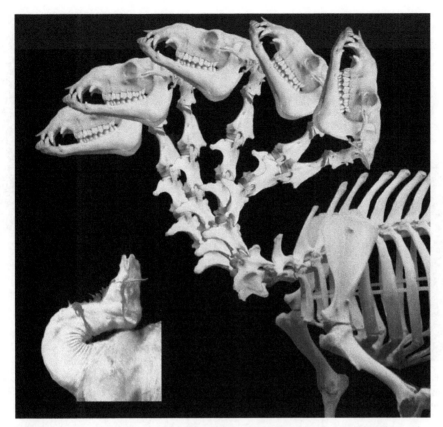

FIGURE 6.11. The remarkable dorsal flexibility of the camel's neck (inset; after Gauthier-Pilters and Daag 1981:fig. 29) might suggest that sauropod necks were also more flexible than might be predicted based on birds, wherein the zygapophyseal synovial capsules limit the travel of the enclosed articular facets prior to disarticulation (cf. Stevens and Parrish 1999; Sereno et al. 1999). The camel's flexibility, however, can be replicated without disarticulation between associated pre- and postzygapophyses. Dorsiflexion is in fact limited osteologically prior to disarticulation (see fig. 6.12). The large angular deflection achieved per joint in the camel's neck is a geometric consequence of its elongate zygapophyses set close to the center of dorsoventral rotation.

analogy, in sauropods. However, the osteological features that limit dorsiflexion in many highly flexible extant taxa are not apparent, as far as we have observed, in sauropods.

The giraffe's restricted dorsoventral motion is also limited osteologically, both dorsally, as just described, and ventrally (B. Curtice, pers. comm. 2001; Parrish and Stevens, pers. obs.). It should be noted that the extreme lateral flexibility of the giraffe is "at the edge," with minimal overlap at the zygapophyses, and not protected by the osteological stops present in avian vertebrae. As in many vertebrates, the range of motion is further delimited by the ligamentous synovial capsules surrounding the zygapophyses, which become taut at a limit of displacement between zygapophyses. In avian cervical vertebrae, at the limits of travel the zygapophyses maintain roughly 25%–50% overlap (Stevens and Parrish 1999; Wedel and Sanders 1999).

The maximum displacement between zygapophyses permitted by the surrounding capsules translates to a maximum angular deflection. The cervical flexibility of the giraffe, the camel, and various avians has been found to be predictably related to the geometry of the zygapophyseal facets and their placement relative to the axes of rotation (figs. 6.11, 6.13). Given the morphological similarities between avian and sauropod cervical vertebrae, Stevens

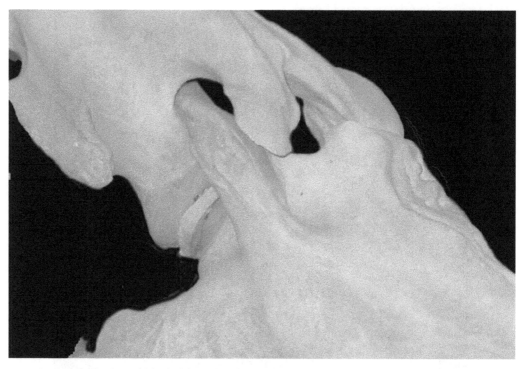

FIGURE 6.12. Dorsiflexion at the base of the neck in the camel is arrested osteologically prior to disarticulation at the zygapophyses. Postzygapophyses lock into depressions just posterior to the associated prezygapophyses at the limit of dorsiflexion.

and Parrish (1999) estimated the flexibility of two diplodocids, using dimensionally accurate models for the articular facets and centra and assuming that the zygapophyses were limited to a maximum displacement (or minimum tolerated overlap) based on that observed in avians (figs. 6.13, 6.14). The result was less overall range of motion than commonly expected, primarily in those regions of the neck where the relatively small zygapophyses would have allowed little excursion.

ESTIMATING FEEDING HEIGHTS

In neutral pose, the neck of every sauropod we have studied thus far has negligible curvature at the base and, at its cranial extent, a tendency to droop or curve ventrally. Determining the mean head height requires understanding the height and slope of the base of the neck, which in turn is determined by the reconstruction of the trunk. The precise degree of arching of the dorsal vertebrae of most sauropods is as yet unknown, and the placement of the pectoral girdles is still a matter of some uncertainty as well. Nonetheless, the maximum effect of these factors on mean feeding height can be estimated for various taxa that have been modeled three-dimensionally.

The variation in head height resulting from differing expectations for pelvic girdle placement can be envisioned by similar triangles. Holding other factors constant, the head can be thought of as cantilevered at the end of the presacral column ahead, pivoted about the acetabular axis, and supported by the glenoid. The variation in head height is proportional to the ratio of rostrum–acetabulum distance to glenoid–acetabulum distance, typically about three times the variation at the shoulder. While there is significant uncertainty in the reconstruction of the pectoral girdles, this effect is compounded by the degree of the arch to the dorsal column (fig. 6.8). Clearly these differences do not

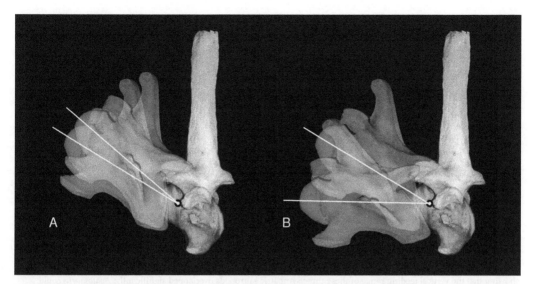

FIGURE 6.13. Giraffe dorsoventral range of motion for cervical vertebra C7 articulating on T1. The center of rotation, as indicated, was estimated by a circular fit to the profile of the condyle of T1. In (A) dorsiflexion of approximately 9° (relative to indicated neutral position slope) is limited osteologically when the postzygapophyses of C7 contact T1. In (B) ventriflexion of roughly 30° is permitted while preserving substantial overlap (about 50%) between zygapophyseal facets.

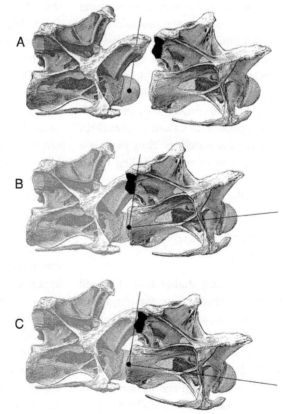

FIGURE 6.14. *Diplodocus carnegii* cervical vertebrae C13 and C14 (after Hatcher 1901: pl. 4) showing estimated limits of the dorsoventral range of motion. In (A) the right postzygapophysis of C13 is shaded, and the midline of prezygapophysis of C14 is indicated along with the center of rotation for dorsoventral flexion at the center of curvature of the condyle of C14. In (B) the two vertebrae are placed in articulation and the dorsiflexion limit of 6° is illustrated, defined as the angular deflection that reduces the zygapophyseal overlap to 50% (note that the anterior margin of the postzygapophysis of C13 is displaced anteriorly to the indicated midline). In (C) ventriflexion is limited to about 8° (at which point the posterior margin of the postzygapophysis is displaced to the prezygapophysis midline).

FIGURE 6.15. DinoMorph models of *Brachiosaurus* in neutral pose and *Apatosaurus* and *Diplodocus* maximally dorsiflexed, showing that the vertical feeding range of these three sympatric sauropods overlapped significantly: *Apatosaurus* could reach the neutral pose head height of *Brachiosaurus*, and *Diplodocus*, while less dorsally flexible, was capable of >4-m elevation. Vertical scale bar represents 5 m.

categorically change the expectation for browse height, but they do have some consequences regarding conventional views on niche partitioning among sauropods. Based on DinoMorph simulations, *Diplodocus* was capable of reaching vegetation at least 4 m high, and *Apatosaurus* had sufficient dorsal flexibility to reach what we estimate to be the mean head height of *Brachiosaurus* (fig. 6.15). On the other hand, *Brachiosaurus* could readily reach down to ground level (without the need to splay its legs giraffe-style), provided it had the modest ability to flex about 8° ventrally in the proximal cervical vertebrae. Thus these two clades shared 4–6 m of vertical feeding range, which also appears to have overlapped those of camarasaurids (fig. 6.15).

FEEDING STRATEGIES AND PALEOECOLOGY

In the last decade, many studies have addressed sauropod feeding, mostly from the standpoint of tooth morphology and microwear (Fiorillo 1998; Christiansen 2000; Upchurch and Barrett 2000), jaw muscle reconstruction and cranial mechanics (Calvo 1994; Barrett and Upchurch 1994; Upchurch and Barrett 2000), or bioenergetics (Farlow 1987; Dodson 1990).

Sauropod feeding is of interest for several reasons. First, as has been widely noted (e.g., Stevens and Parrish 1999), sympatry among sauropod genera is widespread, particularly in the Late Jurassic (although the recent study by Curry Rogers and Forster [2001] indicates that two or more sympatric species of sauropods may have occurred in Madagascar as late as the Late Cretaceous). Second, because sauropods are among the largest terrestrial herbivores that ever lived, understanding their bioenergetics is useful both in considering the scaling of metabolism in vertebrates and in examining the likelihood that sauropods, and other dinosaurs, were endotherms, were ectotherms, or had a unique metabolic physiology.

It is clear from the cranial and dental studies that significant differences existed among sauropod feeding mechanisms. Brachiosaurids and camarasaurids had broad, spatulate teeth that appear to have been optimal for biting off resistant vegetation, whereas the teeth of diplodocids and titanosaurs were more peglike in shape and probably functioned best in cropping or, perhaps, in stripping leaves from branches (Barrett and Upchurch 1994). The recent feeding studies are unanimous in their assertion that no significant processing of food occurred in the mouths of any known sauropods, although the spatulate-toothed forms probably sliced their food when grasping it with their mouths rather than just pulling it off the branches.

Another topic of debate is the range of heights and, in particular, the maximum height at which a given sauropod taxon might have fed. Different workers have based their estimates of maximum feeding height on their assumptions regarding cervical posture. Upchurch and Barrett (2000) estimated the maximum browse height as the sum of shoulder height and neck length. In other words, their maximum browse height assumed a completely vertical neck. The reconstructions in McIntosh et al. (1997) approach this assumption as well. As discussed above, the results of our cervical studies suggest substantially lower maximum browsing heights. Head elevation was far from vertical, reaching only approximately 42° above the horizontal in *Apatosaurus* and about 15° in *Diplodocus* (fig. 6.15). Since the curvature associated with dorsiflexion shortens the effective neck length, maximum browse height would be more accurately estimated as the sum of shoulder height plus roughly 40%–60% of the neck length. Maximum browse height might be of lesser ecological importance than mean browse height, however, given the prevalence of BV feeding in modern herbivores, where larger and taller taxa may capitalize on their ability to ventriflex (i.e., to reach down from their neutral feeding height) to obtain food resources when their preferred fodder becomes scarce (Daag and Foster 1976; Leuthold 1977).

Occasionally some sauropods (usually diplodocids [e.g., Bakker 1986; Paul 2000]) have been depicted as standing tripodally, balanced on their hindlimbs and the proximal end of the caudal vertebral column, and thus as having the potential to reach far higher vegetation. If some sauropods were capable of assuming such a tripodal feeding posture, a far greater range of vertical niche partitioning would have been possible than if all sauropods were obligatory quadrupeds. Again, an accurate body reconstruction is a necessary prerequisite to estimating the additional height that might have been provided by a sauropod rearing up on its hindlimbs. The maximum feeding height, however, would be far harder to estimate, because the acetabulum height would no longer serve as a definitive anchor point.

The hindlimb stance would probably need to be widened to achieve stability, which in turn would lower the center of mass and the fulcrum about which the presacral trunk would pivot. Unless the back were flexed dorsally to compensate for the bending moment induced by the massive gut, the animal's vertical reach would be less than sometimes optimistically depicted. The height of the head at the maximum vertical reach of a tripodal pose, and the width of the feeding envelope in this posture, would depend on how nearly erect the sauropod could stand, on whether it settled some of its weight on the tail kangaroo-style, and on its stability, which would ultimately depend on the acuity of its neuromuscular coordination. The sauropod would need to integrate visual and vestibular signals to movement and to coordinate compensatory movements in order to prevent catastrophic instability, especially when attempting to extend the neck laterally as well as dorsally.

The relationship between sauropod feeding strategies and Mesozoic floral evolution is of interest in at least two ways. First, establishing a correspondence between sauropod tooth, skull, and neck forms and particular plant diets would clarify one of the most important types of energy flow among organisms in Jurassic and Cretaceous ecosystems. Second, at least some paleobotanists (Wing and Tiffney 1987) and vertebrate paleontologists (Bakker 1978) have suggested that "clear-cutting" of Jurassic and Early Cretaceous forests by sauropod herds may have been instrumental in creating ecological conditions that favored the origin of flowering plants.

The lithological unit containing the greatest diversity and abundance of sauropod fossils is the Upper Jurassic (and possibly Lower Cretaceous) Morrison Formation of western

North America. Although floras of the Morrison Formation are not nearly as well preserved and understood as its sauropod fauna, they have been the object of many studies. Miller (1987) surveyed Morrison compression floras, mostly from Montana, and noted that cycadophytes, seed ferns, and ferns were generally more abundant than conifers and ginkgophytes. In their survey of Morrison taphonomy and paleoecology, Dodson et al. (1980) interpreted the Morrison depositional basin as a mostly arid, strongly seasonal, alluvial plain dissected by stream channels and occasional lakes. Rees et al. (2004) surveyed fossil plant distribution and taphonomy in the Late Jurassic Morrison Formation and concluded that the Morrison landscapes were savanna-like, dominated by herbaceous plants and short trees such as ginkgos and cycads, most of which were concentrated near rivers and lakes. Although taller conifers were part of the landscape, they appear to have been rare and confined to riparian corridors and small, isolated patches of forest (Rees et al. 2004).

Comparison of estimated browsing heights of various sauropod taxa with heights of extant members of plant clades that were abundant during the Jurassic and Cretaceous (fig. 6.16) shows a striking correspondence between the mean browse heights of the sauropods and these generalized vegetation dimensions. The only sauropod taxa that appear to have had the capability of grazing on tall gymnosperms were *Brachiosaurus* and perhaps *Camarasaurus*. Studies of cranial anatomy and tooth form in both genera, and of tooth wear in *Camarasaurus*, suggest that they fed on more resistant plant material and possibly utilized more oral processing than was the case for diplodocids (Calvo 1994; Fiorillo 1998, Upchurch and Barrett 1990). A survey of relative abundances of sauropods in the Morrison and Tendaguru sites (Rees et al. 2004) indicates that brachiosaurids are more abundant in the conifer-rich Tendaguru beds than the Morrison deposits. Upchurch and Barrett (1990) also put forth the hypothesis that differential wear on the upper and lower teeth of *Diplodocus* suggested that it utilized both low and high browsing. Diplodocids, along with the titanosaur clade Nemegtosauridae, have distinctive adaptations (including dorsally placed nostrils, a tooth comb restricted to the front of the jaws that was suitable for sieving or cropping, and a ventral inclination of the head relative to the horizontal axis of the braincase) that may have facilitated low browsing and, perhaps, even placement of the greater part of the heads underwater during feeding on aquatic plants (Parrish 2003).

One ecological paradox that has long been posited regarding the Jurassic sauropod-dominated ecosystems is how they could support such a diverse and large assemblage of giant herbivores. One common model proposed involves herds of sauropods moving through forests and denuding them of vegetation as they trampled the trunks (e.g., Wing and Tiffney 1987). Given that the default inclination for most sauropod necks appears to be one that positions the head close to the ground, perhaps it is more plausible that they were also feeding mostly at or near ground level. Lycopods and ferns appear to have been abundant in Mesozoic terrestrial ecosystems and, also, would have served as a rapidly growing, readily renewable food source, much like today's angiospermous grasses. This is not a new speculation, having been proposed for diplodocids (Krasilov 1981; Chatterjee and Zheng 1997), but it does potentially dovetail current thinking about sauropod functional morphology, late Mesozoic paleoecology, and plant physiology.

CONCLUSIONS

Since sauropods were first described, it has gradually become standard to reconstruct most of these dinosaurs as high browsers. The neutral pose of the presacral vertebral column can be reconstructed from the osteology of the articular facets of its component vertebrae. However, for all sauropods thus studied our analysis shows the following conditions: the dorsal portion of the vertebral column is

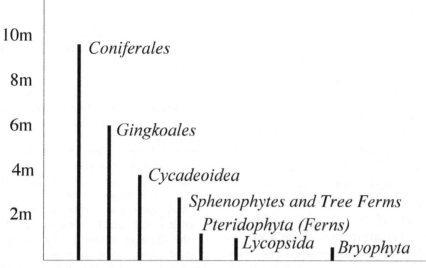

FIGURE 6.16. The bulk of the potential plant biomass in the middle Mesozoic, as exemplified here by the Morrison flora, appears to have been medium to low browse. Even the tallest conifers had the greatest amount of their foliage located in their bottom halves, and some of the fastest-growing and potentially most nutritious plants were ground cover (e.g., lycopods and small ferns). There was little vertical stratification in the flora to drive potential diversification among these herbivores. "Head-down" feeding by sauropods would be particularly apt for accessing the foliage and fructifications of groups like ferns, cycads, and ground cover. Even without reconstructing *Brachiosaurus* with a giraffelike neck, this sauropod was capable of BN feeding at about 5–6 m, while BD feeding by dorsiflexion of merely 3° per joint would potentially raise the head >9 m above ground level. Ventriflexion by about 8° per joint proximally would place the head at ground level, permitting BV feeding that overlaps the mean feeding heights of camarasaurids and diplodocids. The diplodocids, in occasional BD feeding, could have overlapped the camarasaurid vertical range but were unique (with the possible exception of the later, nonsympatric *Euhelopus*) in their ability to feed in BV below mean surface level, a potential adaptation for aquatic feeding (see fig. 6.9). Contrary to the common expectation, vertical differentiation in sauropod feeding envelopes is not evident among these taxa, many of which were coeval and sympatric. The predominantly low mean feeding heights of these sauropods correspond closely to the height of the bulk of available browse. Although dental adaptations suggest clear diversity in feeding preferences among sauropods, there was little vertical stratification of floras to drive potential diversification among these herbivores.

arched, the cervical–dorsal transition is straight, negligible intrinsic curvature occurs in the posterior cervical vertebrae, and a shallow, concave-downward curvature occurs cranially. Most sauropods were medium to low browsers, with generally a "head-down" feeding angle. Although *Brachiosaurus*, possibly the tallest sauropod, is shown to have a nearly horizontal neck, it nonetheless held its head about 6 m above ground level. Therefore, *Brachiosaurus* and, perhaps, *Camarasaurus* (at the upward extent of its feeding envelope) appear to have been the only Jurassic sauropods clearly capable of feeding as "high browsers" on arborescent gymnosperms. Recent paleobotanical studies (e.g., Rees et al. 2004; J. T. Parrish et al., in press) suggest that the greatest abundance of forage in the sauropod-rich, Late Jurassic Morrison Formation was low- to medium-height ground cover, which is consistent with the concept of most Morrison sauropods being low browsers.

ACKNOWLEDGMENTS

The authors are deeply appreciative of the contributions Dr. Jack McIntosh has made to the field of sauropod paleontology. Jack's devotion to sauropods, his encyclopedic familiarity with their fossil record, and his willingness to share that knowledge have enriched our research, and our lives, immensely. Jack has always expressed energetic support for our efforts to define the pose and movements of sauropods quantitatively. Seeing Jack dance a spontaneous Scottish jig in response to our poster at the Pittsburgh SVP quickened our hearts. Since then our work has been driven largely by a desire to induce a repeat performance. Ron Toth and Judy Parrish provided advice and photographs that were instrumental in the estimation of plant height estimates for figure 6.16. Brian Curtice and Christopher McGowan provided photos of the Humboldt Museum sauropods that augmented those taken by J.M.P. Eric Wills provided important contributions to the programming of DinoMorph. Thanks go to Jim Madsen for providing us with the opportunity to manipulate a cast of *Diplodocus carnegii*. Heinrich Mallison provided three-dimensional data for the modeling of *B. brancai*. Matt Bonnan and Phil Platt provided valuable insights into sauropod limb posture. This work was supported in part by National Science Foundation Grant 0093929.

LITERATURE CITED

Alexander, R.M. 1989. Dynamics of Dinosaurs and Other Extinct Giants. Columbia University Press, New York. 167 pp.

Bakker, R.T. 1978. Dinosaur feeding behavior and the origin of flowering plants. Nature 274: 661–663.

———. 1986. Dinosaur Heresies. William Morrow, New York. 481 pp.

Barrett, P.M., and Upchurch, P. 1994. Feeding mechanisms of *Diplodocus*. GAIA 10: 195–204.

Beasley, Walter. 1907. *Diplodocus:* the greatest of all earthly creatures. Scientific American 96(24): 491–492.

Bonnan, M.F. 2001. The Evolution and Functional Morphology of Sauropod Dinosaur Locomotion. Ph.D. dissertation. Northern Illinois University, DeKalb, IL. 722 pp.

Calvo, J.O. 1994. Jaw mechanics in sauropod dinosaurs. GAIA 10: 183–194.

Chatterjee, S., and Zheng, Z. 1997. The feeding strategies in sauropods. J. Vertebr. Paleontol. 17: 37A.

Christian, A., and Heinrich, W.-D. 1998. The neck posture of *Brachiosaurus brancai*. Mittelungen Mus. Naturk. Berlin Geowissenschaften 1: 73–80.

Christian, A., Heinrich, W.-D. and Golder, W. 1999. Posture and mechanics of the forelimbs of *Brachiosaurus brancai* (Dinosauria: Sauropoda). Mittelungen Mus. Naturk. Berlin Geowissenschaften 2: 63–73.

Christiansen, P. 2000. Feeding mechanisms of the sauropod dinosaurs *Brachiosaurus, Camarasaurus, Diplodocus,* and *Dicraeosaurus*. Hist. Biol. 14: 137–152.

Currie, P.J. 1987. New approaches to studying dinosaurs in Dinosaur Provincial Park. In: Czerkas, S.J., and Olson, E.C. (eds.). Dinosaurs Past and Present. Vol. II. University of Washington Press, Seattle. Pp. 100–117.

Curry Rogers, K., and Forster, C.A. 2001. The last of the dinosaur titans: a new sauropod from Madagascar. Nature 412: 530–534

Czerkas, S.A., and Czerkas, S.J. 1991. Dinosaurs: A Global View. Mallard Press, New York. 247 pp.

Daag, A.I., and Foster, J.B. 1976. The Giraffe: Its Biology, Behavior, and Ecology. Van Nostrand Reinhold, New York. 210 pp.

Dodson, P.M. 1990. Sauropod paleoecology. In: Weishampel, D.B., Dodson P., and Osmolska, H. (eds.). The Dinosauria. University of California Press, Berkeley. Pp. 402–407.

Dodson, P.M., Behernsmeyer, A.K., Bakker, R.T., and McIntosh, J.T. 1980. Taphonomy and paleoecology of the dinosaur beds of the Jurassic Morrison Formation. Paleobiology 6: 208–232.

Farlow, J.O. 1987. Speculations about the diet and digestive physiology of herbivorous dinosaurs. Paleobiology 13: 60–72.

Fiorillo, A.R. 1998. Dental microwear patterns of sauropod dinosaurs *Camarasaurus* and *Diplodocus:* evidence for resource partitioning in the Late Jurassic of North America. Hist. Biol. 13: 1–16.

Gauthier-Pilters, H., and Daag, A.I. 1981. The Camel, Its Ecology, Behavior and Relationship to Man. University of Chicago Press, Chicago. 208 pp.

Gilmore, C.W. 1925. A nearly complete articulated skeleton of *Camarasaurus*, a saurischian dinosaur from the Dinosaur National Monument. Mem. Carnegie Mus. 10: 347–384.

Gilmore, C.W. 1936. The osteology of *Apatosaurus* with special reference to specimens in the Carnegie Museum. Mem. Carnegie Mus. 11: 175–300.

Hatcher, J.B. 1901. *Diplodocus* (Marsh): its osteology, taxonomy, and probable habits, with a reconstruction of the skeleton. Mem. Carnegie Mus. 1: 1–63.

Holland, W. J. 1906. The osteology of *Diplodocus* Marsh. Mem. Carnegie Mus. 2: 225–278.

Janensch, W. 1929. Die Wirbelsäule der Gattung *Dicraeosaurus hausemanni*. Palaeontographica 3 (Suppl. 7): 39–133.

Janensch, W. 1950a. Die Wirbelsäule von *Brachiosaurus brancai*. Palaeontographica (Suppl. 7) 3: 27–93.

Janensch, W. 1950b. Die Skelettrekonstrucktion von *Brachiosaurus brancai*. Palaeontographica 3 (Suppl. 7): 95–103.

Janensch, W. 1961. Die Gliedmaszen und Gliedmaszengürtel der Sauropoden der Tendaguru-Schichten. Palaeontographica (Suppl. 7)3: 177–235.

Krasilov, V. A. 1981. Changes of Mesozoic vegetation and the extinction of the dinosaurs. Paleogeog. Paleoclimatol. Palaeoecol. 34: 207–224.

Leuthold, B. 1977. African Ungulates: A Comparative Review of Their Ethology and Behavioral Ecology. Springer Verlag, Berlin. 307 pp.

Leuthold, B., and Leuthold, W. 1972. Food habits of giraffe in Tsavo National Park, Kenya. East Afr Wildl. J. 10: 129–141.

Martin, J. 1987. Mobility and feeding of *Cetiosaurus*: Why the long neck? Occas. Paper Tyrell Mus. Paleontol. 3: 150–155.

McGinnis, H. J. 1982. Carnegie's Dinosaurs. Carnegie Museum of Natural History, Pittsburgh, PA. 129 pp.

McIntosh, J., M. K. Brett-Surman and J. O. Farlow. 1997. Sauropods. In: Farlow, J. O., and Brett-Surman, M. K. (eds.). The Complete Dinosaur. Indiana University Press, Bloomington. Pp. 264–290.

Miller, C. M. 1987. Land plants of the northern Rocky Mountains before the appearance of flowering plants. Ann. Ma. Bot. Garden 74: 692–706.

Osborn, H. F., and Mook, C. C. 1921. *Camarasaurus*, *Amphicoelias*, and other sauropods of Cope. Mem. Am. Mus. Nat. Hist. 3: 247–287.

Owen-Smith, R. N. 1988. Megaherbivores: The Influence of Very Large Body Size on Ecology. Cambridge University Press, Cambridge. 369 pp.

Parrish, J. M. 2003. Mapping ecomorphs onto sauropod phylogeny. J. Vertebr. Paleontol. Suppl. 23(3): 85A–86A.

Parrish, J. M., and Stevens, K. A. 1998. Undoing the death pose: Using computer imaging to restore the posture of articulated dinosaur skeletons. J. Vertebr. Paleontol. Suppl.18: 69A.

———. 2002a. Neck mobility in long-necked vertebrates: from modern mammals to sauropods. J. Morphol. 248: 270.

———. 2002b. Rib angulation, scapular position, and body profiles in sauropod dinosaurs. J. Vertebr. Paleontol. Suppl. 22(3): 95A.

Paul, G. S. 1987. The science and art of restoring the life appearance of dinosaurs and their relatives: a rigorous how-to guide. In: Czerkas, S. J., and Olson, E. C., (eds.). Dinosaurs Past and Present. Vol. II. University of Washington Press, Seattle. Pp. 5–49.

———. 2000. Restoring the life appearance of dinosaurs. In: Paul, G. S. (ed.). The Scientific American Book of Dinosaurs. Bryon Press and Scientific American, New York. Pp. 78–106.

Pellew, R. 1984. The feeding ecology of a selective browser, the giraffe (*Giraffa camelopardalis tippelskirchi*). J. Zool. 202: 57–81.

Pincher, C. 1949. Evolution of the giraffe. Nature 164: 29–30.

Rees, P. M., Noto, C. R., Parrish, J. M., and Parrish, J. T. 2004. Late Jurassic climates, vegetation, and dinosaur distributions. J. Geol. 112: 643–653.

Sereno, P. C., Beck, A. L., Moussa, B., Dutheil, D., Larsson, H. C. E., Lyon, G. H., Sadlier, R. W., Sidor, C. A., Varrichio, D. J., Wilson, G. P., and Wilson, J. A. 1999. Cretaceous sauropods from the Sahara and the uneven rate of skeletal evolution among dinosaurs. Science 286: 1342–1347.

Simmons, R., and Scheepers, L. 1996. Winning by a neck: sexual selection in the evolution of the giraffe. Am. Nat. 148: 771–786.

Stevens, K. A. 2002. DinoMorph: parametric modeling of skeletal structures. Senckenbergiana Lethaea 82(1): 23–34.

Stevens, K. A., and Parrish, J. M. 1999. Neck posture and feeding habits of two Jurassic sauropod dinosaurs. Science, 284: 798–800.

———. 2005a. Neck posture, dentition, and feeding strategies in Jurassic sauropod dinosaurs. In: Carpenter, K., and Tidwell, V. (eds.). Thunder Lizards: The Sauropodomorph Dinosaurs. University of Indiana Press, Bloomington (in press).

———. 2005b. The intrinsic curvature of sauropod necks (Saurischia: Dinosauria) (in preparation).

Stevens, K. A., and Wills, E. D. 2001. Gracile versus robust cervical vertebral designs in sauropods. J. Vertebr. Paleontol. Suppl. 21(3): 104A.

Thulborn, R. A. 1990. Dinosaur Tracks. Chapman and Hall, London. 430 pp.

Upchurch, P., and Barrett, P. M., 2000. The evolution of sauropod feeding mechanisms. In: Sues, H. D. (ed.). Evolution of Herbivory in Terrestrial Vertebrates. Cambridge University Press, Cambridge. pp. 79–122.

Walther, F. 1969. Flight behavior and avoidance of predators in Thompson's gazelle (*Gazella thomsoni* Gunther, 1884). Behaviour 34: 184–221.

Wedel, M. J., and Sanders, R. K. 1999. Comparative morphology and functional morphology of the cervical series in Aves and Sauropoda. J. Vertebr. Paleontol. 19: 83A.

Wilhite, R. 2003. Digitizing large fossil skeletons for three-dimensional applications. Paleontol. Electronica 5(1): 1–10

Wilson, J. A., and Sereno, P. C. 1998. Early evolution and higher-level phylogeny of sauropod dinosaurs. Soc. Vertebr. Paleontol. Mem. 5: 1–68.

Wiman, C. 1929. Die Kriede-dinosaurier aus Shantung. Palaeontol. Sinica (Ser. C) 6: 1–67.

Wing, S. L., and Tiffney, B. H. 1987. The reciprocal interaction of angiosperm evolution and tetrapod herbivory. Rev. Paleobot. Palynol. 50: 179–210.

Woolnough, A. P., and du Toit, J. T. 2001. Vertical zonation of browse quality in tree canopies exposed to a size-structured guild of African browsing ungulates. Oecologia 129: 585–590.

Young, T., and Isbell, L. 1991. Sex differences in giraffe feeding ecology: Energetic and social constraints. Ethology 87: 79–89.

SEVEN

Postcranial Skeletal Pneumaticity in Sauropods and Its Implications for Mass Estimates

Mathew J. Wedel

ONE OF THE SIGNAL FEATURES of sauropods, and one of the cornerstones of our fascination with them, is their apparent efficiency of design. The presacral neural spines of all sauropods have a complex of bony ridges or plates known as vertebral laminae (fig. 7.1; abbreviations used in the figures are listed below). In addition, the vertebral centra of most sauropods bear deep fossae or have large foramina that open into internal chambers. The laminae and cavities of sauropod vertebrae are often considered to be adaptations for mass reduction (Osborn 1899; Hatcher 1901; Gilmore 1925) and have been important in studies of sauropod evolution (McIntosh 1990; Wilson 1999). The possibility that these structures were pneumatic—that they contained or partitioned air-filled diverticula of the lungs or air sacs—has been recognized for over a century (Seeley 1870; Janensch 1947). However, pneumaticity in sauropods has received little attention until recently (Britt 1997; Wilson 1999; Wedel 2003a, 2003b).

My goal here is to review previous work on pneumaticity in sauropods, discuss some outstanding problems, and outline possible directions for future studies. To that end, the chapter is organized around three questions. What criteria do we use to infer pneumaticity in sauropod fossils? What characteristics of pneumatic bones have been (or could be) described? and How can we apply data on skeletal pneumaticity to paleobiological problems, such as estimating the masses of sauropods? Before attempting to answer these questions, it will be useful to review skeletal pneumaticity in living vertebrates.

Institutional abbreviations: BYU, Earth Sciences Museum, Brigham Young University, Provo, Utah; CM, Carnegie Museum, Pittsburgh, Pennsylvania; DGM, Museo de la Divisão de Geologia y Mineralogia, Rio de Janeiro, Brazil; DMNS, Denver Museum of Nature and Science, Denver, Colorado; OMNH, Oklahoma Museum of Natural History, Norman, Oklahoma; USNM, National Museum of Natural History, Smithsonian Institution, Washington, DC.

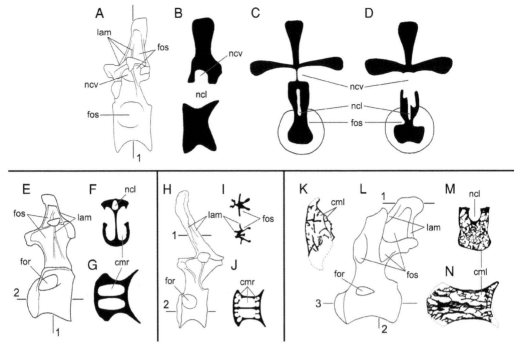

FIGURE 7.1. Pneumatic features in dorsal vertebrae of *Barapasaurus* (A–D), *Camarasaurus* (E–G), *Diplodocus* (H–J), and *Saltasaurus* (K–N). Anterior is to the left; different elements are not to scale. A, A posterior dorsal vertebra of *Barapasaurus*. The opening of the neural cavity is under the transverse process. B, A midsagittal section through a middorsal vertebra of *Barapasaurus* showing the neural cavity above the neural canal. C, A transverse section through the posterior dorsal shown in A (position 1). In this vertebra, the neural cavities on either side are separated by a narrow median septum and do not communicate with the neural canal. The centrum bears large, shallow fossae. D, A transverse section through the middorsal shown in B. The neural cavity opens to either side beneath the transverse processes. No bony structures separate the neural cavity from the neural canal. The fossae on the centrum are smaller and deeper than in the previous example. (A–D redrawn from Jain et al. 1979:pl. 101, 102.) E, An anterior dorsal vertebra of *Camarasaurus*. F, A transverse section through the centrum (E, position 1) showing the large camerae that occupy most of the volume of the centrum. G, a horizontal section (E, position 2). (E–G redrawn from Ostrom and McIntosh 1966:pl. 24.) H, A posterior dorsal vertebra of *Diplodocus*. (Modified from Gilmore 1932:fig. 2.) I, Transverse sections through the neural spines of other *Diplodocus* dorsals (similar to H, position 1). The neural spine has no body or central corpus of bone for most of its length. Instead it is composed of intersecting bony laminae. This form of construction is typical for the presacral neural spines of most sauropods outside the clade Somphospondyli. (Modified from Osborn 1899:fig. 4.) J, A horizontal section through a generalized *Diplodocus* dorsal (similar to H, position 2). This diagram is based on several broken elements and is not intended to represent a specific specimen. The large camerae in the midcentrum connect to several smaller chambers at either end. K, A transverse section through the top of the neural spine of an anterior dorsal vertebra of *Saltasaurus* (L, position 1). Compare the internal pneumatic chambers in the neural spine of *Saltasaurus* with the external fossae in the neural spine of *Diplodocus* shown in J. L, An anterior dorsal vertebra of *Saltasaurus*. M, A transverse section through the centrum (L, position 2). N, A horizontal section (L, position 3). In most members of the clade Somphospondyli the neural spines and centra are filled with small camellae. (K–N modified from Powell 1992:fig. 16.)

Anatomical abbreviations: al, accessory lamina; *cit*, canalis intertransversarius; *cml*, camella; *cmr*, camera; *dsv*, diverticulum supervertebrale; *for*, foramen; *fos*, fossa; *lam*, lamina; *nad*, neural arch diverticulum; *naf*, neural arch fossa; *ncl*, neural canal; *ncs*, neurocentral suture; *ncv*, neural cavity; *nsf*, neural spine fossa; *pcdl*, posterior centrodiapophyseal lamina; *podl*, postzygodiapophyseal lamina; *prdl*, prezygodiapophyseal lamina; *spol*, spinopostzygapophyseal lamina; *sprl*, spinoprezygapophyseal lamina; *vk*, ventral keel.

SKELETAL PNEUMATICITY IN EXTANT TAXA

Pneumatization of the postcranial skeleton in various ornithodiran groups, including sauropods, is just one instance of the more general phenomenon of skeletal pneumatization. Skeletal pneumatization, which includes paranasal, paratympanic, and pulmonary pneumatic spaces, is unique to archosaurs and advanced synapsids (Witmer 1997, 1999). However, diverticula (epithelium-lined outgrowths) of the pharynx or trachea are present in representative taxa from most major lineages of tetrapods, including frogs (Duellman and Trueb 1986), snakes (Young 1991, 1992), birds (King 1966; McClelland 1989a), and primates (Janensch 1947). Pharyngeal and tracheal diverticula are often used to inflate specialized structures used in phonation or visual display. These diverticula do not invade any bones except the hyoid, which is pneumatized by tracheal diverticula in the howler monkey *Alouatta* (Janensch 1947; *Mycetes* of his usage). Diverticula of paranasal and paratympanic air spaces extend down the neck in some species of birds, but these diverticula are subcutaneous or intermuscular and do not pneumatize the postcranial skeleton (King 1966). Extremely rare examples of cervical pneumatization have been reported in humans, but these are pathological cases related to occipitoatlantal fusion (Sadler et al. 1996). Among extant taxa, only birds have extensive postcranial skeletal pneumaticity (PSP).

Extant birds have relatively small, inflexible lungs and an extensive system of air sacs in the thorax and abdomen. The air sacs are flexible and devoid of parenchymal tissue, and their primary function is to ventilate the lungs (King 1966; Duncker 1971; McClelland 1989b). In most birds, the air sacs also give rise to a network of diverticula. Diverticula pass into the viscera, between muscles, and under the skin in various taxa (Richardson 1939; King 1966; Duncker 1971). If a diverticulum comes into contact with a bone, the bone may become pneumatized. Bremer (1940) described the pneumatization of the humerus in the chicken (*Gallus*) as follows. The diverticulum enters the bone because osteoclasts break down the bony tissue ahead of it. The bony tissue immediately adjacent to the diverticulum is replaced by mesenchymal tissue, which degenerates or is resorbed and is in turn replaced by the growing diverticulum. As the diverticulum bores through the cortical bone it produces a pneumatic foramen, which must remain open for pneumatization to proceed normally (Ojala 1957). Once the bone has been penetrated, branches of the diverticulum spread through the marrow cavity by replacing bony trabeculae. The marrow is reduced to small islands of tissue surrounded by the diverticulum. As these islands of marrow degenerate, the branches of the diverticulum anastomose and form a single, epithelium-lined air cavity that occupies most of the internal volume of the bone. The trabecular structure of the bone is greatly reduced, and the inner layers of the cortex are resorbed.

Witmer (1990) pointed out that a pneumatic foramen does not have to be located on the pneumatic bone in question; the intraosseous diverticulum may have spread across a suture from an adjacent pneumatic bone. He called this *extramural pneumatization* and contrasted it with *intramural pneumatization*, in which a diverticulum directly invades a bone and produces a pneumatic foramen. Although Witmer (1990) was concerned with cranial pneumatization, extramural pneumatization also occurs in the postcranial skeleton, for example, between fused vertebrae in the chicken (King 1957; Hogg 1984a).

The term *air sac* has been used by some authors for any reservoir of air in an animal that is lined by epithelium and devoid of parenchymal tissue (e.g., Brattstrom 1959; Cranford et al. 1996). The same term is often used in the ornithological literature to refer specifically to the pulmonary air sacs of birds (e.g., Müller 1907). In this paper, the term air sac is restricted to indicate the pulmonary air sacs of birds. All other epithelium-lined air reservoirs, including those that develop from the lungs and air sacs, are called diverticula. Another important difference is between a pneumatic diverticulum, which is a soft-tissue structure, and the bony recess that it may occupy (Witmer 1999). In many cases, the bony recess is produced by the diverticulum through the process of pneumatization. This causal relationship allows us to infer the presence of diverticula from certain kinds of bony recesses. The study of skeletal pneumaticity in fossil taxa is founded on such inferences.

WHAT CRITERIA DO WE USE TO INFER PNEUMATICITY IN FOSSILS?

How do we recognize skeletal pneumaticity? More specifically, what are the osteological correlates (*sensu* Witmer 1995, 1997) of pneumatic diverticula, such that the presence of the latter can be inferred from the former? Several authors, including Hunter (1774) and Müller (1907), list differences between pneumatic and apneumatic bones. These authors focused on recognizing pneumaticity in extant birds and thus referred to attributes that tend not to fossilize, such as vascularity, oil content, and color. Britt (1993, 1997) provided the most comprehensive list of pneumatic features identifiable in fossil bones: internal chambers with foramina, fossae with crenulate texture, smooth or crenulate tracks (grooves), bones with thin outer walls, and large foramina.

INTERNAL CHAMBERS WITH FORAMINA

The most obvious osteological correlate of pneumaticity is the presence of foramina that lead to large internal chambers. Large chambers, often called "pleurocoels," are present in the presacral vertebrae of most sauropods. They may also be present in the sacral and caudal vertebrae, as in *Apatosaurus* and *Diplodocus* (see Ostrom and McIntosh 1966:pl. 30 and Osborn 1899:fig. 13, respectively). In extant birds, such chambers are invariably associated with pneumatic diverticula (Britt 1993). The presence of similar chambers in the bones of sauropods, theropods, and pterosaurs has been accepted by most authors as prima facie evidence of pneumaticity (Seeley 1870; Cope 1877; Marsh 1877; Janensch 1947; Romer 1966; Britt 1993, 1997; O'Connor 2002). As far as I am aware, no substantive alternative hypotheses have been advanced; as Janensch (1947:10: translated from the German by G. Maier) said, "There is no basis to consider the pleurocentral cavities in sauropod vertebrae as different from similar structures in the vertebrae of birds." In short, no soft tissues other than pneumatic diverticula are known to produce large foramina that lead to internal chambers, and these chambers constitute unequivocal evidence of pneumaticity.

One of the primary differences among the pneumatic vertebrae of different sauropod taxa is the subdivision of the internal chambers. Some taxa, such as *Camarasaurus*, have only a few large chambers, whereas others, such as *Saltasaurus*, have many small chambers (fig. 7.1). Vertebrae with many small chambers have been characterized as "complex" (Britt 1993; Wedel 2003b), in contrast to "simple" vertebrae with few chambers. The concept of "biological complexity" has several potential meanings (McShea 1996). In this paper, complexity refers only to the level of internal subdivision of pneumatic bones; complex bones have more chambers than simple ones. This is "nonhierarchical object complexity" in the terminology of McShea (1996).

EXTRAMURAL PNEUMATIZATION

The only obvious opportunities for extramural pneumatization in the postcranial skeletons of sauropods are between fused sacral and caudal vertebrae and between the sacrum and the ilium. Sacral vertebrae of baby sauropods have deep fossae (Wedel et al. 2000:fig. 14), and at least in *Apatosaurus*, a complex of internal chambers is present before the sacral vertebrae fuse (Ostrom and McIntosh 1966:pl. 30). The co-ossified blocks of caudal vertebrae in *Diplodocus* often include centra with large pneumatic foramina (Gilmore 1932:fig. 3). It is possible that co-ossified centra without foramina could be pneumatized by intraosseous diverticula of adjacent pneumatic vertebrae, although this has not been demonstrated.

Sanz et al. (1999) reported that "cancellous tissue" is present in the presacral vertebrae, ribs, and ilium of *Epachthosaurus* and *Saltasaurus*. The presacral vertebrae of *Saltasaurus* are pneumatic and have a camellate internal structure (fig. 7.1K–N), and pneumatic ribs are known in several titanosaurs (Wilson and Sereno 1998). Further, spongiosa (*sensu* Francillon-Vieillot et al. 1990) are present in apneumatic vertebrae of many—possibly all—sauropods (see "Application to a Paleobiological Problem: Mass Estimates," below), so cancellous bone is not limited to titanosaurs. For these reasons, it seems that the "cancellous tissue" of Sanz et al. (1990) is synonymous with camellate pneumatic bone. If so, then the ilia of some titanosaurs may have been pneumatic. Two possible routes for pneumatization of the ilium are by diverticula of abdominal air sacs and by extramural pneumatization from the sacrum. However, the possibility of ilial pneumatization must remain speculative until better evidence for it is presented.

NEURAL CAVITIES

In many sauropods, the neural spines of the dorsal vertebrae contain large chambers. These chambers communicate with the outside by way of large foramina beneath the diapophyses. Upchurch and Martin (2003) called such chambers neural cavities and discussed their occurrence in *Cetiosaurus*, *Barapasaurus*, and *Patagosaurus*. According to Upchurch and Martin (2003:218), "In *Barapasaurus* and *Patagosaurus*, the neural cavity is linked to the external surface of the arch by a lateral foramen which lies immediately below the base of the transverse process, just in front of the posterior centrodiapophyseal lamina [pcdl]" (see fig. 7.1A). In some dorsal vertebrae of *Barapasaurus*, the neural canal is open dorsally and communicates with the neural cavity (Jain et al. 1979). Upchurch and Martin (2003) mentioned that similar cavities are present in some neosauropods, and Bonaparte (1986:fig. 19.7) illustrated neural cavities in *Camarasaurus* and *Diplodocus*. Jain et al. (1979) and Upchurch and Martin (2003) also described a second morphology (in *Barapasaurus* and *Cetiosaurus*, respectively), in which the neural cavity is divided into two halves by a median septum and does not communicate with the neural canal (fig. 7.1C). Neural cavities are interpreted as pneumatic for the same reason that the more familiar cavities in vertebral centra are: they are large internal chambers connected to the outside through prominent foramina (Britt 1993).

PNEUMATIC RIBS

The dorsal ribs of some sauropods have large foramina that lead to internal chambers. The best-known examples of costal pneumaticity in sauropods are the pneumatic ribs of *Brachiosaurus* (Riggs 1904; Janensch 1950). Pneumatic dorsal ribs are also present in *Euhelopus* and some titanosaurs (Wilson and Sereno 1998). Gilmore (1936) described a foramen that leads to an internal cavity in a dorsal rib of *Apatosaurus*, and pneumatic dorsal ribs have also been reported in the diplodocid *Supersaurus* (Lovelace et al. 2003). Pneumatic dorsal ribs have not been found in *Haplocanthosaurus*, *Camarasaurus*, or any basal diplodocoids, so the character evidently evolved independently in diplodocids and titanosauriforms. Pneumatic ribs are part of a growing list of pneumatic characters that evolved in parallel in diplodocids and titanosauriforms, along with complex vertebral chambers and pneumatic caudal vertebrae (see below).

FOSSAE AND LAMINAE

PNEUMATIC FOSSAE

Fossae are ubiquitous in sauropod vertebrae and are often the sole evidence of pneumaticity. For example, basal sauropods such as *Barapasaurus* have shallow fossae on the presacral centra and neural spines but lack the large internal chambers typical of later sauropods (fig. 7.1). Are these fossae pneumatic? The naive assumption that all fossae are pneumatic will surely lead to the overestimation of pneumaticity. On the other hand, to deny that any fossae are pneumatic unless they contain foramina that lead to large internal chambers is equally false. We need criteria to distinguish pneumatic fossae from nonpneumatic fossae.

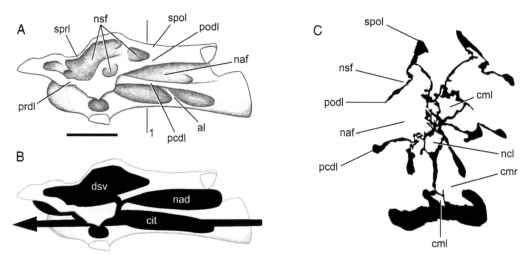

FIGURE 7.2. A cervical vertebra of *Brachiosaurus* and a hypothetical reconstruction of the pneumatic diverticula. A, BYU 12866, a midcervical vertebra of *Brachiosaurus*, in left lateral view. The neural spine fossae are bounded on all sides by the four laminae that connect the pre- and postzygapophyses to the neurapophysis and diapophysis. Some of the neural spine fossae contain large, sharp-lipped foramina. B, Possible appearance of the pneumatic diverticula, shown in black. We can be fairly certain that pneumatic diverticula occupied the fossae on the neural arch, neural spine, and centrum, but the connections between various diverticula and their order of appearance during ontogeny remain speculative. Here the diverticula have been restored based on those of birds, with the canalis intertransversarius running alongside the centrum and the diverticulum supervertebrale occupying the neural spine fossae (see Müller [1907:figs. 3–5, 7, 11, 12] for the appearance of these diverticula in the pigeon). Any connections between the canalis intertransversarius and the diverticulum supervertebrale probably passed intermuscularly, because the laminae bounding the neural spine fossae are uninterrupted by tracks or grooves. C, A transverse section through the midcentrum (A, position 1) traced from a CT image (Wedel et al. 2000:fig. 12C) and corrected for distortion. The volume of air filling the fossae and camellae in the neural arch and spine is unknown, but it may have equaled or exceeded the volume of air in the centrum. Lamina terminology after Wilson (1999). Scale bar equals 20 cm.

The best case for a pneumatic fossa is a fossa that contains pneumatic foramina within its boundaries. The *Brachiosaurus* vertebra shown in figure 7.2 has large, sharply lipped pneumatic foramina in most of the fossae on the lateral sides of the centrum and neural spine (see also Janensch 1950; Wilson 1999). Similar foramina-within-fossae are present in the vertebrae of many other neosauropods, including *Diplodocus* (Hatcher 1901:pl. 3, 7), *Tendaguria* (Bonaparte et al. 2000:fig. 17, pl. 8), and *Sauroposeidon* (Wedel et al. 2000:fig. 8b). The inference that these fossae are pneumatic relies on the presence of unequivocally pneumatic features within the fossae. The inferred presence of pneumaticity is less supported in the case of blind fossae that contain no foramina, such as the large fossae on the dorsal centra of *Barapasaurus* (fig. 7.1).

Wilson (1999) proposed that "subfossae," or fossae-within-fossae, might further support the inference of pneumaticity. "These well defined, smooth-walled depressions are present in many sauropods and seem to be analogous to the more pronounced coels [foramina] that characterize *Brachiosaurus*. Like the coels, these depressions may have housed smaller pneumatic diverticuli [sic] in life" (Wilson 1999:651). This hypothesis is supported by the complex morphology of some pneumatic diverticula in birds. In the ostrich, the large diverticula that lay alongside the cervical vertebrae consist of bundles of smaller diverticula (Wedel 2003a:fig. 2). It seems reasonable to expect that when such a bundle comes into contact with a bone, the aggregate would produce a fossa, within which each diverticulum would produce a subfossa. This hypothesis can and should be tested in future computed tomography (CT) studies. Gower (2001:121) argued that the "multipartite fossae" and "deep multi-chambered concavities" in the dorsal vertebrae of *Erythrosuchus* were more consistent with pneumaticity than with muscular or vascular structures (but see O'Connor 2002).

Britt (1993) proposed that crenulate texture of the external bone is evidence that some fossae are pneumatic. In *Sauroposeidon* the difference in texture between the pneumatic fossae and the adjacent bone is striking, and this allows the boundaries of the fossae to be precisely plotted (Wedel et al. 2000:fig. 7). However, there is little doubt that the fossae of *Sauroposeidon* are pneumatic, because they contain pneumatic foramina. The inference that a blind fossa is pneumatic based on texture alone is less certain. Blind fossae can also contain muscles or adipose tissue (O'Connor 2002). It is not known if these three kinds of fossae can be reliably distinguished on the basis of bone texture. Until this is tested, inferring pneumaticity on the basis of bone texture alone may not be warranted.

For the time being, I know of no test that can definitively determine whether a blind fossa housed a pneumatic diverticulum or some other soft tissue. Pneumatic diverticula often induce bone resorption when they come into contact with the skeleton, and it is possible that external pneumatic features might be recognized by some distinctive aspect of cortical bone histology. I do not suggest that this must be the case, but it is worth investigating.

To determine if a fossa is pneumatic or not, it is worthwhile to consider other potentially pneumatic features on or in the same bone. Consider the fossa bounded by the *podl*, *prdl*, *spol*, and *sprl* in *Haplocanthosaurus* (fig. 7.3). At least in the cervical vertebrae, these fossae do not contain any pneumatic foramina or subfossae, they do not lead to any obvious pneumatic tracks, and the bone texture is smooth rather than crenulate (pers. obs.). In other words, nothing about the fossae themselves indicates that they were pneumatic (as opposed to containing adipose deposits or other soft tissues). However, the centra of the same vertebrae contain deep, sharp-lipped cavities that penetrate to a narrow median septum. By the criteria discussed herein, the cavities in the centra are unequivocally pneumatic. Their presence demonstrates that pneumatic diverticula were in close contact with all of the preserved cervical vertebrae. Because we already know that pneumatic diverticula contacted the cervical vertebrae, it seems safe to infer that the neural spine fossae are pneumatic in origin. At least, the inference of pneumaticity is better founded than it would be based on the neural spine fossae alone.

(As an aside, the nomenclature for vertebral laminae has been thoroughly reviewed and standardized [Wilson 1999], but no standard nomenclature for vertebral fossae exists. It is tempting to propose such a nomenclature, if only to avoid circumlocutions like that used above ["the fossae bounded by the *podl*, *prdl*, *spol*, and *sprl*"]. However, a separate nomenclature for fossae is unnecessary and could be misleading. Hatcher [1901] named several fossae, such as the "infraprezygapophyseal cavity," using the same spatial orientation terms that were commonly used for naming laminae [e.g., Osborn 1899]. Such a position-based nomenclature for fossae shares all of the faults of the old orientation-based systems for naming laminae [for further discussion see Wilson 1999]. Laminae should be defined by the structures they connect [Wilson 1999]. Similarly, I think that fossae should be defined by the laminae that bound them. To list all of the bounding laminae when referring to a fossa may be awkward, but it is also precise.)

VERTEBRAL LAMINAE, HOMOLOGY, AND THE ORIGINS OF POSTCRANIAL SKELETAL PNEUMATICITY

It is tempting to assume that the fossae of basal sauropods are pneumatic because they are homologous to the unequivocally pneumatic features of later sauropods. For example, in *Brachiosaurus* the fossa bounded by the *podl*, *prdl*, *spol*, and *sprl* is clearly pneumatic because it contains pneumatic foramina (fig. 7.2). Does this mean that the equivalent fossa in *Barapasaurus* is also pneumatic? After all, phylogenetic analysis indicates that the bounding laminae are homologous in *Barapasaurus* and *Brachiosaurus* (Wilson 1999, 2002). The answer seems to be

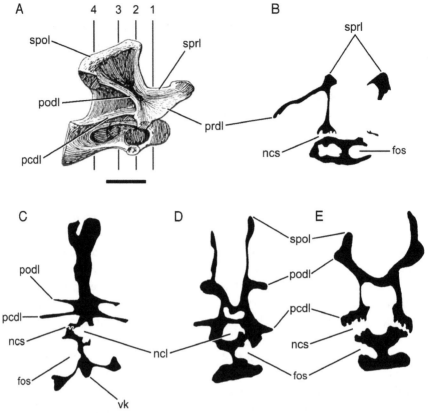

FIGURE 7.3. Pneumatic features in a cervical vertebra of *Haplocanthosaurus*. A, A posterior cervical of *Haplocanthosaurus* in right lateral view (CM 879-7; this specimen was erroneously referred to as CM 572 in Upchurch [1998:fig. 8], and as CM 897-7 in Wedel et al. [2000:fig. 2], Wedel [2003a:fig. 3], and Wedel [2003b:fig. 1]). (Modified from Hatcher (1903:pl. 2.) B–E, Cross sections traced from CT slices. B, Section at A, position 1. C, Section at A, position 2. The opening of the neural canal and the absence of the neurocentral suture on one side are due to a break in the specimen. D, Section at A, position 3. E, Section at A, position 4. The neurocentral sutures are unfused over most of their length, indicating that this animal was not fully mature. Scale bar equals 5 cm.

that the fossae may be homologous, but that is no guarantee that they were produced by the same morphogenetic process. Ontogenetic pathways are themselves subject to evolutionary change. As Hall (1999:347) stated, "A limb built upon one set of rules does not lose its homology with limbs built upon different rules." Conversely, homology does not necessarily indicate identical morphogenetic pathways. The shallow fossae of basal sauropods may have contained deposits of fat such as those identified in birds by O'Connor (2001). It is possible that such adipose deposits were replaced by pneumatic diverticula later in sauropod evolution. In that case, the laminae that bound the fossae would have remained the same, but the tissue that filled the fossae would have changed. The same replacement may also have occurred during ontogeny.

If we order archosaur vertebrae in terms of putatively pneumatic features, the resulting arrangement has no obvious gaps and is roughly congruent with current phylogenies (i.e., Sereno 1991; Wilson 2002). At one end of the spectrum are vertebrae that lack laminae, such as those of extant crocodilians. Very shallow depressions may be present on the neural spines or centra, but these depressions are not bounded by an obvious lip and do not contain subfossae or large foramina. The next grade of vertebral construction is represented by *Marasuchus*, which has low ridges below some of the presacral diapophyses (Sereno and Arcucci 1994); these ridges may represent rudimentary laminae (Wilson 1999). At the next level, a series of diapophyseal and zygapophyseal laminae is primitive for Saurischia (Wilson 1999). These laminae are present in *Herrerasaurus* and prosauropods (Sereno and Novas 1994; Bonaparte 1986), but the fossae they enclose are blind, lack subfossae, and have no obvious textural differences from the adjacent bone (Wedel, pers. obs.). Vertebral centra of these taxa lack fossae. Shallow fossae are present on the centra of early sauropods such as *Isanosaurus*, *Shunosaurus*, and *Barapasaurus*, and neural chambers may be present in the arch and spine (Jain et al. 1979; Zhang 1988; Buffetaut et al. 2000). In *Jobaria* and *Haplocanthosaurus* the central fossae are bounded by a sharp lip and penetrate to a median septum (Sereno et al. 1999; Wedel 2003b, pers. obs.). Finally, most neosauropods have prominent pneumatic foramina that open into chambers that ramify within the centrum, and the fossae of the neural arches and spines contain subfossae or pneumatic foramina.

It is not clear where pneumaticity first appears in the preceding series. At one end of the scale are the vertebrae of crocodiles, which are known to be apneumatic. At the other end are the vertebrae of neosauropods, the pneumatic features of which are virtually identical to those of birds (Janensch 1947). In between, the inference of pneumaticity receives more support as we approach Neosauropoda, but the "break point" between apneumatic and pneumatic morphologies is debatable. The primitive saurischian complex of laminae first appears in small dinosaurs and seems to be structural overkill if pneumatic diverticula were absent (Wilson 1999). An apneumatic interpretation of these laminae requires that a large number of structures that are clearly related to pneumatization in later forms be primitively present for other reasons, and leaves us (at least for now) without a satisfying hypothesis to explain the origin of vertebral laminae. The blind fossae of early saurischians are, at best, equivocal evidence of pneumaticity. However, any explanation that pushes the origin of PSP forward in time will accumulate a corresponding number of ad hoc hypotheses to explain the early appearance of laminae and fossae. For these reasons, I favor Wilson's (1999) hypothesis that laminae are pneumatic in origin and that the appearance of laminae marks the appearance of PSP, although as Wilson (1999:651) pointed out, more work is needed.

Gower (2001) posited widespread pneumaticity in Archosauria based on vertebral fossae. If he is right, PSP originated before the divergence between crocodile- and bird-line archosaurs and was present in virtually all of the noncrocodilian taxa in the series discussed above. O'Connor (2002) questioned the reliability of blind fossae as indicators of pneumaticity, but he did not present evidence to falsify Gower's hypothesis. Indeed, hypotheses of pneumaticity are difficult to falsify; although it is often easy to demonstrate that a bone has been pneumatized, it is difficult to demonstrate that it has not (Hogg 1980). For now, the possibility that the fossae described by Gower are pneumatic cannot be ruled out, but neither can less radical alternative hypotheses.

OTHER OSTEOLOGICAL CORRELATES OF PNEUMATICITY

Pneumatic tracks, thin outer walls, and large foramina are not likely to be falsely interpreted as pneumatic features in sauropods. External tracks are only rarely identified in sauropods. Wedel et al. (2000:fig. 7) illustrated a pneumatic track in *Sauroposeidon*, but the track was not the basis for the pneumatic interpretation; rather, the track was identified as pneumatic because it led away from a deep, sharply lipped pneumatic fossa. Many sauropod vertebrae have thin outer walls, especially those of the aforementioned *Sauroposeidon* (fig. 7.4). However, the thin outer walls of sauropod vertebrae invariably bound large internal chambers that are clearly pneumatic,

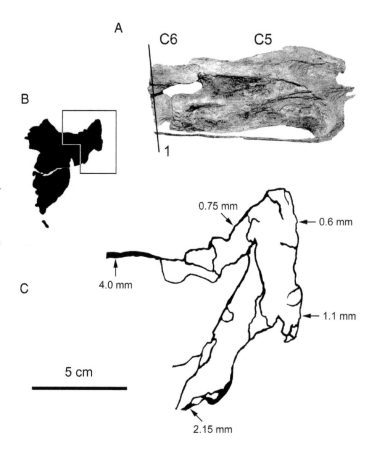

FIGURE 7.4. Internal structure of a cervical vertebra of *Sauroposeidon*, OMNH 53062. A, The posterior two-thirds of C5 and the condyle and prezygapophysis of C6 in right lateral view. The field crew cut though C6 to divide the specimen into manageable pieces. B, Cross section of C6 at the level of the break, traced from a CT image (A, position 1) and photographs of the broken end. The left side of the specimen was facing up in the field and the bone on that side is badly weathered. Over most of the broken surface the internal structure is obscured by plaster or too damaged to trace, but it is cleanly exposed in the ramus of the right prezygapophysis (outlined). C, The internal structure of the prezygapophyseal ramus, traced from a photograph. The arrows indicate the thickness of the bone at several points, as measured with a pair of digital calipers. The camellae are filled with sandstone.

so, again, the inference of pneumaticity does not rest on the equivocal feature. Finally, there is the question of foramina that are not pneumatic, such as nutrient or nervous foramina. Britt et al. (1998) proposed that pneumatic foramina could be distinguished from nutrient foramina on the basis of relative size, with pneumatic foramina typically being about an order of magnitude larger, relative to the length of the centrum. The two kinds of foramina could also be distinguished based on the internal structure of the vertebrae. Pneumatic vertebrae typically lack trabecular bone (Bremer 1940; Schepelmann 1990) and have compact bone in their outer walls and in the septa between pneumatic cavities (Reid 1996). The presence of trabecular bone inside a vertebra is evidence that it is either apneumatic or, at least, incompletely pneumatized (King 1957). Distinguishing pneumatic foramina from nutrient foramina is a potential problem in studies of birds and other small theropods, but most sauropods are simply so large that pneumatic and nutrient foramina are unlikely to be confused. Even juvenile sauropods tend to have large pneumatic fossae rather than small foramina (see Wedel et al. 2000:fig. 14).

DESCRIPTION OF PNEUMATIC ELEMENTS

At least four aspects of skeletal pneumaticity can be described: the external traces of pneumaticity (discussed above), the internal complexity of an element, the ratio of bone to airspace within an element, and the distribution of pneumatic features along the vertebral column.

INTERNAL COMPLEXITY OF PNEUMATIC BONES

This variable has received the most attention in previous studies and is only briefly reviewed here. Longman (1933) recognized that sauropod

TABLE 7.1
Classification of Sauropod Vertebrae into Morphologic Categories Based on Pneumatic Characters

Acamerate	Pneumatic characters limited to fossae; fossae do not significantly invade the centrum.
Procamerate	Deep fossae penetrate to median septum but are not enclosed by ostial margins.
Camerate	Large, enclosed camerae with regular branching pattern; cameral generations usually limited to 3.
Polycamerate	Large, enclosed camerae with regular branching pattern; cameral generations usually 3 or more, with increased number of branches at each generation.
Semicamellate	Camellae present but limited in extent; large camerae may also be present.
Camellate	Internal structure entirely composed of camellae; neural arch laminae not reduced. Large external fossae may also be present.
Somphospondylous	Internal structure entirely composed of camellae; neural arch laminae reduced; neural spine has an inflated appearance.

NOTE: After Wedel et al. (2000:table 3).

vertebrae with internal chambers fall into two broad types, those with a few large chambers and those with many small chambers. Longman called the first type phanerocamerate and the second cryptocamerillan (although he did not explicitly discuss them as products of skeletal pneumatization). Britt (1993, 1997) independently made the same observation and used the terms camerate and camellate to describe large-chambered and small-chambered vertebrae, respectively. Wedel et al. (2000) expanded this terminology to include categories for vertebrae with fossae only and vertebrae with combinations of large and small chambers (table 7.1). Wedel et al. (2000) and Wedel (2003b) also discussed the phylogenetic distribution of different internal structure types. In general, the vertebrae of early diverging sauropods such as *Shunosaurus* and *Barapasaurus* have external fossae but lack internal chambers. Camerae are present in the vertebrae of diplodocids, *Camarasaurus*, and *Brachiosaurus*. Presacral vertebrae of *Brachiosaurus* also have camellae in the condyles and cotyles, and camellae are variably present in the neural spine and apophyses. The vertebrae of *Sauroposeidon* and most titanosaurs lack camerae and are entirely filled with camellae, although some titanosaurs may have vertebral camerae. From published descriptions (Young and Zhao 1972; Russell and Zheng 1994), the vertebrae of *Mamenchisaurus* appear to be camellate.

From the foregoing, it might appear that the internal structures of sauropod vertebrae, their evolution, and their phylogenetic distribution are all well understood. In fact, vertebral internal structure is only known for a small minority of sauropods. Even in those taxa for which the internal structure is known, this knowledge is usually limited to a handful of vertebrae or even a single element, which severely limits our ability to assess serial, ontogenetic, and population-level variation. Despite these limitations, three broad generalizations can be made. First, the vertebrae of very young sauropods tend to have a simple I-beam shape in cross section, with large lateral fossae separated by a median septum (Wedel 2003b). This is true even for taxa in which the vertebrae of adults are highly subdivided, such as *Apatosaurus*. In these taxa the internal complexity of the vertebrae increased during ontogeny. The second generalization is that complex internal structures evolved independently in *Mamenchisaurus* and diplodocids and one or more times in Titanosauriformes (Wedel 2003b). This suggests a general evolutionary trend toward increasing complexity of vertebral internal structure in sauropods, albeit one that took different forms in different lineages (i.e., polycamerate vertebrae in Diplodocidae and somphospondylous vertebrae in Somphospondyli) and that may have been subject to reversals (i.e, camerate vertebrae in some titanosaurs [see Wedel 2003b]). Finally, the largest and longest- necked sauropods, such as *Mamenchisaurus*, the diplodocines, brachiosaurids, *Euhelopus*, and titanosaurs such as *Argentinosaurus* and the unnamed taxon represented by DGM Serie A, all have polycamerate, semicamellate, or fully camellate internal structures. I have previously stated that the complex internal structures were correlated with increasing size and neck length (Wedel 2003a, 2003b). This may or may not be true; I have not performed any phylogenetic tests of character correlation. Nevertheless, the presence of complex internal structures in the vertebrae of the largest and longest-necked sauropods suggests that size, neck length, and internal structure are related.

VOLUME OF AIR WITHIN A PNEUMATIC BONE

The aspect of skeletal pneumaticity that has probably received the least attention to date is the ratio of bone tissue to empty space inside a pneumatic bone. Although many authors have commented on the weight-saving design of sauropod vertebrae (Osborn 1899; Hatcher 1901; Gilmore 1925), no one has quantified just how much mass was saved. The savings in mass could have important paleobiological implications, for example, in determining how much mass to subtract from volumetric mass estimates.

Currey and Alexander (1985) and Cubo and Casinos (2000) reported relevant data on the long bones of birds, which are tubular and may be filled with marrow or air. In both studies, the variable of interest was K, the inner diameter of the element divided by its outer diameter. Both studies found mean values of K between 0.77 and 0.80 for pneumatic bones. The mean for marrow-filled bird bones is 0.65 (Cubo and Casinos 2000), and the mean for terrestrial mammals is 0.53 (calculated from Currey and Alexander 1985:table 1).

The K value is a parameter of tubular bones; it is meaningless when applied to bones with more complex shapes or internal structures, such as sauropod vertebrae. I propose the *airspace proportion* (ASP), or the proportion of the volume of a bone—or the area of a bone section—that is occupied by air spaces, as a variable that can be applied to both tubular and nontubular bones. One problem is that measuring the volumes of objects is difficult and often imprecise. It is usually easier to measure the relevant surface areas of a cross section, but any one cross section may not be representative of the entire bone. For example, the long bones of birds and mammals are usually tubular at midshaft, but the epiphyses mostly consist of marrow-filled trabecular bone or pneumatic camellate bone. Nevertheless, it may be easier to take the mean of several cross sections as an approximation of volume than to directly measure volume, especially in the case of large, fragile, matrix-filled sauropod vertebrae.

For the avian long bones described above, data were only presented for a single cross section located at midshaft. Therefore, the ASP values I am about to discuss may not be representative of the entire bones, but they probably approximate the volumes (total and air) of the diaphyses. For tubular bones, ASP may be determined by squaring K (if r is the inner diameter and R the outer, then K is r/R, ASP is $\pi r^2/\pi R^2$ or simply r^2/R^2, and ASP $= K^2$). For the K of pneumatic bones, Currey and Alexander (1985) report lower and upper bounds of 0.69 and 0.86, and I calculate a mean of 0.80 from the data presented in their table 1. Using a larger sample size, Cubo and Casinos (2000) found a slightly lower mean K of 0.77. The equivalent values of ASP are 0.48 and 0.74, with a mean of 0.64, or 0.59 for the mean of Cubo and Casinos (2000). This means that, on average, the diaphysis of a pneumatic avian long bone is 59%–64% air, by volume.

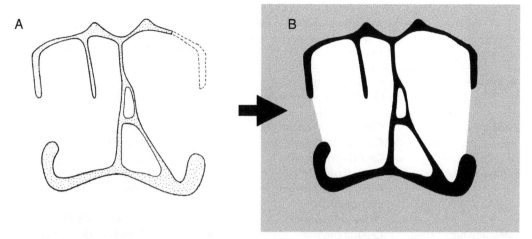

FIGURE 7.5. How to determine the airspace proportion (ASP) of a bone section. (A) A section is traced from a photograph, CT image, or published illustration; in this case, a transverse section of a *Barosaurus africanus* cervical vertebra from Janensch (1947:fig. 3). (B) Imaging software is used to fill the bone, airspace, and background with different colors. The number of pixels of each color can then be counted using Image J (or any program with a pixel count function) and used to compute the ASP. In this case, bone is black and air is white, so the ASP is (white pixels)/(black pixels + white pixels).

How do these numbers compare with the ASPs of sauropod vertebrae? To find out, I measured the area occupied by bone and the total area for several cross sections of sauropod vertebrae (see fig. 7.5 for an example). I obtained the cross-sectional images from CT scans, published cross sections, and photographs of broken or cut vertebrae. For image analysis I used Image J, a free program available online from the National Institutes of Health (Rasband 2003). Some results are presented in table 7.2 (this research is in progress and I will present more complete results elsewhere). The results should be approached with caution: I have only analyzed a few vertebrae from a handful of taxa, and only one or a few cross sections for each bone, so the results may not be representative of either the vertebrae, the regions of the vertebral column, or the taxa to which they belong. The sample is strongly biased toward cervical vertebrae simply because cervicals are roughly cylindrical and fit through CT scanners better than dorsal or sacral vertebrae. Despite these caveats, some regularities emerge.

First, ASP values range from 0.32 to 0.89, with a mean of 0.60. Even though the data may not be truly representative, it seems reasonable to conclude that most sauropod vertebrae contained at least 50% air, by volume, and probably somewhat more. This assumes that the cavities in sauropod vertebrae were entirely filled with air and the amount of soft tissue was negligible. Chandra Pal and Bharadwaj (1971) found that the air spaces in pneumatic bird bones are lined with simple squamous epithelium, so the assumption is probably valid. The ASP values presented here for sauropod vertebrae are similar to the range and mean found for pneumatic long bones of birds (or at least their diaphyses).

Second, although only a handful of measurements are available for each taxon, it is already clear that ASP can vary widely from slice to slice within a single vertebra and probably also between vertebrae of different regions of the skeleton and between individuals of the same species. As we collect more data we may find more predictable relationships, for example, between the ASP values of cervical and dorsal vertebrae or between certain taxa. The system may also be so variable that such relationships will be impossible to detect, if they even exist. Rampant variation seems to be the rule for skeletal pneumaticity in general (e.g., King 1957; Cranford et al. 1996; Weiglein 1999), and it would be surprising if ASP were not also highly variable.

TABLE 7.2
The Airspace Proportion (ASP) of Transverse Sections through Vertebrae of Sauropods and Other Saurischians

	REGION	ASP	SOURCE
Apatosaurus	Cervical condyle	0.69	Wedel (2003b:fig. 6b)
	Cervical midcentrum	0.52	Wedel (2003b:fig. 6c)
	Cervical cotyle	0.32	Wedel (2003b:fig. 6d)
Barosaurus	Cervical midcentrum	0.56	Janensch (1947:fig. 8)
	Cervical, near cotyle	0.77	Janensch (1947:fig. 3)
	Caudal midcentrum	0.47	Janensch (1947:fig. 9)
Brachiosaurus	Cervical condyle	0.73	Janensch (1950:fig. 70)
	Cervical midcentrum	0.67	Wedel et al. (2000:fig. 12c)
	Cervical cotyle	0.39	Wedel et al. (2000:fig.12d)
	Dorsal midcentrum	0.59	Janensch (1947:fig. 2)
Camarasaurus	Cervical condyle	0.49	Wedel (2003b:fig. 9b)
	Cervical midcentrum	0.52	Wedel (2003b:fig. 9c)
	Cervical, near cotyle	0.50	Wedel (2003b:fig. 9d)
	Dorsal midcentrum	0.63	Ostrom and McIntosh (1966:pl. 23)
	Dorsal midcentrum	0.58	Ostrom and McIntosh (1966:pl. 24)
	Dorsal midcentrum	0.71	Ostrom and McIntosh (1966:pl. 25)
Phuwiangosaurus	Cervical midcentrum	0.55	Martin (1994:fig. 2)
Pleurocoelus	Cervical midcentrum	0.55	Lull (1911:pl. 15)
Saltasaurus	Dorsal midcentrum	0.55	Powell (1992:fig. 16)
Sauroposeidon	Cervical prezygapophyseal ramus	0.89	Fig. 7.4
	Cervical midcentrum	0.74	Wedel et al. (2000:fig. 12g)
	Cervical postzygapophysis	0.75	Wedel et al. (2000:fig. 12h)
Theropoda	Cervical prezygapophysis	0.48	Janensch (1947:fig. 16)
	Dorsal midcentrum	0.50	Janensch (1947:fig. 15)

NOTE: Only values for published sections are presented. Much more work will be required to determine norms for different taxa and different regions of the vertebral column, and the values presented here may not be representative of either. Nevertheless, these values suggest that pneumatic sauropod vertebrae were often 50%–60% air, by volume. The mean of these 22 measurements is 0.60.

Third, the lowest ASP values—0.32 in *Apatosaurus* and 0.39 in *Brachiosaurus*—are for slices through the cotyle, or bony cup, at the posterior end of the centrum. Here the cortical bone is doubled back on itself to form the cup, and the wall of the cotyle itself is at an angle to the slice and appears wider in cross section. The cotyle is surrounded by pneumatic chambers in both *Apatosaurus* and *Brachiosaurus*, but these become smaller and eventually disappear toward the end of the vertebra. For these reasons, the cotyle is expected to have a lower ASP than the rest of the vertebra.

Fourth, *Sauroposeidon* has the highest values of ASP, up to a remarkable 0.89. The values for *Sauroposeidon* are even higher than those for the closely related *Brachiosaurus*, and the ranges for the two taxa do not overlap (although they may come to when a larger sample is considered). A very high ASP is probably an autapomorphy of *Sauroposeidon* and may have evolved to help lighten its extremely long (~12-m) neck.

Finally, ASP appears to be independent of the internal complexity of the vertebrae. The *Saltasaurus* vertebra is the most highly subdivided of the sample. The I-beam-like vertebrae of the juvenile *Pleurocoelus* and *Phuwiangosaurus* are the least subdivided; the other taxa fall somewhere in the middle. Nevertheless, most values in the table 7.2, including those for *Saltasaurus*, *Pleurocoelus*, and *Phuwiangosaurus*, fall between 0.50 and 0.60. The means for all taxa other than *Sauroposeidon* also fall within the same range, so there is no apparent trend that relates ASP to internal complexity. Cast in evolutionary terms, this indicates that the evolution of complex internal structures from simple ones involved a redistribution rather than a reduction of bony tissue within the vertebrae. The ASP values of the juvenile *Pleurocoelus* and *Phuwiangosaurus* imply that a similar redistribution was involved in the ontogenetic derivation of complex chambers from juvenile fossae.

The results presented here are preliminary, and the available data are better suited for suggesting hypotheses than for testing them. Much work remains to be done, both in gathering comparative data from extant forms and in exploring the implications of pneumaticity for sauropod biomechanics.

DISTRIBUTION OF PNEUMATICITY ALONG THE VERTEBRAL COLUMN

The two previous sections dealt with the characteristics of a single pneumatic bone. We must also consider the location of pneumatic features in the skeleton, because these features constrain the minimum extent of the diverticular system. For example, in the USNM 10865 skeleton of *Diplodocus*, pneumatic foramina are present on every vertebra between the axis and the nineteenth caudal ([Gilmore 1932; foramina are only present on caudals 1–18 in the skeleton of *Diplodocus* described by Osborn [1899] and on caudals 1–16 in the mounted DMNS skeleton [Wedel, pers. obs.]). This means that in life the pneumatic diverticula reached at least as far anteriorly as the axis and as far posteriorly as caudal vertebra 19 (fig. 7.6). The diverticular system may have been more extensive and simply failed to pneumatize any more bones, but it could not have been any less extensive.

In mapping the distribution of pneumaticity along the vertebral column, it is important to consider where on the vertebrae the pneumatic features are located. In the co-ossified block of *Diplodocus* caudal vertebrae illustrated by Gilmore (1932:fig. 3), the centra of caudals 15–19 bear large pneumatic foramina, but the neural spines lack laminae and do not appear to have been pneumatic. This is in contrast to the presacral, sacral, and anterior caudal vertebrae, which have heavily sculpted neural spines with deep fossae and scattered foramina (see Osborn 1899:figs. 7, 13). In the opposite condition, the neural spines bear laminae and fossae and may have been pneumatic, but the centra lack pneumatic features. Examples include the middle and posterior dorsal vertebrae of *Jobaria* (see Sereno et al. 1999:fig. 3). Sauropod vertebrae can therefore exist in one of four states: (1) both centrum and neural spine pneumatic, as in the presacral vertebrae of most neosauropods; (2) centrum pneumatic but neural spine apneumatic, as in the middle caudals of *Diplodocus*; (3) neural spine pneumatic but centrum apneumatic, as in the posterior dorsals of *Jobaria* (assuming that the laminate neural spines are pneumatic); or (4) no signs of pneumaticity in the centrum or neural spine, as in the distal caudals of most sauropods. Pneumatization of the centrum typically results in large internal cavities with prominent foramina, so the inference of pneumaticity is well supported in conditions 1 and 2. In condition 3 the situation may be less clear. In derived neosauropods such as *Brachiosaurus* and the diplodocids, the neural spine fossae often bear small subfossae and foramina, which indicate that these fossae are pneumatic (see Janensch 1950; Curtice and Stadtman 2001). In more basal sauropods such as *Haplocanthosaurus*, the neural spine fossae are often blind and lack the heavily sculpted texture seen in later forms. The neural spines of these basal sauropods may have been pneumatic, but the inference is less well founded.

The earliest sauropodomorph with distinctly emarginated pneumatic fossae is *Thecodontosaurus caducus* (Yates 2003). In

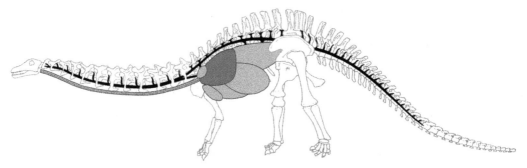

FIGURE 7.6. Hypothetical conformation of the respiratory system of a diplodocid sauropod. The left forelimb, pectoral girdle, and ribs have been removed for clarity. The lung is shown in dark gray, air sacs are light gray, and pneumatic diverticula are black. Only some of the elements shown in this illustration can be determined with certainty: the minimum length of the trachea, the presence of at least some air sacs, and the minimum extent of the pneumatic diverticula. The rest of the respiratory system has been restored based on that of birds, but this remains speculative. The skeleton is modified from Norman (1985:83).

T. caducus, pneumatic fossae are only present on the middle cervical vertebrae. This means that the fossae must have been produced by diverticula of cervical air sacs similar to those of birds (as opposed to diverticula of the lungs proper). A similar pattern of pneumatization in *Coelophysis* indicates that cervical air sacs were present in both sauropodomorphs and theropods by the Norian (Late Triassic), and cervical air sacs are probably primitive for saurischians (Wedel 2004).

In general, more derived sauropods tended to pneumatize more of the vertebral column. Except for the atlas, which is always apneumatic, pneumatic chambers (or prominent fossae) are present in the cervical vertebrae of *Shunosaurus*; in the cervical and anterior dorsal vertebrae of *Jobaria*; in all of the presacral vertebrae of *Cetiosaurus*; in the presacral and sacral vertebrae of most neosauropods; and in the presacral, sacral, and caudal vertebrae of diplodocids and saltasaurids (Wedel 2003a, 2003b, pers. obs.). This caudad progression of vertebral pneumaticity also occurred in the evolution of theropods (Britt 1993) and occurs ontogenetically in extant birds (Cover 1953; Hogg 1984b). At a gross level, the system is both homoplastic and recapitulatory.

In extant birds, diverticula of the cervical air sacs do not extend farther posteriorly than the anterior thoracic vertebrae. If the diverticula of sauropods followed the same pattern of development as those of birds, then the presence of pneumatic sacral vertebrae in most neosauropods indicates the presence of abdominal air sacs (Wedel et al. 2000). There are no strong reasons to doubt that neosauropods had abdominal air sacs. However, the future discovery of a sauropod with a pneumatic hiatus—a gap in the pneumatization of the dorsal vertebrae—would unequivocally demonstrate the presence of abdominal air sacs and their diverticula (Wedel 2003a).

APPLICATION TO A PALEOBIOLOGICAL PROBLEM: MASS ESTIMATES

The implications of PSP for sauropod paleobiology are only beginning to be explored. In particular, skeletal pneumaticity may be an important factor in future studies of the biomechanics and respiratory physiology of sauropods. The most obvious implication of extensive PSP in sauropods is that they may have weighed less than is commonly thought. In this section, the problem of estimating the masses of sauropods is used as an example of how information about PSP may be applied to a paleobiological question.

Two distinct questions proceed from the observation that most sauropod skeletons were highly pneumatic. The first is purely methodological: (How) Should we take pneumaticity into account in estimating the masses of sauropods? The second question is paleobiological: If we find that pneumaticity significantly lightened sauropods, how does that affect our understanding of sauropods as living animals? If pneumaticity did not significantly lighten sauropods, then the second question is moot, so I consider the methodological question first.

METHODS

The masses of dinosaurs are generally estimated using allometric equations based on limb bone dimensions (Russell et al. 1980; Anderson et al. 1985) or volumetric measurements using physical or computer models (Colbert 1962; Paul 1988, 1997; Henderson 1999). If allometric equations are used, then pneumaticity need not be taken into account; the limb bones are assumed to have been as circumferentially robust as they needed to be to support the animal's mass, regardless of how the body was constituted. If an animal with a pneumatic skeleton was lighter than it would have been otherwise, this should already be reflected in its limb bone morphology, and no correction is necessary. On the other hand, if volumetric measurements are used, then it is possible to take skeletal pneumaticity into account and failure to do so may result in mass estimates that are too high.

Volumetric mass estimation is performed in three steps (Alexander 1989). First, the volume of a scale model of the organism is measured. Next, the volume of the model is multiplied by the scale factor to obtain the volume of the organism in life. Finally, the volume of the organism is multiplied by the estimated density to obtain its mass. The presence of air in the respiratory system and pneumatic diverticula can be accounted for in the first two steps, by reducing the estimated volume of model or the organism, or in the third step, by adjusting the density used in the mass calculation. Both methods have been used in published mass estimates of dinosaurs. Alexander (1989) used plastic models in his volumetric study, and he drilled holes to represent the lungs before estimating the center of mass of each model and the proportion of mass supported by the fore- and hindlimbs (see Alexander 1989:figs. 4.6, 5.3). Curiously, he does not seem to have drilled the holes before performing his mass estimates; at least, the holes are only mentioned in conjunction with the center of mass and limb support studies. Henderson (1999) included lung spaces in his digital models for mass estimation purposes and, later, included air sacs and diverticula in a buoyancy study (Henderson 2004). Paul (1988, 1997) used the alternative method of adjusting the density values for the mass calculations. He assigned a specific gravity (SG) of 0.9 to the trunk to account for lungs and air sacs, and an SG of 0.6 to the neck to account for pneumatization of the vertebrae.

Before attempting to estimate the volume of air in a sauropod, it is important to recognize that the air was distributed among four separate regions: (1) the trachea, (2) the "core" respiratory system of lungs and, possibly, pulmonary air sacs, (3) the extraskeletal (i.e., visceral, intermuscular, and subcutaneous) diverticula, and (4) the pneumatic bones. These divisions are important for two reasons. First, the volumes of each region are differently constrained by skeletal remains. The volume of air in the skeleton can be estimated with a high degree of confidence because the sizes of the airspaces can be measured from fossils. In contrast, the volume of the trachea is not constrained by skeletal remains and must be estimated by comparison to extant taxa. The lung/air sac system and extraskeletal diverticula are only partly constrained by the skeleton (see below). This leads to the second point, which is that estimates of all four regions can be made independently, so that skeletal pneumaticity can be taken into account regardless of conformation (birdlike, crocodile-like, etc.) and volume of the core respiratory system.

AN EXAMPLE USING *DIPLODOCUS*

Consider the volume of air present inside a living *Diplodocus*. Practically all available mass

TABLE 7.3
The Volume of Air in Diplodocus

	AIR VOLUME (L)	MASS SAVINGS (KG)
Trachea	104	104
Lungs and air sacs	1,500	1,500
Extraskeletal diverticula	?	?
Pneumatic vertebrae		
Centra		
Cervicals 2–15	136	82
Dorsals 1–10	208	125
Sacrals 1–5	75	45
Caudals 1–19	329	198
Subtotal for centra	748	450
Neural spines		
Cervicals 2–15	136	82
Dorsals 1–10	416	250
Sacrals 1–5	150	90
Caudals 1–19	165	99
Subtotal for spines	867	520
Subtotal for vertebrae	970	1,455
Total	2,574	3,059

NOTE: See the text for methods of estimation. The total volume for vertebrae is 1,615.

estimates for *Diplodocus* (Colbert 1962; Alexander 1985; Paul 1997; Henderson 1999) are based on CM 84, the nearly complete skeleton described by Hatcher (1901). Uncorrected volumetric mass estimates—i.e., those that do not include lungs, air sacs, or diverticula—for this individual range from 11,700 kg (Colbert 1962; as modified by Alexander 1989:table 2.2) to 18,500 kg (Alexander 1985). Paul (1997) calculated a mass of 11,400 kg using the corrected SGs cited above, and Henderson (1999) estimated 14,912 kg, or 13,421 kg after deducting 10% to represent the lungs. For the purposes of this example, the volume of the animal is assumed to have been 15,000 liters. The estimated volumes of various air reservoirs and their effects on body mass are listed in table 7.3.

Estimating the volume of air in the vertebral centra is the most straightforward. I used published measurements of centrum length and diameter from Hatcher (1901) and Gilmore (1932) and treated the centra as cylinders. The caudal series of CM 84 is incomplete, so I substituted the measurements for USNM 1065 from Gilmore (1932); comparison of the measurements of the elements common to both skeletons indicates that the two animals were roughly the same size. I multiplied the volumes obtained by 0.60, the mean ASP of the sauropod vertebrae listed in table 7.2, to obtain the total volume of air in the centra.

The volume of air in the neural spines is harder to calculate. The neural spines are complex shapes and are not easily approximated with simple geometric models. Furthermore, the fossae on the neural arches and spines only partially enclosed the diverticula that occupied them. Did the diverticula completely fill the space between adjacent laminae, did they bulge outward into the surrounding tissues, or did surrounding tissues bulge inward? In the complete absence of in vivo measurements of diverticulum volume in birds, it is impossible to say. Based on the size of the neural spine relative to the centrum in most sauropods (see fig. 7.2), it seems reasonable to assume that in the cervical vertebrae, at least as much air was present in the arch and spine as in the centrum, if not more. In the high-spined dorsal and sacral vertebrae (see fig. 7.1), the volume of air in the neural arch and spine may have been twice that in the centrum. Finally, proximal caudal vertebrae have large neural spines but the size of the spines decreases rapidly in successive vertebrae. On average, the caudal neural spines of *Diplodocus* may have contained only half as much air as their associated centra. These estimates are admittedly rough, but they are probably conservative and so they will suffice for this example.

As they developed, the intraosseous diverticula replaced bony tissue, and the density of that tissue must be taken into account in estimating how much mass was saved by pneumatization of the skeleton. In apneumatic sauropod vertebrae the internal structure is filled with cancellous bone and presumably supported red (erythropoeitic) bone marrow (fig. 7.7). Distal

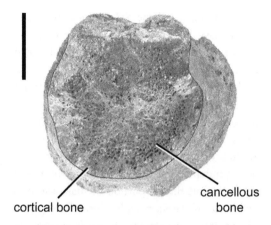

FIGURE 7.7. Internal structure of OMNH 27794, a partial distal caudal vertebra of a titanosauriform. The internal structure is composed of apneumatic cancellous bone, and no medullary cavity is present. Scale bar equals 1 cm.

caudal vertebrae of the theropod *Ceratosaurus* have a large central chamber or centrocoel (Madsen and Welles 2000:fig. 6). This cavity lacks large foramina that would connect it to the outside, so it cannot be pneumatic in origin. The medullary cavities of apneumatic avian and mammalian long bones are filled with adipose tissue that acts as lightweight packing material (Currey and Alexander 1985), and the same may have been true of the centrocoels in *Ceratosaurus* caudals. The presence of a similar marrow cavity in sauropod vertebrae prior to pneumatization cannot be ruled out, but to my knowledge no such cavities have been reported. In birds, the intraosseous diverticula erode the inner surfaces of the cortical bone in addition to replacing the cancellous bone (Bremer 1940), so pneumatic bones tend to have thinner walls than apneumatic bones (Currey and Alexander 1985; Cubo and Casinos 2000). The tissues that may have been replaced by intraosseous diverticula have SGs that range from 0.9 for some fats and oils to 3.2 for apatite (Schmidt-Nielsen 1983:451, table 11.5). For this example, I estimated that the tissue replaced by the intraosseous diverticula had an average SG of 1.5 (calculated from data presented in Cubo and Casinos 2000), so air cavities that total 970 liters replace 1,455 kg of tissue. The extraskeletal diverticula, trachea, lungs, and air sacs did not replace bony tissue

in the body. They are assumed to replace soft tissues (density of 1 g/cm^3) in the solid model.

Extraskeletal diverticula include visceral, intermuscular, and subcutaneous diverticula. None of these leave traces that are likely to be fossilized. The bony skeleton places only two constraints on the extraskeletal diverticula. First, as previously discussed, the distribution of pneumatic bones in the skeleton limits the minimum extent of the diverticular system. Thus, we can infer that the vertebral diverticula in *Diplodocus* must have extended from the axis to the nineteenth caudal vertebra (at least in USNM 1065), but the course and diameter of the diverticula are unknown. The second constraint imposed by the skeleton is that the canalis intertransversarius, if it existed, could not have been larger than the transverse foramina where it passed through them, although it may have been smaller or increased in diameter on either side. I am unaware of any studies in which the in vivo volume of the avian diverticular system is measured. This information vacuum prevents me from including a volume estimate for the diverticular system in table 7.3.

To estimate the volume of the trachea, I used the allometric equations presented by Hinds and Calder (1971) for birds. The length equation, $L = 16.77M^{0.394}$, where L is the length of the trachea (cm) and M is the mass of the animal (kg), yielded a predicted tracheal length of 6.8 m for a 12-ton animal. The cervical series of *Diplodocus* CM 84 is 6.7 m long and the trachea may have been somewhat longer, and I judged the correspondence between the neck length and the predicted tracheal length to be close enough to justify using the equations, especially for the coarse level of detail needed in this example. The volume equation, $V = 3.724M^{1.090}$, yields a volume of 104 liters.

Finally, the volume of the lungs and air sacs must be taken into account. The lungs and air sacs are only constrained by the skeleton in that they must fit inside the ribcage and share space with the viscera. Based on measurements from caimans and large ungulates, Alexander (1989) subtracted 8% from the volume of each of his models to account for lungs. Data presented by King (1966:table 3) indicate that the lungs and air sacs of birds may occupy 10%–20% of the volume of the body. Hazlehurst and Rayner (1992) found an average SG of 0.73 in a sample of 25 birds from 12 unspecified species. On this basis, they concluded that the lungs and air sacs occupy about a quarter of the volume of the body in birds. However, some of the air in their birds probably resided in extraskeletal diverticula or pneumatic bones, so the volume of the lungs and air sacs may have been somewhat lower. In the interests of erring conservatively, I put the volume of the lungs and air sacs at 10% of the body volume.

The results of these calculations are necessarily tentative. The lungs and air sacs were probably not much smaller than estimated here, but they may have been much larger; the trachea could not have been much shorter but may have been much longer, or it may have been of a different or an irregular diameter (see McClelland [1989a] for tracheal convolutions and bulbous expansions in birds); the neural spines may have contained much more or somewhat less air; the ASP of *Diplodocus* vertebrae may be higher or lower; and the tissue replaced by the intraosseous diverticula may have been more or less dense. The extraskeletal diverticula have not been accounted for at all, although they were certainly extensive in linear terms and were probably voluminous as well. Uncertainties aside, it seems likely that the vertebrae contained a large volume of air, possibly 1,000 liters or more if the very tall neural spines are taken into account. This air mainly replaced dense bony tissue, so skeletal pneumatization may have lightened the animal by up to 10%—and that does not include the extraskeletal diverticula or pulmonary air sacs. In the example presented here, the volume of air in the body of *Diplodocus* is calculated to have replaced about 3,000 kg of tissue that would have been present if the animal were solid. If the total volume of the body was 15,000 liters and the density of the remaining tissue was 1 g/cm^3, the body mass would have been about 12 metric tons and the SG of the entire body would have been 0.8. This is lower than the SGs of squamates and crocodilians (0.81–0.89) found by Colbert (1962), higher than the SGs of birds (0.73) found by Hazlehurst and Rayner (1992), and about the same as the SGs (0.79–0.82) used by Henderson (2004) in his study of sauropod buoyancy. Note that the amount of mass saved by skeletal pneumatization is independent of the estimated volume of the body, but the proportion of mass saved is not. Thus if we start with Alexander's (1985) 18,500-liter estimate for the body volume of *Diplodocus*, the mass saved is still 1,455 kg, but this is only 8% of the solid mass, not 10% as in the previous example.

It could be argued that adjusting the estimated mass of a sauropod by a mere 8%–10% is pointless. The mass of the living animal may have periodically fluctuated by that amount or more, depending on the amount of fat it carried and how much food it held in its gut (Paul 1997). Further, the proposed correction is tiny compared to the range of mass estimates produced by different studies, from 11,700 kg (Paul 1997) to 18,500 kg (Alexander 1985). However, there are several reasons for taking into account the mass saved by skeletal pneumatization. The first is that estimating the mass of extinct animals is fraught with uncertainty, but we should account for as many sources of error as possible, and PSP is a particularly large source of error if it is not considered. Also, the range of mass estimates for certain taxa may be very wide, but 8%–10% of the body mass is still a sizable fraction when applied to any one estimate. The entire neck and head account for about the same percentage of mass in volumetric studies (Alexander 1989; Paul 1997), so failing to account for PSP may be as gross an error as omitting the neck and head from the volumetric model. These are the purely methodological reasons for considering the effect of PSP on body mass. There is also the paleobiological consideration, which is that the living animal was 8%–10% lighter because of PSP than it would have been without. Mass reduction of this magnitude almost certainly carries a selective advantage (Currey and Alexander 1985), and this may explain the presence of extensive PSP in many sauropods.

An alternative possibility is that sauropod skeletons weighed as much as they would have in the absence of PSP but that pneumatization allowed the elements to be larger and stronger for the same mass. This hypothesis was first articulated by Hunter (1774) to explain skeletal pneumatization in birds. It is supported by the observation that the skeletons of birds are not significantly lighter than the skeletons of comparably sized mammals (Prange et al. 1979). If this hypothesis is correct, pneumatic elements should be noticeably larger and more voluminous than nonpneumatic elements. The transitions from pneumatic to apneumatic regions of the vertebral column in *Jobaria* (Sereno et al. 1999:fig. 3) and *Diplodocus* (Osborn 1899:fig. 13; Gilmore 1932:fig. 3, pl. 6) are not marked by obvious changes in size or form of the vertebrae. This supports the hypotheses that pneumatic vertebrae were lighter than apneumatic vertebrae and that PSP really did lighten sauropod skeletons.

PALEOBIOLOGICAL IMPLICATIONS

The importance of PSP for sauropod paleobiology is still largely unexplored. To date, Henderson's (2004) study of sauropod buoyancy is the only investigation of the biomechanical effects of PSP. Henderson included pneumatic diverticula in and around the vertebrae in his computer models of sauropods, and found that floating sauropods were both highly buoyant and highly unstable. Pneumaticity may also be important in future studies of neck support in sauropods. Alexander (1985, 1989) calculated that a large elastin ligament would be better suited than muscles to holding up the neck of *Diplodocus*. His calculations were based on a volumetric estimate of 1,340 liters (and, thus, 1,340 kg) for the neck and head. Using the values in table 7.3, one fifth of that volume, or 268 liters, was occupied by airspaces. If Paul (1997) and Henderson (2004) are correct, the SG of the neck may have been as low as 0.6, which would bring the mass of the neck down to about 800 kg (the same result could be obtained by applying the air volumes in table 7.3 to a more slender neck model than that used by Alexander). As the mass of the neck goes down, so to does the perceived need for a large "nuchal" ligament, the existence of which is controversial (see Wedel et al. 2000; Dodson and Harris 2001; Tsuihiji 2004).

Recognition of skeletal pneumaticity in sauropods may also affect physiological calculations. For example, most published studies of thermal conductance in dinosaurs (e.g., Spotila et al. 1973, 1991) have modeled dinosaur bodies using solid cylinders. Air is a better insulator than conductor, but moving bodies of air may cool adjacent tissues by convection or evaporation. The pneumatic diverticula of birds tend to be blind-ended tubes except where they anastomose (Cover 1953), and most are poorly vascularized (Duncker 1971), so there appears to be little potential for evaporative cooling. On the other hand, thermal panting is an important homeostatic mechanism for controlling body temperature in birds and depends on evaporation from nasal, buccopharyngeal, and upper tracheal regions (Lasiewski 1972; Menaum and Richards 1975). At the very least, the inclusion of tracheae, lungs and pneumatic diverticula in thermal conductance models would decrease the effective radius of some of the constituent cylinders. What effect, if any, this would have on the results of thermal conductance studies is unknown, which is precisely the point: it has not been tested.

PROBLEMS AND PROSPECTS FOR FURTHER RESEARCH

Despite a long history of study, research on PSP is, in many ways, still in its infancy. Anyone who doubts the accuracy of this statement is directed to Hunter (1774). In the first published study of PSP, Hunter developed two of the major functional hypotheses entertained today: pneumaticity may lighten the skeleton, or it may strengthen the skeleton by allowing bones of larger diameter for the same mass as marrow-filled bones (see Witmer [1997], for a historical perspective on these and other hypotheses). Although many later authors have documented the presence and extent of PSP in certain birds (e.g., Crisp 1857; King 1957), most have focused on one or a few species (but see O'Connor 2004), some have produced conflicting accounts (reviewed by King 1957), and few have attempted to test functional hypotheses (but see Warncke and Stork 1977; Currey and Alexander 1985; Cubo and Casinos 2000; O'Connor 2004). Evolutionary patterns of PSP in birds are difficult to discern because few species have been studied (King, 1966), usually with little or no phylogenetic context (O'Connor 2002, 2004). Limits of knowledge of PSP in extant vertebrates necessarily limit what can be inferred from the fossil record. For example, disagreements between various published accounts of the development of pneumatization in birds frustrate attempts to infer the ontogenetic development of PSP in sauropods (Wedel 2003a).

Another problem for studies of PSP in fossil organisms is small sample sizes. As mentioned above, few taxa have been intensively studied and the importance of serial, ontogenetic, and intraspecific variation is difficult to assess. Sample sizes are mainly limited by the inherent attributes of the fossils: fossilized bones are rare, at least compared to the bones of extant vertebrates; they may be crushed or distorted; and they are often too large, too heavy, or too fragile to be easily manipulated. Even if these difficulties are overcome, most of the pneumatic morphology is still inaccessible, locked inside the bones.

SOURCES OF DATA

Information on the internal structure of fossil bones comes from three sources: CT studies, cut sections of bones, and broken bones. Although CT studies of fossils are becoming more common, access to scanners is very limited and can be prohibitively expensive. Large fossils, such as sauropod vertebrae, cause logistical problems. Most medical CT scanners have apertures 50 cm or less in diameter, and many sauropod vertebrae are simply too big to fit through the scanners. Furthermore, medical scanners are not designed to image large, dense objects like sauropod bones. The relatively low-energy X rays employed by medical scanners may fail to penetrate large bones, and this can produce artifacts in the resulting images (Wedel et al. 2000). Industrial CT scanners can image denser materials, but the rotating platforms used in many industrial scanners are too small to accept most sauropod vertebrae. For the near-future, CT will likely remain a tool of great promise but limited application.

Cut sections of bones can yield valuable information about pneumatic internal structures. The cuts may be made in the field to break aggregates of bones into manageable pieces, as in the cut *Sauroposeidon* vertebra shown in fig. 7.4. Less commonly, bones may be deliberately cut to expose their cross sections or internal structures, such as the cut specimens illustrated by Janensch (1947:fig. 5) and Martill and Naish (2001:pl. 32). Cutting into specimens is invasive and potentially dangerous to both researchers and fossils. Although cut specimens will continue to appear from time to time, they are unlikely to become a major source of data. In contrast, broken bones are ubiquitous. The delicate structure of pneumatic bones, even large sauropod vertebrae, may make them more prone to breakage than apneumatic bones. For these reasons broken bones are an important resource in studies of PSP and could be exploited more in the future. Published illustrations of broken sauropod vertebrae are numerous; notable examples include Cope (1878:fig. 5), Hatcher (1901:pl. 7), Longman (1933:pl. 16, fig. 3), and Dalla Vecchia (1999:figs. 2, 19). A beautiful example from outside Sauropoda is the broken transverse process of *Tyrannosaurus* illustrated by Brochu (2003:fig. 75).

DIRECTIONS FOR FUTURE RESEARCH

Four attributes of pneumatic bones are listed above under "Description of Pneumatic Elements":

(1) external pneumatic features, (2) internal structure, (3) ASP, and (4) distribution of pneumaticity in the skeleton. Only the second attribute has been systematically surveyed in sauropods (Wedel 2003b), although aspects of the first are treated by Wilson (1999). Knowledge of the fourth is mainly limited to the observation that diplodocines and saltasaurines have pneumatic caudal vertebrae and other sauropods do not (Wedel 2003b). All existing data on the ASPs of sauropod vertebrae are presented in table 7.2. Not only do all four attributes need further study, but the levels of serial, ontogenetic, and intraspecific variation should be assessed whenever possible. Similar data on PSP in pterosaurs, nonavian theropods, and birds are needed to test phylogenetic and functional hypotheses.

The pneumatic diverticula of birds are morphologically and morphogenetically intermediate between the core respiratory system of lungs and air sacs and the pneumatic bones. Understanding the development, evolution, and possible functions of diverticula is therefore crucial for interpreting patterns of PSP in fossil vertebrates. Müller (1907), Richardson (1939), Cover (1953), King (1966), Duncker (1971) and a few others described the form and extent of the diverticular network in the few birds for which they are known, but information on many bird species is lacking or has been inadequately documented (King 1966). The ontogenetic development of the diverticula is very poorly understood; most of what we think we know is based on inferences derived from patterns of skeletal pneumatization (Hogg 1984a; McClelland 1989b). Such inferences tell us nothing about the development of the many visceral, intermuscular, and subcutaneous diverticula that do not contact the skeleton or pneumatize any bones. These diverticula could not have evolved to pneumatize the skeleton. Most diverticula that pneumatize the skeleton must grow out from the core respiratory system before they reach their "target" bones, so they probably also evolved for reasons other than skeletal pneumatization (Wedel 2003a). Those reasons are unknown, in part because the physiological functions—or exaptive effects (sensu Gould and Vrba 1982)—of diverticula remain obscure. Three important physiological questions that could be answered with existing methods are: (1) What volume of air is contained in the diverticula in life? (2) What is the rate of diffusion of air into and out of blind-ended diverticula? and (3) In cases where diverticula of different air sacs anastomose, is air actively circulated through the resulting loops?

Finally, more work is needed on the origins of PSP; if nothing else, Gower's (2001) unconventional hypothesis has drawn attention to this need. Potential projects include histological and biomechanical studies to assess the structure and functions of vertebral laminae (Wilson 1999). In addition, criteria for distinguishing the osteological traces of adipose deposits, muscles, vascular structures, and pneumatic diverticula are badly needed for the interpretation of potentially pneumatic features in fossil bones. This problem is the subject of ongoing research by O'Connor (1999, 2001, 2002).

CONCLUSIONS

The best evidence for pneumaticity in a fossil element is the presence of large foramina that lead to internal chambers. Based on this criterion, pneumatic diverticula were present in the vertebrae of most sauropods and in the ribs of some. Vertebral laminae and fossae were clearly associated with pneumatic diverticula in most eusauropods, but it is not clear whether this was the case in more basal forms. Measurements of vertebral cross sections indicate that, on average, pneumatic sauropod vertebrae were 50%–60% air, by volume. Taking skeletal pneumaticity into account may reduce mass estimates of sauropods by up to 10%. Although the functional and physiological implications of pneumaticity in sauropods and other archosaurs remain largely unexplored, most of the outstanding problems appear tractable, and there is great potential for progress in future studies of pneumaticity.

ACKNOWLEDGMENTS

This work is dedicated to Jack McIntosh, the dean of sauropod workers, whose generosity and enthusiasm continually inspire me. I am grateful to my former and current advisors, Richard Cifelli, Bill Clemens, and Kevin Padian, for their encouragement and sound guidance over many years. This work was completed as part of a doctoral dissertation in the Department of Integrative Biology, University of California, Berkeley. In addition to my advisors, I thank the other members of my dissertation committee, F. Clark Howell, David Wake, and Marvalee Wake, for sound advice and enlightening discussions. Portions of this work are based on a CT study conducted in collaboration with R. Kent Sanders, without whose help I would be nowhere. I thank the staff of the University of Oklahoma Hospital, Department of Radiology, for their cooperation, expecially B. G. Eaton for access to CT facilities and Thea Clayborn, Kenneth Day, and Susan Gebur for performance of the scans. I thank David Berman, Michael Brett-Surman, Brooks Britt, Sandra Chapman, Jim Diffily, Janet Gillette, Wann Langston, Jr., Paul Sereno, Derek Siveter, Ken Stadtman, Dale Winkler, and the staff of the Leicester City Museum for access to specimens in their care. Translations of critical papers were made by Will Downs, Nancie Ecker, and Virginia Tidwell and obtained courtesy of the Polyglot Paleontologist Web site (http://www.uhmc.sunysb.edu/anatomicalsci/paleo). A translation of Janensch (1947) was made by Gerhard Maier, whose effort is gratefully acknowledged. I thank Pat O'Connor for many inspiring conversations and for gracious access to unpublished data and papers in press. Lauryn Benedict, Rebecca Doubledee, Andrew Lee, Anne Peattie, and Michael P. Taylor read early drafts of this chapter and made many helpful suggestions. I am especially grateful to Pat O'Connor and an anonymous reviewer for thoughtful comments that greatly improved the manuscript. Funding was provided by grants from the University of Oklahoma Graduate College, Graduate Student Senate, and Department of Zoology and the University of California Museum of Paleontology and Department of Integrative Biology. This is University of California Museum of Paleontology contribution no. 1841.

LITERATURE CITED

Alexander, R. McN. 1985. Mechanics of posture and gait of some large dinosaurs. Zool. J. Linn. Soc. 83: 1–25.

Alexander, R. McN. 1989. Dynamics of Dinosaurs and Other Extinct Giants. Columbia University Press, New York. 167 pp.

Anderson, J. F., Hall-Martin, A., and Russell, D. A. 1985. Long-bone circumference and weight in mammals, birds and dinosaurs. J. Zool. 207: 53–61.

Bonaparte, J. F. 1986. The early radiation and phylogenetic relationships of the Jurassic sauropod dinosaurs, based on vertebral anatomy. In: Padian, K. (ed.). The Beginning of the Age of Dinosaurs. Cambridge University Press. Cambridge. Pp. 247–258.

Bonaparte, J. F., Heinrich, W.-D., and Wild, R. 2000. Review of *Janenschia* Wild, with the description of a new sauropod from the Tendaguru beds of Tanzania and a discussion on the systematic value of procoelous caudal vertebrae in the Sauropoda. Palaeontographica A 256: 25–76.

Brattstrom, B. H. 1959. The functions of the air sac in snakes. Herpetologica 15: 103–104.

Bremer, J. L. 1940. The pneumatization of the humerus in the common fowl and the associated activity of theelin. Anat. Rec. 77: 197–211.

Britt, B. B. 1993. Pneumatic Postcranial Bones in Dinosaurs and Other Archosaurs. Ph.D. dissertation. University of Calgary, Calgary. 383 pp.

Britt, B. B. 1997. Postcranial pneumaticity. In: Currie, P. J., and Padian, K. (eds.). The Encyclopedia of Dinosaurs. Academic Press, San Diego, CA. Pp. 590–593.

Britt, B. B. Makovicky, P. J., Gauthier, J., and Bonde, N. 1998. Postcranial pneumatization in *Archaeopteryx*. Nature 395: 374–376.

Brochu, C. A. 2003. Osteology of *Tyrannosaurus rex*: insights from a nearly complete skeleton and high-resolution computed tomographic analysis of the skull. Soc. Vertebr. Paleontol. Mem. 7: 1–138.

Buffetaut, E., Suteethorn, V., Cuny, G., Tong, H., Le Loeuff, J., Khansuba, S., and Jongautchariyakal, S. 2000. The earliest known sauropod dinosaur. Nature 407: 72–74.

Chandra Pal and Bharadwaj, M. B. 1971. Histological and certain histochemical studies on the respiratory system of chicken. II. Trachea, syrinx, brochi and lungs. Indian J. Anim. Sci. 41: 37–45.

Colbert, E. H. 1962. The weights of dinosaurs. Am. Mus. Novitates 2076: 1–16.

Cope, E. D. 1877. On a gigantic saurian from the Dakota Epoch of Colorado. Palaeontol. Bull. 25: 5–10.

Cope, E. D. 1878. On the saurians recently discovered in the Dakota beds of Colorado. Am. Nat. 12: 71–85.

Cover, M. S. 1953. Gross and microscopic anatomy of the respiratory system of the turkey. III. The air sacs. Am. J. Vet. Res. 14: 239–245.

Cranford, T. W., Amundin, M., and Norris, K. S., 1996. Functional morphology and homology in the odontocete nasal complex: implications for sound generation. J. Morphol. 228: 223–285.

Crisp, E. 1857. On the presence or absence of air in the bones of birds. Proc. Zool. Soc. London 1857: 215–200.

Cubo, J., and Casinos, A. 2000. Incidence and mechanical significance of pneumatization in the long bones of birds. Zool. J. Linn. Soc. 130: 499–510.

Currey, J. D., and Alexander, R. McN. 1985. The thickness of the walls of tubular bones. J. Zool. 206: 453–468.

Curtice, B., and Stadtman, K. 2001. The demise of *Dystylosaurus edwini* and a revision of *Supersaurus vivianae*. Mesa Southwest Mus. Bull. 8: 33–40.

Dalla Vecchia, F. M. 1999. Atlas of the sauropod bones from the Upper Hauterivian–Lower Barremian of Bale/Valle (SW Istria, Croatia). Natura Nacosta 18: 6–41.

Dodson, P., and Harris, J. D. 2001. Necks of sauropod dinosaurs: Support for a nuchal ligament? J. Morphol. 248: 224.

Duellman, W. E., and Trueb, L. 1986. Biology of Amphibians. McGraw-Hill, New York. 670 pp.

Duncker, H.-R. 1971. The lung air sac system of birds. Adv. Anat., Embryol. Cell Biol. 45: 1–171.

Francillon-Vieillot, H., de Buffrénil, V., Castanet, J., Géraudie, J., Meunier, F. J., Sire, Y., Zylberberg, L., and de Ricqlés, A. 1990. Microstructure and mineralization of vertebrate skeletal tissues. In: Carter, J. G. (ed.). Skeletal Biomineralization: Patterns, Processes and Evolutionary Trends. Vol. 1. Van Nostrand Reinhold, New York. Pp. 471–548.

Gilmore, C. W. 1925. A nearly complete articulated skeleton of *Camarasaurus*, a saurischian dinosaur from the Dinosaur National Monument, Utah. Mem. Carnegie Mus. 10: 347–384.

Gilmore, C. W. 1932. On a newly mounted skeleton of *Diplodocus* in the United States National Museum. Proc. U. S. Natl. Mus. 81: 1–21.

Gould, S. J., and Vrba, E. S. 1982. Exaptation—A missing term in the science of form. Paleobiology 8: 4–15.

Gower, D. J. 2001. Possible postcranial pneumaticity in the last common ancestor of birds and crocodilians: evidence from *Erythrosuchus* and other Mesozoic archosaurs. Naturwissenschaften 88: 119–122.

Hall, B. K. 1999. Evolutionary Developmental Biology. 2nd ed. Kluwer Academic, Norwell, MA. 491 pp.

Hatcher, J. B. 1901. *Diplodocus* (Marsh): its osteology, taxonomy, and probable habits, with a restoration of the skeleton. Mem. Carnegie Mus. 1: 1–63.

Hatcher, J. B. 1903. Osteology of *Haplocanthosaurus*, with a description of a new species, and remarks on the probable habits of the Sauropoda, and the age and origin of *Atlantosaurus* beds. Mem. Carnegie Mus. 2: 1–72.

Hazlehurst, G. A., and Rayner, J. M. V. 1992. Flight characteristics of Triassic and Jurassic Pterosauria: an appraisal based on wing shape. Paleobiology 18: 447–463.

Henderson, D. M. 1999. Estimating the masses and centers of mass of extinct animals by 3-D mathematical slicing. Paleobiology 25: 88–106.

Henderson, D. M. 2004. Tipsy punters: sauropod dinosaur pneumaticity, buoyancy and aquatic habits. Proc. Biol. Sci. 271 (Suppl.): S180–S183.

Hinds, D. S., and Calder, W. A. 1971. Tracheal dead space in the respiration of birds. Evolution 25: 429–440.

Hogg, D. A. 1980. A comparative evaluation of methods for identification of pneumatization in the avian skeleton. Ibis 122: 359–363.

Hogg, D. A. 1984a. The distribution of pneumatisation in the skeleton of the adult domestic fowl. J. Anat. 138: 617–629.

Hogg, D. A. 1984b. The development of pneumatisation in the postcranial skeleton of the domestic fowl. J. Anat. 139: 105–113.

Hunter, J. 1774. An account of certain receptacles of air, in birds, which communicate with the lungs, and are lodged both among the fleshy parts and in the hollow bones of those animals. Philos. Trans. Roy. Soc. London 64: 205–213.

Jain, S. L., Kutty, T. S., Roy-Chowdhury, T. K., and Chatterjee, S. 1979. Some characteristics of *Barapasaurus tagorei,* a sauropod dinosaur from the Lower Jurassic of Deccan, India. Proceedings of the IV International Gondwana Symposium, Calcutta, Vol. 1. Pp. 204–216.

Janensch, W. 1947. Pneumatizitat bei Wirbeln von Sauropoden und anderen Saurischien. Palaeontographica (Suppl. 7) 3: 1–25.

Janensch, W. 1950. Die Wirbelsaule von *Brachiosaurus brancai.* Palaeontographica (Suppl. 7) 3: 27–93.

King, A. S. 1957. The aerated bones of *Gallus domesticus.* Acta Anat. 31: 220–230.

King, A. S. 1966. Structural and functional aspects of the avian lungs and air sacs. Int. Rev. Gen. Exp. Zool. 2: 171–267.

Lasiewski, R. C. 1972. Respiratory function in birds. In: Farmer, D. S., and King, J. R. (eds.). Avian Biology. Vol. II. Academic Press, New York. Pp. 287–342.

Longman, H. A. 1933. A new dinosaur from the Queensland Cretaceous. Mem. Queensland Mus. 10: 131–144.

Lovelace, D., Wahl, W. R., and Hartman, S. A. 2003. Evidence for costal pneumaticity in a diplodocid dinosaur (*Supersaurus vivianae*). J. Vertebr. Paleontol. 23: 73A.

Lull, R. S. 1911. Systematic paleontology of the Lower Cretaceous deposits of Maryland: Vertebrata. Lower Cretaceous Volume, Maryland Geological Survey. Pp. 183–211.

Madsen, J. H., and Welles, S. P. 2000. *Ceratosaurus* (Dinosauria, Theropoda): a revised osteology. Utah Geol. Surv. Misc. Publ. 00-2: 1–80.

Marsh, O. C. 1877. Notice of new dinosaurian reptiles from the Jurassic Formation. Am. J. Sci. 14: 514–516.

Martill, D. M., and Naish, D. 2001. Dinosaurs of the Isle of Wight. Palaeontological Association, London. 433 pp.

Martin, V. 1994. Baby sauropods from the Sao Khua Formation (Lower Cretaceous) in northeastern Thailand. GAIA 10: 147–153.

McClelland, J. 1989a. Larynx and trachea. In King, A. S., and McClelland, J. (eds.). Form and Function in Birds. Vol. 4. Academic Press, London. Pp. 69–103.

McClelland, J. 1989b. Anatomy of the lungs and air sacs. In King, A. S., and McClelland, J. (eds.). Form and Function in Birds. Vol. 4. Academic Press, London. Pp. 221–279.

McIntosh, J. S. 1990. Sauropoda. In Weishampel, D. B., Dodson, P., and Osmolska, H. (eds.). The Dinosauria. University of California Press, Berkeley, CA. Pp. 345–401.

McShea, D. W. 1996. Metazoan complexity: Is there a trend? Evolution 50: 477–492.

Menaum, B., and Richards, S. A. 1975. Observations on the sites of respiratory evaporation in the fowl during thermal panting. Resp. Physiol. 25: 39–52.

Müller, B. 1907. The air-sacs of the pigeon. Smithson. Misc. Collect. 50: 365–420.

Norman, D. 1985. The Illustrated Encyclopedia of Dinosaurs. Crescent Books, New York. 208 Pp.

O'Connor, P. M. 1999. Postcranial pneumatic features and the interpretation of respiratory anatomy from skeletal specimens. J. Vertebr. Paleontol. 19: 67A.

O'Connor, P. M. 2001. Soft-tissue influences on archosaurian vertebrae: interpreting pneumatic and vascular features. J. Vertebr. Paleontol. 21: 84A.

O'Connor, P. M. 2002. Pulmonary pneumaticity in non-dinosaurian archosaurs with comments on *Erythrosuchus* and distal forelimb pneumaticity in pterosaurs. J. Vertebr. Paleontol. 22: 93A.

O'Connor, P. M. 2004. Pulmonary pneumaticity in the postcranial skeleton of extant Aves: a case study examining Anseriformes. J. Morphol. 261: 141–161.

Ojala, L. 1957. Pneumatization of the bone and environmental factors: experimental studies on chick humerus. Acta Oto-Laryngol. Suppl. 133: 1–28.

Osborn, H. F. 1899. A skeleton of *Diplodocus*. Mem. Am. Mus. Nat. Hist. 1: 191–214.

Ostrom, J. H., and McIntosh, J. S. 1966. Marsh's Dinosaurs: The Collections from Como Bluff. Yale University Press, New Haven, CT. 388 pp.

Paul, G. S. 1988. The brachiosaur giants of the Morrison and Tendaguru with a description of a new subgenus, *Giraffatitan*, and a comparison of the world's largest dinosaurs. Hunteria 2(3): 1–14.

Paul, G. S. 1997. Dinosaur models: the good, the bad, and using them to estimate the mass of dinosaurs. Dinofest Int. Proc. 1997: 129–154.

Powell, J. E. 1992. Osteología de *Saltasaurus loricatus* (Sauropoda–Titanosauridae) del Cretácico Superior del noroeste Argentino. In: Sanz, J. L., and Buscalioni, A. D. (eds.). Los Dinosaurios y Su Entorno Biotico: Actas del Segundo Curso de Paleontología en Cuenca. Instituto Juan de Valdes, Cuenca, Argentina Pp. 165–230.

Prange, H. D., Anderson, J. F., and Rahn, H. 1979. Scaling of skeletal mass to body mass in birds and mammals. Am. Nat. 113: 103–122.

Rasband, W. 2003. Image J. National Institutes of Health, Bethesda, MD (http://rsb.info.nih.gov/ij/).

Reid, R. E. H. 1996. Bone histology of the Cleveland-Lloyd dinosaurs and of dinosaurs in general Part 1: Introduction: introduction to bony tissues. Brigham Young Univ. Geol. Stud. 41: 25–72.

Richardson, F. 1939. Functional aspects of the pneumatic system of the California brown pelican. Condor 41: 13–17.

Riggs, E. S. 1904. Structure and relationships of the opisthocoelian dinosaurs. Part II: The Brachiosauridae. Field Columbian Mus. Publ. Geol. 2: 229–247.

Romer, A. S. 1966. Vertebrate Paleontology. 3rd ed. University of Chicago Press, Chicago, IL. 491 pp.

Russell, D. A., and Zheng, Z. 1994. A large mamenchisaurid from the Junggar Basin, Xinjiang, People's Republic of China. Can. J. Earth Sci. 30: 2082–2095.

Russell, D. A., Beland, P., and McIntosh, J. S. 1980. Paleoecology of the dinosaurs of Tendaguru (Tanzania). Mem. Soc. Geol. France. 59: 169–175.

Sadler, D. J., Doyle, G. J., Hall, K., and Crawford, P. J. 1996. Craniocervical bone pneumatisation. Neuroradiology 38: 330–332.

Sanz, J. L., Powell, J. E., LeLoeuff, J., Martinez, R., and Pereda Superbiola, X. 1999. Sauropod remains from the Upper Cretaceous of Laño (northcentral Spain). Titanosaur phylogenetic relationships. Estud. Mus. Cie. Nat. Alava 14(Numero Esp. 1): 235–255.

Schepelmann, K. 1990. Erythropoietic bone marrow in the pigeon: development of its distribution and volume during growth and pneumatization of bones. J. Morphol. 203: 21–34.

Schmidt-Nielsen, K. 1983. Animal Physiology: Adaptation and Environment. Cambridge University Press, Cambridge. 619 pp.

Seeley, H. G. 1870. On *Ornithopsis*, a gigantic animal of the pterodactyle kind from the Wealden. Ann. Mag. Nat. Hist. (Ser. 4) 5: 279–283.

Sereno, P. C. 1991. Basal archosaurs: phylogenetic relationships and functional implications. Soc. Vertebr. Paleontol. Mem. 2: 1–53.

Sereno, P. C., and Arcucci, A. B. 1994. Dinosaurian precursors from the Middle Triassic of Argentina: *Marasuchus lilloensis*, gen. nov. J. Vertebr. Paleontol. 14: 33–73.

Sereno, P. C., and Novas, F. E. 1994. The skull and neck of the basal theropod *Herrerasaurus ischigualastensis*. J. Vertebr. Paleontol. 13: 451–476.

Sereno, P. C., Beck, A. L., Dutheil, D. B., Larsson, H. C. E., Lyon, G. H., Moussa, B., Sadleir, R. W., Sidor, C. A., Varricchio, D. J., Wilson, G. P., and Wilson, J. A. 1999. Cretaceous sauropods and the uneven rate of skeletal evolution among dinosaurs. Science 286: 1342–1347.

Spotila, J. R., Lommen, P. W., Bakken, G. S., and Gates, D. M. 1973. A mathematical model for body temperatures of large reptiles: implications for dinosaur ecology. Am. Nat. 107: 391–404.

Spotila, J. R., O'Connor, P. M., Dodson, P., and Paladino, F. V. 1991. Hot and cold running dinosaurs: body size, metabolism and migration. Modern Geol. 16: 203–227.

Tsuihiji, T. 2004. The ligament system in the neck of *Rhea americana* and its implication for the bifurcated neural spines of sauropod dinosaurs. J. Vertebr. Paleontol. 24: 165–172.

Upchurch, P. 1998. The phylogenetic relationships of sauropod dinosaurs. Zool. J. Linn. Soc. 124: 43–103.

Upchurch, P., and Martin, J. 2003. The anatomy and taxonomy of *Cetiosaurus* (Saurischia, Sauropoda) from the Middle Jurassic of England. J. Vertebr. Paleontol. 23: 208–231.

Warncke, G., and Stork, H.-J. 1977. Biostatische und thermoregulatorische Funktion der Sandwich-Strukturen in der Schädeldecke der Vögel. Zool. Anzeiger 199: 251–257.

Wedel, M. J. 2003a. Vertebral pneumaticity, air sacs, and the physiology of sauropod dinosaurs. Paleobiology 29: 243–255.

Wedel, M. J. 2003b. The evolution of vertebral pneumaticity in sauropod dinosaurs. J. Vertebr. Paleontol. 23: 344–357.

Wedel, M. J. 2004. The origin of postcranial skeletal pneumaticity in dinosaurs. Proceedings of the 19th International Congress of Zoology, Beijing, China Zoological Society. Pp. 443–445.

Wedel, M. J. Cifelli, R. L., and Sanders, R. K. 2000. Osteology, paleobiology, and relationships of the sauropod dinosaur *Sauroposeidon*. Acta Palaeontol. Polonica 45: 343–388.

Weiglein, A. H. 1999. Development of the paranasal sinuses in humans. In: Koppe, T., Nagai, H., and Alt, K. W., (eds.). The Paranasal Sinuses of Higher Primates. Quintessence, Chicago, IL. Pp. 35–50.

Wilson, J. A. 1999. A nomenclature for vertebral laminae in sauropods and other saurischian dinosaurs. J. Vertebr. Paleonotol. 19: 639–653.

Wilson, J. A. 2002. Sauropod dinosaur phylogeny: critique and cladistic analysis. Zool. J. Linn. Soc. 136: 217–276.

Wilson, J. A., and Sereno, P. C. 1998. Early evolution and higher-level phylogeny of sauropod dinosaurs. Soc. Vertebr. Paleontol. Mem. 5: 1–68.

Witmer, L. M. 1990. The craniofacial air sac system of Mesozoic birds (Aves). Zool. J. Linn. Soc. 100: 327–378.

Witmer, L. M. 1995. The extant phylogenetic bracket and the importance of reconstructing soft tissues in fossils. In: Thomason, J. J., (ed.). Functional Morphology in Vertebrate Paleontology. Cambridge University Press, Cambridge. Pp. 19–33.

Witmer, L. M. 1997. The evolution of the antorbital cavity of archosaurs: a study in soft-tissue reconstruction in the fossil record with an analysis of the function of pneumaticity. Soc. Vertebr. Paleontol. Mem. 3: 1–73.

Witmer, L. M. 1999. The phylogenetic history of paranasal air sinuses. In: Koppe, T., Koppe, Nagai, H., and Alt, K. W. (eds.). The Paranasal Sinuses of Higher Primates. Quintessence, Chicago, IL. Pp. 21–34.

Yates, A. M. 2003. A new species of the primitive dinosaur *Thecodontosaurus* (Saurischia: Sauropodomorpha) and its implications for the systematics of basal dinosaurs. J. Syst. Palaeontol. 1: 1–42.

Young, B. A. 1991. Morphological basis of "growling" in the king cobra, *Ophiophagus hannah*. J. Exp. Zool. 260: 275–287.

Young, B. A. 1992. Tracheal diverticula in snakes: possible functions and evolution. J. Zool. 227: 567–583.

Young, C. C., and Zhao, X.-J. 1972. [*Mamenchisaurus hochuanensis*, sp. nov.] Inst. Vertebr. Paleontol. Paleoanthropol. Monogr. A 8: 1–30. (In Chinese.)

Zhang, Y. 1988. [The Middle Jurassic dinosaur fauna from Dashanpu, Zigong, Sichuan.] J. Chengdu Coll. Geol. 3: 1–87. (In Chinese.)

EIGHT

The Evolution of Sauropod Locomotion

MORPHOLOGICAL DIVERSITY OF A SECONDARILY QUADRUPEDAL RADIATION

Matthew T. Carrano

SAUROPOD DINOSAUR LOCOMOTION, like that of many extinct groups, has historically been interpreted in light of potential modern analogues. As these analogies—along with our understanding of them—have shifted, perspectives on sauropod locomotion have followed. Thus early paleontologists focused on the "whalelike" aspects of these presumably aquatic taxa (e.g., Osborn 1898), reluctantly relinquishing such ideas as further discoveries began to characterize sauropod anatomy as more terrestrial. Although this debate continued for over a century, the essentially terrestrial nature of sauropod limb design was recognized by the early 1900s (Hatcher 1903; Riggs 1903). Aside from a few poorly received attempts (e.g., Hay 1908; Tornier 1909), comparisons have usually been made between sauropods and terrestrial mammals, rather than reptiles. Particular similarities were often noted between the limbs of sauropods and those of elephants and rhinos (Holland 1910; Bakker 1971).

With respect to sauropod locomotion, these comparisons with proboscideans have been fruitful but also have tended to become canalized. In this regard, the words of paleontologist W. C. Coombs (1975:23) remain particularly apt, as much for their still-relevant summary of the status quo in sauropod locomotor research as for their warning to future workers:

> It is a subtle trap, the ease with which an entire reptilian suborder can have its habits and habitat preferences deduced by comparison not with all proboscideans, not with the family Elephantidae, not with a particular genus or even a single species, but by comparison with certain populations of a single subspecies.
>
> Deciding that a particular modern animal is most like sauropods is no guarantee of solving the problem of sauropod behavior.

Similarly, modern analogues play a limited role in illuminating the evolution of sauropod locomotion. What information may be gleaned from terrestrial mammals and applied to sauropods is more likely to inform aspects of general limb design, for the simple reason that such information is likely to be rather general in scope. For example, it has been suggested that sauropods employed a limited locomotor repertoire relative to other, smaller dinosaurs in order to maintain limb safety factors (e.g., Wilson and Carrano 1999). This was based on structural limits and behavior changes that had been observed between large- and small-bodied extant mammals (Biewener 1990).

For finer-scale patterns of locomotor evolution to be understood, data on locomotor morphology must first be brought into a phylogenetic context. Fortunately, several recent cladistic studies have greatly clarified systematic relationships within Sauropoda (e.g., Upchurch 1995, 1998; Salgado et al. 1997; Wilson and Sereno 1998; Curry Rogers and Forster 2001; Wilson 2002), although greater resolution is still needed, and many taxa have yet to be studied in detail. Furthermore, the changes apparent within sauropod evolution will almost certainly be subtler than those between sauropods and other dinosaurs. Therefore interpretations cannot rely solely on general analogies with extant taxa, but must also address the specific differences perceived between different taxa. In this regard, not all aspects of sauropod locomotor morphology may be interpretable, or even explicable, although their presence may still be noteworthy.

In this chapter, I use a phylogenetic framework to analyze sauropod locomotor evolution. This aspect of sauropod biology is approached quantitatively with measurements taken directly from specimens and is integrated with qualitative morphological observations. I interpret the resultant patterns in light of the evolution of body size and quadrupedalism, and note several sauropod locomotor specializations. Finally, three large ingroups (diplodocoids, basal macronarians, and titanosaurians) are described in greater detail, to illustrate a portion of the smaller-scale diversity evident in sauropod locomotor morphology.

BODY SIZE AND BODY-SIZE EVOLUTION

Body size likely played a central role in sauropod evolution. As Dodson (1990:407) noted, "Large size with all its biological implications was intrinsic to sauropod biology and was established at the outset of sauropod history." Even the most primitive known sauropods were large relative to other dinosaurs, usually reaching at least 5 metric tons; this is often close to the largest size attained by most other dinosaur clades (e.g., Anderson et al. 1985; Peczkis 1994). Ultimately, some sauropods achieved maximum body sizes exceeding 40 metric tons, with some estimates suggesting even higher masses (e.g., Colbert 1962).

Not surprisingly, several morphological features associated with early sauropod evolution appear to be size-related. This implies that the morphological hallmarks of the earliest sauropods (and, by extension, the diagnosis of Sauropoda itself) cannot be entirely separated from the acquisition of large body size.

These features include (1) a columnar, graviportal limb posture, (2) increased limb bone robusticity, (3) shortened distal limb segments (Carrano 2001), and (4) increased femoral midshaft eccentricity (Wilson and Carrano 1999; Carrano 2001). In other groups of terrestrial amniotes, these features are correlated with increased body size (e.g., Alexander et al. 1979; Scott 1985; Carrano 1998, 1999, 2001). Most are similarly correlated within other dinosaur clades and across Dinosauria as a whole (Carrano 2001). Additionally, the acquisition of a graviportal limb posture may be tied to reduction of lower-limb flexion and extension, implying a functional correlation with reduction in the lower-limb extensor attachments sites such as the olecranon and cnemial crest (see "Sauropod Locomotor Specializations," below).

The simultaneous appearance of these characters at the base of Sauropoda (Upchurch 1995, 1998; Wilson and Sereno 1998; Wilson 2002) is probably an artifact of an insufficiently resolved and recorded early sauropod history. Nonetheless, their close ties to large size in other vertebrate groups suggest that they may be genuinely intercorrelated. In other words, the appearance of more than one of these characters within a single clade or large-bodied taxa may be predictable.

Limb scaling relationships in dinosaurs were examined by Carrano (2001), although sauropods were not highlighted in that study. The results of that analysis demonstrated that dinosaur hindlimb bones scale strongly linearly regardless of posture, with the only evident differences being attributed to changes in body size. Specifically, although hindlimb elements of quadrupedal dinosaurs often appear to be more robust than those of bipedal forms, they are similarly robust at any given body size (fig. 8.1). Indeed, quadrupedalism is associated with an increase in body size over the primitive condition in all dinosaur clades in which it evolves. Dinosaur forelimb elements showed a trend opposite to that expected, becoming relatively longer and less robust in quadrupeds. This shift in scaling was interpreted as a response to the need for a longer forelimb to accompany the hindlimb in generating stride length.

Among thyreophorans and ceratopsians, scaling trends are consistent with this overall pattern, indicating that hindlimb bones did not undergo significant proportional changes during the transition to quadrupedalism apart from those associated with size increases. Forelimb changes are similar to those seen in quadrupeds generally (fig. 8.2). "Semibipedal" dinosaurs, such as prosauropods and derived ornithopods, tend to scale intermediately between bipeds and quadrupeds in both fore- and hindlimb elements (fig. 8.1). They do not show significant differences from either but, instead, overlap the two postures in forelimb scaling. The general picture available from these analyses is one of remarkable uniformity among all dinosaurs regardless of posture, and of the overwhelming influence of body size (versus posture) on limb bone dimensions. These patterns are apparent whether comparisons are made using anteroposterior, mediolateral, or circumferential femoral dimensions (Carrano 2001).

Other features of sauropod limbs are also best viewed as responses to increased body size. Femoral midshaft eccentricity increases in all dinosaurs with increasing body size (Carrano 2001). The apparent disparity between sauropod femora and those of other dinosaurs is an artifact of the size discrepancy between them; extraordinarily large ornithopod and thyreophoran femora show eccentricities comparable to those of sauropods. Similarly, the proportionally short distal limb segments of sauropods are also mirrored among large thyreophorans, ceratopsians, and ornithopods, as well as large terrestrial mammals. Furthermore, the length of the distal limb decreases with increasing body mass in dinosaurs and terrestrial mammals generally (Carrano 1999, 2001). Thus, sauropod limb scaling properties are again entirely consistent with size-related trends.

Beyond this, body-size evolution has not received specific attention in sauropods, except in the larger context of body-size evolution in dinosaurs (Carrano 2005). Here I used squared-change parsimony to reconstruct ancestral body size values using terminal taxon body sizes and a composite phylogeny (fig. 8.3). Femoral length and midshaft diameters were used as proxies for body size because they scale strongly linearly with it (see appendix 8.1). Once ancestral values are reconstructed, ancestor–descendant comparisons can be made throughout Sauropoda. Specifically, I analyzed evolutionary patterns by comparing each ancestor to its descendants (including all internal nodes), as well as by comparing the reconstructed ancestral value for Sauropoda with each terminal taxon (table 8.1). In addition, I used Spearman rank correlation to investigate correlations with patristic distance from the base of the phylogeny, testing whether more derived taxa tend to be larger than more primitive forms (table 8.2).

Not surprisingly, early sauropod evolution is characterized by a steady increase in body size from the condition seen in *Vulcanodon* and larger prosauropods to that in basal neosauropods, representing at least a doubling in size over some 40 million years (Fig. 8.4). This pattern predominates throughout basal sauropods and into Neosauropoda, but the pattern within Neosauropoda is more complex.

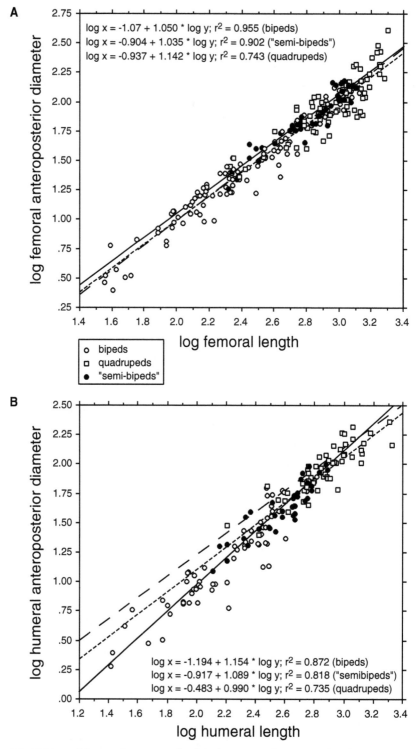

FIGURE 8.1. Limb bone scaling associated with limb posture in dinosaurs. Reduced major axis regressions showing bipeds (open circles, solid lines), "semi-bipeds" (filled circles, dotted lines), and quadrupeds (open squares, dashed lines) from all dinosaur clades. (A) Femoral anteroposterior diameter versus length; (B) humeral anteroposterior diameter versus length. There are few scaling differences among the femora of different groups, but increasingly negative scaling of the humerus accompanies quadrupedalism.

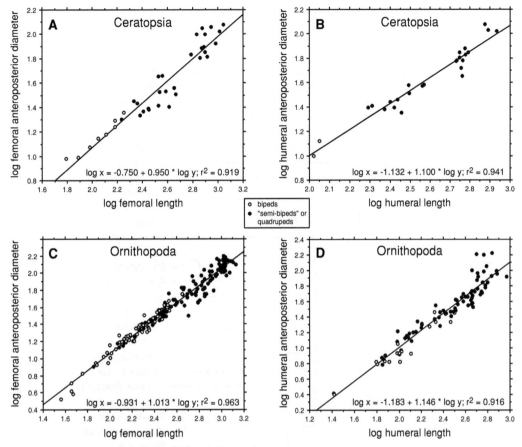

FIGURE 8.2. Limb bone scaling associated with changes in limb posture in nonsauropod dinosaurs. Reduced major axis regressions showing bipeds (open circles) and either "semibipeds" or quadrupeds (filled circles) in two clades for which the transition to quadrupedalism is represented. (A) Femoral diameter versus length in Ceratopsia; (B) humeral diameter versus length in Ceratopsia; (C) femoral diameter versus length in Ornithopoda; (D) humeral diameter versus length in Ornithopoda. Different postural types scale very similarly within each group.

Diplodocoids and macronarians both show divergences in body-size evolution, with members of each group having increased and decreased in size from the primitive condition. The largest members of both clades (*Apatosaurus* among diplodocoids, *Brachiosaurus*, *Argyrosaurus*, and *Argentinosaurus* among macronarians) are similar in size, but macronarians reach substantially smaller adult body sizes (*Saltasaurus*, *Neuquensaurus*, *Magyarosaurus*; ~1.5–3 metric tons) than the smallest diplodocoids (*Dicraeosaurus*, *Amargasaurus*; ~5–10 tons). More notable is the steady, consistent decrease in body size among derived macronarians, from some of the largest (e.g., *Argentinosaurus*; ~50 tons) to some of the smallest (e.g., *Saltasaurus*; ~3 tons) sauropod body sizes over about 30 million years. This trend includes titanosaurians and reaches its nadir among saltasaurines.

This analysis supports previous assertions of body-size increase in sauropods, emphasizing its persistence throughout much of the clade. However, it also highlights an unappreciated complexity to this pattern, demonstrating that most body-size increases occur early in sauropod evolution and were largely complete by the Upper Jurassic neosauropod radiation. Subsequent decreases are also evident, particularly among macronarians. These may be a response to having reached an upper bound on body size or represent size-based diversification in later sauropod lineages. In this study,

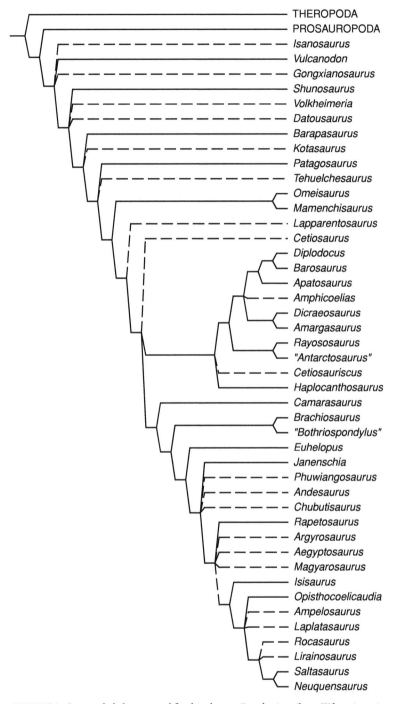

FIGURE 8.3. Sauropod phylogeny used for this chapter. Based primarily on Wilson (2002), but also Salgado et al. (1997), Wilson and Sereno (1998), Upchurch (1998), and Curry Rogers (pers. comm.). Taxa with dashed lines were not analyzed cladistically but are placed according to their presumed position in table 13 of Wilson (2002).

TABLE 8.1
Results of Body-Size Analyses for Sauropoda

	MEAN	SUM	SKEW	MEDIAN	N	+	−
A. All ancestor–descendant comparisons							
FL	0.014	1.203	0.771	0.006	89	47	42
FAP	0.009	0.641	0.331	0.011	70	38	32
FML	0.013	0.911	0.239	0.008	70	41	29
B. Basal node–terminal taxon comparisons							
FL	0.209	10.879	−0.588	0.239	52	46	8

NOTE: Results using squared-change parsimony reconstructions based on measurements of femoral length. In A, comparisons are between each reconstructed ancestral node and each descendant taxon; in B, they are between the basal reconstructed ancestral node for each clade and all its descendant terminal taxa. These results are drawn from a larger study (Carrano 2005), in which all available dinosaur taxa were included. Abbreviations: skew, skewness; +, number of positive ancestor–descendant changes; −, number of negative ancestor–descendant changes; FAP, femoral anteroposterior diameter; FL, femoral length; FML, femoral mediolateral diameter.

TABLE 8.2
Spearman Rank Correlations of Body Size and Patristic Distance for Sauropoda

	ρ	z	p	ρ†	z†	p†
All Sauropoda						
FL	−0.104	−0.973	0.331	−0.106	−0.998	0.318
FAP	−0.201	−1.671	0.095*	−0.204	−1.697	0.090*
FML	−0.219	−1.833	0.067*	−0.223	−1.863	0.063*
Diplodocoidea						
FL	0.526	2.409	0.016*	0.519	2.377	0.018*
FAP	0.456	1.766	0.078*	0.449	1.738	0.082*
FML	0.705	2.639	0.008*	0.700	2.618	0.009*
All non-macronarian Sauropoda						
FL	0.663	4.499	<0.001*	0.662	4.491	<0.001*
FAP	0.415	2.490	0.013*	0.413	2.478	0.013*
FML	0.379	2.208	0.027*	0.376	2.193	0.028*
Macronaria						
FL	−0.718	−3.515	<0.001*	−0.727	−3.526	<0.001*
FAP	−0.205	−0.869	0.385	−0.211	−0.894	0.371
FML	−0.552	−2.467	0.014*	−0.563	−2.519	0.012*

NOTE: Abbreviations as in Table 8.1, Note. Daggers (†) indicate values corrected for ties; asterisks (*) indicate significant p values.

Magyarosaurus—which has been described as a "dwarf" sauropod (Jianu and Weishampel 1999)—is not a unique taxon but instead represents the size endpoint for an entire clade of relatively small taxa. Because smaller sauropods tend to be the end products of their respective lineages (and therefore have relatively high patristic distance values), there is a weak negative size correlation within Sauropoda. However, when macronarians are analyzed separately, they show a significant size decrease, while other sauropod groups (diplodocids and all nonmacronarians) exhibit significant size increases (table 8.2).

Trend analyses (McShea 1994; Alroy 2000a, 2000b; Carrano 2005) suggest that sauropod

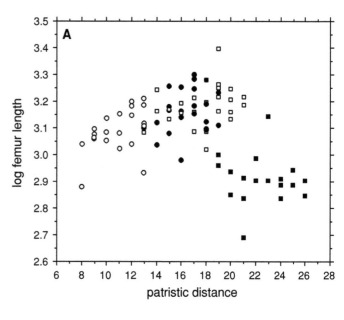

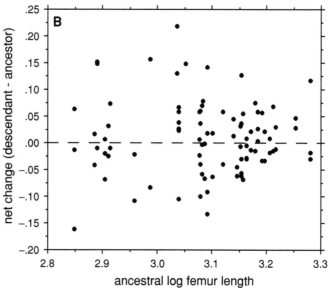

FIGURE 8.4. Body-size evolution in Sauropoda, based on measurements of femur length. (A) Body size versus patristic distance, showing increases throughout most of sauropod evolution, particularly from basal sauropods (open circles) to diplodocoids (open squares) and basal macronarians (filled circles). Note the steady size decrease in derived macronarians (filled squares). (B) Net change (descendant size minus ancestral size) versus ancestral body size, showing that changes during sauropod evolution are nearly random (i.e., no correlation).

body-size evolution is largely characterized by "active" macroevolutionary processes. Two details of this pattern support such an interpretation: (1) the range of sauropod body sizes expands through time (and through the phylogeny), with concomitant loss of smaller taxa as larger taxa appear; and (2) there is a weak positive correlation between ancestor–descendant changes and ancestral sizes. The apparent upper size bound for sauropods (which may well be altered by subsequent discoveries) is minimally about 50 metric tons, substantially greater than those for other dinosaur groups (which tend to be between 5 and 10 tons). Their apparent lower size bound, approximately 1–3 metric tons, is also considerably larger than that for other dinosaurs (usually between 0.05 and 0.5 tons) (Carrano 2005).

QUADRUPEDALISM

Quadrupedalism represents a second fundamental, dominant characteristic of all sauropods, having been achieved in the earliest known forms and retained in all known descendants. Much of the appendicular skeleton shows adaptations for quadrupedal posture, particularly in the manus and forelimb (McIntosh 1990; Upchurch 1995, 1998; Wilson and Sereno 1998). These changes include lengthening of the forelimb elements relative to those of the hindlimb, as well as morphological modifications of the distal forelimb elements associated with locomotion and weight bearing (see below).

Based on such criteria, the primitive sauropod *Vulcanodon* appears to have been fully quadrupedal (Cooper 1984). Its more primitive relatives *Blikanasaurus* and *Isanosaurus* entirely lack forelimb materials, so their postural status cannot be assessed. The recently described basal sauropod *Antetonitrus* (Yates and Kitching 2003) has a relatively long forelimb but retains prosauropod-like manual elements, suggesting that quadrupedalism may have been facultative in this taxon. More complete forelimb materials are needed to better elucidate its postural and locomotor capabilities.

Other outgroup sauropodomorph taxa (prosauropods and stem-sauropodomorphs) all appear to have retained at least facultative bipedalism (e.g., Galton 1990). Thus the transition to quadrupedalism in sauropods is only hinted at by the known fossil record. Nonetheless, it can be examined indirectly through comparison with other dinosaur groups in which the transition is recorded (Ceratopsia, Thyreophora), along with "semibipedal" taxa such as prosauropods and derived ornithopods.

Among thyreophorans and ceratopsians, the shift to quadrupedalism is also accompanied by marked changes in forelimb morphology. Not surprisingly, hindlimb morphology is usually little altered, instead retaining many of the locomotor features already present in the bipedal members of each clade. Most differences in the hindlimb can be attributed to those associated with concomitant body size increases (Carrano 2001; see also "Body Size and Body-Size Evolution," above). In the forelimb, however, the manus is modified to become weight-bearing through consolidation of the carpus (via fusion and/or loss) and reduction in its mobility, incorporation of the metacarpals into a bound unit, and transformation of the unguals into blunt, hooflike structures. Aside from relative lengthening of the forelimb, these are the primary features used to infer quadrupedalism in dinosaur taxa.

Most of these changes are already present in basal sauropods (e.g., *Shunosaurus*, *Barapasaurus*), indicating that these animals were already fully (and probably obligatorily) quadrupedal. The manus of *Vulcanodon* is incomplete, but the long humerus and forearm suggest that this taxon was also probably fully quadrupedal (Raath 1972; Cooper 1984). Nevertheless, the transition to permanent quadrupedalism likely involved a series of morphological changes that occurred across a range of facultatively quadrupedal taxa. *Antetonitrus* may represent one such intermediate, combining a relatively long forelimb with a primitive manus (Yates and Kitching 2003).

More derived basal sauropods record the completion of this evolutionary process (fig. 8.5). Despite our fragmentary understanding of its early stages, relics of the transition to quadrupedalism remain in basal sauropods such as *Shunosaurus*, *Barapasaurus*, *Omeisaurus*, *Mamenchisaurus*, and "cetiosaurs" (e.g., *Lapparentosaurus*, *Patagosaurus*, *Cetiosaurus*). Although these forms were fully quadrupedal, they document the continued acquisition of additional characters related to this posture. Their phylogenetic positions help to clarify the original sequence of character acquisition in earlier forms by "removing" later-appearing features from the early stages of the evolution of sauropod quadrupedalism.

The primitive saurischian manus was probably digitigrade, as suggested by the posture of manus prints of otherwise bipedal dinosaurs (e.g., Lull 1953; Baird 1980). Additionally, other dinosaur clades that underwent a transformation to quadrupedalism also appear to have passed through (and sometimes remained at) a phase involving a digitigrade manus. The major

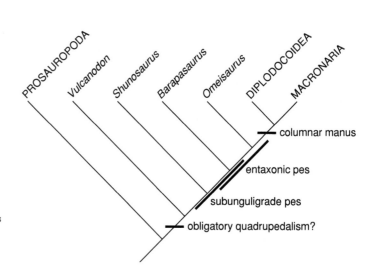

FIGURE 8.5. The transition to quadrupedalism in Sauropoda. Although basal sauropods were likely obligate quadrupeds, numerous additional modifications for this posture were developed in basal taxa. Phylogeny from Wilson and Sereno (1998).

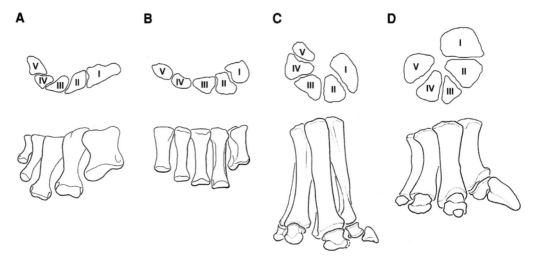

FIGURE 8.6. Changes in the metacarpus in Sauropoda. The metacarpals acquire a semicircular, columnar arrangement within sauropods, with a reduction in length disparities between elements. Proximal (top row) and dorsal (bottom row) views of the articulated right metacarpus; proximal views are shown with the palm toward the top of the page. (A) *Massospondylus carinatus* (modified from Cooper 1981). (B) *Shunosaurus lii* (modified from Zhang 1988). (C) *Brachiosaurus brancai* (modified from Janensch 1922; Wilson and Sereno 1998). (D) *Janenschia robusta* (modified from Janensch 1922; Wilson and Sereno 1998). Numbers refer to metacarpals I through V. Not to scale.

changes in manus morphology, therefore, began from a primitively digitigrade structure in which the individual metacarpals are most firmly interconnected proximally, arranged in a dorsoventral arch, and held at an oblique angle to the substrate (fig. 8.6). This morphology would have served to resist long-axis bending during flexion and extension at the carpus by allowing the metacarpals to act together in distributing and deflecting these loads. In the most primitive sauropods for which complete manus remains are known (e.g., *Shunosaurus*, *Omeisaurus*), the intermetacarpal facets are already somewhat elongated distally, indicating that the bones were arranged into a tight, amphiaxonic arch while being held nearly vertically relative to the substrate (Osborn 1904; Gilmore 1946; Christiansen 1997). Coombs (1975) described this arrangement as mimicking columns beneath a supported floor (i.e., the carpus), an analogy that is probably mechanically appropriate. Compressional loading was likely increased relative to bending in such a structure.

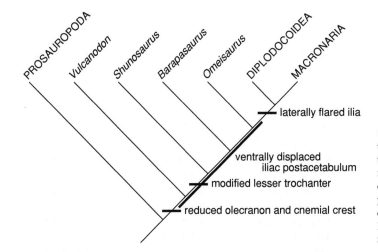

FIGURE 8.7. Sauropod locomotor specializations. The acquisition of unique morphological features within sauropods occurred throughout basal taxa, prior to the divergence of diplodocoids and macronarians. Phylogeny from Wilson and Sereno (1998).

Meanwhile, the sauropod pes also underwent significant changes in morphology and posture. Primitively, the saurischian (and dinosaurian) pes was a digitigrade structure with four primary weight-bearing digits, an arrangement retained by prosauropods, primitive theropods (*Herrerasaurus, Eoraptor*), and basal ornithischians (*Pisanosaurus, Lesothosaurus*). Basal sauropods enlarged digit V into an additional support element (Raath 1972; Cruickshank 1975), representing a reversal to a more primitive archosaurian condition. These same taxa also show a shift in metatarsal orientation from relatively upright (subvertical) to more horizontal. Sauropod footprints show evidence that these low-angled metatarsals were supported by a fleshy pad (Coombs 1975; Gallup 1989), analogous to those of modern proboscideans and rhinocerotids, indicating that the sauropod pes was subunguligrade rather than strictly digitigrade.

SAUROPOD LOCOMOTOR SPECIALIZATIONS

Other changes documented in basal sauropod evolution do not appear to be related to either quadrupedalism or large body size but, instead, represent unique components of the sauropod locomotor apparatus. Like the changes discussed previously, these modifications occurred early in sauropod evolution and were essentially in place by the appearance of Neosauropoda (fig. 8.7).

All sauropods show evidence of reduction in the lower-limb muscle attachments. This occurs in both the fore- and the hindlimbs, where the insertion areas for these muscles are reduced relative to the condition in other saurischians (and other dinosaurs). In the forelimb, the ulnar olecranon process is reduced to a flat, rugose surface in primitive sauropods; this condition is reversed only in titanosaurians (Christiansen 1997; Wilson and Sereno 1998). This process represents the insertion of musculi (mm.) triceps, which are the major forearm extensors (e.g., Christiansen 1997). The hindlimb shows reduction of the cnemial crest on the tibia, representing the insertion for the knee extensors (mm. iliotibiales, musculus [m.] ambiens, and mm. femorotibiales). These changes suggest a shift in basal sauropods away from significant use of the lower limb during locomotion and toward relying primarily on protraction and retraction at the hip and shoulder to generate stride length.

As in other quadrupedal dinosaurs, both the manus and the pes of sauropods show phalangeal reduction (Osborn 1904; Coombs 1975; Upchurch 1995, 1998; Wilson and Sereno 1998). Usually reduction occurs primarily in length, with individual phalanges becoming

TABLE 8.3
Loss of Manual Phalanges in Sauropoda

	SPECIMEN	PRESERVED	RECONSTRUCTED
Theropoda			2-3-4-2*-0
Prosauropoda			2-3-4-3*-2*
Shunosaurus lii	ZDM T5402	2-2-2*-2*-2*	2-2-2*-2*-2*
Omeisaurus tianfuensis	ZDM T	2-2-?-?-1	2-2-2-2-2*
Jobaria tiguidensis	MNN TIG3	2-2-2-2-2	2-2-2-2-2
Apatosaurus louisae	CM 3018	2-2*-1-1-1	2-2*-1-1-1
Apatosaurus excelsus	CM 563	2-1-1-1-1	2-2*-1-1-1
Camarasaurus sp.	FMNH 25120	2-1-1-1-1	2-2*-1-1-1
Camarasaurus sp.	AMNH 823	2-1-1-1-1	2-2*-1-1-1
Camarasaurus grandis	GMNH-VP 101	2-1-1-1-1	2-2*-1-1-1
Brachiosaurus brancai	HMN S II	2-1-1-1-1	2-2*-1-1-1
Janenschia robusta	HMN Nr. 5	2-2*-1-1-1	2-2*-1-1-1
Opisthocoelicaudia skarzynskii	ZPal MgD-I/48	0-0-0-1*-0	0-0-0-1*-0
Alamosaurus sanjuanensis	USNM 15660	0-0-0-0-0	0-0-0-0-0

NOTE: Formulas list numbers of phalanges in order from digit I through digit V. Asterisks (*) indicate vestigial phalanges; question marks (?) indicate questionable numbers.

compact and often disclike. However, sauropods continue this trend to its extreme, eliminating many of the manual and pedal phalanges altogether. In the manus particularly, this reduction is carried to the point that all the unguals, and nearly all the phalanges, are eventually lost (Osborn 1904, 1906; Gilmore 1946; Wilson and Sereno 1998; table 8.3). Phalanges typically serve as points of flexion and extension for the manus, and their loss indicates the loss of these functions within this portion of the limb. This represents the end point of the transformation from a propulsive, digitigrade manus to one that was almost entirely dedicated to columnar support.

Tracking this set of changes is problematic, however, because phalanges are often lost prior to discovery of the specimen. This is particularly true of vestigial, nubbinlike phalanges such as those in the manus of *Shunosaurus* (Zhang 1988). Conservatively, manual phalangeal reduction appears to occur primarily at two points in sauropod evolution. First, most of the phalanges on digits II–V were lost in Neosauropoda, and then the remainder were lost in derived titanosaurs. However, some evidence suggests that this pattern is more complex. For example, *Apatosaurus louisae* (CM 3018) retains two phalanges on digit II (Gilmore 1936). Many other neosauropods (perhaps most nontitanosaurs) retain a distal articular facet on II-1 that implies the presence of an additional phalanx, albeit a small one. The vestigial IV-1 in *Opisthocoelicaudia* (Borsuk-Bialynicka 1977) might also be a genuine relic. Thus, additional stages of reduction may have existed between Neosauropoda and titanosaurs (table 8.3).

In addition, the primitive dinosaurian pes is a mesaxonic structure, with the weight-bearing axis passing down the central digit (III) (fig. 8.8). This condition, in which the central metatarsals are the longest and most robust, is evident in prosauropods and in basal sauropods (*Vulcanodon*). However, it is altered in more derived sauropods (e.g., *Shunosaurus, Omeisaurus*) such that the medial metatarsals are larger than the central elements. In particular, metatarsal I becomes the most robust, and the weight-bearing axis shifts to a more medial, entaxonic position (Coombs 1975). This type of pedal structure is highly unusual among tetrapods, being evident elsewhere among

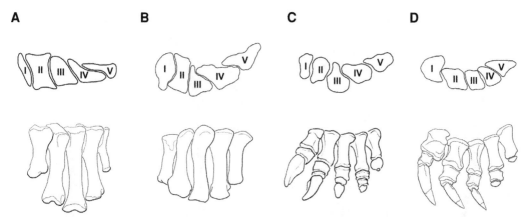

FIGURE 8.8. Changes in the metatarsus in Sauropoda. The metatarsals change from mesaxonic to entaxonic, as the medial digits become relatively larger. Proximal (top row) and dorsal (bottom row) views of the articulated left metatarsus; the proximal views are shown plantar-upward. (A) *Massospondylus carinatus* (modified from Cooper 1981) . (B) *Vulcanodon karibaensis* (modified from Cooper 1984). (C) *Shunosaurus lii* (modified from Zhang 1988; Wilson and Sereno 1998). (D) *Apatosaurus louisae* (modified from Gilmore 1936). Numbers refer to metatarsals I through V. Not to scale.

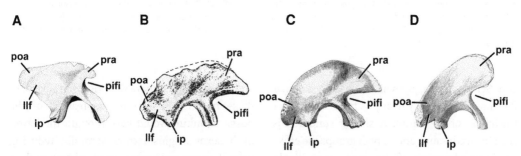

FIGURE 8.9. Changes in the ilium in Sauropoda. The ilium is expanded anteroposteriorly, and the postacetabulum is depressed ventrally to the level of the ischial peduncle. Right pelves are shown in lateral view. (A) *Massospondylus carinatus* (modified from Cooper 1981). (B) *Shunosaurus lii* (modified from Zhang 1988). (C) *Dicraeosaurus hansemanni* (modified from Janensch 1961). (D) *Brachiosaurus brancai* (modified from Janensch 1961). Abbreviations: ip, ischial peduncle; llf, origination area for lower limb flexors; pifi, space anterior to ilium for passage of m. puboischiofemoralis internus 2; poa, postacetabulum; pra, preacetabulum. Not to scale.

megatheriid xenarthrans. This shift may have accompanied a more general change from narrow-bodied, predominantly bipedal taxa in which the limbs were positioned well under the body to wider-bodied quadrupedal taxa in which the limbs were more laterally positioned. In such wider-bodied taxa, the weight-bearing axis would pass through the medial portion of the pes instead of the central portion, as reflected in the modified proportions of metatarsals I and II.

The ilium is substantially modified in sauropods (fig. 8.9). Primitively (in prosauropods and basal theropods), this bone is dorsoventrally narrow and relatively short anteroposteriorly (Carrano 2000). In all known sauropods, both the preacetabular and the postacetabular processes are expanded anteroposteriorly and dorsoventrally. The preacetabulum is large and lobate (Raath 1972; Wilson and Sereno 1998), arching above a large space anteroventral to the ilium through which m. puboischiofemoralis internus 2 likely passed (Romer 1923; Carrano and Hutchinson 2002). This space is larger than that seen in neotheropods, implying that m. puboischofemoralis internus 2 did not undergo the alterations in size or position seen in that group (Carrano and Hutchinson 2002).

The preacetabular ilium is also flared laterally in most neosauropods, with the blade curving outward from a point just dorsal to the pubic peduncle. This is carried to an extreme in titanosaurians (Powell 1986, 1990, 2003; Wilson

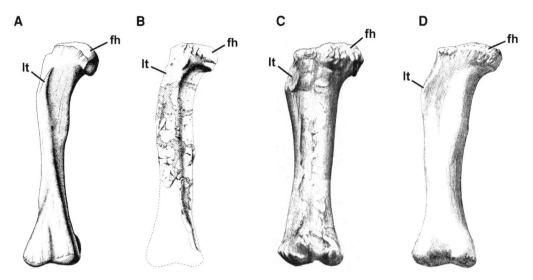

FIGURE 8.10. Changes in the femoral lesser trochanter within Sauropoda. The lesser trochanter is reduced from a distinct process to a rugose bump and shifts in position from anterolateral to lateral. Right femora are shown in anterior view. (A) *Massospondylus carinatus* (modified from Cooper 1981). (B) *Vulcanodon karibaensis* (modified from Cooper 1984). (C) *Apatosaurus excelsus* (modified from Riggs 1903). (D) *Brachiosaurus brancai* (modified from Janensch 1961). Abbreviations: fh, femoral head; lt, lesser trochanter. Not to scale.

and Sereno 1998), in which the preacetabular ilium reaches laterally nearly as far as the greater trochanter of the femur. A similar condition is seen in ankylosaurs, and in both groups the effect would have been to shift the origins of m. iliotibialis 1 and m. iliofemoralis externus laterally. As a result, these muscles would have had reduced mediolateral actions relative to their anteroposterior and/or dorsoventral ones.

The iliac postacetabulum is also large but extends much farther ventrally than the preacetabulum. In *Vulcanodon* (Raath 1972; Cooper 1984) and *Shunosaurus* (Zhang 1988), the postacetabulum is depressed close to the level of the ischial peduncle, much more so than the condition in prosauropods and theropods. It reaches the level of the ischial peduncle in many diplodocoids, and actually exceeds it in many titanosaurians. Several lower-limb flexors originated on the iliac postacetabulum, including mm. flexores tibiales internii, m. flexor tibialis externus, and m. iliofibularis (Cooper 1984), although the exact placement and bounds of these muscles are not clear (Carrano and Hutchinson 2000). In most neosauropods, this portion of the ilium often bears complex fossae and rugosities that mark the origins of several of these muscles. Therefore, depression of the postacetabulum would have brought the lower-limb flexor origins farther ventrally, reducing their dorsoventral components relative to their anteroposterior actions.

On the femur, the lesser trochanter was primitively part of the dinosauriform trochanteric shelf, a ridge running around the anterolateral corner of the proximal femur that served as the insertion for m. iliofemoralis externus (Hutchinson 2001). In most dinosaurs, this shelf is differentiated into a distinct process (the lesser trochanter, for portions of mm. iliotrochanterici) as well as a rugose mound (for m. iliofemoralis externus) (Hutchinson 2001; Carrano and Hutchinson 2002). This occurs independently in several dinosaur clades (Carrano 2000), but sauropods are a persistent exception to this trend (fig. 8.10). Instead, although the trochanteric shelf is modified in sauropods, the lesser trochanter is reduced to a rounded anterolateral ridge in *Vulcanodon* (Cooper 1984) and a rugose lateral bump in other sauropods. The m. iliofemoralis insertion was not drawn proximally as in other dinosaurs (Carrano 2000), instead remaining well below the femoral head.

This transition would have accomplished two things. First, m. iliofemoralis externus would have been placed laterally, not anteriorly as in most dinosaurs (Carrano 2000). Thus this muscle would have retained a role in femoral abduction (and possibly rotation), rather than femoral protraction. Second, it remained relatively distal in position and was (apparently) never elaborated into separate muscles; no analogues to mm. iliotrochanterici of theropods and birds were present. Therefore sauropods likely relied on m. puboischiofemoralis internus for femoral protraction, as did most non-theropod dinosaurs.

LOWER-LEVEL PATTERNS IN SAUROPOD LOCOMOTOR EVOLUTION

Although acquired stepwise in basal sauropods, the morphological specializations discussed above characterize nearly all known sauropod taxa. Their appearances mark the emplacement of the sauropod locomotor apparatus, which remained fundamentally unaltered throughout subsequent neosauropod evolution for the next 100 million years. Yet on this functional and morphological foundation was built a diversity of sauropod locomotor habits that has gone mostly unappreciated. This is largely thanks to the perceived uniformity of the sauropod appendicular skeleton, which, while genuine at one scale (and in that sense remarkable), has also helped to obscure smaller-scale patterns in sauropod locomotor evolution (e.g., Hatcher 1903:55). Discovering and interpreting lower-level evolutionary patterns require an understanding of smaller-scale differences between sauropod groups.

Here I discuss the differences among three major sauropod groups—diplodocoids, basal macronarians, and titanosaurians. I do not seek to present a detailed study of these taxa; my goal is rather to demonstrate that existing information is sufficient to illustrate marked morphological (and, by inference, functional) differences among these taxa. Future work in this area is likely to be quite fruitful.

DIPLODOCOIDEA

Diplodocoids are the sister-group to Macronaria within Neosauropoda (fig. 8.3). The group includes three major clades (Diplodocidae, Dicraeosauridae, and Rebbachisauridae) and over 25 taxa, of which at least 15 are well known. They are generally perceived as smaller and more slender than other sauropods, based largely on reconstructions of *Diplodocus*, *Amargasaurus*, and *Dicraeosaurus*. However, specimens of *Apatosaurus* (YPM 1860) and *Seismosaurus* (NMMNH P-3690; possibly synonymous with *Diplodocus* [Lucas and Heckert 2000]) rival the largest macronarians in overall size. (As noted earlier, this gigantism was achieved independently in these two clades.)

Diplodocoids have gracile limbs that are short relative to the trunk, making them somewhat low-slung animals (fig. 8.11A). Compared to basal macronarians, diplodocoids would have had shorter strides relative to their trunk length. A marked discrepancy between the fore- and the hindlimb lengths exaggerates the condition typically present in other quadrupedal dinosaurs. The manus remains relatively short, similar to that in more primitive sauropods. A few diplodocoids (*Diplodocus* and particularly *Amphicoelias*) are notable in having a reduced femoral midshaft eccentricity (Osborn and Mook 1921) compared to other, similar-sized sauropods.

In addition, diplodocoids have a modified metatarsus in which metatarsals III and IV are "twinned" in morphology. This contrasts (albeit subtly) with the condition in other sauropods, where these two bones are similar in length but distinct from one another in shape. The functional implications of such a pedal structure are unclear; it may belie some minor locomotor specialization or simply be a product of increased limb bone slenderness. In other respects, however, the diplodocoid appendicular skeleton is relatively conservative, retaining

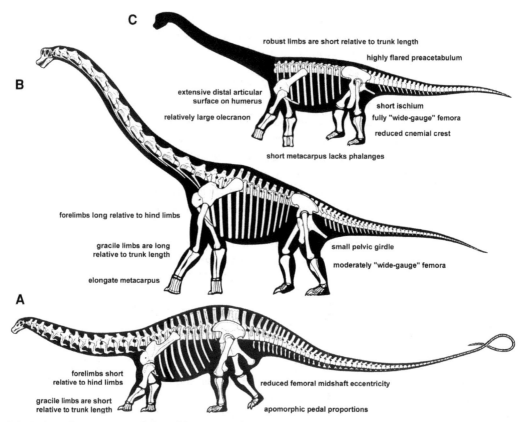

FIGURE 8.11. Comparative morphology of three sauropod groups. (A) Diplodocoidea (*Apatosaurus louisae*); (B) basal Macronaria (*Brachiosaurus brancai*); (C) Titanosauria (*Opisthocoelicaudia skarzynskii*). Skeletal reconstructions modified from Wilson and Sereno (1998). Approximately to scale.

many sauropod and neosauropod symplesiomorphies.

BASAL MACRONARIA

"Basal macronarians" are a paraphyletic assemblage of more than 20 species of taxa including the Brachiosauridae, *Camarasaurus*, *Euhelopus*, *Janenschia*, and the most basal titanosaurians (e.g., *Aeolosaurus*, *Pleurocoelus*) (Wilson and Sereno 1998; Wilson 2002). Many of these forms are large even among sauropods, with taxa such as *Brachiosaurus*, *Sauroposeidon*, and *Argentinosaurus* representing some of the largest known terrestrial vertebrates.

Basal macronarians are characterized by relatively gracile limbs that are long relative to the trunk (fig. 8.11B). As a result, these taxa would have had longer stride lengths compared to diplodocoids. The forelimbs are long relative to the hindlimbs (again, opposite to the diplodocoid condition), a tendency most evident in *Brachiosaurus* (FMNH P25107, HMN D) but also seen in other taxa ("*Bothriospondylus madagascariensis*," MNHN specimen [Lapparent 1943]; *Euhelopus*, PMU R234 [Wiman 1929]; *Lusotitan atalaiensis*, MIGM holotype [Lapparent and Zbyszewski 1957]). In addition, the metacarpus is elongate relative to that of diplodocids and primitive sauropods, allowing it to contribute increasingly to stride length. Overall, most of these changes would have allowed the forelimb to approach or exceed the hindlimb in its ability to contribute to forward progression. This represents a reversal of the typical (and primitive) dinosaurian condition.

Many of these forms also show a reduction in the size of the pelvic girdle relative to the pectoral girdle. This would have reduced the origination areas for most of the proximal hindlimb musculature, again suggesting a shift in locomotor dominance toward the forelimb. Some of the "wide-gauge" features that exemplify titanosaurians (see below) are present in these taxa, but to a much lesser degree. For example, brachiosaurids show a medial offset of the femoral head, moderate flaring of the preacetabular ilium, and reduction of the pollex ungula—all changes eventually carried further by titanosaurians (Wilson and Carrano 1999).

Recently, Stevens and Parrish (1999) suggested that sauropods generally did not raise the neck above the level of the shoulder. The resulting clade of long-necked but low-browsing animals stands in marked contrast to conventional reconstructions of sauropods with giraffe-like neck postures (e.g., Paul 1988). I do not attempt to evaluate this hypothesis here, but it is worth noting that an additional effect of forelimb lengthening in basal macronarians would be to raise the height of the shoulder, thereby elevating the maximum potential height of the neck in animals otherwise constrained to low browsing. Thus basal macronarians may have been relatively high-browsing sauropods thanks to their forelimb adaptations.

TITANOSAURIA

Titanosaurian sauropods comprise approximately 50 taxa of derived macronarians (Wilson and Sereno 1998; Curry Rogers and Forster 2001; Wilson 2002) such as *Rapetosaurus*, *Ampelosaurus*, and the Saltasaurinae. Of these, approximately 35 are known from diagnostic materials, and many fewer from relatively complete specimens. As noted above, titanosaurians are unusual in being markedly smaller than most other sauropods, but in addition, they display a suite of features that readily distinguish them from other neosauropods (fig. 8.11C). These include an outward-canted femur with a correspondingly angled distal articular surface, a widened sacrum, and a significantly more elliptical femoral midshaft relative to body size (Wilson and Sereno 1998; Wilson and Carrano 1999). Such features are consistent with an interpretation of titanosaurians as the authors of "wide-gauge" sauropod trackways (Wilson and Carrano 1999).

Titanosaurians are also characterized by highly flared iliac preacetabular processes (Powell 1986, 1990, 2003; Christiansen 1997), somewhat analogous to the condition seen in ankylosaurs. This flaring would have displaced the origins of m. iliofemoralis and m. iliotibialis 1 more laterally, as noted above (Carrano 2000). The pelvic elements are relatively small, as in basal macronarians. In addition, the ischium is very short relative to the pubis (Wilson and Sereno 1998), and the anterior caudals have ball-in-socket articulations, perhaps associated with an increased predominance of bipedal rearing in these taxa (Borsuk-Bialynicka 1977; Wilson and Carrano 1999).

Reversing the primitive condition for sauropods, titanosaurian humeri have anteroposteriorly long distal articular surfaces (Wilson and Sereno, 1998; table 8.4), indicating a wider range of flexion–extension at the elbow than in other sauropods (Osborn 1900). Although the limb joint surfaces are not well modeled and are now missing what were probably substantial cartilage caps (Hatcher 1901; Coombs 1975), it is reasonable to assume that the true articular cartilage would have been at least as anteroposteriorly extensive as the bony attachment surface. I measured the linear height of the articular surface on the anterior face of the distal humerus and compared it among sauropod groups relative to total humerus length. These differences are significant when compared using both the unpaired, two-group t-test ($t = -5.346$, df = 16; $p < 0.001$) and the Mann–Whitney U-test ($U = 2.000$, $Z = -3.184$; $p = 0.015$).

Both fore- and hindlimbs are short relative to the trunk. The particularly short titanosaurian forefoot is the most derived among sauropods in having entirely lost the carpus and phalanges (including the manual ungual) or having both reduced to cartilage (Gilmore 1946; Salgado et al. 1997; Wilson and Sereno 1998). Whatever function the enlarged pollex had originally evolved

TABLE 8.4
Humeral Articulation Angles in Sauropods

	SPECIMEN	HL	AHD	%
Shunosaurus lii	T5401	648	37	5.7
Omeisaurus tianfuensis	T5701	790	27	3.4
Mamenchisaurus constructus	IVPP V.947	1,080	52	4.8
"Cetiosaurus" mogrebiensis	Tamguert n'Tarit	1,370	42	3.1
Barosaurus africanus	HMN k 37	970	16	1.6
Dicraeosaurus hansemanni	HMN Q 11	740	24	3.2
Camarasaurus grandis	YPM 1901	1,288	90	7.0
Brachiosaurus altithorax	FMNH P25107	2,040	105	5.1
Brachiosaurus brancai	HMN S II	2,130	104	4.9
Euhelopus zdanskyi	GSC specimen	1,090	40	3.7
Tehuelchesaurus benitezii	MPEF-PV 1125	1,140	41	3.6
Janenschia robusta	HMN P 8	890	40	4.5
Gondwanatitan faustoi	MN 4111-V	729	71	9.7
Aegyptosaurus baharijensis	BSP 1912 VIII 61	1,000	83	8.3
Alamosaurus sanjuanensis	USNM 15560	1,360	90	6.6
Opisthocoelicaudia skarzynskii	ZPAL MgD-I/48	1,000	58	5.8
Saltasaurus loricatus	PVL 4017-67	611	66	10.8
Isisaurus colberti	ISIR 335/59	1,480	148	10.0

NOTE: Measurements are in millimeters. Abbreviations: AHD, anterior height of distal articulation; HL, humeral length. % = (AHD/HL) × 100.

for in ancestral bipedal saurischians, its conversion into an entirely locomotor structure was complete in titanosaurians. These barrel-chested taxa also exhibit a modified shoulder joint (Powell 1986, 2003; Wilson and Sereno 1998) that, in conjunction with changes at the elbow, suggests increased forelimb mobility and flexibility compared to other sauropods.

CONCLUSIONS

The basic sauropod appendicular morphology, including its functional relationships to locomotion, is fundamentally tied to the constraints imposed by large size on a vertebrate in a terrestrial environment. Many sauropod appendicular synapomorphies are also found in other large dinosaurs and in large terrestrial mammals, and their appearance is likely size related. Body size within sauropods increases steadily throughout the clade, with subsequent size decreases in both diplodocoids (dicraeosaurids) and macronarians (saltasaurines). Trend analyses suggest that this conforms to a "passive" rather than an "active" trend (*sensu* McShea 1994).

Features associated with the acquisition of obligate quadrupedalism—including a relatively long forelimb and supportive manus—are already present in the primitive sauropod *Vulcanodon*, but the transition is completed in successively more derived basal sauropods. Few scaling changes are inferred to have accompanied this postural shift aside from lengthening of the forelimb.

Other aspects of sauropod limb morphology are unique to the clade, including the development of a columnar, tightly arched manus with a reduced phalangeal count. The subunguligrade pes is entaxonic, indicating a shift in primary weight support toward the medial aspect of the pes. The femoral lesser trochanter is fully lateral and reduced to a rugose bump, representing the least elaborate such structure within Dinosauria. The iliac preacetabulum (and associated muscle origins) are flared laterally, whereas the postacetabulum (and associated muscle origins) are depressed ventrally.

These basic aspects of sauropod appendicular design are never radically altered, but sauropod locomotor evolution nevertheless includes an unappreciated morphological diversity. Diplodocoids, basal macronarians, and titanosaurians each show locomotor specializations that distinguish them from one another. Although many of these differences are subtle relative to those that distinguish sauropods from other dinosaurs, they nonetheless suggest that all sauropods did not employ identical locomotor styles or habits. Future studies should focus on these smaller-scale morphological differences to fully interpret and appreciate the true locomotor diversity within Sauropoda.

ACKNOWLEDGMENTS

I would like to thank J. A. Wilson, K. A. Curry Rogers, and D. Chure for inviting me to contribute to this volume honoring Jack McIntosh. I would also like to thank J. A. Wilson and K. A. Curry Rogers for their many comments and advice on preparing this chapter, particularly regarding many of the more esoteric (and unpublished) aspects of sauropod phylogeny. The manuscript was markedly improved thanks to additional comments from J. A. Wilson and G. R. Erickson. The translations of Lapparent (1943) and Lapparent and Zbyszewski (1957) are available at The Polyglot Paleontologist Web site (http://ravenel.si.edu/paleo/paleoglot).

LITERATURE CITED

Alexander, R. M., Jayes, A. S., Maloiy, G. M. O., and Wathuta, E. M. 1979. Allometry of limb bones of mammals from shrews (*Sorex*) to elephant (*Loxodonta*). J. Zool. London 189: 305–314.

Alroy, J. 2000a. Understanding the dynamics of trends within evolving lineages. Paleobiology 26(3): 319–329.

———. 2000b. New methods for quantifying macroevolutionary patterns and processes. Paleobiology 26(4): 707–733.

Anderson, J. F., Hall-Martin, A., and Russell, D. A. 1985. Long-bone circumference and weight in mammals, birds and dinosaurs. J. Zool. London (A) 207: 53–61.

Baird, D. 1980. A prosauropod dinosaur trackway from the Navajo Sandstone (Lower Jurassic) of Arizona. In: Jacobs, L. L. (ed.). Aspects of Vertebrate History: Essays in Honor of Edwin Harris Colbert. Museum of Northern Arizona Press, Flagstaff. Pp. 219–230.

Bakker, R. T. 1971. Ecology of the brontosaurs. Nature 229: 172–174.

Biewener, A. A. 1990. Biomechanics of mammalian terrestrial locomotion. Science 250: 1097–1103.

Borsuk-Bialynicka, M. 1977. A new camarasaurid sauropod *Opisthocoelicaudia skarzynskii* gen. n., sp. n. from the Upper Cretaceous of Mongolia. Palaeontol. Polonica 37: 5–64.

Carrano, M. T. 1998. Locomotion in non-avian dinosaurs: integrating data from hindlimb kinematics, in vivo strains, and bone morphology. Paleobiology 24(4): 450–469.

———. 1999. What, if anything, is a cursor? Categories versus continua for determining locomotor habit in dinosaurs and mammals. J. Zool. 247: 29–42.

———. 2000. Homoplasy and the evolution of dinosaur locomotion. Paleobiology 26: 489–512.

———. 2001. Implications of limb bone scaling, curvature and eccentricity in mammals and non-avian dinosaurs. J. Zool. 254: 41–55.

———. 2005. Body-size evolution in the Dinosauria. In: Carrano, M. T., Gauden, T. J., Blob, R. W. and Wible, J. R. (eds.). Amnioti Paleobiology: Perspectives on the Evolution of Mammals, Birds and Reptiles. University of Chicago Press, Chicago.

Carrano, M. T., and Hutchinson, J. R. 2002. The pelvic and hind limb musculature of *Tyrannosaurus rex*. J. Morphol. 253(3): 201–228.

Christiansen, P. 1997. Locomotion in sauropod dinosaurs. GAIA 14: 45–75.

Colbert, E. H. 1962. The weights of dinosaurs. Am. Mus. Novitates 2076: 1–16.

Coombs, W. P., Jr. 1975. Sauropod habits and habitats. Palaeogeogr. Palaeoclimatol. Palaeoecol. 17: 1–33.

Cooper, M. R. 1981. The prosauropod dinosaur *Massospondylus carinatus* Owen from Zimbabwe: its biology, mode of life and phylogenetic significance. Occas. Papers Natl. Mus. Monum. Rhodesia Ser. B Nat. Sci. 6(10): 689–840.

———. 1984. A reassessment of *Vulcanodon karibaensis* Raath (Dinosauria: Saurischia) and the origin of the Sauropoda. Palaeontol. Afr. 25: 203–231.

Cruickshank, A. R. I. 1975. The origin of sauropod dinosaurs. So. Afr. J. Sci. 71: 89–90.

Curry Rogers, K. A., and Forster, C. A. 2001. The last of the dinosaur titans: a new sauropod from Madagascar. Nature 412: 530–534.

Dodson, P. 1990. Sauropod paleoecology. In: Weishampel, D. B., Dodson, P., and Osmólska, H. (eds.). The Dinosauria. University of California Press, Berkeley. Pp. 402–407.

Gallup, M. R. 1989. Functional morphology of the hindfoot of the Texas sauropod *Pleurocoelus* sp. indet. In: Farlow, J. O. (ed.). Paleobiology of the Dinosaurs. Geological Society of America Special Paper 238, Boulder, CO. Pp. 71–74.

Galton, P. M. 1990. Basal Sauropodomorpha—Prosauropoda. In: Weishampel, D. B., Dodson, P., and Osmólska, H. (eds.). The Dinosauria. University of California Press, Berkeley. Pp. 320–344.

Gilmore, C. W. 1936. Osteology of *Apatosaurus*, with special reference to specimens in the Carnegie Museum. Mem. Carnegie Mus. 11(4): 175–278.

———. 1946. Reptilian fauna of the North Horn Formation of central Utah. U. S. Dept. Interior Geol. Surv. Prof. Paper 210-C: 29–53.

Hatcher, J. B. 1901. *Diplodocus* (Marsh): its osteology, taxonomy, and probable habits, with a restoration of the skeleton. Mem. Carnegie Mus. 1(1): 1–63.

———. 1903. Osteology of *Haplocanthosaurus*, with description of a new genus, and remarks on the probable habits of the Sauropoda and the age and origin of the *Atlantosaurus* beds. Mem. Carnegie Mus. 2(1): 1–72.

Hay, O. P. 1908. On the habits and pose of the sauropodous dinosaurs, especially of *Diplodocus*. Am. Nat. 42: 672–681.

Holland, W. J. 1910. A review of some recent criticisms of the restorations of sauropod dinosaurs existing in the museums of the United States, with special reference to that of *Diplodocus carnegiei* in the Carnegie Museum. Am. Nat. 44: 259–283.

Hutchinson, J. R. 2001. The evolution of femoral osteology and soft tissues on the line to extant birds (Neornithes). Zool. J. Linn. Soc. 131: 169–197.

Janensch, W. 1922. Das Handskelett von *Gigantosaurus robustus* und *Brachiosaurus brancai* aus den Tendaguru-Schichten Deutsch-Ostafrikas. Centralblatt Mineral. Geol. Paläontol. 1922: 464–480.

———. 1961. Die Gliedmaszen und Gliedmaszengürtel der Sauropoden der Tendaguru-Schichten. Palaeontogr. Suppl. VII(1) (teil 3, leif 4): 177–235.

Jianu, C.-M. and Weishampel, D. B. 1999. The smallest of the largest: a new look at possible dwarfing in sauropod dinosaurs. Geol. Mijnbouw 78: 335–343.

Lapparent, A. F. de. 1943. Les dinosauriens jurassiques de Damparis (Jura). Mem. Soc. Géol. France Nouvelle Ser. 47: 1–21.

Lapparent, A. F. de, and Zybszewski, G. 1957. Les dinosauriens du Portugal. Serv. Geol. Portugal Nova Ser Mem. 2: 1–63.

Lucas, S. G., and Heckert, A. B. 2000. Jurassic dinosaurs in New Mexico. In: Lucas, S. G., and Heckert, A. B. (eds.). Dinosaurs of New Mexico. New Mexico Museum of Natural History and Science Bulletin, Albuquerque. Pp. 43–45.

Lull, R. S. 1953. Triassic life of the Connecticut Valley (revised). State Conn. State Geol. Nat. Hist. Surv. Bull. 81: 1–336.

McIntosh, J. S. 1990. Sauropoda. In: Weishampel, D. B., Dodson, P., and Osmólska, H. (eds.). The Dinosauria. University of California Press, Berkeley. Pp. 345–401.

McShea, D. W. 1994. Mechanisms of large-scale evolutionary trends. Evolution 48(6): 1747–1763.

Osborn, H. F. 1898. Additional characters of the great herbivorous dinosaur *Camarasaurus*. Bull. Am. Mus. Nat. Hist. 10(12): 219–233.

———. 1900. The angulation of the limbs of Proboscidia, Dinocerata, and other quadrupeds, in adaptation to weight. Am. Nat. 34: 89–94.

———. 1904. Manus, sacrum, and caudals of Sauropoda. Bull. Am. Mus. Nat. Hist. 20(14): 181–190.

———. 1906. The skeleton of *Brontosaurus* and skull of *Morosaurus*. Nature 1890(73): 282–284.

Osborn, H. F., and Mook, C. C. 1921. *Camarasaurus, Amphicoelias*, and other sauropods of Cope. Mem. Am. Mus. Nat. Hist. New Ser. 3: 247–387.

Paul, G. S. 1988. The brachiosaur giants of the Morrison and Tendaguru with a description of a new subgenus, *Giraffatitan*, and a comparison of the world's largest dinosaurs. Hunteria 2(3): 1–14.

Peczkis, J. 1994. Implications of body-mass estimates for dinosaurs. J. Vertebr. Paleontol. 14(4): 520–533.

Powell, J. E. 1986. Revision de los titanosauridos de America del Sur. Ph.D. thesis. Universidad Nacional de Tucumán, Facultad de Ciencias Naturales.

———. 1990. *Epachthosaurus sciuttoi* (gen. et sp. nov.), un dinosaurio saurópodo del Cretácico de Patagonia (Provincia de Chubut, Argentina). Actas del V Congreso Argentino de Paleontologia y Bioestratigrafia, Tucumán, Argentina. Pp. 123–128.

———. 2003. Revision of South American titanosaurid dinosaurs: palaeobiological, palaeobiogeographical and phylogenetic aspects. Rec. Queen Victoria Mus. 111: 1–173.

Raath, M. A. 1972. Fossil vertebrate studies in Rhodesia: a new dinosaur (Reptilia: Saurischia) from near the Tria-Jurassic boundary. Arnoldia 5(30): 1–37.

Riggs, E. S. 1903. Structure and relationships of opisthocoelian dinosaurs. Part I: *Apatosaurus* Marsh. Field Columbian Mus. Geol. Ser. 2(4): 165–196.

Salgado, L., Coria, R. A., and Calvo, J. O. 1997. Evolution of titanosaurid sauropods. I: Phylogenetic analysis based on the postcranial evidence. Ameghiniana 34(1): 3–32.

Scott, K. M. 1985. Allometric trends and locomotor adaptations in the Bovidae. Bull. Am. Mus. Nat. Hist. 179(2): 197–288.

Stevens, K. A., and Parrish, J. M. 1999. Neck posture and feeding habits of two Jurassic sauropods. Science 284: 798–800.

Tornier, G. 1909. Wie war der *Diplodocus carnegii* wirklich gebaut? Sitz. Berliner Gesselschaft Naturf. Fr. Berlin 1909: 193–209.

Upchurch, P. 1995. The evolutionary history of sauropod dinosaurs. Philos. Trans. Roy. Soc. London B 349: 365–390.

———. 1998. The evolutionary history of sauropod dinosaurs. Philos. Trans. Roy. Soc. London B 349: 365–390.

Wilson, J. A. 2002. Sauropod dinosaur phylogeny: critique and cladistic analysis. Zool. J. Linn. Soc. 136: 217–276.

Wilson, J. A., and Carrano, M. T. 1999. Titanosaurs and the origin of "wide-gauge" trackways: a biomechanical and systematic perspective on sauropod locomotion. Paleobiol. 25: 252–267.

Wilson, J. A., and Sereno, P. C. 1998. Early evolution and higher-level phylogeny of sauropod dinosaurs. J. Vertebr. Paleontol. Mem. 5, 18(2; Suppl.): 1–68.

Wiman, C. 1929. Die Kreide-Dinosaurier aus Shantung. Palaeontol. Sinica Ser. C 6(1): 1–67.

Yates, A. M., and Kitching, J. W. 2003. The earliest known sauropod dinosaur and the first steps towards sauropod locomotion. Proc. Roy. Soc. London B270(1525): 1753–1758.

Zhang, Y. 1988. The Middle Jurassic Dinosaur fauna from Dashanpu, Zigong, Sichuan. Vol. I. Sauropod Dinosaurs (1). *Shunosaurus*. Sichuan Publishing House of Science and Technology, Chengdu. Pp. 1–89.

APPENDIX 8.1. TAXA AND MEASUREMENTS USED IN THIS CHAPTER

Taxa are listed in a general phylogenetic hierarchy, but see figure. 8.3 for details of their hypothesized interrelationships. All measurements are in millimeters and were log-transformed prior to analysis. Abbreviations as in table 8.1. Institutional abbreviations: AMNH, American Museum of Natural History, New York, New York; BMNH, The Natural History Museum, London, UK; BSP, Bayerische Staatssammlung für Paläontologie, München, Germany; CH/P.W., Department of Mineral Resources, Bangkok, Thailand; CM, Carnegie Museum of Natural History, Pittsburgh, Pennsylvania; CMNH, Cleveland Museum of Natural History, Cleveland, Ohio; DINO, DINOLab, Fruita, Colorado; FMNH, Field Museum of Natural History, Chicago, Illinois; GMNH, Gunma Museum of Natural History, Gunma, Japan; HMN, Humboldt Museum für Naturkunde, Berlin, Germany; ISI, Indian Statistical Institute, Kolkata, India; IVPP, Institute of Palaeoanthropology and Palaeontology, Beijing, China; MACN, Museo Argentino de Ciencias Naturales "Bernardino Rivadavia," Buenos Aires, Argentina; MCNA, Museo de Ciencias Naturales de Alava, Vitoria-Gasteiz, Spain; MDE, Musée des Dinosaures, Esperaza, France; MIGM, Museu Geológico do Instituto Geológico e Mineiro, Lisboa, Portugal; MLP, Museo de La Plata, La Plata; Argentina; MN, Museu Nacional, Rio de Janeiro, Brazil; MNHN, Muséum National d'Histoire Naturelle, Paris, France; MPCA, Museo Provincial "Carlos Ameghino," Cipoletti, Argentina; MPEF, Museo Paleontológico "Egidio Feruglio," Trelew, Argentina; MUCP, Museo de Ciencias Naturales de la Universidad Nacional del Comahue, Neuquén, Argentina; OUM, Oxford University Museum, Oxford, UK; PMU, Palaeontological Museum, University of Uppsala, Uppsala, Sweden; PVL, Instituto Miguel Lillo, Tucumán, Argentina; QG, Queen Victoria Museum, Harare, Zimbabwe; State University of New York (SUNY), Stony Brook, New York; YPM, Peabody Museum of Natural History, Yale University, New Haven, Connecticut; ZDM, Zigong Dinosaur Museum, Zigong, China; ZPAL, Muzeum Ziemi Polska Akademia Nauk, Warsaw, poland.

	SPECIMEN	FL	FAP	FML
Sauropoda				
Gongxianosaurus shibeiensis	Holotype	1,164.0		270.0
Isanosaurus attavipachi	CH4-1	760.0	71.5	121.6
Vulcanodon karibaensis	QG 24	1,100.0	140.0	174.0
Eusauropoda				
Barapasaurus tagorei	Holotype	1,365.0	131.0	187.0
Cetiosaurus oxoniensis	OUM	1,626.0		305.0
Datousaurus bashanensis	IVPP V.7262	1,057.0	147.0	
Kotasaurus yamanpalliensis	111/S1Y/76	1,130.0	80.0	160.0
Lapparentosaurus madagascariensis	"Individu taille max"	1,590.0	240.0	
Mamenchisaurus constructus	IVPP V.948	1,280.0	207.0	
Mamenchisaurus hochuanensis	IVPP holotype	860.0		
Omeisaurus junghsiensis	IVPP holotype		103.0	
Omeisaurus tianfuensis	ZDM T5701	1,310.0		206.0
Patagosaurus fariasi	PVL 4076	1,542.0	135.5	
	PVL 4170			255.0
Shunosaurus lii	IVPP V.9065	1,250.0		188.0
Tehuelchesaurus benitezii	MPEF-PV 1125	1,530.0		243.0
Volkheimeria chubutensis	PVL 4077	1,156.0	148.0	75.1
Neosauropoda				
Diplodocoidea				
Amargasaurus cazaui	MACN-N 15	1,050.0	128.8	180.0
Amphicoelias altus	AMNH 5764	1,770.0	210.0	216.0
Apatosaurus ajax	YPM 1860	2,500.0		
Apatosaurus excelsus	FMNH 7163	1,830.0	310.0	310.0
Apatosaurus louisae	CM 3018	1,785.0	174.0	332.3
Barosaurus africanus	HMN NW 4	1,361.0	150.6	204.2

APPENDIX 8.1. (continued)

	SPECIMEN	FL	FAP	FML
Barosaurus lentus	AMNH 6341	1,440.0	120.2	204.3
Cetiosauriscus stewarti	BMNH R.3078	1,360.0	190.0	195.0
Dicraeosaurus hansemanni	HMN m	1,220.0	142.5	
	HMN dd 3032			192.3
Diplodocus carnegii	CM 84	1,542.0	174.0	
	CM 94			186.0
Diplodocus longus	YPM 1920	1,645.0		
	AMNH 223		143.0	
Haplocanthosaurus priscus	CMNH 10380	1,745.0		
	CM 572		207.0	
Rayososaurus tessonei	MUCPv-205	1,440.0		220.0
Macronaria				
"Bothriospondylus madagascariensis"	MNHN uncat.	1,460.0	110.0	
Brachiosaurus altithorax	FMNH P25107	2,000.0		365.0
Brachiosaurus brancai	HMN St	1,913.0	151.7	299.0
Camarasaurus lentus	DINO 4514	1,470.0		252.0
	CM 11338		86.5	
Camarasaurus supremus	AMNH 5761a	1,800.0	255.0	
	GMNH-PV 101			228.0
Euhelopus zdanskyii	PMU R234	955.0	100.0	142.0
Titanosauriformes				
Aegyptosaurus baharijensis	BSP 1912 VIII 61	1,290.0	75.0	223.0
Ampelosaurus atacis	MDE uncat. 1	802.0		157.5
	MDE uncat. 2		66.0	
Andesaurus delgadoi	MUCPv-132	1,550.0		226.0
"Antarctosaurus wichmannianus"	FMNH 3019	1,770.0		
	MACN 6904		77.5	217.0
Argyrosaurus superbus	PVL 4628	1,910.0	160.0	300.0
Chubutisaurus insignis	MACN 18222	1,715.0	265.0	
Janenschia robusta	HMN IX	1,330.0		
	HMN P		131.5	188.8
Laplatasaurus araukanicus	MLP-Av 1047/1128	1,000.0		
Lirainosaurus astibiae	MCNA 7468	686.0		97.0
Magyarosaurus dacus	BMNH R.3856	488.0	43.7	66.8
Neuquensaurus australis	MLP CS 1121/1103	700.0	110.0	
Neuquensaurus robustus	MLP CS 1094	799.0		134.5
	MLP CS 1480		120.0	
Opisthocoelicaudia skarzynskii	ZPAL MgD-I/48	1,395.0	108.0	280.0
Phuwiangosaurus sirindhornae	CH/P.W. 1-1/1-21	1,250.0	85.0	215.0
Rapetosaurus krausei	SUNY uncat.	687.0	91.2	63.1
Rocasaurus muniozi	MPCA-Pv 56	768.0		117.0
Saltasaurus loricatus	PVL 4017-80	875.0		164.6
	PVL 4017-79		90.0	
Titanosaurus indicus	BMNH R.5934	865.0	67.3	80.9

NINE

Steps in Understanding Sauropod Biology

THE IMPORTANCE OF SAUROPOD TRACKS

Joanna L. Wright

SAUROPOD TRACKS ARE SPECTACULAR and have captured public imagination since they were first recognized and described in the first half of the twentieth century. Roland T. Bird publicized the bathtub-sized tracks in the Glen Rose Formation (Texas) at sites such as Paluxy River and Davenport Ranch in *Natural History* (Bird 1939, 1944) soon after he had first described them, and dinosaur tracks were as popular then as they have ever been. Bird's research on sauropod tracks led him to a number of conclusions that were radical for the time. He noticed that these trackways seldom showed tail drag marks (Bird 1944), and despite some rather sprightly sauropod reconstructions around the turn of the twentieth century, sauropods had always been depicted dragging their long tails on the ground. He came to the conclusion that sauropods, then thought to be semiaquatic, like long-necked hippopotamuses, had left these tracks when they were walking half-submerged in water and floating their tails behind them. While not the dynamic pose seen in modern reconstructions (Wilson & Sereno 1998), this could be construed as a step in the right direction.

Little progress was made in documenting sauropod tracks for many years after this, until the late 1980s, when both the Glen Rose tracksites (Farlow et al. 1989) and the Purgatoire site (MacClary 1938; Lockley et al. 1986; fig. 9.1) were redescribed. These studies are indicative of an increased interest in dinosaur tracks in general, including sauropod tracks, which has continued until the present day.

This study is the first comprehensive review of sauropod tracks since 1994. It differs from previous reviews (Farlow 1992; Lockley et al. 1994b) not only with the addition of tracksite information published since 1994, but also with the construction of predicted sauropod track morphologies for comparison with fossilized sauropod tracks. In addition, synapomorphies from recent phylogenetic studies of sauropods (Salgado et al. 1997; Upchurch 1998; Wilson and Sereno 1998) are used to determine sauropod track characteristics and to predict differences between the trackways of different sauropod clades (see also Carrano and Wilson 2001). Data from published sauropod tracksites are used herein to investigate patterns in global and temporal distribution, lithological and paleoenvironmental associations

FIGURE 9.1. The Purgatoire tracksite in southeastern Colorado. (Photo courtesy of USDA Forest Service.)

and to test hypotheses about the paleoenvironmental and paleolatitudinal preferences of sauropods (Lockley 1991; Farlow 1992; Lockley et al. 1994b). Recently reported Late Triassic sauropod tracks conform to sauropod track synapomorphies.

SAUROPOD FOOT MORPHOLOGY AND PREDICTIVE ICHNOLOGY

As with many dinosaurs, pedal elements of sauropods are rare and complete feet and hands even rarer. However, patterns do emerge from what is known about sauropod feet and certain generalizations can be made. Despite the commonly held view that sauropod feet are essentially conservative in morphology (Farlow 1992), osteological features of the manus and pes have been used in recent phylogenies (Upchurch 1998; Wilson and Sereno 1998).

Fifteen of Upchurch's 204 characters (7%) might be reflected in sauropod trackways, and 24 of Wilson and Sereno's 109 characters (22%): 9 of these characters are used by both studies. In addition, Salgado et al. (1997) used three additional characters in their phylogeny of the titanosaurs. In total 33 osteological features might directly affect trackway morphology. However, some of these characters combine to produce the same results in trackways so that 33 separate trackway changes will not be apparent (see appendix 9.1).

A simplified sauropod cladogram with special reference to these features is shown in figure 9.2. Omission of clades and taxa not distinguished by any pedal characters removes most of the differences between Upchurch's (1998) and Wilson and Sereno's (1998) cladograms, yielding a "podial consensus cladogram" showing the clades that can be recognized

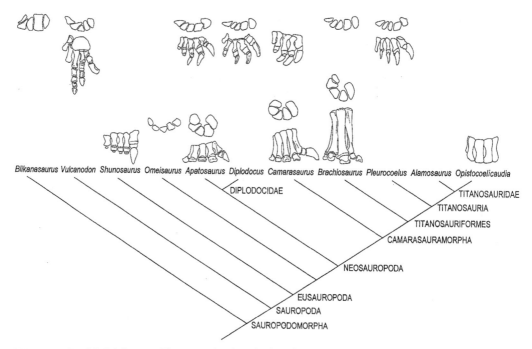

FIGURE 9.2. Simplified cladogram of the sauropods (after Salgado et al. 1997; Upchurch 1998; Wilson and Sereno 1998), showing only taxa that can be discriminated partially based on podial characters of representative sauropods, with pes (above) and manus (below) in anterior and proximal views: *Vulcanodon* (after Raath 1972); *Shunosaurus* (after Zhang et al. 1984); *Omeisaurus* (after Wilson and Sereno 1998); *Apatosaurus* (after Gilmore 1936); *Camarasaurus* (after Farlow 1992); *Brachiosaurus* (after Wilson and Sereno 1998); *Diplodocus* (after Farlow, 1992); *Brachiosaurus* (after Janensch 1961); *Opistocoelicaudia* (after Wilson and Sereno 1998). Figures not to scale but drawn to similar foot widths for ease of morphological comparisons.

based on synapomorphies. These are numbered consecutively and are referred to as (#n) in the following sections.

Traditionally, fossil vertebrate tracks are assigned to producers by attempting to fit known manus and pes skeletons to tracks. Such an approach has allowed identification of several different types of tracks, including those of sauropods. However, it is limited by the desire to fit skeletons into their purported tracks. The predictive approach (Unwin 1989) uses the structure of the pedal skeleton of potential trackmakers to construct hypothetical tracks that can then be compared to fossil tracks. This method constrains the morphological range of tracks made by the animal in question and has the potential to exclude certain trackmaker candidates from consideration. However, if none of the tracks fit, then there are two possible conclusions: (a) none of the animals considered is a possible trackmaker, which might mean that the producer of that track morphotype has not yet been discovered; or (b) the skeletal reconstructions are incorrect. Sauropods had massive bones and their fossilized remains are usually relatively uncrushed so it is possible to reconstruct the feet with a reasonable degree of accuracy. It should be possible to predict the types of tracks that different clades of sauropods would produce using the characteristics listed above. Hypothetical sauropod tracks have been constructed using the method outlined by Unwin (1989; fig. 9.3). The predicted tracks are "ideal"; they show the maximum amount of detail possible. Many sauropod tracks are found in sediments that did not preserve the features of the trackmaker's foot with such fidelity and so few tracks will show all the features of the predicted tracks.

The cladistic approach (Olsen 1995; Carrano and Wilson 2001) is similar to the predictive approach (Unwin 1989), but rather than constructing "ideal" tracks based on the osteological

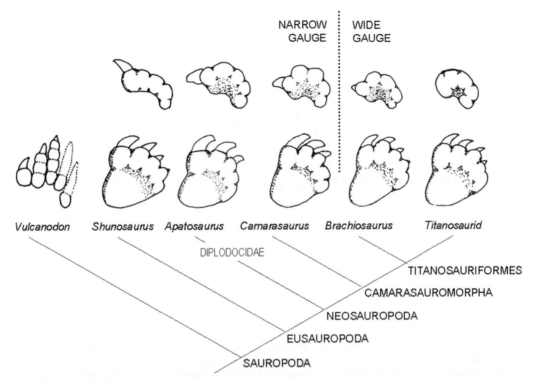

FIGURE 9.3 Predicted tracks of sauropods in phylogenetic context based on skeletal morphology. The exact relative positions of manus–pes pairs cannot be predicted. Not to scale. Drawn to same pes print length for ease of morphological comparison.

structure of known sauropod pedes, it relies exclusively on the recognition of synapomorphies revealed in tracks (e.g. Carrano and Wilson 2001). The approach outlined here is a combination of the two methods (predicted tracks based on synapomorphies).

Basal sauropod (e.g., *Vulcanodon*) tracks should have been made by an animal progressing quadrupedally (podial characteristic #1), with five weight-bearing digits on the manus and pes (#2, 3, 6). These are characteristic of all sauropod tracks. The pes was digitigrade and the long axis would probably have been approximately parallel to the trackway midline. The first pedal digit impression should show a large sickle-shaped claw inclined at an oblique plane to the ground (#4, 5), with the axis of the claw directed anterolaterally (fig. 9.3). Digit I is shorter than digits II and III and the lengths of digits IV and V are unknown. The manual digits of *Vulcanodon* are unknown. The reconstruction looks superficially very like some purported prosauropod tracks, except that five digits contacted the ground and digit I is very short, with a laterally rather than a medially directed claw impression. No tracks like these are known.

Eusauropod tracks would show five short digits subequal in length on the manus, with no discrete phalangeal pads (#7,8). Manual digit I bears a large claw. The arrangement of the manus digits is disputed; it is uncertain whether they should be arranged in a crescent (Wilson and Sereno 1998) or a semicircle (Upchurch 1998). Pes impressions should show five short digit impressions (#12–14), and no separate phalangeal pads (#13). The first three digits should show narrow claw impressions (#15) decreasing in size from digit I (#16). In most eusauropods these claws would lie at an angle to the substrate and be directed anterolaterally (#18). The claw impression of digit IV, if present, would be small (#17). The pes prints would toe outward from the trackway midline (#11,19) and bear a large fleshy heel pad (#9,10) making them elongate (fig. 9.3).

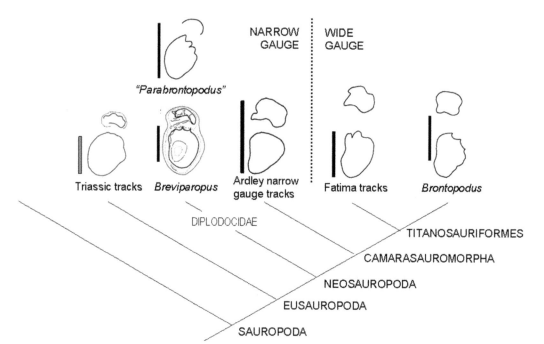

FIGURE 9.4. Examples of sauropod manus and pes fossilized footprints in phylogenetic context. Scale bars equal 1 m except for the Triassic tracks, where the scale bar equals 20 cm. Track illustrations (left to right): Triassic tracks from Peacock Canyon, New Mexico (after Lockley and Hunt 1995); *Parabrontopodus* (after Lockley et al. 1986, 1994a); *Breviparopus* (after Ishigaki 1989); narrow gauge trackways from Ardley Quarry, UK (after Day et al. 2002); trackways from Fatima, Portugal (after Santos et al. 1994); *Brontopodus* (after Farlow et al. 1989).

Neosauropod manus impressions would be semicircular (#20, 21), but see the discussion below. Both manus and pes tracks would have even shorter digits (#22, 23). Neosauropod pes impressions are likely to be very similar to one another—the main differences being in the relative sizes of the pedal claws and how many there are: two, three, or four (fig. 9.4).

Diplodocid pes impressions should only show three claw impressions; pes impressions of *Barosaurus* would only show two (fig. 9.2).

The shape of camarasauromorph manus impressions would probably be the same as those of neosauropods (i.e., semicircular) but they might be slightly deeper relative to the pes impressions (#25) and the anteromedial part of the manus impression might be deeper than the posterolateral portion (#26).

The lateral deflection of the femoral head in titanosauriforms (#29) may indicate that they would produce a wider trackway than other sauropods (Wilson and Carrano 1999). This is probably the greatest change in the track morphology for this clade (fig. 9.3), but in addition, the manus impressions should show only a small claw on digit I (#27, 28).

Titanosaurids should produce even wider trackways than titanosauriforms (#33), the anteromedial part of the manus impression might be even deeper (#30), and no digit or claw impressions should be visible in manus tracks (#31,32), although pads on the distal ends of the metacarpals might leave impressions (fig. 9.3).

Comparison of predicted sauropod tracks with the sauropod track record (fig. 9.5) yields some correspondence, indicating that (a) these tracks were made by sauropods; (b) no basal sauropod (e.g., *Vulcanodon*) tracks have been discovered; and (c) tracks predicted to be made by titanosaurs can be found from at least the middle Jurassic onwards (Day et al. 2002).

The predicted tracks show substantial digit I manus claw impressions for all sauropods except titanosaurs, as well as claw impressions

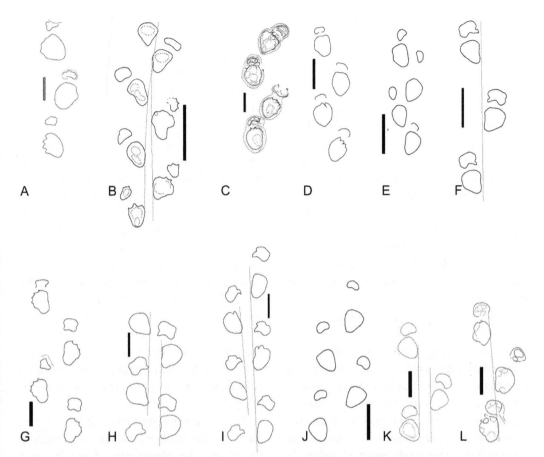

FIGURE 9.5. Representative sauropod trackways. (A) Peacock Canyon, New Mexico (after Lockley and Hunt 1995). (B) Trackway from the Lias of Italy (after Avanzini and Leonardi 1994) changes from wide to narrow gauge along the trackway. (C) *Breviparopus taghbloutensis* (after Ishigaki 1989). (D) and (E), *Parabrontopodus mcintoshi* (after Lockley et al. 1986, 1994a). (F) Large manus narrow gauge trackways from Ardley Quarry, UK; note the large manus claw impression (after Day et al. 2002). (G) *Brontopodus birdi* (after Farlow et al. 1989). (H) and (I) Trackways from Fatima, Portugal; note the very large manus and claw impression (after Santos et al. 1994). (J) *Rotundichnus muenchehagensis* (after Lockley et al. 2004). (K) *Sauropodichnus giganteus* (after Calvo 1999). (L) Large manus narrow gauge trackways from China; the wide gauge trackways this smaller trackway was associated with showed the same morphology, and the narrowness of this trackway may reflect the youth of the trackmaker (after Lockley et al. 2002b). Scale bars equal 1 m except for (A) and (L) where scale bars equal 20 cm. All tracks drawn to the same pes length for ease of morphological and trackway pattern comparison. Note the differences in internal trackway width (gauge), stride lengths, and manus position and orientation.

on all pes prints. However, few sauropod trackways preserve manus claw impressions, and many pes impressions show no signs of claw marks. Thus, the absence of manus claw impressions cannot by itself indicate a titanosaur origin, but is more likely to be a preservational artifact. Manus impressions with claw marks indicate a nontitanosaur origin. The first metacarpal of many sauropods is shorter than the rest and thus the digital pad of digit I might be smaller than shown and the claw might only impress at its distal end. Such claw impressions would be more susceptible to sediment collapse and correspondingly less likely to be preserved. Pittman and Gillette (1989) suggested that sauropod pedal "claws were 'wrapped' along the lateral edge of the foot by sauropods walking across firm or moist substrates, or extended forward for braking, turning or traction in muddy areas" (Pittman and Gillette 1989:331), providing a functional reason for the absence of these claw marks in many sauropod tracks.

The ideal tracks would also seem to indicate that the great majority of tracks were made by titanosauriforms or basal eusauropods and that nontitanosauriform neosauropods left very few tracks. This is because, according to the characters presented above, all neosauropods, possibly all eusauropods, had metacarpals arranged in a semicircular column, which would have produced horseshoe-shaped tracks. Yet many sauropod tracks show crescentic manus impressions so a question arises over the reconstruction of the sauropod manus. It seems most likely that some neosauropods (diplodocoids), while having vertically oriented columnar ligament-bound metacarpals, did not have them arranged in a tight 270° arc but in a wider 210° crescentic arc (Hand 1999). This is only a slightly tighter configuration than that of *Omeisaurus* (fig. 9.2). This would then allow neosauropods to be considered potential trackmakers of crescentic manus impressions (figs. 9.3, 9.4), which are very common in the Late Jurassic, when neosauropods are the most common sauropods known from osteological remains. Most sauropod tracks are very difficult to place in a phylogenetic context, either because they are too poorly preserved to show the relevant details or because their morphology does not fit predicted morphology.

The most notable predicted differences in sauropod tracks are: (a) basal sauropods to eusauropods, where the pes should change from digitigrade to semidigitigrade with a heel pad and be toed out from the trackway axis; (b) the transition from a crescentic to a semicircular manus, which may occur in camarasauromorphs; and (c) the change from narrow to wide gauge trackways, which should occur in titanosauriforms. Most other transitions would only be evident in very well-preserved trackways.

SAUROPOD TRACK MORPHOLOGY AND CLASSIFICATION

Manus tracks are crescentic to horseshoe-shaped, with digit III directed anterolaterally (figs. 9.4, 9.5). Manus claw impressions are seldom preserved, and even more rarely are details of claw shape or vertical orientation visible in trackways. Notable exceptions are trackways of *Brontopodus birdi* at the type locality (Bird 1944; Farlow et al. 1989) (fig. 9.5) and trackways at Ardley Quarry, UK (Day et al. 2002) (fig. 9.5), and Galinha, Portugal (Santos et al. 1994). The absence of manual claw impressions has incited speculation about the function and position of the sauropod manus claw and several hypotheses have been proposed (see Upchurch 1994). Impressions of other manual digits are even more uncommon but have been reported (Leonardi and Avanzini 1994; Farlow et al. 1989; Farlow 1992; Ishigaki 1988; Lockley et al. 2002b). Manus impressions are always preserved in front of the pes impressions but the exact position may vary. The manus may be closer or farther away from the trackway midline than the pes and it may be very close to the anterior margin of the pes or up to a pes length away (fig. 9.5). Manus impressions are angled outward from the axis of the trackway at angles of 5–75°.

Pes impressions are triangular to oval, with the axis of the print toed outward at an angle of 10–30° and the widest part of the footprint directed anterolaterally (figs. 9.4–9.6). The back of the footprint narrows to the posterior margin; it may be rounded or pointed (fig. 9.6). Pedal claw impressions are more often preserved than those of the manus but many trackways still do not preserve them, for instance, most trackways at the Purgatoire locality (Lockley et al. 1986) and pedal claw impressions are not preserved in many of the trackways at Ardley Quarry (Day et al. 2002). Where preserved, pedal claw impressions are directed laterally or anterolaterally and decrease in size from digit I. Pes tracks may show two to four claw impressions (figs. 9.4–9.6). Non-claw-bearing digits may only impress as a slight bulge on the lateral side of the footprint immediately anterior to the "heel" pad, although some pes tracks preserve separate impressions of rounded toes (Lockley and Hunt 1995).

Sauropod tracks range in pes length from 20 cm (Lim et al. 1994; Lockley et al. 2002b) to

FIGURE 9.6. Well-preserved sauropod pes tracks showing claw impressions from the Late Jurassic. (A) CU-MWC 194.2, Lost Springs site, Morrison Formation, Utah; (B) CU-MWC 188.28, Summerville Formation, Arizona.

more than 1 m. Even the smallest sauropod tracks show the characteristic subtriangular toed-out pes impressions with anterolaterally directed claw impressions. The relative sizes of manus and pes impressions are considered important in sauropod track classification but it is difficult to predict this from osteology because of the soft tissue "heel" pad.

SAUROPOD TRACKWAY GAUGE

Sauropod trackways are termed narrow gauge when the inside margins of the pes impressions overlap the trackway midline and wide gauge when they do not. When this division was proposed (Farlow 1992; Lockley et al. 1994a) an intermediate category, medium gauge, was illustrated (Meyer et al. 1994) but not described. The illustration seemed to be a difference of scale (the internal width of the trackway relative to pes print size was the same as in the wide gauge example) and this category has not been used subsequently in trackway descriptions; occasionally modifiers such as "moderately" and "slightly" are applied to wide gauge trackways (e.g., Lockley et al. 2001).

Despite the proposed link between wide gauge trackways and titanosauriforms (Wilson and Carrano 1999), trackway gauge does not seem to reflect the anatomy of the trackmaker; other characteristics such as locomotor style and ontogeny contribute, e.g. some change from one to the other along their length (Leonardi and Avanzini 1994; fig. 9.5B) and some sauropods may have made a narrow gauge trackway as juveniles and a wide gauge one when older (Lockley et al. 2001, 2002a; fig.9.5L). The pes impressions of some other trackways are very close to the trackway midline so that they do not qualify as narrow gauge in a strict sense, yet to classify them as wide gauge seems excessive (figs. 9.5, 9.6). The distinction is thus not as clear-cut as often implied but it may reveal useful information on the identity of the trackmaker. Wide gauge trackways are known from rocks as old as the Sinemurian and Pliensbachian (Ishigaki 1988; Farlow 1992; Leonardi and Avanzini 1994) (fig. 9.5), but the earliest sauropod trackways (Lockley et al. 2002b) are narrow gauge.

Brontopodus birdi is the classic ichnotaxon to which the description wide gauge was first applied (Farlow et al. 1989; Farlow 1992). The ichnogenus *Parabrontopodus* was erected to provide a similar standard for narrow gauge trackways (Lockley et al. 1994a). Sauropod trackways are most often referred to one of these two ichnogenera based on the internal width of the trackway. These two ichnogenera are also used to exemplify the other characteristics of these tracks, such as the relative size of the manus. Some Paluxy *Brontopodus* show narrow gauge locomotion along some trackway segments (Farlow, pers. comm., 2004) so gauge is partially behavioral in even derived sauropods. Thus sauropod tracks should be classified primarily on the basis of footprint morphology and only secondarily on the internal width of the trackway.

The type trackway of *Brontopodus* has a manus-pes area ratio of 1:3; in *Parabrontopodus* or *Breviparopus* this ratio is 1:5 (Lockley et al. 1994a). Other wide gauge trackways have been reported with a ratio as high as 1:2 (Santos et al., 1994). While an association of a relatively large manus with a wide gauge is most common, there are exceptions: both large manus, narrow gauge (Day et al. 2002; Lockley et al. 2001) and small manus, wide gauge (Leonardi and Avanzini 1994; Day et al. 2002; fig. 9.5). In addition, a continuum of manus print sizes relative to the pes impression has been reported (from 1:2 to 1:6), which indicates that the division between large and small manus print trackways may not be as clear-cut as often portrayed. Insufficient numbers of manus and pes skeletons are known to be able to determine patterns of manus and pes relative sizes but even very closely related sauropods can show significant differences in these ratios; for instance, the manus of *Diplodocus* is relatively small in comparison to that of the pes, whereas that of *Apatosaurus* is relatively large (pers. obs.).

Wide gauge trackways have been attributed to titanosauriforms on the basis of their deflected femoral head (# 29 above), which may have given them a wider stance and hence a wider trackway (Wilson and Carrano 1999). Titanosauriforms are not known from osteological remains before the Late Jurassic. However, the first purported titanosaur tracks are reported in the Middle Jurassic (Day et al. 2002), prior to their appearance in the osteological record. It has been suggested that wide gauge trackways with large manus claw impressions from the Middle Jurassic of Portugal (Santos et al. 1994) were made by camarasauromorphs, although they do not have a deflected femoral head (Wilson and Sereno 1998).

It has been claimed that wide gauge trackways are the dominant type of sauropod trackway in the Cretaceous and that narrow gauge ones are more common before the Late Jurassic; the Late Jurassic being a transitional period (Lockley, Meyer, Hunt and Lucas 1994; Wilson and Carrano 1999). Lockley, Meyer, Hunt and Lucas (1994) plotted sauropod trackways, indicating numbers of narrow, wide, and unknown gauge, but on a logarithmic scale. This was in order to accommodate the large range in data (anywhere from one to 100 trackways plotted per stage), but it masks differences in numbers of known trackways per interval because stages with one trackway are shown as over a quarter of the length of one with 10 trackways, or an eighth of the length of one with a hundred. Wilson and Carrano (1999) plotted wide and narrow gauge sauropod trackways as percentages of total known sauropod trackways which allows easier comparison of relative proportions. However, a plot of tracksites, rather than trackways, by stage (fig. 9.7) illustrates the inadequacy of the database on which this conclusion is based. Most tracksites preserve either narrow or wide gauge trackways so tracksites have been plotted in order that large sites like Purgatoire or Paluxy, where many trackways are preserved on a single bedding plane, do not overwhelm data from sites with less areal exposure. Sites which preserve both wide and narrow gauge trackways are plotted as one of each. Figure 9.7 shows the distribution of tracksites

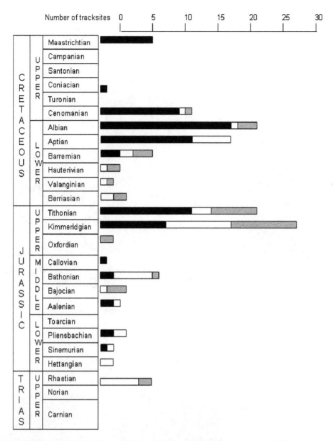

FIGURE 9.7 Stratigraphic distribution of sauropod tracksites by stage. (■) Wide gauge, (▨)narrow gauge, (□) unknown. Based on 64 sites reported in the literature up to 2003 plus the 102 sites from South Korea, here counted as 10 (Cenomanian). Tracksites from the Morrison Formation have been split evenly between the Kimmeridgian and the Tithonian. Data in Table 9.1.

through time. There are insufficient data to determine much about the distribution of narrow and wide gauge trackmakers through time. The Late Kimmeridgian has been suggested as the interval when wide gauge trackways became more numerous than narrow gauge trackways. However, the picture is not that simple. For instance, in Switzerland, seven early Kimmeridgian sites preserve wide gauge trackways and only in the Late Kimmeridgian have narrow gauge trackways been reported (Meyer and Thuring 2003). In most time periods (e.g. in the Maastrictian seven of the eight tracksites are in Bolivia) the data are concentrated in small geographic areas (see Table 9.1) and great caution must be exercised in drawing global conclusions from the small amount of data that exist.

Assuming a crescentic configuration for the metacarpals of diplodocoids (Hand 1999), a comparison with the sauropod osteological record (Wilson and Sereno 1998) might suggest that Late Jurassic narrow gauge trackways were made by diplodocoids, Late Cretaceous wide gauge trackways by titanosaurs, and Late Jurassic to Early Cretaceous wide gauge trackways by titanosauriforms.

ICHNOGENERA ATTRIBUTED TO SAUROPODS

About a dozen sauropod ichnogenera have been named, although only a few (three to five) are considered valid. Sauropod ichnogenera were reviewed by Farlow et al. (1989) and Lockley et al. (1994a); unless otherwise stated, their assessment of the validity of sauropod ichnotaxa is herein considered correct.

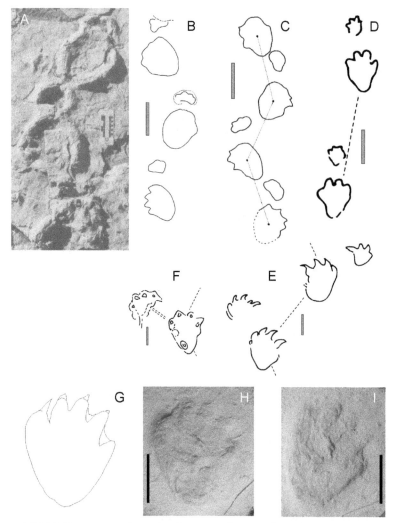

FIGURE 9.8. Triassic age tracks attributed to sauropods. (A) Trackway from Furnish Canyon, Colorado (Lockley et al. 2001). (B) Trackway from Peacock Canyon, New Mexico; (C) trackway from Cub Creek, Utah (after Lockley and Hunt 1995). (D) *Sauropodopus antiquus*; (E) *Tetrasauropus unguiferous*; (F) *Pseudotetrasauropus jaquesi* (Ellenberger 1972; after Lockley and Meyer 1999). (G)–(I) CU-MWC 172.22, pes impression showing five digits, four with claw impressions. Lighting from the right in (H) and the left in (I). Scale bars in (B)–(I) equal 20 cm. (B)–(F) drawn to same pes print length for ease of comparison. (G)–(I) drawn to same scale.

EUSAUROPODA

Tetrasauropus (Ellenberger 1972). Sauropod trackways from the Late Triassic were described and referred to *Tetrasauropus* (Lockley et al. 2001; fig. 9.8). The recognition of these trackways is consistent with the recent discovery of Triassic sauropods (Buffetaut et al. 2000, 2002). The Triassic sauropod trackways have been met with a certain amount of skepticism, but with the enumeration of sauropod podial diagnostic characters (appendix 9.1) and their extrapolation to diagnostic features of sauropod trackways, these claims can be revisited and evaluated.

These trackways have four anterolaterally directed claw impressions with a fifth non-claw-bearing digit, (fig. 9.8). Digit I is the leading digit in the pes but is shorter than the other digits, although the claw impression is larger. The deepest part of the track is the inside margin of the foot, which is oriented anteromedially and forms the leading edge. The long axis of the pes impression is angled outward at an angle of 20–30° to the trackway midline (fig. 9.8A–C). The "heel" pad is substantial and the posterolateral part of the track is also well impressed. Digits I and V are impressed most deeply, although all five digits are clear. Well-preserved manus impressions show five digits arranged in a gentle crescent; none shows any indications of claws. The manus impressions are preserved in front of and slightly inside the pes impressions and angled outward at a variable angle of 5–45° (fig. 9.8). Manus impressions are 15%–20% of the area of the pes. The trackways are narrow to slightly wide gauge (see Lockley et al. 2001:fig. 3). These trackways fit the sauropod track diagnostic criteria and are most likely to have been made by Late Triassic sauropods. Similar tracks have been described from other Late Triassic sites including Wales (Lockley et al. 1996), Greenland (Jenkins et al. 1994), Italy (Dalla Vecchia 1996), and Switzerland (Meyer and Thuring 2003). Confirmation of these occurrences would indicate a Laurasian distribution of sauropods in the Late Triassic. Some tracks from the Pliensbachian of Morocco (Ishigaki 1988; morphological type 2 of Farlow [1992]) appear very similar to these tracks and might also be congeneric.

The small size of these tracks indicates a hip height of 0.8–1.5 m. This is consistent with the fact that some of these Late Triassic sauropods were small (Buffetaut et al. 2000), although the tracks could also have been made by a juvenile of a larger Late Triassic sauropod (Buffetaut et al. 2002). However, the consistent small size of these tracks at several widely spaced localities might argue for small adult trackmakers.

Lockley et al. (2001) referred these trackways to *Tetrasauropus*; despite the differences in foot and claw impression orientations, both morphotypes were considered sufficiently similar to be referred to the same ichnogenus. However, osteological features (see #5 and 18 above) indicate that pedal claw orientation in sauropods is a function of the structure of the foot skeleton and therefore an important difference in footprint morphology. Thus, the morphological differences between the tracks from North America and the United Kingdom, which can be attributed to sauropods, and those of the type *Tetrasauropus* trackway from South Africa, which is excluded from a sauropod origin, indicate that these two types of trackways can no longer be considered congeneric. When more well-preserved trackways of this morphotype are discovered it may be possible to erect a new ichnogenus.

Tracks from the Late Triassic attributed to sauropods were first described by Ellenberger (1972), although this trackmaker identification was questioned (Olsen and Galton 1984). These four-toed tracks, *Sauropodopus antiquus* (fig. 9.8D), do not fit the sauropod track diagnostic criteria outlined above and have been attributed to chirotheres (Lockley and Meyer 2000). Other tracks from these sites named *Tetrasauropus unguiferous* (fig. 9.8E) were attributed to prosauropods (Ellenberger 1972; Lockley and Meyer 1999). These rounded tracks show the impressions of four large anteromedially directed claws on the pes and three on the manus. The axis of the pes impressions is parallel to the trackway midline. This trackway does not correspond to predicted sauropod track morphology. *Pseudotetrasauropus* includes a number of ichnospecies, some showing four or five short toes, with rounded or pointed terminations, a large heel pad, and manus impressions and others showing separated toes with discrete phalangeal pads made by animals progressing bipedally. *P. jaquesi* is the only one of these that could be interpreted as of sauropod origin; it has a five-fingered manus that faces anterolaterally and lies outside the line of the pes impressions and a four-toed triangular pes oriented parallel to the trackway midline (fig. 9.8F). The toes are short, with blunt terminations; one may show a claw impression, which is oriented anteromedially. The four toes and claw orientation, in addition to the pes orientation, indicate that the maker of this trackway is unlikely to have been a sauropod.

These Triassic sauropod trackways seem to have been made not by basal sauropods such as *Vulcanodon*, but by eusauropods; they show evidence of a semidigitigrade foot and four laterally compressed sickle-shaped unguals lying at an angle to the substrate (see podial synapomorphies #5 and 7–15 and the summary of eusauropod predicted track features; Figs. 9.3, 9.4A, B, 9.5, 9.6). These tracks provide the first evidence of eusauropods in the Late Triassic.

UNNAMED SAUROPOD TRACK MORPHOTYPE A
This was recovered from the Early Jurassic of Morocco (Ishigaki 1985, 1988; Farlow 1992) and Italy (Leonardi and Avanzini 1994; fig. 9.5). The trackway is characterized by long pes impressions showing four clear toes and suggestions of a fifth.

NEOSAUROPODA–DIPLODOCIDAE
Breviparopus taghbaloutensis (Dutuit and Ouazzou 1980) was originally described from the Middle Jurassic of Morocco. The trackway was narrow gauge, with medium to small (manus:pes ratio, 1:3.5–5) crescentic manus prints (fig. 9.5C). Well-preserved pes impressions show three claws directed anterolaterally with two pads, presumably from digits IV and V behind the claw of digit III (fig. 9.5). The pes impressions are approximately 1 m long, and the manus impressions 50 cm wide. Well-preserved manus impressions show indications of a single medially directed claw (Ishigaki 1989). Manus impressions are farther from the trackway midline than pes impressions. The manus impressions are very close to the pes impressions and it has been suggested that the pes impressions truncated those of the manus (Dutuit and Ouazzou 1980). However, manus-only trackways from the same area show a similar morphology (Farlow et al., 1989), the manus area, at one-fourth to one-fifth that of the pes, is typical for a narrow gauge trackway and the presence of raised rims around both types of tracks indicates that little overstepping is likely to have taken place. Dutuit and Ouazzou (1980) suggested the name *Breviparopus taghbaloutensis* for ease of referral to the footprints at this site (Demnat), but they illustrated the trackway with photographs and diagrams and the name has been used since to refer to tracks from both this site and others (e.g., Ishigaki 1988, 1989; Calvo 1999). Both Farlow et al. (1989) and Lockley et al. (1994a) considered it a valid ichnotaxon.

Parabrontopodus mcintoshi (Lockley et al. 1994) was erected on the basis of abundant ichnofossils from the Purgatoire tracksite of southeastern Colorado, although the holotype is a manus–pes pair (CU-MWC 190.5). All trackways from this site were referred to *Parabrontopodus* because they are all narrow gauge. These tracks are not as well preserved as those of *Brontopodus*; pedal claw impressions are rarely preserved and the manus tracks are crescentic. *Parabrontopodus* was diagnosed as a narrow gauge sauropod trackway with outwardly rotated pes impressions, three laterally directed pedal claw impressions, and small semicircular manus impressions (Lockley et al. 1994a: 140). A diplodocid has been suggested as a possible trackmaker for the tracks from the type locality, based on the smallish size of the tracks and their preservation in the Morrison Formation.

The original diagnosis did not state how *Parabrontopodus* could be differentiated from *Breviparopus*. *Breviparopus* is considered a valid ichnotaxon (Farlow et al. 1989; Lockley et al. 1994a) and the type trackway of *Breviparopus* is well preserved (Ishigaki 1985, 1989). It is not clear how *Parabrontopodus* can be distinguished from *Breviparopus*, and thus *Parabrontopodus* may be a junior synonym of *Breviparopus*. Since the erection of *Parabrontopodus*, few trackways have been referred to *Breviparopus* (exceptions include Calvo 1999), however, if *Parabrontopodus* is indeed a junior synonym of *Breviparopus*, tracks referred to *Parabrontopodus* should be referred to *Breviparopus*. The trackway from which the type specimen was taken (CU-MWC 190.5; fig. 9.5D) is very similar in manus and pes morphology to *Breviparopus* (fig. 9.5C), but the topotype trackway (CU-MWC 190.3) shows a rather different trackway pattern (fig. 9.5E).

In this trackway the manus impressions not only are generally farther in front of the pes, but also tend to be closer to the midline, than those of the type trackway, and rather than being oriented at approximately 20–25° to the midline, they make an angle of 45–70° to it (fig. 9.7B, C). This may indicate an important difference in locomotion styles between the two types of sauropods. CU-MWC 190.3 also has more elongate pes impressions and a slightly shorter relative stride length: 2.2–2.5 times the pes print length (although stride length is related to speed rather than taxonomy), in comparison to the stride length 3 times the pes length in the type trackway and *Breviparopus*. Farlow et al. (1989) previously suggested that there might be two sauropod track morphotypes at Purgatoire.

TITANOSAURIFORMES

Brontopodus birdi (Farlow et al. 1989) is based on well-preserved and substantial material from Texas (fig. 9.5G). The length and width of *Brontopodus* manus tracks are approximately equal, clawless, and U-shaped, with digit impressions I and V slightly separated from conjoined digit impressions II–IV. Manus track centers are closer to the trackway midline than pes track centers. Well-preserved pes tracks show three large, laterally directed claw marks, decreasing in size from digit I to digit III, a small claw on digit IV, and a pad on digit V. Manus and pes impressions are well away from the trackway midline (more than 0.5 pes print widths)—the classic wide gauge pattern. Manus impressions are about one-third the area of the pes impressions. *Brontopodus* stride length at the type locality varies from approximately two to five times the pes print length, and the pace angle (related to stride length) is 100–120°; stride length should increase with increased speed and so should not be expected to be a constant within trackways or ichnogenera. *Brontopodus* differs from all other sauropod ichnotaxa in that the manus length and width are approximately equal; this indicates that the trackmaker possessed metacarpals arranged in a tight arc, rather than a crescent. *Brontopodus* has been attributed to a titanosauriform such as the contemporary *Pleurocoelus* (Farlow et al. 1989; Gallup 1989; Wilson and Carrano 1999).

POSSIBLE TITANOSAURIFORMES (WIDE GAUGE TRACKWAYS)

Rotundichnus muenchehagensis (Hendricks 1981) was erected on the basis of the sauropod tracks at Münchehagen (Berriasian), Germany. The size of these tracks alone indicates that they were made by sauropods, but the arrangement and orientation of manus and pes impressions are also consistent with a sauropod origin (fig. 9.5J). No claw impressions or other morphological details are preserved and the trackways have a high manus:pes ratio (ca. 1:3) and are wide gauge. The lack of morphological detail means that these tracks cannot be diagnostic of any particular type of sauropod or trackway and should not be applied to tracks from other sites. The stride length is 3.2 times the pes print length. In contrast to *Brontopodus* (see above), the manus is semicircular in shape; the manus length is half the dimension of its width, and there is no indication of the separation of the lateral digits from the central digits typical of *Brontopodus*.

Sauropodichnus giganteus (Calvo 1991) was erected on the basis of a trackway from Argentina, consisting of large (90-cm maximum dimension) rounded impressions interpreted as having been made by sauropod pedes. The manus impressions were presumably overstepped completely by those of the pes (Mazzetta and Blanco 2001). This ichnogenus was considered a nomen dubium by Lockley et al. (1994a). Calvo (1999) revised the diagnosis based on undoubted sauropod trackways with manus prints (fig. 9.5K) from the localities of Cerrito del Bote and El Chocón (Albian–Cenomanian). These trackways are wide gauge, with crescentic manus impressions almost one quarter the area of the pes impressions (range, 1:3–1:5; average, 1:3.7). This spans the range of manus heteropody from large to small manus. No claw impressions are preserved but the crescentic manus morphology may mean that these trackways do not fall into *Brontopodus*, although less well-preserved manus impressions of *Brontopodus* from the type locality are crescentic in shape (Farlow et al. 1989). *Sauropodichnus* is more similar to *Rotundichnus* than to *Brontopodus*. The main differences are that *Sauropodichnus* has a relatively smaller manus and slightly narrower trackway, although it may eventually turn out to be a junior synonym of *Rotundichnus*. Trackways referred to *Breviparopus* were also reported from these localities (Calvo 1999). Titanosaurs have been suggested as trackmakers for both types of trackways (Calvo 1999), despite discrepancies between the known titanosaur skeletal morphology and the trackways (crescentic manus impressions and narrow gauge trackways).

Elephantopoides barkhausensis (Kaever and Lapparent 1974; Freise and Klassen 1979) was established on the basis of sauropod tracks from Barkhausen (Late Jurassic), Germany. These tracks are wide gauge, with a crescentic manus one-third the area of the pes and the trackway width half the width of the pes. No morphological details such as claw impressions are preserved. The lack of morphological detail makes it difficult to refer tracks from other sites to this ichnotaxon.

SAUROPOD TRACK DISTRIBUTION

Sauropod tracks are now known from the Late Triassic to the latest Cretaceous (fig. 9.7) and from every continent except Antarctica (fig. 9.9), although some intervals and areas are better documented than others. Table 9.1 is a summary of tracksite data used in figures. 9.7 and 9.9–9.11.

SPATIOTEMPORAL DISTRIBUTION

The patchy distribution of sauropod tracksites (fig. 9.9) places limitations on interpretation of the data. However, bearing this in mind, and using the phylogenetic correspondences from figure 9.5, the track record is generally consistent with the osteological record.

Lockley (1991:123, 124) suggests that "from their first appearance ... right through to Cretaceous times brontosaur tracksites are associated with limy substrates in tropical and subtropical latitudes." Farlow (1992) tested this hypothesis by examining the distribution of sauropod tracks by environment and paleolatitude and comparing this to the distribution of other dinosaur tracks and skeletal remains. He concluded that current knowledge of sauropod tracksites was "inadequate for determining latitudinal and habitat preferences of sauropods" (Farlow 1992:90). Lockley et al. (1994b) performed an analysis similar to that of Farlow but added recently discovered tracksites, raising the number of known tracksites from 22 to 190, of which 102 were in a single formation in one small area of South Korea. Lockley et al. (1994b) defined a tracksite as a specific geographic location on a particular geographic area and thus split up several of Farlow's (1992) sites; for example, the Purgatoire site has four track-bearing levels. Lockley et al. (1994b) only considered sauropod tracks and concurred with Lockley's (1991) original hypothesis. The method of Lockley et al. (1994b) is followed in this report.

It is possible to test the contention that sauropods preferred paleolatitudes of 0–30°N (Lockley et al. 1994b). However, most sauropod tracks are found at paleolatitudes of 30–45°N (figs. 9.9, 9.10). The second-hightest number of tracksites is found at paleolatitudes of 15–30°N, which happens to be the paleolatitudinal range of much of North America and Europe during the Mesozoic where prospecting for tracksites has been most concentrated. Tracks from other areas are usually discovered by local residents, with subsequent documentation by scientists from developed countries (Lockley 2002b), and are generally concentrated near population centers (Lockley 2002a). Sauropod tracksite distribution is more informative about the state of modern countries than about the habitat preferences of sauropods.

To test if sauropod tracks are found at different paleolatitudes than other types of dinosaur tracks, especially ornithopods, as suggested by Lockley (1991; Lockley et al. 1994b), it would be necessary to plot other dinosaur track-

TABLE 9.1
Summary of Sauropod Tracksite Publications Around the World

AGE	COUNTRY	NO. SITES	LITHOLOGY	PALEOENVIRONMENT	GAUGE	REFERENCE(S)
Late Triassic						
Norian-Rhaetian	Western USA	5	Clastic	Fluvial	n	Lockley et al. 2001
Norian-Rhaetian	Wales	1	Clastic, redbeds		n	Lockley et al. 1996
Norian-Rhaetian	Switzerland	1	Carbonate	Coastal		Meyer & Thuring 2003
	Greenland	1				Jenkins et al. 1994
Early Jurassic						
Pliensbachian	Northeastern Italy	1	Carbonate	Coastal	w	Avanzini et al. 2001
Pliensbachian	Morocco	3	Carbonate	Coastal	b	Ishigaki 1988
Early Sinemurian	Northern Italy	1	Carbonate	Coastal	b	Dalla Vecchia 1994; Avanzini et al. 1997
Hettangian	Poland	2	Sandstone	Fluvial	n	Gierlinski & Sawicki 1998
Middle Jurassic						
Aalenian–Bathonian	Yorkshire, UK	9	Terrigenous clastics	Fluvial	b	Romano & Whyte 2003
Baj–Bath	Portugal	1	Limestone	Coastal	w	Antunes & Mateus 2003; Santos et al. 1994
Baj–Bath	Morocco	4				Ishigaki 1989; Dutuit & Ouazzou 1980
Bathonian	Oxford, UK	1		Coastal	b	Day et al. 2002
Callovian	UT, USA	1	Sandstone	Coastal	w	Foster et al. 2000
	Southeastern Mexico	1	Arenitic siltstone	Fluvial		Ferrusquía-Villafranca et al. 1996
	Portugal	1				Lockley & Meyer 1999; Antunes & Mateus 2003
Late Jurassic						
Oxfordian	Chile	1	Sandstone	Marine platform		Leonardi 1989
Oxfordian	Tadjikistan	1				Novikov et al. 1993
Early Kimmeridgian	Switzerland	4	Mudstone	Coastal	w	Meyer & Thuring 2003
Early Kimmeridgian	Switzerland	3			w	Meyer & Thuring 2003
Kimmeridgian	Portugal	1	Limestone		b	Lockley & Meyer 1999
Kimmeridgian	Portugal	1	Limestone			Antunes & Mateus 2003

TABLE 9.1 (continued)

AGE	COUNTRY	NO. SITES	LITHOLOGY	PALEOENVIRONMENT	GAUGE	REFERENCE(S)
Kimmeridgian	CO, USA	1	Sandstone	Floodplain		Barnes & Lockley 1994
Kimmeridgian		1	Sandstone		w	Meyer & Thuring 2003
Late Kimmeridgian	Switzerland	4			n	Meyer & Thuring 2003
Late Kimmeridgian	Switzerland	1			w	Meyer & Thuring 2003
Tithonian–Kimmeridgian	CO, USA	6			n	Prince & Lockley 1989
Tithonian–Kimmeridgian	CO, USA	3	Limestone	Lacustrine		
Tithonian–Kimmeridgian	WY, USA	2	Limestone	Coastal	w	
Tithonian	UT, USA	5				
Tithonian	UT, USA	1	Muddy sandstone			
Tithonian	Switzerland	3			w	Meyer & Thuring 2003
Tithonian	France	1				Allain & Pereda-Suberbiola 2003
Tithonian	Portugal	8	Limestone	Coastal	w	Lockley & Meyer 2000; Lockley et al. 1994c
	Portugal	6				Antunes & Mateus 2003
	Niger	1				Weishampel et al. 1990
Early Cretaceous						
Berriasian	UK	1	Limestone	Coastal		Wright et al. 1997
Berriasian	Germany	1	Sandstone	Coastal	n	Fischer 1998
Berriasian	Spain	1	Calcareous		n	Lockley & Meyer 2000
Berriasian	Chile	1		Coastal swamp		Leonardi 1989
Berriasian	Thailand	1	Sandstone	Coastal		Le Loeff et al. 2002
Hauterivian	Portugal	2				Antunes & Mateus 2003
Haut–Barr	Italy	1	Limestone	Coastal		Dal Sasso 2003
Neocomian	Brazil	2		Lacustrine		Leonardi 1989
Neocomian	Australia	1	Sandstone	Coastal	n	Thulborn et al. 1994
Neocomian	Australia	1	Sandstone	Swamp/forest	n	Thulborn et al. 1994
Barremian–Aptian	UK	1	Terrigenous clastics	Fluvial		Radley 1994; Insole & Hutt 1994
Barremian–Aptian	Spain	1			w	Moratalla et al. 1994
Barremian–Aptian	Spain	1	Silty sandstone		w	Moratalla et al. 1994

Age	Country	N	Lithology	Environment	Gauge	Reference
Barremian–Aptian	China	4	Sand/mudstone, clastics	Fluvial	b	Lockley et al. 2002b
Aptian	Spain	10	Carbonates	Coastal	b	Casanovas et al. 1997
Albian	Spain	2	Conglomerate, Sandstone	Fluvial		Moratalla et al. 1994
Albian	Argentina	3	Mudstone	Fluvial	w	Calvo 1999
Albian	TX, USA	8	Limestone	Coastal	w	Pittman 1989
Albian	AR, USA	1			b	Pittman & Gillette 1989
Late Albian	Croatia	1	Carbonate	Marine platform		Dalla Vecchia 1994
	Brazil	1	Sandstone	Fluvial		Leonardi 1989
	Tadjikistan	2			w	
Late Cretaceous						
Cenomanian	Croatia	1			w	Lockley & Meyer 2000
Cenomanian	Portugal	1				Antunes & Mateus 2003
Cenomanian	South Korea	102	Shales, siltstone	Lacustrine	w	Lim et al. 1994
Cenomanian	TX, USA	1	Limestone	Coastal	w	Farlow et al. 1989
late Cenomanian	Croatia	1	Carbonate	Marine platform, humid	n	Dalla Vecchia 1994
Turonian–Coniacian	China	1	Terrigenous clastics	Subtropical	w	Lockley et al. 2002b
Maastrichtian	Spain	1	Silty limestone	Coastal	w	Schulp & Brokx 1999
Maastrichtian	Bolivia	1	Sandstone	Fluvial	w	Lockley et al. 2002a
Maastrichtian	Bolivia	5	Limestone	Lacustrine	w	Lockley et al. 2002a
Maastrichtian	Bolivia	1	Limestone	Marine platform	w	Lockley et al. 2002 (Leonardi & Avanzini 1994)

Note: Abbreviations: n, narrow gauge; w, wide gauge; b, both wide and narrow gauge.

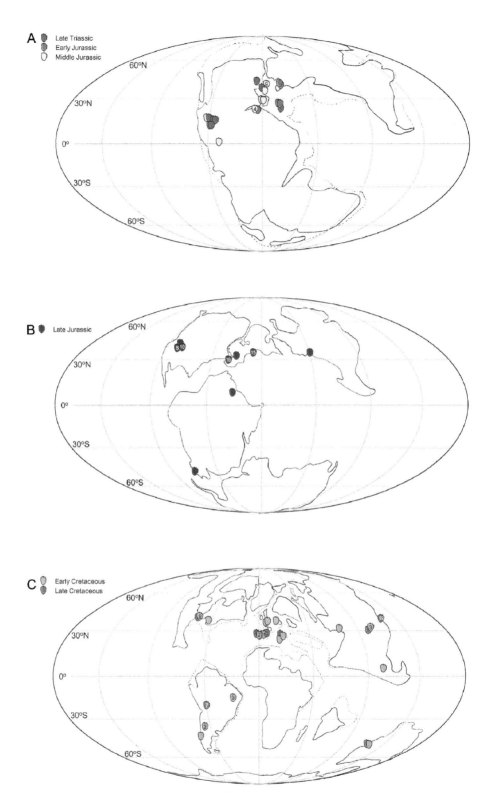

FIGURE 9.9. The global distribution of sauropod tracksites. (A) Late Triassic to Middle Jurassic tracksites shown on an Early Jurassic paleogeographic map; (B) Late Jurassic tracksites; (C) Early and Late Cretaceous tracksites plotted on a Maastrichtian paleogeographic map. Data in Table 9.1.

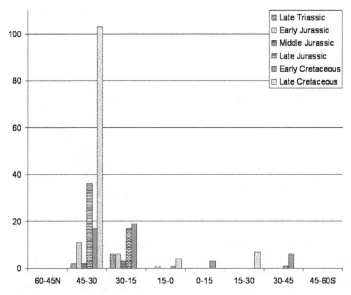

FIGURE 9.10. Paleolatitudinal distributions of sauropod tracksites. Tracksite occurrences plotted as conservatively as possible, given the available data (i.e., where tracksites fell between two categories, they were placed in the one nearer the equator).

site distributions against paleolatitude. Farlow (1992) did this, not only for tracks but also for skeletal remains, and found few differences; this has not been done since. In South Korea, where more than a hundred tracksites have been reported, there is obviously no latitudinal separation, even if sauropod and ornithopod tracks are seldom found on the same bedding plane.

PALEOENVIRONMENTAL DISTRIBUTION

As biogenic sedimentary structures produced by the behavior of animals, tracks reveal where living animals roamed. Tracks may have the potential to reveal habitat preferences of their producers if favored habitats are as likely to be preserved as any other and if these habitats have track preservation potential. However, sediments and tracks in lowland and coastal environments are much more likely to be preserved than those in more upland or inland environments, and certain other environments, such as forests, are much less likely to preserve tracks.

Lockley et al. (1994b:242) suggested that "sauropod tracks were most commonly associated with marine carbonate platform sequences and carbonate-evaporite lake basins." Their diagrams illustrating the relationship between sauropod tracksites and sedimentary facies show that the majority of sauropod tracks are found in carbonate environments. However, their diagrams show that they conflated lithology and paleoenvironments, counting all environments as carbonates apart from "fluviolacustrine." Some sauropod tracks from lacustrine paleoenvironments are preserved in siliciclastic sediments (e.g., Paik et al. 2001). Figure 9.11 shows sauropod tracksites plotted by lithology and by paleoenvironment, with the South Korean localities counted as 10 (rather than 102). If the total number of Korean tracksites were inserted, lacustrine environments would be boosted to 52%, marine and coastal environments would decrease to 28%, and other inland environments, mainly fluvial, would decrease to 20%. This demonstrates how this one small geographic area, during one time period, can overwhelm the whole data set. The Korean lacustrine deposits are not evaporitic and therefore do not necessarily indicate a semiarid climate. These data show no support for the idea that sauropods preferred carbonate substrates. The

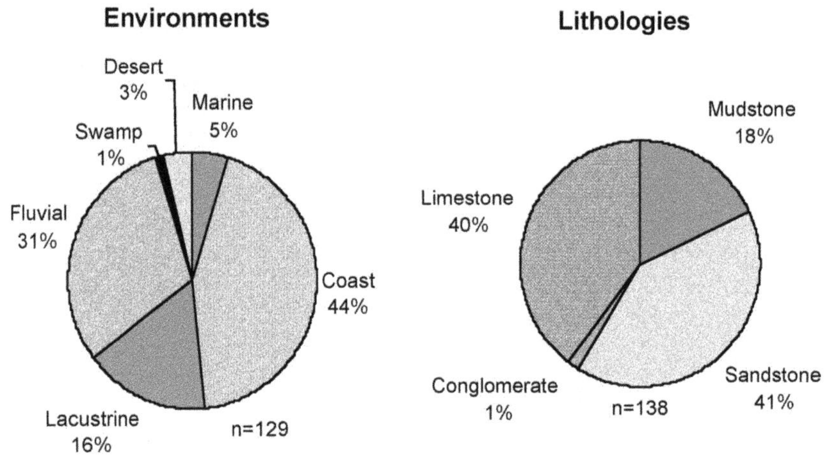

FIGURE 9.11. Paleoenvironmental and lithological associations of sauropod tracksites. South Korean localities counted as 10 (rather than 102), but because they are siliciclastic (Paik et al. 2001), the addition of the total number would shift the balance away from carbonate localities, to only 25% carbonates. Number of sites in each plot is not the same because published reports do not always give both sets of data.

65% of tracksites from coastal or lacustrine environments may simply indicate that tracks in such localities are more likely to be preserved.

There is no major shift in environmental preferences through time (table 9.2), nor do the makers of narrow or wide gauge trackways seem to have had different habitat preferences (table 9.3). Overall, most sauropod tracksites from the Late Jurassic and Early Cretaceous are preserved in carbonates, and more sauropod tracksites from other times are preserved in siliclastic sediments. This may reflect the predominant types of terrestrial depositional environments in prospected areas during these time periods, rather than trackmaker habitat preferences. A predominance of coastal tracksites may correspond with an increase in coastal plain environments related to the breakup of Pangaea.

Associations of sauropod tracks with other types of fossils are not reexamined herein so the assertion that sauropod tracks are preferentially preserved with theropod tracks is not tested. However, sauropod tracks have been found in association with ornithopod tracks (Thulborn et al. 1994; Calvo 1999; Paik et al. 2001), although an association with no tracks or theropod tracks seems to be more common. However, as ornithopod tracks only become common in the Cretaceous, perhaps only Cretaceous sites should be considered in determining track association preferences (fig. 9.12).

No attempt has been made herein to assess paleoclimate of the tracksites, so claims that many sauropod tracksites are found in semiarid to seasonal climates cannot be assessed, except to note that this covers a wide range of climates and could be interpreted as describing the majority of climates around the world.

DIFFERENCE IN DISTRIBUTION OF SAUROPOD TRACKS AND BODY FOSSILS

One of the most potentially interesting areas of sauropod track research is the contrasting information that is revealed by tracks and skeletal remains. Most sauropod bones represent individuals at least 80% of adult body size, while many tracks are those of very small individuals, probably juveniles, as no known sauropods are this small as adults (Lockley et al. 1994c). This is especially true of sauropod tracks from Asia; recent discoveries from China have continued this trend (Lockley et al. 2002b). The relatively high proportion of juvenile sauropod trackways may be interpreted in several ways. Juveniles may have inhabited areas where footprints have a higher preservation potential, for instance,

TABLE 9.2
Lithology and Paleoenvironments by Trackway Gauge Through Time

AGE	GAUGE	LITHOLOGY	PALEOENVIRONMENT
Late Triassic	n	Terrigenous clastics	Fluvial
Early Jurassic	2 n	Terrigenous clastics	River floodplain
	5 n & w	Carbonates	Marginal marine
Middle Jurassic	2 w	1 sandstone	Coastal
		1 limestone	Coastal
	5 n & w	1 limestone	Coastal
		4 clastics	Fluvial
Late Jurassic	w	Clastics or limestone	Mainly coastal
	n	Limestone or clastics	Coastal or inland
Early Cretaceous	w & n	Ubiquitous	Ubiquitous
Late Cretaceous	w	Siliciclastic	Lacustrine
	w	Siliciclastic	Marginal marine

NOTE: In the Late Cretaceous two geographic areas preserved the majority of these sites. Abbreviations: n, narrow; w, wide.

lakeshores, riverbanks, and lagoonal shorelines, whereas the adults spent more time in areas where bones are more likely to be preserved, such as low-lying alluvial plains. Sauropod clutch sizes are thought to have been large, and many more sauropods would hatch out than survive to adulthood. Their greater numbers might account for the high proportion of small sauropod trackways.

LOCOMOTION AND BEHAVIOR

Trackways preserve direct evidence of behavior; they are the nearest we are likely to get to a "movie" of a dinosaur. Trackways can be used to constrain locomotor hypotheses of their producers and have the potential to reveal details of the animal's locomotion such as weight distribution, step cycle, or gait.

Sauropod trackways unequivocally show that the trackmakers had an upright stance and a generally parasagittal gait. All known sauropod trackways show slow walking speeds, in keeping with their graviportal anatomy. Sauropod pes impressions range in length from 20 cm to more than 1 m, which gives estimated hip heights of 1–5 m (Thulborn 1990).

Well-preserved sauropod trackways show depth variations within footprints, which may indicate the sequence of weight distribution in the sauropod step cycle (Wilson and Carrano 1999). Where footprint depth variation has been described the patterns are consistent. The anteromedial side of the pes print is the deepest; this forms the leading edge of the track and would have been the push-off point for the recovery stroke. The next deepest part of the pes impression is the lateral side of the track (mainly the "heel"); this would have been the first part of the foot to contact the substrate for the beginning of the power stroke. *Brontopodus* tracks from the type locality, the topotype trackway of *Parabrontopodus* (CU-MWC 190.3 [Lockley et al. 1994a]), and the Late Triassic sauropod track CU-MWC 172.22 (fig. 9.8) all show this depth distribution pattern.

Pittman and Gillette (1989) suggested from footprint evidence that sauropod pedal claws might be wrapped along the lateral edge of the foot when walking on firm substrates and extended for added traction when turning, braking, or walking on more slippery substrates. Wrapping of the claws along the edge of the foot might help explain the lack of pedal claw impressions in many sauropod trackways.

Manus impressions seem to be deepest at the anterior side (digits I–III) and shallowest at

TABLE 9.3
Sauropod Track Morphotypes with Possible Trackmakers

MORPHOTYPE	MANUS:PES	GAUGE	CLAW/DIGIT IMPRESSIONS	AGE	GEOGRAPHIC LOCATION	POSSIBLE TRACKMAKER	REFERENCE(S)
1 "*Tetrasauropus*"	Small manus	Narrow	Distinct digit impressions & 4 pedal claw impressions	L Tr–E Jr	USA, Wales, (Greenland, Italy, Switzerland, Morocco)	Eusauropod	Lockley et al. 1996, 2001 (Meyer & Thuring 2003; Dalla Vecchia 1996; Jenkins et al. 1994; Ishigaki 1988) (fig. 9.5A)
2	Small manus	Wide or variable	Indications of manus digits; up to 4 pedal claw impressions	E Jr	Italy, Morocco	Eusauropod	Ishigaki 1988; Leonardi & Avanzini 1994 (fig. 9.5B)
3 *Breviparopus* (*Parabrontopodus*)	Small manus (1:5)	Narrow	May have 1 manus claw impression; up to 3 pedal claw impressions	M–L Jr	USA, N. Africa	Neosauropod ?Diplodocid	Lockley et al. 1994. (fig. 9.5C–E)
4	Large manus	Narrow	Large digit I claw impression	M Jr	UK	???	Day et al. 2002 (fig. 9.5F)
5	Small manus	Wide	No manual claw impressions	M Jr	UK	Titanosaurid	Day et al. 2002
6 *Brontopodus*, (*?Rotundichnus*, *?Sauropodichnus*)	Large manus (1:3)	Wide	Very small manus claw impression if present; 4 pedal claw impressions	E K	USA, Germany, S. America	Titanosauriform	Farlow et al. 1989 (fig. 9.5G)
7	Very large manus (1:2)	Wide	Large claw on manual digit I; 4 large pedal claw impressions	L Jr	Portugal	?Camarasauromorph	Santos et al. 1994 (fig. 9.5H, I)

NOTE: These do not necessarily correspond to ichnogenera; several valid ichnogenera could be contained within one of these broad morphotypes. Abbreviations: E, Early; M, Middle; L, Late; J, Jurassic; K, Cretaceous; Tr, Triassic.

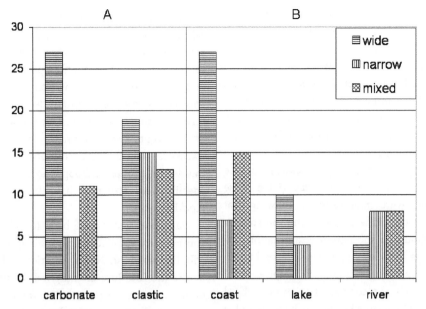

FIGURE 9.12. Sauropod tracksites preserving narrow, wide or both gauges of trackways plotted by (A) lithology (n = 90) and (B) environment (n = 83). Numbers of sites are not the same because not all publications give both sets of data.

the back, regardless of the orientation of the manus impressions. This may indicate that most of the weight on the manus was borne by the middle three digits. This might also partly explain the lack of manus claw impressions in most sauropod tracks.

Occasionally extra traces are preserved in sauropod trackways, which provide extra evidence of limb motion. Drag marks posterior to manus impressions have been described for the New York slab of *Brontopodus* from the Paluxy River site (Farlow et al. 1989) and for tracks from the Purbeck Limestone Group (Wright et al. 1997). These drag marks are slightly curved and show that the forelimb described a slight lateral arc during the recovery stroke.

Finer morphological details of sauropod tracks, such as skin impressions, are not often preserved, although they have been reported (Lockley and Hunt 1995). These tracks showed small polygonal scales in the sole of the foot; this is similar to the pattern reported for the soles of the feet of other dinosaurs such as hadrosaurs, iguanodontids, and theropods. The rarity of sauropod skin impressions may be because of their large size, although large hadrosaur and iguanodontid tracks with skin impressions seem to be more common, or it may simply be that the substrates in which sauropod tracks have so far been preserved do not provide sufficient resolution to reveal such details.

Manus-only sauropod trackways have been interpreted as sauropod swimming tracks by Ishigaki (1989), although because of the weight distribution of the sauropod body, it is considered more likely that these are a preservational artifact and the manus prints are undertracks punched through from a level above (Lockley and Rice 1990).

Several sauropod tracksites preserve a number of parallel trackways (Bird 1944; Lockley et al. 1986; Farlow et al. 1989). At some sites, such as Davenport Ranch, Texas, in the Glen Rose Formation, gregarious behavior has been demonstrated (Bird 1944; Lockley 1991). At others such as the Purgatoire tracksite, which is thought to have been an ancient lake shore, a geographic control is strongly indicated (Lockley et al. 1986). Many sites with multiple sauropod trackways either do not preserve a large enough area to assess gregarious behavior (e.g., Lockley et al. 2002b) or are not preserved in sufficient detail.

Multiple sauropod trackways heading in the same direction on a single bedding plane are preserved in numbers up to 25. This may be an indication of "herd" size, or we may not be finding larger "herd" sizes because bedding planes extensive enough to show larger groups are so seldom exposed. Sauropod trackways at the Davenport Ranch site in Texas (Glen Rose Formation) were interpreted as showing that the larger animals were protecting the smaller ones, although a subsequent examination disproved this (Lockley 1991). No sauropod tracksites have shown evidence of complex social behavior such as a consistent or strategic herd structure. Really large herds may have trampled the ground so extensively that no discrete trackways are preserved. There is no indication that sauropods traveled in herds of hundreds or thousands as has been inferred from bonebeds of ceratopsians or hadrosaurs, even though tracks of these ornithischians do not give any indication of the huge size of their herds.

SUMMARY AND PROSPECTUS

Despite the purported conservative nature of the sauropod foot, characteristic features of sauropod tracks can be identified and used to determine if footprints were indeed produced by sauropods. These characteristics allow us to identify certain Late Triassic tracks as being of sauropod origin and allow us to distinguish a number of other sauropod footprint morphotypes. In addition to predicted morphotypes, a couple of other sauropod ichnomorphotypes can be distinguished. These show sauropod footprint characteristics but do not correspond to the podial anatomy of any known sauropod. However, the feet of many sauropods are unknown.

Sauropod trackways show that sauropods progressed quadrupedally with a parasagittal gait. Slow walking speeds are preserved at all tracksites. Sauropods walked with their feet toed-out at up to 30° from the trackway midline and the dorsal surface of the manus directed anterolaterally at angles of up to 75° (figs. 9.5, 9.8). The gauge of sauropod trackways is mainly a product of pelvic and hindlimb anatomy (Wilson and Carrano 1999) of some sauropod groups (e.g., titanosauriforms), although in some instances it seems to be related to ontogeny or locomotory style. Sauropod tracksites indicate that sauropods sometimes traveled in groups, although no indications of specific herd structure have been identified (Lockley 1991).

Sauropod tracks have been found on every continent except Antarctica. Their apparent distribution solely in North America and Europe before the Late Jurassic may well be the result of more intensive prospecting there, because sauropod skeletal remains have been found in Asia in the Late Triassic (Buffetaut et al. 2001, 2002) and in southern Africa in the Early Jurassic (Raath 1972). The tracks and bone fossil records appear to sample different assemblages and reveal complementary information.

This synthesis of currently known sauropod tracksites indicates that sauropod tracksites are preserved in approximately equal numbers in siliciclastic and carbonate lithologies. Marginal marine and inland environments are also represented in approximately equal numbers. The patterns of tracksite preservation in these lithologies and environments through time may reflect predominant depositional environments with a high preservation potential at any given time. If sauropods preferred mountainsides or forests as a living environment, their fossils (tracks or bones) would be unlikely to be preserved, so only statements about differing frequencies of occurrence in paleoenvironments with a high preservation potential can be made, and the data show no such difference. There are currently insufficient data in both the skeletal and the ichnological records to make any general statements about sauropod habitat preferences. Sauropods were a globally distributed, long-lived group probably capable of exploiting a wide range of habitats, lithologies, and climates (Farlow 1992, Thulborn et al. 1994).

Increasingly sophisticated computer programs now allow us to begin to investigate sauropod tracks in new ways. Computer animation programs allow the use of trackways as a control on models of dinosaur locomotion (Henderson 2002). Statistical modeling allows more detailed analysis of trackway patterns independent of variables related to the speed of the trackmaker, and photogrammetric techniques can produce contour maps of footprints, which can be used in more detailed locomotor studies than have previously been possible. Laser surveying can be used to make detailed maps accurate to a few millimeters (Wright et al. 1997). Such technology is likely to become ubiquitous and less expensive and will allow tracks to be used in increasingly more sophisticated ways to investigate sauropod biology further.

ACKNOWLEDGMENTS

I thank Jeff Wilson, Kristi Curry Rogers, and Dan Chure for asking me to participate in this symposium and to contribute a chapter to this book. Thanks are also due to Liz Cook, Jordon Hand, and Deborah Thomas for reading drafts of the manuscript. This chapter was greatly improved by the comments of an anonymous referee.

LITERATURE CITED

Allain, R., and Superbiola, X. P. 2003. Dinosaurs of France. C. R. Palevol. 2: 27–44.

Antunes, M. T., and Mateus, O. 2003. Dinosaurs of Portugal. C. R. Palevol. 2: 77–95.

Avanzini, M., Frisia, S., van den Driessche, K., and Keppens, E. 1997. A dinosaur tracksite in an early Liassic tidal flat in northern Italy: paleoenvironmental reconstruction from the sedimentology and geochemistry. Palaios 12: 538–551.

Avanzini, M., Leonardi, G., Tomasoni, R., and Campolongo, M. 2001. Enigmatic dinosaur trackways from the Lower Jurassic (Pliensbachian) of the Sarca Valley, northeast Italy. Ichnos 8: 235–242.

Barnes, F. A., and Lockley, M. G. 1994. Trackway evidence for social sauropods from the Morrison Formation, eastern Utah (USA). GAIA 10: 37–42.

Bird, R. T. 1939. Thunder in his footsteps. Nat. His. 43: 254–261

Bird, R. T. 1944. Did *Brontosaurus* ever walk on land? Nat. Hist. 53: 61–67

Buffetaut, E., Suteethorn, V., Cuny, G., Tong, H., Le Loeff, J., Khansubha, S., and Jongautcharlyaku, S. 2000. The earliest known sauropod dinosaur. Nature 407: 72–74.

Buffetaut, E., Suteethorn, V., Le Loeff, J., Cuny, G., Tong, H., and Khansubha, S. 2002. The first giant dinosaurs: a large sauropod from the Late Triassic of Thailand. C. R. Palevol 1: 103–109.

Calvo, J. 1991. Huellas de dinosaurios en la Formacion Río Limay (Albiano-Cenomaniano?), Picun Leufu, Provincia de Neuquen, República Argentina. Ameghiniana 28: 241–258.

———. 1999. Dinosaurs and other vertebrates of the Lake Ezequiel Ramos Mexia area, Neuquen-Patagonia, Argentina. In: Tomida, Y., Rich, T., and Vickers-Rich, P. (eds.). Proceedings of the Second Gondwanan Dinosaur Symposium. National Science Museum Monographs No. 15, Tokyo. Pp. 13–45.

Carrano, M. T., and Wilson, J. A. 2001. Taxon distributions and the tetrapod track record. Paleobiology 27: 564–582.

Casanovas, M., Fernandez, A., Pérez-Lorente, F., and Santafé, J. V. 1997. Sauropod trackways from site el Sobaquilla (Muniulla, La Rioja, Spain) indicate amble walking. Ichnos 5: 101–107.

Dalla-Vechia, F. M. 1994. Jurassic and Cretaceous sauropod evidence in the Mesozoic carbonate platforms of the southern Alps and Dinards. GAIA 10: 65–74

———. 1996. Archosaurian trackways in the Upper Carnian of Dogna Valley (Udine, Fruili, NE Italy). Natura Nascosta 12: 5–17.

Dal Sasso, C. 2003. Dinosaurs of Italy. C. R. Palevol. 2: 45–66.

Day, J. J., Upchurch, P., Norman, D. B., Gale, A. S., and Powell, H. P. 2002. Sauropod trackways, evolution and behavior. Science 296: 1659.

Dutuit, J.-M., and Ouazzou, A. 1980. Découverte d'une piste de Dinosaure sauropode sue le site d'empreintes de Demnat (Haut-Atlas marocain). Mem. Soc. Geol. France 139: 95–102.

Ellenberger, P. 1972. Contribution à la classification des Pistes de Vertébrés du Trias: Les types du Stormberg d'Afrique du Sud Palaeovertebrata, Mémoire Extraordinaire. Laboratoire de Paléontologie de Vertébrés, Montpellier. 104 pp.

Farlow, J. O. 1992. Sauropod tracks and trackmakers: integrating the ichnological and skeletal records. Zubia 10: 89–138.

Farlow, J. O., Pittman, J. G., and Hawthorne, J. M. 1989. *Brontopodus birdi*, Lower Cretaceous sauropod footprints from the US Gulf coastal plain. In:

Gillette, D. D., and Lockley, M. G. (eds.). Dinosaur Tracks and Traces. Cambridge University Press, Cambridge. Pp. 371–394.

Ferrusquía-Villafranca, I., Jiménez-Hidlago, E., and Bravo-Cuevas, V. M. 1996. Footprints of small sauropods from the Middle Jurassic of Oaxaca, southeastern Mexico. In: Morales, M. (ed.). The Continental Jurassic. Mus. North. Ariz. Bull. 60: 119–126.

Fischer, R. 1998. Die Saurierfährten im Naturdenkmal Münchehagen. Mitteilung. Inst. Geol. Paläontol. Univ. Hannover 37: 3–59.

Foster, J., Hamblin, A., and Lockley, M. G. 2000. The oldest evidence of a sauropod dinosaur in the western United States and other important vertebrate trackways from Grand Staircase– Escalante National Monument, Utah. Ichnos 7: 169–181.

Freise, H., and Klassen, H. 1979. Die Dinosaurierfahrten von Barkhausen im Wiehengebirge. Veroffentl. Landkr. Osnabruck 1: 1–36.

Gallup, M. R. 1989. Functional morphology of the hindfoot of the Texas sauropod *Pleurocoelus* sp. Indet. In: Farlow, J. O. (ed.). Paleobiology of the Dinosaurs. Geol. Soc. Am. Spec. Paper 238: 71–74.

Gierlinski, G., and Sawicki, G. 1998. New sauropod tracks from the Lower Jurassic of Poland. Geol. Q. 42: 477–480.

Gilmore, C. W. 1936. Osteology of *Apatosaurus*, with special reference to specimens in the Carnegie Museum. Mem. Carnegie Mus. 9(4): 175–272.

Hand, J. D. 1999. A fully articulated Apatosaurus manus from the Morrison Formation of Wyoming. Geological Society of America, Abstracts with Programs. Vol. 30.

Henderson, D. M. 2002. Wide and narrow gauge sauropod trackways as a consequence of body mass distribution and the requirement for stability. J. Vertebr. Paleontol. 22: 64A.

Hendricks, A. 1981. Die Saurierfährten von Münchehagen bei Rehburg-Loccum (NW-Deutschland). Abhandlung. Landesmus. Naturkunde Münster 43: 1–22.

Insole, A. N., and Hutt, S. 1994. The palaeoecology of the dinosaurs of the Wessex Formation (Wealden Group, Early Cretaceous), Isle of Wight, southern England. Zool. J. Linn. Soc. 112: 197–215.

Ishigaki, S. 1985. Nature Study 31: 5–8. (In Japanese.)

———. 1988. Les empreintes des dinosaures du Jurasique inferieur du Haut Atlas central marocain. Notes Serv. Geol. Maroc 44: 79–86.

———. 1989. Footprints of swimming sauropods from Morocco. In: Gillette, D. D., and Lockley, M. G. (eds.). Dinosaur Tracks and Traces. Cambridge University Press, Cambridge. Pp 83–86.

Janensch, W. 1922. Das Handskelett von *Gigantosaurus* und *Brachiosaurus* aus den Tendaguru-Schichten Deutsch-Ostafrikas: Centralblatt für Mineralogie, Geologie, und Paläontologie. Pp. 464–480.

Janensch, W. 1961. Die Gliedmaszen und Gliedmaszengürtel der Sauropoden der Tendaguru-Schiechten. Palaeontographica (Suppl. 7[1]) 3(4): 177–235.

Jenkins, F. A., Shubin, N. H., Amaral, W. W., Gatesey, S. M., Schaff, C. R., Clemensen, L. B., Downs, W. R., Davidson, N. R., Bonde, B., and Osbaeck, F. 1994. Late Triassic continental vertebrates and depositional environments of the Fleming Fjord Formation, Jameson Land East Greenland. Meddelelser Grønland Geosci. 32: 3–25.

Kaever, M., and Lapparent, A. F. de. 1974. Les traces des pas de dinosaures du Jurasique de Barkhausen (Basse Saxe, Allemagne). Bull. Soc. Geol. France 16: 516–525.

Le Loeff, J., Khansubha, S., Buffetaut, E., Suteethorn, V., Tong, H., and Souillat, C. 2002. Dinosaur footprints from the Phra Wihan Formation (Early Cretaceous of Thailand). C. R. Palevol. 1: 287–292.

Leonardi, G., 1989. Inventory and statistics of the South American dinosaurian ichnofauna and its paleobiological interpretation. In Gillette, D. D. and M. G. Lockley, (eds) Dinosaur Tracks and Traces, Cambridge University Press, Pp. 371–394.

Leonardi, G. and Avanzini, M. 1994. Dinosauri in Italia. Le Scienze. Quaderni 76: 69–81, Milan.

Lim, S. K., Lockley, M. G., Yang, S.-Y., Fleming, R. F., and Houck, K. A. 1994. Preliminary report on sauropod tracksite from the Cretaceous of Korea. GAIA 10: 109–117.

Lockley, M. G. 1991. Tracking Dinosaurs. Cambridge University Press, Cambridge. 238 pp.

Lockley, M. G., and Hunt, A. P. 1995. Dinosaur Tracks and Other Fossil Footprints of the Western United States. Columbia University Press, New York. 338 pp.

Lockley, M. G., and Meyer, C. A. 1999. Dinosaur Tracks and Other Fossil Footprints of Europe. Columbia University Press, New York. 323 pp.

Lockley, M. G., and Rice, A. 1990. Did *Brontosaurus* ever swim out to sea? Ichnos 1: 81–90.

Lockley, M. G., Houck, K., and N. K. Prince. 1986. North America's largest dinosaur tracksite: implications for Morrison Formation paleoecology. Geol. Soc. Am. Bull. 57: 1163–1176.

Lockley, M. G., Farlow, J. O., and Meyer, C. A. 1994a. *Brontopodus* and *Parabrontopodus* ichnogen nov. and the significance of wide and narrow gauge sauropod trackways. GAIA 10: 126–34.

Lockley, M. G., Meyer, C. A., Hunt, A. P., and Lucas, S. G. 1994b. The distribution of sauropod tracks and trackmakers. GAIA 10: 233–248.

Lockley, M. G., Meyer, C. A., and Santos, V. F. 1994c. Trackway evidence for a herd of juvenile sauropods from the Late Jurassic of Portugal. GAIA 10: 43–48.

Lockley, M. G., King, M., Howe, S., and Sharp, T. 1996. Dinosaur tracks and other archosaur footprints from the Triassic of South Wales. Ichnos 5: 23–41.

Lockley, M. G., Wright, J. L., Hunt, A. P., and Lucas, S. G. 2001. The Late Triassic sauropod track record comes into focus: old legacies and new paradigms. New Mexico Geological Society Guidebook, 52nd Field Conference, Geology of the Llano Estacado.

Lockley, M. G., Schulp, A. S., Meyer, C. A., Leonardi, G., and Mamani, D. K. 2002a. Titanosaur trackways from the Upper Cretaceous of Bolivia: evidence for large manus, wide gauge locomotion and gregarious behaviour. Cretaceous Res. 23: 383–400.

Lockley, M. G., Wright, J. L., White, D., Li., J., Feng, L., Li, H., and Matsukawa, M. 2002b. The first sauropod trackways from China. Cretaceous Res. 23: 363–381.

Lockley, M. G., Wright, J. L., and Thies, D. 2004. Some observations on the dinosaur tracks at Münchehagen (Lower Cretaceous), Germany. Ichnos 11: 261–274.

MacClary, J. S. 1938. Dinosaur trails of Purgatory. Sci. Am. 158: 72.

Mazzetta, G. V., and Blanco, R. E. 2001. Speeds of dinosaurs from the Albian–Cenomanian of Patagonia and sauropod stance and gait. Acta Palaeontol. Polonica 46: 235–246.

Meyer, C.A., Lockley, M.G., Robinson, J.W. and V. F. Santos, V.F. 1994. A comparison of well-preserved sauropod tracks from the Late Jurassic of Portugal and the Western United States: evidence and implications. Gaia: Revista de Geociencias, Museu Nacional de Historia Natural, Lisbon, Portugal. 10: 57–64.

Meyer, C.A. and Thuring, B., 2003. Dinosaurs in Switzerland, Comptes Rendu Palevol 2: 103–117.

Moratalla, J. J., Garcia-Mondejar, J., Lockley, M. G., Sanz, J. L. and Jimenez, S. 1994. Sauropod trackways from the Lower Cretaceous of Spain. GAIA 10: 75–83.

Novikov, V. P., Suprichev, V. V., and Salikhov, F. S. 1993. Dinosaur tracksites in the Obikhingou River Basin (Peter the Great range). Rep. Acad. Sci. Tadjikistan Repub. 1(36): 55–58.

Olsen, P. E. 1995. A new approach for recognizing track makers. Geol. Soc. Am. Abstr. Progr. 27: 86.

Olsen, P. E., and Galton, P. 1984. A review of the reptile and amphibian assemblages from the Stormberg from Southern Africa with special emphasis on the footprints and the Stormberg. Paleontograp. Afr. 25: 87–110.

Paik, I. S., Kim, H. J., and Lee, Y. I. 2001. Dinosaur track-bearing deposits in the Cretaceous Jindong Formation, Korea: occurrence, palaeoenvironments and preservation. Cretaceous Res. 22: 79–92.

Pittman, J. G. 1989. Stratigraphy, lithology, depositional environment and track type of dinosaur trackbearing beds of the Gulf Coastal Plain. In: Gillette, D. D., and Lockley, M. G. (eds.). Dinosaur Tracks and Traces. Cambridge University Press, Cambridge. Pp. 135–154.

Pittman, J. G., and Gillette, D. D. 1989. The Briar Site: a new sauropod dinosaur tracksite in Lower Cretaceous beds of Arkansas, USA. In: Gillette, D. D., and Lockley, M. G. (eds.). Dinosaur Tracks and Traces. Cambridge University Press, Cambridge. Pp. 313–332.

Prince, N. K., and Lockley, M. G. 1989. The sedimentology of the Purgatoire Tracksite region, Morrison Formation of southeastern Colorado. In: Gillette, D. D., and Lockley, M. G. (eds.). Dinosaur Tracks and Traces. Cambridge University Press, Cambridge. Pp. 155–164.

Raath, M. A. 1972. Fossil vertebrate studies in Rhodesia: a new dinosaur (Reptilia: Saurischia) from near the Trias-Jurassic boundary. Arnoldia 5: 1–37.

Radley, J. D. 1994. Field meeting, 24–5 April 1993: the Lower Cretaceous of the Isle of Wight. Proc. Geol. Assoc. 105: 145–152.

Romano, M., and Whyte, M. A. 2003. Jurassic dinosaur tracks and trackways of the Cleveland Basin, Yorkshire: preservation, diversity and distribution. Proc. Yorkshire Geol. Soc. 54: 185–215.

Salgado, L., Coria, R.A., and Calvo, J.O. 1997. Evolution of titanosaurid sauropods. I: Phylogenetic analysis based on the postcranial evidence. Ameghiniana 34: 3–32.

Santos, V. F., Lockley, M. G., Meyer, C. A., Carvalho, J., Galopim de Carvalho, A. M., and Moratalla, J. J. 1994. A new sauropod tracksite from the Middle Jurassic of Portugal. GAIA 10: 5–14.

Schulp, A. S., and Brokx, W. A. 1999. Maastrichtian sauropod tracks from the Fumanya site, Berguedà, Spain. Ichnos 6: 239–250.

Thulborn, R. A. 1990. Dinosaur Tracks. Cambridge University Press, Cambridge. 410 pp.

Thulborn, T., T., Hamley, and Foulkes, P. 1994. Preliminary report on sauropod dinosaur tracks in the Broome Sandstone. GAIA 10: 85–94.

Unwin, D. M. 1989. A predictive method for the identification of vertebrate ichnites and its application to pterosaur tracks. In: Gillette, D. D., and Lockley, M. G. (eds.). Dinosaur Tracks and Traces. Cambridge University Press, Cambridge. Pp. 259–274.

Upchurch, P. 1994. Manus claw function in sauropod dinosaurs. GAIA 10: 161–72.

———. 1998. The phylogenetic relationships of sauropod dinosaurs. Zool. J. Linn. Soc. 124: 43–103.

Weishampel, D., Dodson, P., and Osmolska, H. 1990. The Dinosauria. University of California Press, Berkeley. 733 pp.

Wilson, J. A., and Carrano, M. T. 1999. Titanosaurs and the origin of "wide gauge" trackways: a biomechanical and systematic perspective on sauropod locomotion. Paleobiology 25: 252–267.

Wilson, J. A., and Sereno, P. C. 1998. Early evolution and higher-level phylogeny of sauropod dinosaurs. Society of Vertebrate Paleontology Memoir 5.

Wilson, J.A. and Upchurch, P., 2003. A revision of *Titanosaurus* Lydekker (Dinosauria–Sauropoda), the first dinosaur genus with a 'gondwanan' distribution. Journal of Systematic Palaeontology 1: 125–160.

Wright, J. L., Radley, J. D., Upchurch, P., and Wimbledon, W. A. 1997. Keates' Quarry dinosaur footprint site, Intermarine Member, Purbeck Limestone Group (Berriasian). UK National Trust Site Report.

Zhang, Y., Yang, D., and Peng, G. 1984. [New materials of *Shunosaurus* from the Middle Jurassic of Dashanpu, Zigong, Sichuan.] J. Chengdu Coll. Geol. Suppl. 2: 1–12. (In Chinese.)

APPENDIX 9.1. PODIAL SYNAPOMORPHIES OF SAUROPODOMORPHS AND THEIR POTENTIAL EXPRESSION IN TRACKS

Abbreviations: C, Upchurch (1998) numbered characters; WS, Wilson and Sereno (1998) numbered characters; Sa, Salgado et al. (1997) numbered characters. See also Carrano and Wilson (2001).

	CHARACTER NO.	PREVIOUS NO.	SYNAPOMORPHY	ICHNOLOGICAL EXPRESSION
Sauropoda: sauropodomorphs more closely related to Saltasaurus *than* Plateosaurus *(Wilson & Sereno 1998)*				
Posture	1	C 186, C 158, WS 1	vQuadrupedal posture	Trackway should show manus and pes impressions
Pes	2	WS 14	Proximal ends of metatarsals I & V subequal in area to those of metatarsals II & IV: metatarsal I may even be larger in area than metatarsal II (fig. 9.3)	Tracks showing that fewer than five digits contacted the ground cannot have been made by sauropods
	3	WS 15	Length of metatarsal V >70% metatarsal IV (fig. 9.3)	
	4	WS 16	Entaxonic pes structure	Strong decreasing gradient in ungual size from digit I to digit V should be visible in footprints
	5	WS 17	Ungual of pedal digit I deep and narrow	Digit I pedal claw impressions are predicted to be visible, and sickle-shaped and oblique to the ground rather than triangular
Manus	6	C 168	Metacarpal V robust and >90% of the length of the longest metacarpal; all five manual digits participated in weight bearing	Digital impressions, if present, should show five digits.
Eusauropoda (Upchurch 1995): sauropods more closely related to Saltasaurus *than to* Vulcanodon				
Manus	7	C 170, WS 43	Phalanges on manual digits II & III reduced; all digits subequal in length apart from digit I, which bears a very large claw	Digital impressions should be of a similar depth. Large claw impression should be apparent on the medial side of the manus print
	8		Manual phalanges (other than unguals) broader than long, without well-formed collateral ligament pits (fig. 9.3).	Short, wide digit impressions, with no discrete phalangeal pads
	9	WS 50	Metatarsal III length <25% that of tibia; overall reduction in relative length of metatarsals	Eusauropod pes impressions would be relatively larger, in length and width, than those of more basal sauropods
	10	WS 52	Metatarsals with spreading configuration; indicates a sloping orientation for the metatarsals; semidigitigrade pedal structure	Presence of fleshy pad (Janensch 1922)—pes impressions would be widest anteriorly, narrowing to the rear.

APPENDIX 9.1. (continued)

CHARACTER NO.	PREVIOUS NO.	SYNAPOMORPHY	ICHNOLOGICAL EXPRESSION
11	WS 51, C 198	Metatarsal I thicker than all other metatarsals, short, and very robust; asymmetric pes	Sauropod pes tracks are turned outward at up to 30° from the trackway midline; digit I is the leading digit in the foot and bears the most force during push-off. Leading margin of sauropod pes impressions (anteromedial side of the pes) often deepest part of print (Wilson & Sereno 1998)
12	WS 53	Nonterminal phalanges of pedal digits broader than long	Reduced digit lengths; pedal digit impressions will be very short and show no discrete phalangeal pads
13	C 202	Loss/extreme reduction of collateral ligament pits on pedal phalanges	
14	WS 55, C 200	Penultimate phalanges of pedal digits II–IV rudimentary or absent	
15	WS 56	Pedal digits I–III have sickle-shaped unguals	Deep narrow claws for all three digits should be evident
16	WS 54	Ungual of pedal digit I longer than metatarsal I	Ungual I very large and should be evident
17	WS 57	Ungual of pedal digit IV rudimentary or absent	Tracks should show three large claws, decreasing in size from I to III (see no. 4 above) and either no or a small claw impression for digit IV; digit V should not bear a claw, although it might leave a short blunt digit impression. Any track with five pedal claw impressions could not have been made by a eusauropod
18	WS 64	Pedal unguals asymmetrical. All pedal unguals lay with part of their medial surface on the ground and with the distal end of the ungual directed laterally	Footprints should show pedal claw impressions pointing anterolaterally with respect to the pes axis. Individual claw impressions oblique to the ground, flattened, shallow, and curved
19	WS 73	Metatarsals III & IV with minimum transverse shaft diameters <65% of	May indicate that even more weight is being placed on the inner side

APPENDIX 9.1. (continued)

	CHARACTER NO.	PREVIOUS NO.	SYNAPOMORPHY	ICHNOLOGICAL EXPRESSION
			metatarsals I & II	of the pes, but such a subtle difference is unlikely to be discernible in footprints

Eusauropoda (Upchurch 1995): sauropods more closely related to Saltasaurus *than to* Vulcanodon.

Manus	20	WS 80	Long intermetacarpal articulations; metacarpals bound by ligaments into a rigid column; digitigrade manus	Shapes and size of manus impressions within trackways should be very consistent; more consistent morphology than those of basal sauropods
	21	WS 81	Metacarpal proximal ends subtriangular; arc of ca. 270a°	Horseshoe-shaped rather than crescentic manus impressions
	22	C 171	Manual phalangeal formula reduced to 2-2-1-1; further reduction of manus digit length	Unlikely to be evident in footprints
Pes	23	C 201	Pedal phalangeal formula reduced to 2-3-4-2-1; further reduction of pes digit length	Unlikely to be evident in footprints.
	24		Proximal pedal phalanges narrow toward their lateral and palmar margins.	

Camarasauromorpha (Salgado et al. 1997): the most recent common ancestor of Camarasauridae and Titanosauriformes

Manus	25	Sa 12, C167 WS 93	Lengthened metacarpals relative to radius; forelimb relatively longer; Shift in weight distribution?	May have deeper manus than pes impressions
	26	WS 94	Metacarpal I subequal in length to metacarpal IV; relative increase in length; shift from metacarpals II–IV being the longest to I–III; Related to weight distribution?	As above

Titanosauriformes: *Brachiosaurus* + titanosaurs

Manus	27	WS 98	Distal condyle of metacarpal I undivided, phalangeal articular surface reduced; Shift in body weight distribution?	Even deeper manus impressions?—unlikely to be evident in footprints
	28	Sa 16(1)	Claw on manual digit I reduced	Claw impressions in titanosauriform trackways should be small
Pes	29	C 187, WS 100	Femoral head deflected medially	Linked to production of wide gauge trackways (Wilson & Carrano 1999)

APPENDIX 9.1. (continued)

	CHARACTER NO.	PREVIOUS NO.	SYNAPOMORPHY	ICHNOLOGICAL EXPRESSION
			Titanosauria: (Wilson & Upchurch 2003)	
Manus	30	C 166	Metacarpal I longer than metacarpal III and subequal to or longer than metacarpal II; completion of the transition of the main weight-bearing digits of the manus from II–IV to I–III	Even deeper manus impressions?—unlikely to be evident in footprints
	31	Sa 16(2)	Claw on manual digit I reduced	No claw impressions should be visible in the manus tracks
	32	Sa 27	Manual phalanges absent, related to no. 31 above; there can be no claws if there are no phalanges	No phalangeal impressions should be visible in manus tracks of titanosaurids, although distal ends of the metacarpals may have borne pads that left impressions visible in tracks
Pes	33		Deflected femoral axis and beveled knee joint (Wilson & Carrano 1999); development of no. 29 above	Even wider gauge trackways

TEN

Nesting Titanosaurs from Auca Mahuevo and Adjacent Sites

UNDERSTANDING SAUROPOD REPRODUCTIVE BEHAVIOR AND EMBRYONIC DEVELOPMENT

Luis M. Chiappe, Frankie Jackson, Rodolfo A. Coria, and Lowell Dingus

THOUSANDS OF SAUROPOD EGG clutches, some containing eggs with exquisitely preserved embryonic bone and integument, have been discovered in the Late Cretaceous nesting site of Auca Mahuevo (Chiappe et al. 1998, 2000, 2001, 2004; Dingus et al. 2000; Chiappe and Dingus, 2001; Coria et al. 2002) and adjacent localities in northwestern Patagonia, Argentina (fig. 10.1). Five expeditions to this extraordinary area (1997, 1999, 2000, 2001, and 2002) have yielded a wealth of information for understanding the prehatching development, the nesting structure, the egg morphology and malformation, and the reproductive behavior of these dinosaurs. Cranial characters of the in ovo embryos allowed the identification of the eggs as those of titanosaurs (Chiappe et al. 1998, 2001; Salgado et al. 2005). Textural differences in the sediments containing some clutches have illuminated aspects of the nest structure of these animals (Garrido et al. 2001; Chiappe et al. 2004). Microstructural studies have expanded our understanding of the eggshell variation (Grellet-Tinner et al. 2004) and incidence of egg malformation (Jackson et al. 2001, 2004) within a titanosaur population. Detailed mapping of clutch spatial distribution and egg-bed stratigraphic position, together with studies of their sedimentary context, have provided the basis for inferring aspects of the nesting behavior of these dinosaurs (Chiappe et al. 2000; Chiappe and Dingus 2001). Combining taxonomic constraint with extensive sampling, research at Auca Mahuevo and its adjacent localities offers the clearest picture to date of sauropod reproduction and embryonic development. In this chapter, we summarize the major developments of this research program and discuss their significance in light of previous interpretations of the reproductive biology of these colossal dinosaurs.

GEOLOGICAL SETTING AND ASSOCIATED DINOSAUR FAUNA

Auca Mahuevo lies approximately 120 km northwest of the city of Neuquén in the homonymous Argentine province (fig. 10.1). Two adjacent nesting sites, Barreales Norte and Barreales Escondido, are 15 and 22 km south of Auca Mahuevo, respectively. These three sites occur within an 85-m-thick sequence of sandstone,

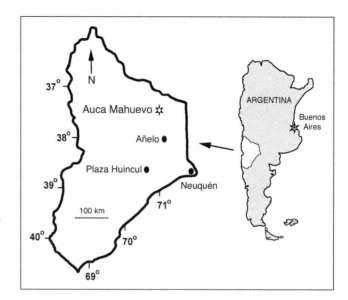

FIGURE 10.1. Map of the province of Neuquén (Argentina) indicating the location of Auca Mahuevo. Barriales Norte and Barreales Escondido are 15 and 22 km south of Auca Mahuevo, respectively.

siltstone, and mudstone of the Anacleto Formation (fig. 10.3), one of the lithostratigraphic units of the fossiliferous Cenomanian–Campanian Neuquén Group (Ramos 1981; Legarretta and Gulisano 1989; Ardolino and Franchi 1996; Leanza 1999; Dingus et al. 2000). Recent paleomagnetic analysis of rocks from the lower portion of the Auca Mahuevo section containing egg-beds 1–3 established the presence of a Reversed magnetozone in the Anacleto Formation (Dingus et al. 2000). In conjunction with earlier biochronologic correlations, this magnetozone was tentatively correlated with C33R, in the early–middle Campanian, between 83.5 and 79.5 million years ago (Dingus et al. 2000).

Exposures at Auca Mahuevo include at least four distinct egg-bearing layers (figs. 10.2, 10.3; egg-beds 1–4), which occur in uniform mudstones representing overbank deposits on a fluvial plain (Chiappe et al. 2000). Two of these layers (egg-beds 2 and 3) can be subdivided into two horizons of eggs, separated by a few centimeters of sediments in the case of egg-bed 3 and about 1 m in the case of egg-bed 2. Egg-beds 3 and 4 are laterally continuous for at least several kilometers (Chiappe et al. 2000; fig. 10.2). Most egg clutches exhibit no discernible evidence of nest structure. However, thin sandstones representing abandoned channel and crevasse splay deposits occur within the Auca Mahuevo section, and in egg-bed 4 they occasionally preserve nesting trace fossils (Chiappe et al. 2004).

Less than 1 m above the nesting structures in egg-bed 4 is a sandy red mudstone that contains a great number of sauropod tracks (Loope et al. 2000). The sauropod tracks are recognizable as thin (1-cm-thick), laterally discontinuous limy deposits that measure up to 80 cm in diameter and are oval to circular in shape. These platter-shaped features are interpreted to contain precipitates of calcium carbonate within the track depression, possibly from evaporation of standing water (Loope et al. 2000). Traceable over several kilometers, the track horizon provides an index layer useful for verifying that the underlying nesting traces occur on the upper surface of a single sandstone stratum. Additional layers of calcium carbonate precipitates, also interpreted as footprints, occur elsewhere in the Auca Mahuevo section.

Clutches from egg-bed 3 occur in paleovertisols (Chiappe and Dingus 2001), recognizable by the abundance and widely varying orientation of slickensides—striated surfaces produced by soil movement within the nesting ground. Vertisols today are associated with clay-rich parent materials and are widespread in regions that experience wet–dry climatic

FIGURE 10.2. Aerial photograph showing egg-beds 3 and 4 at Auca Mahuevo. These egg-beds can be traced laterally for several kilometers.

cycles under semiarid to subhumid environmental conditions. Similar depositional conditions have been inferred for the Late Jurassic–Late Cretaceous of the northern third of Patagonia, where the climate regime has been reconstructed as warm, arid to semiarid, and with a distinct dry season (Andreis 2001).

Several egg-beds also occur at Barreales Norte and Barreales Escondido. Although stratigraphic correlations between these localities and Auca Mahuevo are still preliminary, the available data suggest that these egg-beds can be traced across the vast distances that separate the three sites. Abundant geodes of pale blue celestite crystals that occur in a discrete horizon above Auca Mahuevo's egg-bed 3 are also found at approximately the same stratigraphic position above an egg-bed at Barreales Norte. A footprint layer similar to those of Auca Mahuevo also occurs at this locality at comparable stratigraphic positions. In addition, comparable thicknesses separate egg-beds at these three localities.

Several fossils of adult sauropod and theropod dinosaurs have been found at Auca Mahuevo. Remains of titanosaurs have been collected from egg-bed 4 and from strata between this egg layer and egg-bed 3 (fig. 10.3). These remains are yet to be studied in detail. Among the theropods found at this site is the nearly complete skeleton of *Aucasaurus garridoi* (Coria et al. 2002), an abelisaurid collected from a laminated mudstone unit some 25 m above egg-bed 4. Isolated teeth comparable in

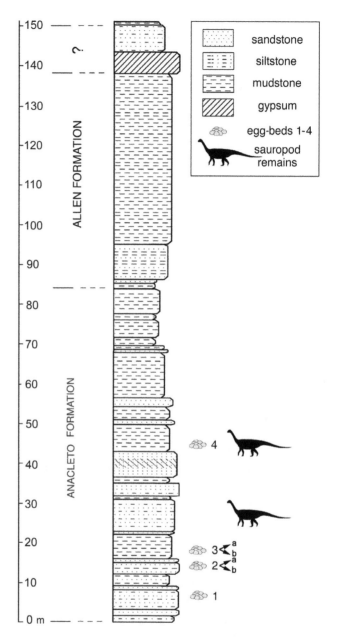

FIGURE 10.3. Composite stratigraphic section at Auca Mahuevo. Note the presence of four stratigraphically distinct beds of essentially identical titanosaur eggs. Egg-beds 2 and 3 can each be subdivided into two layers.

morphology with those of dromeosaurids and the fragmentary remains of a large indeterminate theropod the size of a charcharodontosaurid were also collected from Auca Mahuevo (Coria and Arcucci 2005).

EGGS, CLUTCHES, AND NESTS

Eggs from all these localities exhibit similar size, shape, microstructure, and surface ornamentation as the eggs that contain diagnostic titanosaur remains. The eggs are spherical to subspherical and approximately 13–15 cm in diameter, with a tubercular surface ornamentation consisting of single, rounded nodes (fig. 10.4). The eggshell consists of a single structural layer of calcite—approximately 1.3 mm thick in well-preserved samples—pierced by a pore network of vertical and horizontal canals that intersect one another at the bases of the eggshell units (Grellet-Tinner et al. 2004). The overall microstructure of the eggshell is similar to that

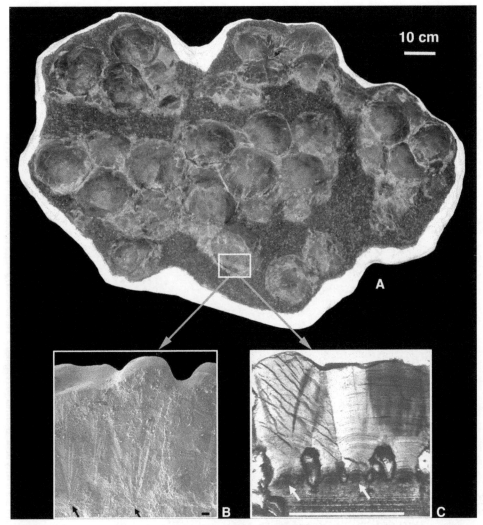

FIGURE 10.4. (A) Auca Mahuevo clutch (Museo Carmen Funes, Plaza Huincul, Argentina; MCF-PVPH-258; quarry of egg-bed 3 illustrated in fig. 10.5) containing nearly 40 eggs. (B) and (C) Scanning electron micrograph and thin section of the Auca Mahuevo eggshell. Scale bars in B and C equal 10 micron and 1 mm, respectively.

described as the ootaxon *Megaloolithus patagonicus* from the Anacleto Formation at Neuquén City (Calvo et al. 1997; fig. 10.4). *Megaloolithus patagonicus* has recently been proposed as a possible junior synonym of *Megaloolithus jabalpurensis* (Vianey-Liaud et al. 2003), a Late Cretaceous (Maastrichtian) oospecies from India. *Megaloolithus patagonicus* is also very similar to Late Cretaceous eggshells from Perú identified by Vianey-Liaud et al. (1997) as *Megaloolithus pseudomamillare* (Grellet-Tinner et al. 2004). Future studies are likely to synonymize many of the megaloolithid oospecies that have been named from various localities around the world. Vianey-Liaud and others (2003) took an initial step in this direction and greatly reduced the number of valid megaloolithid oospecies from India.

Although it is difficult to ascertain the completeness of a fossil egg-clutch, Auca Mahuevo's clutches are composed of numerous eggs, most typically from about 20 to nearly 40 eggs (Chiappe et al. 2004; Jackson et al. 2004; fig. 10.4). Eggs are stacked one on top of the other without any internal spatial arrangement (up to three stacked layers of eggs have been described

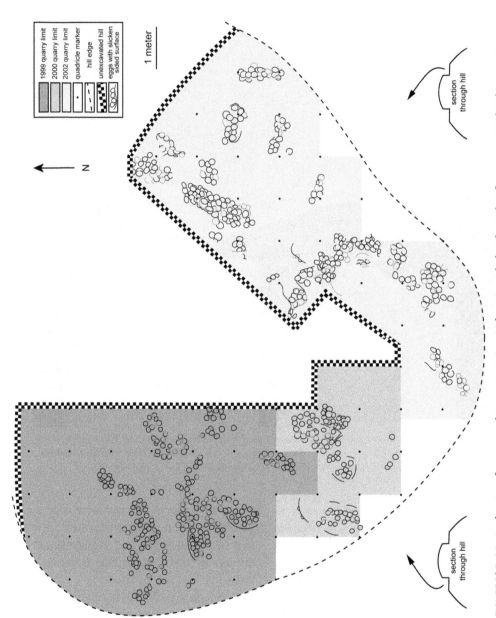

FIGURE 10.5. Map in plan view of eggs exposed at a quarry in Auca Mahuevo's egg-bed 3. The total surface area excavated in this quarry is approximately 65 m².

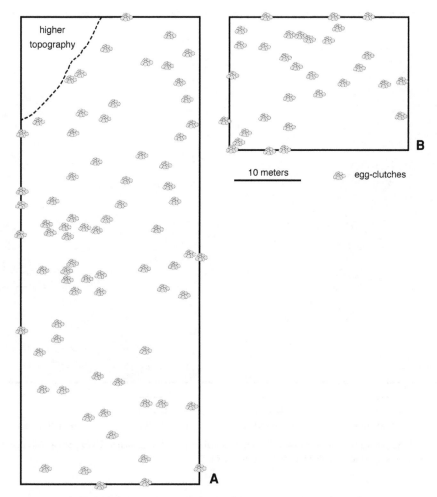

FIGURE 10.6. Clutch maps of two sites contained within erosional surfaces of Auca Mahuevo's egg-bed 3. Seventy-four and 31 randomly distributed egg clutches were mapped within 1,701 m^2 (A) and 486 m^2 (B). Even if recognition of egg-clutch boundaries is sometimes difficult (see text), these numbers provide a minimal estimate of the clutches laid on these surfaces.

for some clutches). More than 500 whole eggs were quarried over a 65-m^2 surface of egg-bed 3 during the 1999, 2002, and 2002 field seasons (fig. 10.5). Spatial analysis of the eggs exposed during 1999, in conjunction with a three-dimensional map constructed from field data, revealed an unexpectedly high egg density (11 eggs/m^2) in this quarry. At this quarry, the boundaries of individual egg clutches are sometimes difficult to determine. In some cases, eggs occur as large accumulations that clearly represent more than one clutch (fig. 10.5). To a certain degree, such a pattern can be explained as the consequence of postburial egg displacement due to edaphic processes—a substantial displacement along the friction planes of slickensides has been observed on several instances. It is also possible that the eggs were somewhat displaced by flotation prior to their final burial. Detailed stratigraphy and mapping of egg depths within this quarry also revealed two distinct levels of eggs, separated by several centimeters of sediment, which were interpreted as different egg-laying events (Chiappe et al. 2000).

Mapping of egg clutches exposed on erosional surfaces of egg-bed 3 produced a concentration of 74 and 31 randomly distributed egg clutches within 1,701 and 486 m^2, respectively

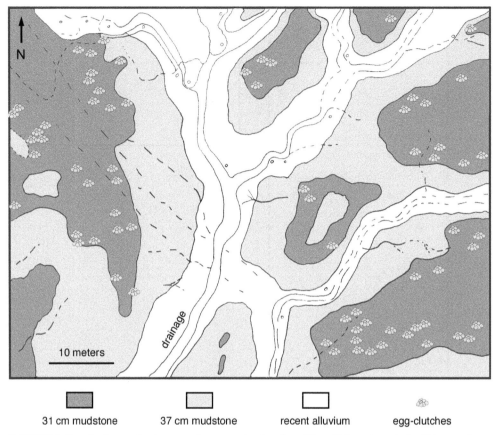

FIGURE 10.7. Egg-clutch map of a site exposed on an erosional surface of Auca Mahuevo's egg-bed 4. This map shows a density comparable to that recorded for egg-bed 3 (fig. 10.6).

(Chiappe et al. 2000; fig. 10.6). Egg clutches on erosional surfaces are often recognized as accumulations of partially weathered eggs whose periphery (eggshell) is still vertically oriented within the substrate. Heavily weathered clutches are recognized as large accumulations of broken eggshells that fan out from a core and that are separated from other egg clutches by areas with minimal eggshell. If any, maps constructed with these criteria for recognition underestimate the number of egg clutches that were laid within these paleosurfaces because of the above-mentioned causes of egg accumulation and displacement. The maximum stratigraphic thickness of strata containing egg clutches within these areas was less than 70 cm. A similar map of an erosional surface of egg-bed 4 showed a comparable high density of randomly distributed egg clutches (fig. 10.7). This high concentration of clutches provides opportunities for studying aspects of sauropod reproductive biology that are otherwise difficult to assess. For example, the discovery of abnormal, multilayered eggs in 6 of nearly 400 in situ clutches surveyed at Auca Mahuevo's egg-beds 2 and 3 (Jackson et al. 2001, 2004) provided the first assessment of the incidence of egg malformation in a sauropod population. This study detected three different types of abnormal eggshells and the fact that pathologic eggs were laid in clutches containing a majority of normal eggs.

Several egg clutches from egg-bed 4 preserve evidence of nest architecture. The eggs of these clutches are similar in size, shape, and microstructure to other Auca Mahuevo eggs containing embryonic remains of titanosaur sauropods (Chiappe et al. 2001). The clutches are contained in large, subcircular to subellipti-

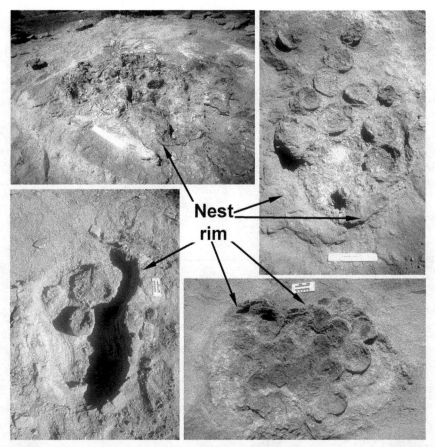

FIGURE 10.8. Four titanosaur nests from Auca Mahuevo's egg-bed 4. These nests consist of surface depressions whose periphery is defined by an elevated rim of structureless sand. Textural differences between the depression fill and the substrate indicate that titanosaurs laid their eggs on the surface.

cal to kidney-shaped depressions in sandstone, although the depression and interstitial spaces between the eggs are filled with mudstone (fig. 10.8). All of the depressions truncate primary stratification of the host substrate and are encircled by a rim of structureless sandstone. The maximum axis of the depressions ranges in length between 100 and 140 cm and their depth is about 10 to 18 cm. These egg-filled depressions are interpreted as excavated nests on the basis of lithologic criteria (Garrido et al. 2001; Chiappe et al. 2004): (1) the depressions truncate the primary stratification of the host sandstone, (2) a massive rim surrounds the perimeter of the depression, and (3) textural differences exist between the host sandstone and the in-filling sediment.

We have interpreted the structureless sand of the depression's rim as the piled debris produced during the construction of the nest and the mudstone surrounding the eggs as the result of a flooding event that is similar to most overbank deposition of the Anacleto Formation in the Auca Mahuevo section. The fact that the eggs in the recognized nests are not entombed by sandstone but by mudstone resulting from the flooding event indicates that the nesting sauropod did not bury the eggs after they were laid (Chiappe et al. 2004). All other clutches from Auca Mahuevo are likely to have been laid in similarly constructed surface nests, which are not recognizable due to the lack of textural differences between the host mudstone and the in-filling sediment. These nests could have

FIGURE 10.9. Titanosaur embryonic remains from Barreales Norte, 15 km south of Auca Mahuevo. Note the remains of limb bones covered by the eggshell of tuberculate ornamentation.

been lined and covered with vegetation (Grellet-Tinner et al. 2004), a suggestion that several researchers have made for other megaloolithid-type eggs from the Late Cretaceous of France (Erben 1970; Kérourio 1981; Cousin 1997).

EMBRYONIC MORPHOLOGY AND SYSTEMATICS

In ovo embryonic remains have been found in two situations: (1) encased in highly cemented fragments of eggs that occur as "float" on erosional surfaces and (2) compressed against the bottom inner shell of in situ eggs. Despite the abundance of embryos, no remains of hatchlings or early juveniles have been discovered. So far, most embryonic remains have been collected from Auca Mahuevo's egg-bed 3 (Chiappe et al. 1998, 2001), although some have also been found at Barreales Norte (fig. 10.9) and Barreales Escondido.

Patches of integument, preserved as calcitic impressions (negative and positive), are common in highly cemented egg fragments that

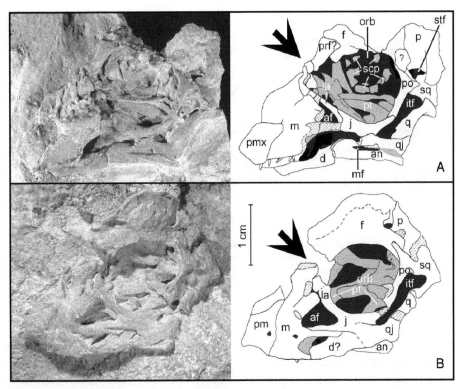

FIGURE 10.10. Photos and interpretive drawings of embryonic titanosaur skulls from Auca Mahuevo's egg-bed 3 in left lateral view. (A) MCF-PVPH-272; (B) MCF-PVPH-263. Arrows point to the approximate location of external nares. Abbreviations: af, antorbital fenestra; an, angular; d, dentary; f, frontal; itf, infratemporal fenestra; j, jugal; la, lacrimal; m, maxilla; mf, mandibular fenestra; orb, orbit; p, parietal; pmx, premaxilla; po, postorbital; prf, prefrontal; pt, pterygoid; q, quadrate; qj, quadratojugal; scp, scleral plates; sq, squamosal; stf, supratemporal fenestra.

occur on erosional surfaces. These skin impressions display a range of nonoverlapping tubercular patterns including rosettes, flowerlike arrangements, and rows of larger tubercles (Chiappe et al. 1998).

Osteological remains are most typically found flattened against the bottom inner shell of in situ eggs. Cranial material is better ossified than limb material, which typically lacks ends (fig. 10.9). Some embryos are clearly larger (by as much as 25%) than others. Although it is difficult to estimate the degree of development of the embryos in comparison to the embryonic stages of extant reptiles, the basic sauropod morphogenetic plan is readily visible in the available skulls. A general "Haeckelian pattern" is also evident in these skulls, which in some respects resemble conditions found outside Sauropoda (e.g., jugal forming part of the ventral margin of the skull, minimally retracted nares).

The anatomical information available in the handful of embryos initially prepared supported the identification of these embryos as neosauropod dinosaurs (Chiappe et al. 1998), the clade originating from the common ancestor of the Late Jurassic *Diplodocus longus* and the Late Cretaceous titanosaur *Saltasaurus loricatus* (Wilson and Sereno 1998). Originally, although synapomorphies of Sauropoda were identified among the disarticulated skulls (e.g., jugal process of the postorbital much longer than the rostrocaudal extension of the dorsal end of this bone [Chiappe et al. 1998]), the dental morphology played a key role in the taxonomic identification of the embryos: the smooth enamel (devoid of denticles) of the crowns and the pencil shape (straight margins and tapering

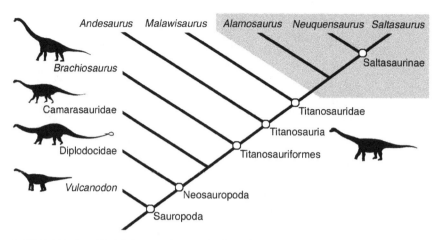

FIGURE 10.11. Simplified cladogram of sauropod relationships. Derived dental characters present in the Auca Mahuevo embryos support their inclusion in a Subgroup of Titanosauria that excludes the African *Malawisaurus dixeyi*. (After Salgado et al. 1997.) Icons from Wilson and Sereno (1998).

crowns) of the teeth, derived characters of neosauropods and diplodocoids + titanosaurs, respectively (Wilson and Sereno 1998; Wilson 2002), supported the placement of the embryos within Neosauropoda. However, because pencil-shaped teeth are commonly interpreted as independently evolved in both diplodocoids and titanosaurs (e.g., Salgado and Calvo 1997; Wilson and Sereno 1998; Curry Rogers and Forster 2001; Wilson 2002), the presence of this condition alone was insufficient for placing the Auca Mahuevo embryos within either of these two neosauropod clades. Subsequent discoveries of embryos yielding more complete and articulated skulls (Chiappe et al. 2001; Salgado et al. 2005) (fig. 10.10) revealed additional synapomorphies of neosauropods and of the more inclusive taxon Eusauropoda (fig. 10.11)—taxa more closely related to *Saltasaurus loricatus* than to *Vulcanodon karibaensis* (Wilson and Sereno 1998). For example, the new embryos display the eusauropod condition of a stepped snout, an absence of a fossa surrounding the antorbital fenestra, a rostral expansion of the quadratojugal, and a lack of contact between this bone and the squamosal (Chiappe et al. 2001; fig. 10.10). Likewise, the new embryos exhibit the neosauropod condition of a postorbital bar that is broader transversely than rostrocaudally (Chiappe et al. 2001). Most importantly, comparisons between these more recently discovered embryonic skulls and titanosaur cranial remains revealed several apparent synapomorphies of this sauropod clade. For example, the embryos exhibit the same ventral notch of the dentigerous margin, between the maxilla, jugal, and quadratojugal, as the titanosaurs *Rapetosaurus krausei* (Curry Rogers and Forster 2001, 2004) and *Nemegtosaurus mongoliensis* (Nowinski 1971), from the Late Cretaceous of Madagascar and Mongolia, respectively. They also share with this taxon and with other titanosaurs from the Late Cretaceous of Argentina (i.e., *Antarctosaurus wichmannianus* and an undescribed titanosaur from northwestern Patagonia) the low rostral portion of the dentary (Chiappe et al. 2001). The remarkable width of the skull roof, as inferred from the size of the frontals and parietals, and the presence of a large mandibular fenestra are other derived features shared by the embryos and the above-mentioned undescribed titanosaur from northwestern Patagonia (Coria and Salgado 1999). These embryonic skulls represent the most complete titanosaur crania. Furthermore, the embryonic material suggests that previous skull reconstructions of these dinosaurs as "camarasauroid" (Salgado and Calvo 1997) are incorrect. Neither a vertical orientation of the quadrate nor the presence of broad nares separated by an elevated premaxillary–nasal arch, conditions typical of "camarasauroid" skulls, is present in the well-preserved skulls of the Auca Mahuevo embryos (Chiappe et al. 2001).

Although the recently discovered embryos have provided support for identification of the Auca Mahuevo eggs as those of titanosaurs, systematic placement of the embryos beyond Titanosauria (i.e., all sauropods more closely related to *Saltasaurus loricatus* than to either *Brachiosaurus brancai* or *Euhelopus zdanskyi* [Wilson and Sereno 1998]) remains problematic. This is primarily because of the paucity of cranial anatomical information available for adult titanosaurs (Salgado and Calvo 1997; Curry Rogers and Forster 2001). The presence of pencil-like teeth, so far known only for *Nemegtosaurus mongoliensis* (Nowinski 1971), *Alamosaurus sanjuanensis* (Kues et al. 1980), and saltasaurines (most recent common ancestor of *Neuquensaurus australis* and *Saltasaurus loricatus* plus all its descendants [Salgado and Calvo 1997]) among titanosaurs, provides support for the placement of the embryos within a subgroup of titanosaurs that excludes the most primitively toothed *Malawisaurus dixeyi* (Jacobs et al. 1993; fig. 10.11). Nonetheless, this interpretation becomes more complex in light of recent phylogenetic inferences indicating that this specialized dental condition could have evolved more than once within titanosaurs (Curry Rogers and Forster 2001).

Despite these reservations, comparisons between the embryos and the best-preserved skulls of adult titanosaurs, those of *Nemegtosaurus mongoliensis* and *Rapetosaurus krausei*, suggest that dramatic transformations must have occurred during the ontogeny of these dinosaurs. The frontals and parietals became greatly reduced in size and they migrated to the dorsocaudal and caudal region of the orbit. The latter became ventrally constricted and adopted an inverted tear-shaped appearance. The rostrum became substantially enlarged, probably as a consequence of maxillary expansion, and the maxilla developed a connection with the quadratojugal, thus excluding the jugal from the ventral margin of the skull. In addition, the external nares expanded in size and migrated backward, to be relocated on top of the orbits.

In addition, the Auca Mahuevo embryos have provided evidence that may potentially clarify the sequence of transformations that occurred during the long evolution of sauropods (Chiappe et al. 2001). An example of how these new developmental data could potentially elucidate aspects of sauropod evolution is provided by examining two salient features of their cranial architecture: the narial retraction and the forward rotation of the braincase. Salgado and Calvo (1997) suggested that the partial to extreme retraction of the eusauropod nares could have been evolutionarily coupled to the forward rotation of the braincase, best exemplified in diplodocids (McIntosh 1997) and some titanosaurs (Salgado and Calvo 1997; Curry Rogers and Forster 2001, 2004). Because the best-preserved embryos are exposed in lateral view, crushed against the inner shell (fig. 10.10), the exact location of the external nares cannot be directly observed. However, the location of the external nares can be inferred from the orientation of the lacrimals, which in eusauropods mark the approximate caudal end of the nares. The rostrodorsal orientation of the lacrimal suggests that the nares of the embryos opened in front of the orbit, dorsorostral to the antorbital fenestra. This position is also supported by the rostral extension of the frontals, which in the Auca Mahuevo embryos nearly reach the rostral margin of the orbit (fig. 10.10). The paucity of cranial material of adult titanosaurs has prevented determination of the location of the nares in this group of sauropods—however, *Rapetosaurus krausei* (Curry Rogers and Forster 2004) and *Nemegtosaurus mongoliensis* appear to have fully retracted nares (this condition is best observed in a yet undescribed specimen of *Nemegtosaurus mongoliensis* housed at the Mongolian Natural History Museum in Ulaanbataar). Although it is likely that the minimally retracted nares of the embryos migrated backward during postnatal allometric development, the rostroventral orientation of the quadrate and squamosal suggests that at that particular stage of development, the embryonic braincase was partially rotated (see Salgado and Calvo [1997] for correlations between braincase rotation and quadrate orientation). This evidence is contrary to Salgado and Calvo's (1997) hypothesis of a concerted evolution of the narial retraction and braincase rostral rotation of eusauropods (Chiappe et al. 2001). These conditions are likely to have evolved independently, although confirmation of this awaits the discovery of adult skulls with conditions resembling those of the Auca Mahuevo embryos—unretracted nares and already rotated braincase.

BEHAVIORAL INFERENCES

Although the behavior of extinct organisms cannot be directly observed, it can be inferred when the product of an organism's activity is preserved in the fossil record (Clark et al. 1999). The in situ eggs of Auca Mahuevo and adjacent localities are the preserved physical evidence of the sauropods' egg-laying behavior. The mapping and collection of eggs, clutches, and nests at these localities have led us to infer several aspects of the reproductive behavior of titanosaur sauropods (Chiappe et al. 2000; Chiappe and Dingus 2001).

The high concentration of egg clutches distributed in a relatively narrow stratigraphic horizon (e.g., 70 cm in Auca Mahuevo's egg-bed 3) suggests a gregarious nesting behavior. Even if each egg-bed could preserve eggs laid during more than one closely occurring nesting season, and taking into consideration that the specifics of the gregarious behavior we have envisioned (the number of females nesting at approximately the same time and in a given season, the frequency of reproductive seasons, and other similar questions) remained unanswered, the density of eggs contained in these layers is such that the conclusion of gregariousness seems unavoidable. It is highly implausible that solitary females laid the thousands of egg clutches contained in these localities. Based on clutches that have been quarried in situ, many eggs were preserved whole, suggesting that they were buried quickly and did not sit out on the paleosurface for very long, rendering them vulnerable to natural processes of disintegration and trampling during subsequent breeding seasons.

The six stratigraphically distinct egg layers (egg-beds 1–4, with egg-beds 2 and 3 each consisting of two egg levels) containing eggs of similar morphology suggest that one sauropod species nested at this site at least six separate times. A minimum of two beds of morphologically similar eggs also occurs at Barreales Norte and Barreales Escondido, supporting site fidelity for these localities as well. The most parsimonious assumption, therefore, is that all eggs were laid by the same titanosaur species. Discovery of additional embryonic remains in eggs from these egg-bearing layers will provide a means for testing this hypothesis.

The discovery at Auca Mahuevo of well-preserved nest traces provided indisputable evidence of nest construction and architecture (Chiappe et al. 2004). Sedimentological evidence showed that, contrary to most modern reptiles, titanosaurs laid eggs in excavated depressions without burying them. Although nest attendance by titanosaurs may be inferred by phylogenetic bracketing (all living archosaurs attend their nests), adult size and proximity between clutches (fig. 10.5) suggest little or no parental care of their clutches, a conclusion again supported by the lack of evidence of trampling in our quarry of egg-bed 3, where most eggs show minimal crushing.

COMPARISONS TO OTHER NESTING SITES

A large number of dinosaur egg localities contain clutches of eggs similar to those from Auca Mahuevo (e.g., subspherical eggs with a relatively thick eggshell comprised of a single structural layer of calcite, with shell units well separated from each other and a tuberculate surface ornamentation) (Sahni et al. 1990; Powell 1992; Vianey-Liaud et al. 1990; Mohabey 1996; Calvo et al. 1997). Traditionally, these eggs have been classified within the Megaloolithidae category of eggshell parat4xonomy and considered to have been laid by sauropod dinosaurs (Zhao 1979; Mikhailov 1991, 1997). However, Auca Mahuevo and its adjacent localities are the only sites in the world where diagnosable remains of sauropod dinosaurs have been found inside eggs, and although these embryos support the identification of some megaloolithid-type eggs (e.g., *Megaloolithus patagonicus*) as sauropod eggs, it would be risky to extrapolate such a conclusion to all other eggs of similar morphology.

Despite this paucity of eggs containing identifiable embryonic remains at other sites, several assumptions have been made regarding the nest construction, egg-laying behavior, and physiology of sauropod dinosaurs. Inference of nest architecture typically is based on clutch geometry rather than primary lithologic attributes of the surrounding sediment (Dughi and Sirugue 1966; Kérourio 1981; Williams et al. 1984; Faccio 1990; Sahni et al. 1990; Powell 1992; Sanz et al. 1995; Mohabey 1996; Calvo et al. 1997). For example, a single layer of megaloolithid eggs from India that occurred in a <1-m² area was used to infer a saucer-shaped sauropod nest (Mohabey 1996, 2000). However, the margins of the "nest" are described as homogeneous with the host rock, with no observable lithological differences. Fossil eggs such as these provide no evidence of nest structure that results from excavation by adult dinosaurs and are, therefore, more appropriately called clutches. To the best of our knowledge, the Auca Mahuevo sauropod nests provide the only documentation of sauropod nest architecture based on lithologic attributes (Chiappe et al. 2004).

A number of egg-laying behaviors have been attributed to sauropod dinosaurs (Moratalla and Powell 1990). For example, it has been suggested that arcs comprised of 15 to 20 eggs (radii, 1.3–1.7m) resulted from the turning radius of the egg-laying female (Cousin et al. 1990). By comparing published limb dimensions of the European Late Cretaceous titanosaur *Hypselosaurus* to radii of the arcs, Cousin et al. (1990) proposed a crouching position for the adult sauropod during egg laying. However, none of the eggs assumed to be of *Hypselosaurus* contain embryonic remains, thus rendering their identification as indeterminate. Such ad hoc assumptions regarding the taxonomy of the egg-laying dinosaur can potentially bias interpretations of data and obscure the true taphonomic picture.

Colonial nesting and/or site fidelity have also been hypothesized for sauropod dinosaurs (Sanz et al., 1995; Figueroa and Powell 2000; López-Martínez 2000; Mohabey 2000). One study reported abundant fragmented eggshell and 24 megaloolithid nests arranged in three clusters in a 6,000-m² area. Extrapolation of the data, however, extended the number of eggs to 300,000, purportedly laid by sauropods nesting on a seashore (Sanz et al. 1995; López-Martínez 2000). Inferences made from these calculations included territorial behavior, high population density, site fidelity, and site preference (Sanz et al. 1995). These interpretations were challenged on the basis of time-averaging of the deposit, nonsynchronous deposition of egg horizons, pedogenesis, and other sedimentological/taphonomic evidence (Sander et al. 1998). Without detailed taphonomic analysis, such paleobiological inferences appear unwarranted (Sander et al. 1998).

Taxonomically unidentified eggs have also been used in studies of sauropod physiology (Case 1978; Erben et al. 1979; Bakker 1986; Paul 1990). For example, egg size data and estimated sauropod hatchling weight were used for determining lifetime reproductive potential as a function of sauropod body mass (Paul 1990) and to estimate the age of sauropods at sexual maturity (Case 1978). In light of the taxonomic uncertainty of the eggs used for these studies, their conclusions are unwarranted.

Many megaloolithid eggs found worldwide share a similar structural morphology with specimens from Auca Mahuevo and most likely represent eggs laid by sauropod dinosaurs. However, taxonomic identification of eggs based on bones within the same stratigraphic unit carries a high potential for error, as shown with the misidentification of both oviraptorid and troodontid eggs (Norell et al. 1994; Horner and Weishampel 1996). In the case of megaloolithid eggs, this situation is further complicated by the discovery of neonate remains of the hadrosaurid *Telmatosaurus transylvanicus* in the proximity of megaloolithid egg-clutches from the late Cretaceous of Romania (Grigorescu et al. 1994, 2003), an association that, if confirmed, would highlight the paraphyletic nature of this egg category. Egg studies that depend on taxonomic or ontogenetic comparisons (e.g., embryo to adult size), therefore, should await definitive identification based on embryonic remains within the egg. Thus far, only the in ovo sauropod remains discovered at Auca Mahuevo provide this crucial evidence.

ACKNOWLEDGMENTS

We thank Michelle Schwengle for rendering the illustrations and Richard Aspinall for assisting with the statistical analyses of egg distribution of Auca Mahuevo's egg-bed 3. We are also grateful to Kristi Curry Rogers and Jeff Wilson for inviting us to contribute to this book and to David Varricchio for his thorough review of the original manuscript. Field and postfield research for this research was supported by the Ann and Gordon Getty Foundation, the Charlotte and Walter Kohler Charitable Trust, the Dirección General de Cultura de Neuquén, the Fundación Antorchas, the Infoquest Foundation, and the Municipalidad de Plaza Huincul.

LITERATURE CITED

Andreis, R. R. 2001. Paleoecology and environments of the Cretaceous sedimentary basins of Patagonia (southern Argentina). VII International Symposium on Mesozoic Terrestrial Ecosystems. Publ. Espec. Asoc. Paleontol. Argentina 7: 7–14.

Ardolino, A. A., and Franchi, M. R. 1996. Geología y Recursos Minerales del Departamento Añelo. Provincia del Neuquén, Rep. Argentina. Publicación Conjunta de la Dirección Provincial de Minería de la Provincia del Neuquén y Dirección Nacional del Servicio Geológico. Anaies (Geol.) 25: 9–106.

Bakker, R. T. 1986. The Dinosaur Heresies. William Morrow, New York. 482 pp.

Calvo, J. O., Engelland, S., Heredia, S. E., and Salgado, L. 1997. First record of dinosaur eggshells (?Sauropoda–Megaloolithidae) from Neuquén, Patagonia, Argentina. GAIA 14: 23–32.

Case, T. J. 1978. Speculations on the growth rate and reproduction of some dinosaurs. Paleobiology 4: 320–328.

Chiappe, L. M., and Dingus, L. 2001. Walking on Eggs. Scribner, New York. 219 pp.

Chiappe, L. M., Coria, R. A., Dingus, L., Jackson, F., Chinsamy, A., and Fox, M. 1998. Sauropod dinosaur embryos from the Late Cretaceous of Patagonia. Nature 396: 258–261.

Chiappe, L. M., Dingus, L., Jackson, F., Grellet-Tinner, G., Aspinall, R., Clarke, J., Coria, R. A., Garrido, A., and Loope, D. 2000. Sauropod eggs and embryos from the Upper Cretaceous of Patagonia. 1st Symposium of Dinosaur Eggs and Embryos, Isona, Spain. Pp. 23–29.

Chiappe, L. M., Salgado, L., and Coria, R. A. 2001. Embryonic skulls of titanosaur sauropod dinosaurs. Science 293: 2444–2446.

Chiappe, L. M., Schmitt, J. G., Jackson, F., Garrido, A., Dingus, L., and Grellet-Tinner, G. 2004. Nest structure for sauropods: Sedimentary criteria for recognition of dinosaur nesting traces. Palaios 19: 89–95.

Clark, J. M., Norell, M. A., and Chiappe, L. M. 1999. An oviraptorid skeleton from the Late Cretaceous of Ukhaa Tolgod, Mongolia, preserved in an avian-like brooding position over an oviraptorid nest. Am. Mus. Novitates 3265: 1–36.

Coria, R. A., and Arcucci, A. 2005. Nuevos dinosaurios terópodos de Auca Mahuevo, Provincia del Neuquén (Cretácico tardío, Argentina). Ameghiniana (in press).

Coria, R. A., and Salgado, L. 1999. Nuevos aportes a la anatomia craneana de los sauropodos titanosauridos. Ameghiniana 36: 98.

Coria, R. A., Chiappe, L. M., and Dingus, L. 2002. A new close relative of *Carnotaurus sastrei* Bonaparte 1985 (Theropoda: Abelisauridae) from the Late Cretaceous of Patagonia. J. Vertebr. Paleontol. 22: 460–465.

Cousin, R., 1997. Les gisements d'oeufs de dinosauriens des Hautes Corbiéres et des Corbiéres Orientales (Aude): Ponte, nidification, microstructre des coquilles. Bull. Soc. Etudes Sci. Aude 97: 29–46.

Cousin, R., Bréton, G., Fournier, R., and Watté, J. 1990. Dinosaur egglaying and nesting in France. In: Carpenter, K., Hirsch, K. F., and Horner, J. R. (eds.). Dinosaur Eggs and Babies. Cambridge University Press, New York. Pp. 56–74.

Curry Rogers, K., and Forster., C. 2001. The last of the dinosaur titans: a new sauropod from Madagascar. Nature 412: 530–534.

———. 2004. The skull of *Rapetosaurus krausei* (Sauropoda: Titanosauria) from the Late Cretaceous of Madagascar. J. Vertebr. Paleontol. 24: 121–144.

Dingus, L., Clarke, J., Scott, G. R., Swisher, C. C., III, Chiappe, L. M., and Coria, R. A. 2000. First magnetostratigraphic/faunal constraints for the age of sauropod embryo-bearing rocks in the Neuquén Group (Late Cretaceous, Neuquén Province, Argentina). Am. Mus. Novitates 3290: 1–11.

Dughi, R., and Sirugue, F. 1966. Sur la fossilization des oeufs de dinosaurs. C. R. Sea. Acad. Sci. 262: 2330–2332.

Erben, H. 1970. Ultrastrukturen und Mineralisation rezenter und fossiler Eischalen bei Vögeln und Reptilien. Biomineral. Forschungsber. 1: 1–66.

Erben, H., Hoefs, J., and Wedepohl, K. H. 1979. Paleobiological and isotopic studies of eggshells from a declining dinosaur species. Paleobiology 5: 380–414.

Faccio, G. 1990. Dinosaurian eggs from the Upper Cretaceous of Uruguay. In: Carpenter, K., Hirsch, K. F., and Horner, J. R. (eds.). Dinosaur Eggs and Babies. Cambridge University Press, New York. Pp. 47–55.

Figueroa, C., and Powell, J. E. 2000. Structure and permeability properties of thick dinosaur eggshells from the Upper Cretaceous of Patagonia, Argentina. Results of a computer simulation analysis. In: Bravo, A. M., and Reyes, T. (eds.). 1st International Symposium of Dinosaur Eggs and Embryos, Isona, Spain. Pp. 51–60.

Garrido, A. C., Chiappe, L. M., Jackson, F., Schmitt, J., and Dingus, L. 2001. First sauropod nest structures. J. Vertebr. Paleontol. 21: 53A.

Grellet-Tinner, G., Chiappe, L. M., and Coria, R. A. 2004. Eggs of titanosaurid sauropods from the Upper Cretaceous of Auca Mahuevo (Argentina). Can. J. Earth Sci. 41: 949–960.

Grigorescu, D., Weishampel, D., Norman, D., Seclamen, M., Rusu, M., Baltres, A., and Teodorescu, V. 1994. Late Maastrichtian dinosaur eggs from the Hateg Basin (Romania). In: Carpenter, K., Hirsh, K. F., and Horner, J. R. (eds.). Dinosaur Eggs and Babies, Cambridge University Press, Cambridge. Pp. 75–87.

———. 2003. The puzzle of Tustea incubation site (Upper Maastrichtian, Hateg Basin, Romania): Hadrosaur hatchlings close to Megaloolithidae type of eggshell. 2nd International Symposium on Dinosaur Eggs and Babies, Abstracts, 16.

Horner, J. R., and Weishampel, D. B. 1996. A comparative embryological study of two ornithischian dinosaurs—A correction. Nature 332: 256–257.

Jackson, F. D., Garrido, A. C., and Loope, D. B., 2001. Sauropod egg clutches containing abnormal eggs from the ate Cretaceous of Patagonia: clues to reproductive biology. J. Vertebr. Paleontol. 21: 65A.

Jackson, F. D., Garrido, A. C., Schmitt, J., Chiappe, L. M., Dingus, L., and Loope, D. B. 2004. Abnormal, multilayered titanosaurid (Dinosauria: Sauropoda) eggs from in situ clutches at the Auca Mahuevo locality, Neuquén Province, Argentina. J. Vertebr. Paleontol. 24: 913–922.

Jacobs, L. L., Winkler, D. A., Downs, W. R., and Gomani, E. 1993. New material of an Early Cretaceous titanosaurid sauropod dinosaur from Malawi. Palaeontology 26: 523–534.

Kérourio, P. 1981. Nouvelles observations sur le mode de nidification et de ponte chez les dinosauries du Cretace terminal du Midi de la France. C. R. Sommaire Soc. Geol. France 1: 25–28.

Kues, B. S., Lehman, T., and Rigby, J. K. 1980. Teeth of *Alamosaurus sanjuanensis*, a Late Cretaceous sauropod. J. Paleontol. 54: 864–869.

Leanza, H. A. 1999. The Jurassic and Cretaceous terrestrial beds from southern Neuquén Basin, Argentina. Inst. Super. Correl. Geol. Ser. Misc. 4: 1–30.

Legarretta, L., and Gulisano, C. 1989. Analisis estratigrafico secuencial de la Cuenca Neuquina (Triasico Superior–Terciario Inferior). In: Chebli, G., and Spalletti, L. (eds.). Cuencas Sedimentarias Argentinas, Serie de Correlacion Geologica Nro. 6. Universidad Nacional de Tucumán, Tucumán, Argentina. Pp. 221–243.

Loope, D., Schmitt, J., and Jackson, F. 2000. Thunder platters: thin discontinuous limestones in Late Cretaceous continental mudstones of Argentina may be chemically infilled sauropod tracks. Geological Society of America, Abstracts with Programs: A450.

López-Martínez, N. 2000. Eggshell sites from the Creatceous-Tertiary transition in south–central Pyrenees (Spain). In: Bravo, A. M., and Reyes, T. (eds.). 1st International Symposium of Dinosaur Eggs and Embryos, Isona, Spain. Pp. 95–115.

McIntosh, J. 1997. Sauropoda. In: Currie, P. J., and Padian, K. (eds.). Encyclopedia of Dinosaurs, Academic Press, New York. Pp. 654–658.

Mikhailov, K. E. 1991. Classification of fossil eggshells of amniotic vertebrates. Acta Paleontol. Polonica 36: 193–238.

———. 1997. Fossil and recent eggshell in amniotic vertebrates: fine structure, comparative morphology and classification. Special Papers in Paleontology, No. 56. Paleontological Association, London. 80 pp.

Mohabey, D. M. 1996. A new oospecies, *Megaloolithus matleyi*, from the Lameta Formation (Upper Cretaceous) of Chandrapur district, Maharashtra, India, and general remarks on the palaeoenvironment and nesting behavior of dinosaurs. Cretaceous Res. 17: 183–196.

———. 2000. Indian Upper Cretaceous (Maestrichtian) dinosaur eggs: their parataxonomy and implication in understanding the nesting behavior. In: Bravo, A. M., and Reyes, T. (eds.). 1st International Symposium of Dinosaur Eggs and Embryos, Isona, Spain. Pp. 95–115.

Moratalla, J. J., and Powell, J. E. 1990. Dinosaur nesting patterns. In: Carpenter, K., Hirsch, K. F., and Horner, J. R. (eds.). Dinosaur Eggs and Babies. Cambridge University Press, Cambridge. Pp. 37–46.

Norell, M. A., Clark, J. M., Dashzeveg, D., Barsbold, R., Chiappe, L. M., Davidson, A., McKenna, M. C., Altangerel, P., and Novacek, M. J. 1994. A theropod dinosaur embryo and the affinities of the Flaming Cliffs dinosaur eggs. Science 266: 779–782.

Nowinski, A. 1971. *Nemegtosaurus mongoliensis* n. gen., n. sp. (Sauropoda) from the Uppermost Cretaceous of Mongolia. Palaeontol. Polonica 25: 57–81.

Paul, G. 1990. Dinosaur reproduction in the fast lane: implications for size, success, and extension. In: Carpenter, K., Hirsch, K. F., and Horner, J. R. (eds.). Dinosaur Eggs and Babies. Cambridge University Press, Cambridge. Pp. 244–255.

Powell, J. 1992. Hallazgo de huevos asignables a dinosaurios titanosáuridos (Saurischia, Sauropoda) de la Provincia de Río Negro, Argentina. Acta Zool. Lilloana 41: 381–389.

Ramos, V. A. 1981. Descripción geológica de la hoja 33c, Los Chihuidos Norte, Provincia del Neuquén. Servicio Nacional de Minería y Geología. Boletín 182.

Sahni, A. S. K., Tandon, A., Jolly, S., Bajpai, S., Sood, A., and Srinivasan, S. 1990. Upper Cretaceous dinosaur eggs and nesting sites from the Deccan volcano-sedimentary province of peninsular India. In: Carpenter, K., Hirsch, K. F., and Horner, J. R. (eds.). Dinosaur Eggs and Babies. Cambridge University Press, Cambridge. Pp. 204–226.

Salgado, L., and Calvo, J. O. 1997. Evolution of titanosaurid sauropods. II: The cranial evidence. Ameghiniana 34: 33–48.

Salgado, L., Coria, R. A., and Chiappe, L. M., 2005. Osteology of the sauropod embryos from the Upper Cretaceous of Argentina. Acta Paleontol Polonica 50: 79–92.

Sander, P. M., Peitz, C., Gallemi, J., and Cousin, R. 1998. Dinosaurs nesting on a red beach? Earth Planet. Sci. 327: 67–74.

Sanz, J. L., Moratalla, J. J., Díaz-Molina, M., López-Martínez, N., Kälin, O., and Vianey-Liaud, M. 1995. Dinosaur nests at the sea shore. Nature 376: 731–732.

Vianey-Liaud, M., Mallan, P., Buscail, O., and C. Montgelard. 1990. Review of French dinosaur eggshells: morphology, structure, mineral, and organic composition. In: Carpenter, K., Hirsch, K. F., and Horner, J. R. (eds.). Dinosaur Eggs and Babies. Cambridge University Press, Cambridge. Pp. 151–183.

Vianey-Liaud, M., Hirsch, K. F., Sahni, A., and Sigé, B. 1997. Late Cretaceous Peruvian eggshells and their relationships with Laurasian and Eastern Gondwanian material. Geobios 30: 75–90.

Vianey-Liaud, M., Khosla, A., and Garcia, G. 2003. Relationships between European and Indian dinosaur eggs and eggshells of the oofamily Megaloolithidae. J. Vertebr. Paleontol. 23: 575–585.

Williams, D. L. G., Seymour, R. S., and Kérourio, P. 1984. Structure of fossil dinosaur eggshell from the Aix Basin, France. Palaeogeogr. Palaeoclimatol. Palaeoecol. 45: 23–37.

Wilson, J. A. 2002. Sauropod dinosaur phylogeny: critique and cladistic analysis. Zool. J. Linn. Soc. 136: 217–276.

Wilson, J. A., and Sereno, P. C. 1998. Early evolution and higher-level phylogeny of sauropod dinosaurs. J. Vertebr Paleontol 18 (Suppl. 2): 1–68.

Zhao, Z. 1979. The advancement of researches on the dinosaurian eggs in China. South China Mesozoic and Cenozoic "Red Formation" Science, Beijing. Pp. 329–340.

ELEVEN

Sauropod Histology

MICROSCOPIC VIEWS ON THE LIVES OF GIANTS

Kristina Curry Rogers and Gregory M. Erickson

SAUROPOD HATCHLINGS MAY have been only a meter long from head to tail (Chiappe et al. 1998, 2001) and weighed in at less than 10 kg (Breton et al. 1985; Weishampel and Horner 1994), while many adult sauropods attained sizes rivaling those of extant whales (Appenzeller 1994; Seebacher 2001; Erickson et al. 2001). This range of sizes is greater than for any other dinosaurian lineage, and includes the largest terrestrial vertebrates ever to inhabit the planet. In addition to this dramatic range of ontogenetic size variability, sauropod sizes vary interspecifically (fig. 11.1). The Romanian titanosaur *Magyarosaurus* has been dubbed the "smallest of the largest" because of its minute adult stature (~5 m [Jianu and Weishampel 1999]). Similarly, the saltasaurids *Saltasaurus* and *Neuquensaurus* are on the "small" end of sauropod size, reaching adult lengths of only 7 m and weights of ~10,000 kg (Powell 1986, 1990, 2003). Conversely, *Argentinosaurus*, *Paralititan*, and *Seismosaurus* push the envelope of size for land-dwelling vertebrates (Anderson et al. 1985; Coe et al. 1987; Gillette 1991; Appenzeller 1994; Seebacher 2001; Smith et al. 2001). As adults, these giants are estimated to have reached lengths of more than 30 m and weighed between 30 and 100 tons.

The growth strategies that permitted sauropods to attain such massive proportions have historically been a topic of great interest, but until recently they remained among the greatest of biological mysteries. In early investigations, workers extrapolated from reptilian growth rates and hypothesized that giant sauropods would have taken decades to reach sexual maturity and well over a century to attain their enormous adult sizes (e.g., Case 1978b; Calder 1984). More recent workers have focused specifically on evidence gleaned from bones and have revealed astounding new data that indicate that sauropod growth rates soared through most of ontogeny (e.g., Ricqlès 1968a,b; Reid 1981, 1990, 1997; Rimblot-Baly et al. 1995; Curry 1998; Sander 2000, 2003; Erickson et al. 2001; Curry Rogers et al. 2003; Sander and Tückmantel 2003).

On the outside, fossilized bones provide the raw data on which we base our interpretation of sauropod anatomy, relationships, function, and biomechanics, and our confidence in the

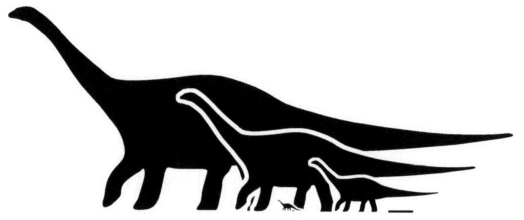

FIGURE 11.1. Interspecific and ontogenetic variation in sauropod size. At hatching, sauropods were likely < 1 m long and under 10 kg. As adults, sauropod sizes ranged from 7 m long and 10,000 kg (e.g., *Saltasaurus*) to 30 m long and approaching 100 tons (e.g., *Argentinosaurus*). Most sauropod adults (e.g., *Diplodocus, Camarasaurus*) fall in the middle of this range. Scale bar equals 1 m.

restored version of long-extinct sauropods rests almost exclusively on evidence obtained from fossil bones. The sizes and body proportions outlined above are readily reconstructed from the skeleton, even when remains are relatively fragmentary. Muscle scars on bones help determine where muscles originate and insert, and the resultant pattern of musculature dictates reconstructed body shape, as well as our interpretations of bone and muscle synergy and biomechanical function (chapters 6 and 8). Similarly, osteological correlates of soft tissues preserved in fossil bones (e.g., pneumatic fossae; chapter 7) make it possible to reconstruct unpreserved soft parts that may never be found in the fossil record. For example, Witmer (2001) reconstructed *Diplodocus* with terminally positioned external nares on the basis of soft tissue correlates in the skull. The gross morphology of bones thus enriches the story of sauropod evolution. Through osteology we document the debut and demise of major lineages and the evolutionary tale told by the trail of characters observed in fossil skeletons.

The internal architecture of fossil bones documents patterns of bone deposition and remodeling and can sometimes allow access to the portions of the bone growth record throughout ontogeny. The clues preserved in fossil bone tissue thus provide the necessary comparative data for choosing among extant analogues, and allow for sauropod growth rate quantification on microstructural and organismal levels. Bone histology breathes new life into fossil bones and offers a rigorous, testable means of addressing hypotheses of sauropod growth rates and life history. Did sauropods grow indeterminately? Did sauropods grow at constant rates throughout ontogeny, or did they experience regular cycles of relative growth rate variation? When did sauropods attain maturity? How long did it take sauropods to reach their massive adult sizes? Did all sauropods exhibit the same basic growth strategy? Along the same lines, is sauropod histology relevant to sauropod phylogeny? The answers to these questions lie under microscopes and deep within sauropod bone tissue.

In this chapter we consider sauropod growth throughout ontogeny from both histological and whole-body perspectives. We first provide a primer on relevant bone histological patterns, followed by a brief overview of the evidence from living and fossil vertebrate taxa relevant to qualitative interpretations of sauropod bone tissue. We outline a method for garnering quantified tissue growth rates directly from sauropod bones and then pull away from microstructural analysis to a broader focus on sauropod whole-body growth rates through developmental mass extrapolation. We present a case study for *Apatosaurus* that incorporates all methods discussed in the text and conclude the chapter with a prospectus for future work on sauropod growth rates.

BONE HISTOLOGY: STUDYING SAUROPOD GROWTH FROM THE INSIDE OUT

In extant vertebrates, distinctive bone tissue types are associated with different growth rates and trajectories. Fortunately, patterns of vertebrate bone tissue organization are readily retrieved from fossils, in which the mineral constituents of bone persist, as do vacuities left by bone's vascular and nervous supply. Thus, when analogous tissue patterns occur in extinct and extant taxa, bone growth rates and overall growth strategies in extinct taxa may be inferred (Enlow and Brown 1956, 1957, 1958; Ricqlès 1977, 1980; Ricqlès et al. 1991; Chinsamy 1990, 1993; Curry 1998, 1999; Sander 2000, 2003a, 2003b; Padian et al. 2001; Chinsamy and Elzanowski 2001). Sauropod bone tissue exhibits a range of typologies comparable to the range of tissue types found in extant vertebrate lineages. Several different levels of bone tissue organization are relevant to discussions of growth rates in sauropods, including the depositional pattern of primary bone protein and mineral, as well as the pattern and extent of bone vasculature (also termed porosity, because of the inclusion of lymphatic and nervous tissue in bone canals Margerie et al. 2002; Starck and Chinsamy 2002) and the degree and nature of remodeling (Francillon-Vieillot et al. 1990). The combination of these characteristics in different patterns allows us to draw conclusions regarding sauropod bone formation rate, age, and longevity, and can even be used to develop sigmoidal growth curves that allow comparison of whole-body growth rates among extinct dinosaurs and living vertebrates. The bone classificatory scheme for descriptive tissue histology developed by Francillon-Vieillot et al. (1990) is employed in the following discussion.

ORGANIZATION OF PRIMARY BONE TISSUE

Primary bone organization is among the most important aspects for determining relative bone depositional rates. As a bone begins an active phase of growth, osteoblasts secrete osteoid, which contains the elastic protein, collagen. The mineral component of bone is composed of hydroxyapatite crystals that align in parallel with the deposited collagen fibers. In living vertebrates, the degree of primary bone organization thus reflects the rate of collagen/hydroxyapatite deposition. Though collagen is only rarely preserved in the fossil record (Wykoff 1975; Currey 1987; Schwietzer et al. 1999, 2005; Carter and Beaupré 2001), hydroxyapatite crystal alignment persists and, thus, indicates relative rates of primary bone deposition.

There are three general categories of bone fibrillar organization, each thought to be the outcome of its formative rate. When the bone depositional rates are relatively slow, collagen fibers and hydroxyapatite crystals show a tight parallel alignment and there are endogenous pulses to the mineralization front. The latter gives the bone a striated texture under polarized light and is thus commonly called *lamellar* bone (fig. 11.2A); each increment in bone growth with parallel fiber alignment is a single lamella. Stronger, endogenous biorhythms (usually annual) often cause physiological disruptions to bone formation and lead to the formation of lines of arrested growth (LAG; see below). The second type of primary bone tissue is known as *parallel fibered* and is deposited at slightly faster rates (fig. 11.2B). In the third case, bone is formed at more rapid rates, and collagen fibers and adhering hydroxyapatite crystals align haphazardly. This type of rapidly growing bone tissue is specified as *woven*, or fibrous, bone (fig. 11.2C). Its genesis is rarely disrupted by endogenous biorhythms in living taxa, although LAG are not uncommon in dinosaurs (Reid 1981, 1984, 1990; Chinsamy 1990; Varricchio, 1993).

In actively growing taxa that generally exhibit rapid growth rates (e.g., eutherian mammals and birds), the woven and lamellar types

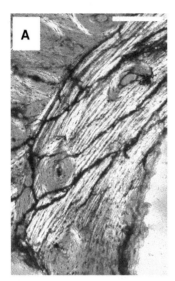 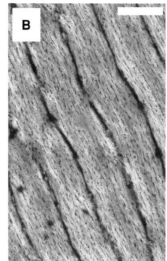 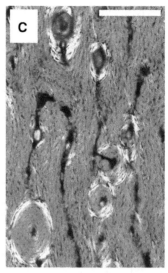

FIGURE 11.2. Bone fiber organization. (A) Endosteal, lamellar zonal bone presents a parallel alignment of hydroxyapatite crystals and collagen fibers, and results in a striated texture under polarized light. (B) Periosteal, parallel-fibered bone and flattened, circumferential vascular canals. (C) Periosteal, woven-fibered bone, with collagen fibers and hydroxyapatite crystals arranged haphazardly. This example includes several scattered secondary osteons. Scale bars equal 100 μm.

of bone tissue occur in combination. The gross size and shape of the bone are dictated by appositional woven bone growth at the outer surfaces of elements. Later, vascular canals trapped within the woven bone are partially infilled by layers of slower growing lamellar bone. The resultant, complexly organized tissue is termed the "fibrolamellar complex" (fig. 11.3).

ORGANIZATION OF BONE VASCULATURE

Like the organization of primary bone tissue, patterns of primary bone vascular supply are thought to vary with bone depositional speeds. As vascular canals are trapped by newly deposited bone, they are termed primary osteons. Primary osteons in a single direction may be oriented longitudinally, circularly, radially, or obliquely relative to the long axis of the bone under examination (fig. 11.4). In more rapidly growing bone tissue, primary vascular canals run in multiple directions and commonly intercalate in three major patterns (fig. 11.5). *Plexiform* vascularization (Enlow 1963) occurs when longitudinal, circular, and radial vascular canals interweave to form a three-dimensional vascular plexus (fig. 11.5A). *Laminar* vascularization (Foote 1913) is characterized by primary vascular canals oriented circularly and longitudinally in superimposed laminae (fig. 11.5B). *Reticular* vascularization is an obliquely and irregularly oriented arrangement of the vascular canals (fig. 11.5C).

Haversian bone tissue results from the centrifugal obliteration of primary vascular canals and cortices and subsequent centripetal redeposition of bone to form secondary osteons. Multiple cycles of remodeling result in "dense Haversian bone," in which little primary bone remains. Primary bone remodeling and the formation of secondary osteons are a time-dependent process. It is most frequently linked to age, localized mechanical stress (e.g., muscle attachment), mineral homeostasis, and fatigue damage repair (Amprino 1948; Lacroix 1971; Martin and Burr 1989). Haversian bone is most common in the bone tissue of mammals (Ricqlès et al. 1991), but also occurs in other vertebrates including birds, "reptiles," and dinosaurs (Mantell 1850; Queckett 1855; Owen 1859; Foote 1913; Enlow and Brown 1956, 1957, 1958).

Recent experimental work on extant birds (Castanet et al. 1996, 2000; Margerie et al. 2002; Starck and Chinsamy 2002) indicates that bones characterized by laminar and retic-

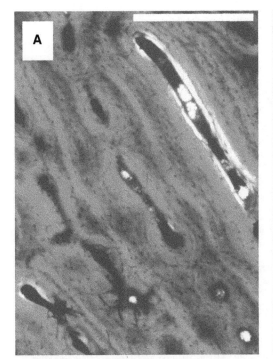 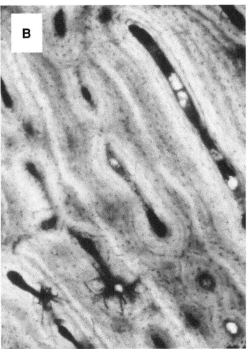

FIGURE 11.3. The fibrolamellar complex with lamellar–zonal bone deposited centrifugally around primary vascular canals in a woven bone matrix to form primary osteons. (A) *Apatosaurus* radius in polarized light. Lamellar–zonal bone is bright; woven matrix is darker. (B) *Apatosaurus* radius in normal light. Scale bar equals 100 μm.

ular vascular organization overlap in their growth rate ranges and are the fastest-growing tissue types (centripetal deposition in bony calli can actually be even faster, at +1 mm/day [Lanyon and Rubin 1984]). Plexiform vascularity occurs at significantly lower rates (Castanet et al. 2000). Bones exhibiting longitudinal vascular organization are typically the slowest-growing type of vascularized primary bone. These findings are particularly significant for studies of dinosaurian bone growth rates, because they provide a means of constraining the estimates for local tissue formation rates in extinct taxa (table 11.1).

THE PERIODICITY OF BONE GROWTH

Growth marks in bone tissue represent a periodicity in bone depositional rate and can be differentiated on the basis of general morphology and biological significance. Zones correspond to periods of active growth and osteogenesis. They are thicker than other growth marks (except perhaps "polish marks" [Sander 2000])

and normally consist of woven and/or parallel-fibered matrices. In the fibrolamellar complex, zones are comprised of superimposed laminae. Annuli are narrower than zones and reflect periods of relatively slow growth rates. They are normally composed of parallel and/or lamellar fibered matrices and may occur as thin rings in localized regions of cortical bone. LAG (also known as "rest lines") represent periods of temporary, complete, or near-complete cessation of appositional growth and represent the external surface of the bone at the time the LAG were deposited. The length of time represented by an individual LAG varies, and may span from a few days to as long as six months (Castanet and Smirina 1990).

Bones characterized by LAG, annuli, zones, or any combination of these is typically termed zonal bone (fig; 11.6). Bone tissue occurring in zones may be lamellar and poorly vascularized if an individual deposits bone slowly throughout its active growth period (e.g., wild crocodilians; fig. 11.6A). Alternatively, bone within

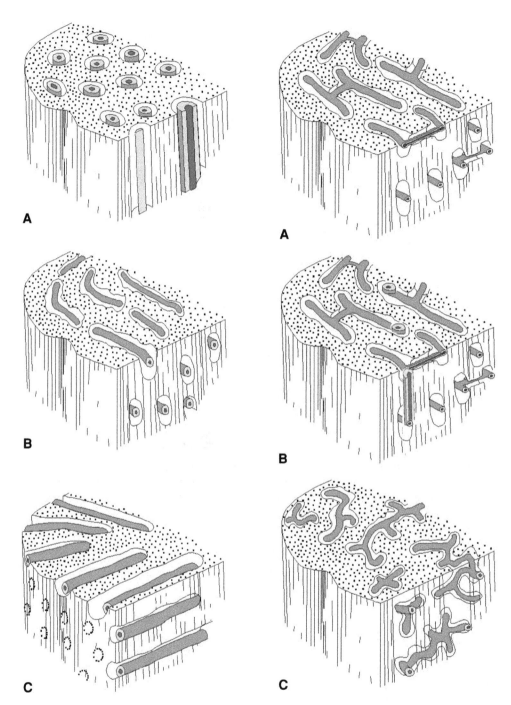

FIGURE 11.4. Organization of primary osteons. (A) Longitudinal vascular canals parallel the long axis of the bone; (B) circular vascular canals ring the circumference of the bone; (C) radial vascular canals radiate from the endosteal regions of the bone. (Modified from Francillon-Viellot et al. 1990.)

FIGURE 11.5. Intercalating primary osteons. (A) Laminar vasculature, with a predominance of circular and radial interwoven canals. (B) Plexiform bone, with circular, radial, and longitudinal interwoven canals. (C) Reticular bone, with an oblique and irregular orientation of primary osteons. Depositional rates of reticular and laminar bone generally overlap and reflect the most rapid rates of primary bone growth.

TABLE 11.1
Relationship Between Bone Growth Rates and Bone Tissue Types

	AVERAGE GROWTH RATE (μM/DAY)		
	LAMINAR BONE	RETICULAR BONE	SUBRETICULAR/ LONGITUDINAL CANALS
Ostrich	24.7	33.2	4.9
Emu	27.1	31.6	5.4
Mink	13.7	—	—
Duck	10.0	—	—

NOTE: Rates outlined below are average rates as determined by Castanet et al. (1993, 1996, 2000) and Ricqlès et al. (1991).

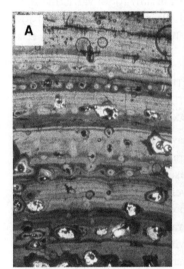 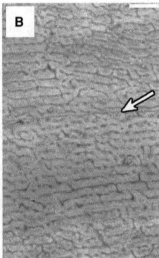 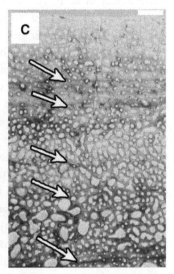

FIGURE 11.6. The periodicity of bone growth. (A) Cortical bone with lamellar–zonal, poorly vascularized zones of active growth punctuated by numerous lines of arrested growth (LAG) indicating periods of dramatic decrease in bone deposition. (B) Cortical bone of highly vascularized, fibrolamellar bone punctuated by one LAG (indicated by arrow). (C) Zonal bone not punctuated by LAG but demarcated by a regular decrease in the size and density of primary vascular canals. Arrows indicate the onset of each new cycle. Scale bars in (A) and (C) equal 500 μm. Scale bar in (B) equals 100 μm.

zones may be fibrolamellar, with abundant vasculature, and thus tell the tale of relatively rapid growth during an ontogenetic stage (fig. 11.6B). This pattern is typical among diverse dinosaurian taxa, including *Maiasaura* (Horner et al. 2000), *Hypacrosaurus* (Cooper and Horner 1999), *Syntarsus* (Chinsamy 1990), and *Troodon* (Varricchio 1993). In some taxa, including humans, LAG are deposited at the external margins of a bone once the majority of skeletal growth has ceased. In these taxa, minor thickening of the bone may still occur by accretion of small amounts of poorly vascularized lamellar bone tissue (the "external fundamental system" [Francillon-Vieillot et al. 1990; Ricqlès et al. 1991]). A low porosity and lamellated structure predispose accretionary bone to resist loads in vivo, and may serve as a significant structural reinforcement in maturing animals (Francillon-Viellot 1990). This avascular, accretionary tissue is frequently marked with LAG and resorption lines and has been associated with growth rate plateaus in extant taxa (e.g., Curry 1998; Sander 2000, 2003a; fig. 11.6C).

An additional type of primary bone stratification has been observed in extant cetaceans (Altman and Ditmer 1972), sirenians (Fawcett 1942), leatherback turtles (Rhodin 1985), ichthyosaurs (Buffrénil and Mazin 1990), and some sauropods (Ricqlès 1968b, 1983; Reid 1981; Rimblot-Baly et al. 1995; Curry 1999; Sander 2000, 2003; Sander and Tückmantel 2003). In these taxa, continuous fibrolamellar bone deposition occurs, but a regular variation in this continual depositional pattern is indicated by variable vascular patterns arranged as superimposed cycles of growth (Ricqlès 1968a,b, 1983; Rimblot-Baly et al. 1995; Curry 1999; Sander 2000, 2003; Sander and Tückmantel 2003). Regions of relatively slower growth exhibit longitudinal vascularity, while areas of faster growth exhibit more extensive canal networks. Once again, cycles are not marked by LAG or clear annuli. Rhodin (1985) concluded that such structures reflect cyclical growth of primary cortices, with faster and slower growth periods. Such cycles as observed in sauropods record mild modulations of bone deposition in a fast (to very fast) context. They are not equivalent to cycles resulting from a steep drop in bone formation (where annuli or LAG are common), as regularly observed in extant poikilotherms (Castanet and Naulleau 1985; Castanet and Smirina 1990) and some other dinosaurs (e.g., *Syntarsus* [Chinsamy 1990], *Troodon* [Varricchio 1993], *Massospondylus* [Chinsamy 1993]). Sander (2000) provided an alternative explanation for the presence of cortical stratification in *Janenschia robusta*. Rather than a specific change in growth rate, he postulated that cycles observed in low magnification resulted from an abruptly lower mineralized bone tissue, which causes a decrease in reflectivity in bright-field illumination.

Cortical stratification of all sorts, including LAG, annuli, and general cycles, may reflect cyclical annual growth, seasonal stress, and/or other endogenous rhythms (Castanet and Smirina 1990; Erickson and Tumanova 2000). Thus, it is most commonly utilized in skeletochronology. These cyclical patterns of bone deposition can essentially provide a minimum age estimate for an individual, particularly when the biological meaning of the growth mark is considered, and an effort is made to account for growth cycles effaced by medullar expansion (e.g., Chinsamy 1990; Varricchio 1993; Rimblot-Baly et al. 1995; Curry 1999; Sander 2000; Erickson et al. 2001; Sander and Tückmantel 2003). Phylogenetic parsimony has demonstrated that annual cycle/LAG deposition is plesiomorphic for vertebrates (e.g., Castanet and Smirina 1990). Thus, in most studies of dinosaurian growth and longevity, growth marks are regarded as annual increments (Ricqlès 1983; Chinsamy 1990; Varricchio 1993; Erickson and Brochu 1999; Curry 1999; Erickson and Tumanova 2000; Erickson et al. 2001).

POTENTIAL PITFALLS OF EQUATING BONE MICROSTRUCTURE AND GROWTH RATES IN EXTINCT TAXA

Overall, we have had much success in documenting general patterns of tissue morphology that can be used for qualifying bone growth rates. However, most dinosaur bones are uniquely composed of tissues with slow-growing reptilian and rapid-growing avian/mammalian attributes, leaving results somewhat inconclusive. Several characteristics of bone biology must be considered in any analysis of growth strategy in fossil vertebrates.

The destructive nature of histological analysis, and difficulty in processing extremely robust sauropod elements, has traditionally restricted investigations of sauropod bone tissue to single elements. Single-element analyses are problematic when the goals are determination of generalized growth patterns and quantification of overall body growth rates because bone microstructure is influenced by ontogeny, environment, mechanics, and phylogeny (e.g., Padian et al. 2001; Starck and Chinsamy 2002). Different bones in the same skeleton may have differing numbers of growth lines due to variable remodeling rates (Chinsamy 1993; Horner et al. 1999, 2000; Padian et al. 2001). Similarly, bone tissue architecture varies with the position of the thin section, with bone at middiaphyses typically composed of compacta and spongiosa, bone at metaphyses comprised of hypertrophied calcified cartilage, and bone at epiphyses a mosaic of tissue types, all of which may have variable growth rates. The diversity of possible formative influences coupled with the fact that tissue structure varies among elements and within single elements prohibits attempts to assess generalized life history attributes (e.g., longevity, growth rates, ages) from single bones. Thus, studies that cut across some of these constraints and allow for a comparison of intraskeletal variation are the best approach for determining general histological and growth patterns (Horner et al. 1999).

Remodeling of primary bone tissue over the course of ontogeny is also problematic for documenting longevity, particularly in extinct vertebrates. The record of the earliest stages of ontogenetic growth can be lost either through medullar expansion or through remodeling of primary bone tissue through the formation of secondary tissue. Thus, arguments for overall growth strategy in extinct taxa must be framed in the context of an ontogenetic series. Younger bones capture growth marks and bone tissue types that may be remodeled/obliterated in later life. Ontogenetic series thus afford a unique opportunity to document a more complete record of osteogenic variation during the life of a particular taxon (Chinsamy 1990, 1993; Curry 1999; Horner et al. 1999; Erickson and Tumanova 2000; Sander 2000; Erickson et al. 2001; Sander and Tückmantel 2003).

With regard to bone growth rates and their realistic expression of overall skeletal growth, the following must be considered. The same bone tissue can form at a range of different rates (Ricqlès et al. 1991; Castanet et al. 2000; Margerie et al. 2002; Starck and Chinsamy 2002) and depositional rates for different tissue types (as in the case of laminar and reticular bone) actually overlap. Recent work on bone tissue growth rates in extant duck (*Anas platyrynchos* [Castanet et al. 1996; Margerie et al. 2002]), ostrich (*Struthio camelus* [Castanet et al. 2000]), emu (*Dromaius novaehollandiae* [Castanet et al. 2000]), and Japanese quail (*Coturnix japonica* [Starck and Chinsamy 2002]) allowed for the correlation of bone growth rates with observed patterns of fibrillar and vascular organization. The first quantitative data on growth dynamics of the primary cortical bone of juvenile ratites demonstrated that high growth rates are linked to highly vascularized bone tissue but that rates for the same tissue type could be as diverse as 10–80 μm/day (Castanet et al. 2000). Detailed work on *Anas* indicated that growth rates for the same tissue type vary from 20 to 110 μm/day (Margerie et al. 2002). The growth rates of fibrolamellar bone in living mammals have only been documented in one study (Ricqlès et al. 1991), at a rate of 2.5–4.5 μm/day. Despite this growth rate overlap among vascular patterns, transitions between bone fibrillar types do correspond to specific changes in formative rates (table 11.1). Margerie et al. (2002) and Starck and Chinsamy (2002) independently cautioned that the presence and proportion of primary porosity may be more relevant in the interpretation of bone growth rates than the primary vascular pattern.

In any case, under simple structurofunctional assumptions and using the principle of parsimony, when a particular tissue type is preserved in a fossil taxon and quantitative growth data on similar tissues exist for living animals, application of bone growth rates in living vertebrates to similar bone tissues in extinct taxa is possible and potentially informative. Investigation of tissue growth rates in this manner can help to restrict estimates of age at various growth stages, especially when skeletochronologic indicators are absent (e.g., Curry 1999; Sander 2000; Sander and Tückmantel 2003).

The relationship between localized measures of tissue formation rates and overall body growth is untested for virtually all skeletal elements, even in extant taxa (Erickson et al. 2001). This lack of data can lead to problems in drawing conclusions from single elements in fossil taxa. These results emphasize the importance of examining homologous bones, standardizing thin sections, and comparing a multielement sample in an ontogenetic context.

WHAT CAN BONE HISTOLOGY REALLY TELL US?

Qualification of the mechanisms and speed of deposition recorded by primary bone and vasculature in extant taxa assists us in (1) registering general growth patterns in extinct taxa and determining appropriate modern analogues; (2) identifying variation of bone growth patterns among individual skeletal elements to gain a more detailed understanding of how function, phylogeny, ontogeny, biorhythms, and environment influence bone development; and (3) allowing for quantification of bone tissue depositional rates. Each of these questions can be addressed through the analysis of the broad-scale histological patterns and trends observed in an ontogenetic series of extinct taxa from the same taxon and locality.

GENERAL SAUROPOD GROWTH PATTERN

Most studies agree that sauropods commonly deposited abundant, fibrolamellar bone. For most of ontogeny, this bone was highly vascularized in laminar, reticular, and/or plexiform patterns. Late in ontogeny, bone vasculature drops off and bone matrices are most frequently comprised of parallel- and lamellar-fibered tissue, thus indicating that in late ontogeny growth substantially slowed. LAG are not common among sauropods until late in ontogeny, although some taxa exhibit cortical stratification in the form of cycles or "polish lines" (e.g., Sauropoda indet. [Reid 1981], *Lapparentosaurus* [Ricqlès 1983; Rimblot-Baly et al. 1995], *Apatosaurus* [Curry 1999], *Janenschia* [Sander 2000; Sander and Tückmantel 2003]). All sauropod bone tissue, including that where "polish lines" or growth cycles are common, is inconsistent with the use of reptilian growth analogues. Instead, histological analyses among sauropods reveal a fibrillar and vascular organization most similar to that of rapidly growing mammals and birds. These results suggest that prolonged ontogeny was not required for the attainment of gigantic body sizes in sauropods (Ricqlès 1968a, 1980, 1983; Rimblot-Baly et al. 1995; Curry 1998, 1999; Sander 2000; Erickson et al. 2001). Counts of cortical stratification supports this hypothesis, indicating that sauropods might have grown to half-size by the age of 5 years and to their adult size by between 10 and 30 years of age (Curry 1999; Sander 2000; Sander and Tückmantel, 2003).

HISTOLOGICAL VARIABILITY AND SAUROPOD BIOLOGY

Significantly, not all sauropod bones demonstrate the same depositional patterns, indicating that formation of different tissue patterns is influenced by functional environment or interelement allometry (Curry 1999; Sander 2000). In the case of sauropods and other dinosaurs, the slightly different histological stories revealed by different elements can provide the key data for addressing more specific life history, quantitative, and whole-body growth rate questions (Varricchio 1993; Castanet et al. 2000; Curry 1998, 1999; Padian et al. 2001). For example, Sander (2000) documented dimorphism in his histological sample of *Barosaurus* and hypothesized that the difference was due to sexual dimorphism and not related to taxonomy. In one morph, bone growth was rapid and continuous, with relatively little haversian remodeling. In the second morph, remodeling was extensive and occurred in the context of slower, discontinuous bone deposition. Sander hypothesized that the highly remodeled bone was likely from a female, with the discontinuities the result of reproduction. Sander ruled out local tissue variability in his sample because of the lack of intermediate tissue types deep in the cortical bone. This methodology highlights the potential for histology to reveal details of dinosaur lives, when little other evidence is available.

QUANTITATIVE SAUROPOD BONE GROWTH RATES

Bone histology has been the basis for interpreting growth strategies and life histories in dinosaurs since their earliest discoveries.

Amprino (1947) was the first to recognize the functional relationship between primary bone organization and the speed of its deposition. "Amprino's Rule" thus predicts that given tissue typologies are deposited within a specific set of growth rate parameters, even if that typology occurs in different bones, individuals, or taxa. Determination of the quantity of primary tissue types in fossil bones thus allows a unique insight into quantified bone growth rates in extinct or long-dead taxa.

When skeletochronologic data are not available, we can bound our assessments of the maximum and minimum number of days that an individual grew through the application of "Amprino's Rule". For example, if we measure the quantity of laminar bone tissue present in a sauropod femur, and we know the range of possible daily depositional rates for laminar tissue in a variety of living vertebrates, we can use the highest and lowest laminar tissue growth rates in living vertebrates to provide a quick look at the sauropod lifespan. The robustness of these methods has been poorly tested among living vertebrates (Castanet et al. 2000; Chinsamy and Starck 2002; Margerie et al. 2002), and more work is necessary in large-bodied mammals, reptiles, and birds before clear answers may be assessed, especially in light of the fact that fibrolamellar bone can be deposited at a 30-fold range of rates.

When skeletochronologic data exist in the form of LAG or growth cycles, we can determine the average daily growth rate for highly vascularized primary tissue forming zones by calculating the average bone apposition rate per growth cycle. To compare this number to values obtained for living vertebrates, we convert our value for the dinosaur to micrometers per day, using the number of days in the time period of interest (i.e., 377 for the Jurassic [Wells 1963]). Of course, these rates must be considered in light of the fact that some days of nondeposition occur during LAGs, and it is impossible to account for this time when calculating rates. Depositional rates obtained for incremental tissue deposits can then be applied to noncyclic bone, as long as vascular and fibrillar constitution is consistent. This growth rate calculation only relies on the assumption that cycles represent a year of bone deposition and does not require the application of growth rates from the poorly sampled modern fauna. To date, these methods have been applied in a handful of sauropod taxa (Curry 1998; Sander 2000; Sander and Tückmantel 2003) and indicate that sauropods might have grown at rates on the conservative end of the laminar, fibrolamellar bone spectrum (table 11.1).

BONE MICROSTRUCTURE AND OVERALL BODY GROWTH RATES

As outlined above, bone histology provides an exceptional tool for studying bone growth in dinosaurs at the microscopic level. But is it possible to extrapolate from the microstructural level to a more inclusive view of overall sauropod body growth? Even more significantly, is it possible to compare somatic growth rates at the macroscopic level we calculate in sauropods to the rates among extant vertebrates? The most common and physiologically informative measures of developmental growth rates in living taxa are not derived directly from bone microstructure, but from some more easily obtained unit (e.g., mass increase). In most vertebrates, mass changes with respect to age can be represented by sigmoidal curves (Case 1978a; Sussman 1964; Purves et al. 1992) with three characteristic growth stages (fig. 11.7). An initial slow growth lag phase is followed by the rapid growth exponential stage when the majority of mass is typically accrued. The ontogenetic trajectory culminates in a reduction or cessation of development during the stationary phase.

Mass standardized comparisons of maximum overall body growth rates among phylogenetically and morphologically diverse taxa can be made using values from the exponential stage of development. These comparisons indicate that rates absolutely increase with respect to body mass and that individual clades have characteristic whole-body growth rates (Case 1978a, 1978b; Calder

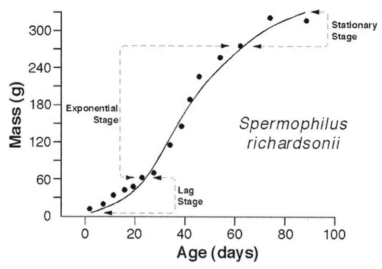

FIGURE 11.7. Growth curve for *Spermophilus richardsonii* ("Richardson's ground squirrel" [modified from Sussman 1964]). Early ontogeny is characterized by a slow-growth, lag stage during infancy. This is followed by an exponential growth stage midway through development, when maximum growth rates are obtained and most mass is accrued. Development culminates with a stationary stage, when growth slows or comes to a standstill.

1984; Erickson et al. 2001). Retrieving these data from fossil bones is impossible without the quantitative record of age at various ontogenetic stages provided by histological analysis. The monumental scale of most sauropods begs the question, What is the typical sauropod body growth trajectory? Erickson and Tumanova (2000) uniquely combined skeletochronologic data with a unique scaling principle they termed developmental mass extrapolation (DME) and quantified whole-body growth rates for dinosaurs. DME requires a reliable age determination for bones and mass quantifications at different ontogenetic stages.

AGE ASSESSMENT

Age is most commonly assessed from growth marks in histologically prepared specimens, as described above. Growth marks/bone tissue lost due to medullar expansion and bone remodeling with increased age must be accounted for by sequentially superimposing subadult specimens on those from larger individuals.

BODY MASS ASSESSMENT: DEVELOPMENTAL MASS EXTRAPOLATION

Body mass for adult animals in a sample may be estimated using the minimal long-bone diaphyseal circumferences and the interspecific allometric equation established by Anderson et al. (1985). Because these equations are not applicable throughout ontogeny, Erickson and Tumanova (2000) utilized DME, an allometric scaling principle. The principle is based on the fact that if a linear measurement from an anatomical structure scales with body mass throughout ontogeny, and the body masses of more than one individual are known, then comparable linear measurements taken throughout ontogeny can be used to extrapolate individual body masses. Femur length is one of the best possible measures for this method because it scales isometrically during ontogeny in extant archosaurs (Dodson 1975; Carrier and Leon 1990; Erickson and Tumanova 2000).

The first step in applying DME to dinosaurs is to take the cube of femur length (l^3) for each specimen in an ontogenetic series. Each value is then taken as a percentage of the adult value (the value for the largest known bone of interest). The percentages are then converted to fractional values and multiplied by the adult mass estimate (see above) to obtain body masses for each of the included growth stages (Erickson and Tumanova 2000; Erickson et al. 2001;

Curry Rogers et al. 2003). The accuracy of this method in predicting exponential stage growth rates has been tested in humans and wild alligators and is accurate within 5% in each case (Erickson et al. 2001). Following application of DME, growth curves can be reconstructed for the extinct taxon of interest.

In our recent analyses of dinosaurian growth trajectories, we have demonstrated that sigmoidal equations accurately describe growth data for all dinosaurs (Erickson et al. 2001; Curry Rogers et al. 2003). All members of Dinosauria grow at rates 2–56 times faster than those of any living reptile. As a clade, dinosaurs did not grow at rates intermediate to those of reptiles, birds, and mammals, or at rates equivalent to those of living altricial birds. Instead, our results indicate that Dinosauria exhibited its own, unique growth trajectory and was unique among vertebrate clades in having members with growth rates below, equal to, or above typical mammalian/avian rates. Interestingly, the growth rates of dinosaurs depended on the size of the taxon of interest: the larger the dinosaur, the more rapid the overall body growth rate. The regression equation we established for Dinosauria included six taxa spanning the phylogenetic, size, and temporal range for the clade. It allowed us to make quantified predictions of dinosaurian growth rates beyond the range of included taxa (Erickson et al. 2001). Sauropod growth rates vary from taxon to taxon and approach those known for extant whales (absolutely and relatively some of the fastest-growing eutherians). For example, a fully grown *Apatosaurus* grew at about 14,460 g/day, compared with 20,700 g/day for a 30,000-kg gray whale (*Eschrichtius robustus* [Case 1978a]). Though this rapid growth rate may seem astounding, it is upheld by bone histological data and still does not hold a candle to the growth rates observed in the extant blue whale (*Balaenoptera musculus*, the fastest-growing animal ever, at 66,000 g/day, [Case 1978a]). In fact, even the enormous sauropod *Argentinosaurus* (unconservatively estimated to weigh ~100,000 kg at adult size) is still predicted to have lower growth rates (55,638 g/day) than *B. musculus* (Case 1978a).

OVERVIEW OF SAUROPOD GROWTH STRATEGIES

Though sauropods were traditionally regarded as long-lived animals with extremely prolonged ontogenies (Case 1978b; Reid 1981), recent years have witnessed a shift in our understanding of sauropod growth patterns and mechanisms. These new results have been based primarily on bone histological data. A handful of diverse sauropods have been studied. Ricqlès (1968a) was the first to conduct an explicit analysis of sauropod bone histological patterns in "*Bothriospondylus*" (=*Lapparentosaurus*) *madagascariensis*. In two papers (1968a, 1983), Ricqlès documented the qualitative histology of humeri and noted the presence of vascular cycles deposited in the context of highly vascularized fibrolamellar bone. Reid (1981) observed numerous LAG in the pubis of an unidentified sauropod and challenged Ricqlès' interpretation of sauropods as rapidly growing organisms.

Rimblot-Baly et al. (1995) extended the Ricqlès qualifications through the inclusion of a *Lapparentosaurus* ontogenetic series. Rimblot-Baly et al. (1995) confirmed Ricqlès' (1968a, 1983) interpretations of cortical stratification and also noted lamellar, accretionary tissue at the cortical periphery in their largest samples. The work by Rimblot-Baly et al. (1995) is also significant because it constitutes the first attempt at quantifying bone growth rates directly from sauropod bone tissue: they proposed a bone appositional rate of ~7 μm/day. This rate is similar to rates observed in living cows and other large mammals and birds, which typically deposit laminar, fibrolamellar bone.

Sander (2000, 2003) and Sander and Tückmantel (2003) provided histological descriptions for the Jurassic sauropods *Brachiosaurus*, *Barosaurus*, *Dicraeosaurus*, *Janenschia*, and *Apatosaurus*. In each taxon, highly vascularized fibrolamellar bone dominates throughout ontogeny. In addition, Sander (2000) distinguished taxa on the basis of the relative degree of bone remodeling and spatial characteristic. He concluded that *Brachiosaurus* attained sexual maturity at 40% maximum size, while *Barosaurus* may not have reached this plateau until 70% of body size had been obtained. Polish lines present in the single *Janenschia* sample yielded age estimates of sexual maturity at 11 years, maximum size at 26 years, and death at 38 years of age. In addition, Sander utilized differing histological patterns in *Barosaurus* individuals to develop a hypothesis of sexual dimorphism related to reproduction.

Analyses such as those performed by Rimblot-Baly et al. (1995) and Sander (2000, 2003) have provided abundant data and the requisite raw material for furthering our understanding of sauropod growth trajectories. Both consider taxa in an ontogenetic series, and Sander (2000) considers multiple skeletal elements. The building of our base of knowledge about sauropod growth dynamics relies on such detailed analytical studies, and we now have the tools to take this work even further.

APPLICATION AND EXAMPLE: *APATOSAURUS* AS A CASE STUDY

To date, *Apatosaurus* is the only sauropod for which all the methods outlined above have been utilized to document patterns in bone and life history strategy. Here we apply each method to an ontogenetic sample of *Apatosaurus* to illustrate (1) the kinds of data that can be generated, (2) the utility and shortcomings of each method, and (3) how different methods outlined in this chapter can be used to build on and test answers derived from their alternatives.

THE SAMPLE

Curry (1999) histologically analyzed five ulnae, four radii, and three scapulae from at least four *Apatosaurus* individuals at various stages of ontogenetic development. The available sample was divided into four relative age classes on the basis of overall bone length, degree of ossifica-

TABLE 11.2
Apatosaurus Relative Age Classes, Derived from Maximum Length, Muscle Scar Development, and Bone Surface Finish

	ELEMENT	MAXIMUM LENGTH (CM)	PERCENTAGE OF ADULT SIZE
I	Radius R1	45.5	61
	Ulna U1	47.9	61
	Scapula S1	70.7	34
II	Radius R2	54.5	73
	Ulna U2	52.2	67
	Scapula S2	117.4	56
III	Radius R3	67.5	91
	Ulna U3	59.4	76
IV	Radius R4	74.2	100
	Ulna U4	78.5	100
	Scapula S3	210.9	100

tion, and muscle scar development (table 11.2). To rule out the effects of environment on the interpretation of histological patterns, all but the largest adult specimen were from a single locality. Similarly, multiple elements from different functional environments were included to negate the influence of biomechanics on the overall histological signal. Transverse thin sections were made at standardized locations in each element (fig. 11.8). Comparable histological organization of bones of similar sizes and inferred relative ages served as an independent test of *a priori* age class assessments. Age classes I and II represented juveniles up to ~75% of adult size, age class III represented subadults between 76% and 91% of adult size, and age class IV represented fully grown adults. Qualitative data on interelement histological variation and ontogenetic bone depositional patterns served as the basis for the first quantifications of *Apatosaurus* skeletal growth strategy.

HISTOLOGICAL CHARACTERIZATIONS, APPLICATIONS, AND INFERENCES

Laminar and plexiform fibrolamellar bone predominates during 90% of ontogenetic growth in all *Apatosaurus* elements studied. Laminar–plexiform deposits are present in each element

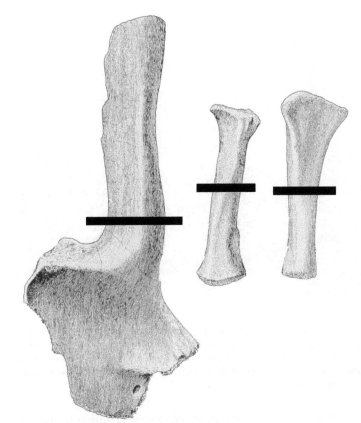

FIGURE 11.8. Histological sample of *Apatosaurus excelsus*, from left to right: Scapula, radius, and ulna. Scapulae were sectioned at the proximal end of the scapular blade. Radii and ulnae were sectioned at middiaphysis in areas devoid of muscle scars. (Modified from Curry 1999: fig. 1.)

but have different gross morphologies in radii, ulnae, and scapulae of similar relative ages (fig. 11.9). Radii and ulnae exhibit noncyclic, continuous deposition of laminar bone throughout most of ontogeny (fig. 11.9A). In contrast, cyclicity of laminar bone deposition is evident at low magnifications in *Apatosaurus* scapulae (fig. 11.9B), and cycles are not demarcated by annuli or LAG. Instead, each scapular cycle is demarcated only by a regular variation in vascular canal density and width, always in the context of continuous laminar bone deposition. Despite this histovariability between forelimb and girdle elements, all three elements maintained rapid rates of bone growth throughout ontogeny, slowing only after subadult sizes were reached.

After ~90% of adult size was attained, the highly vascularized fibrolamellar primary bone was replaced by the first deposits of slower-growing avascular, lamellar-zonal bone with numerous annuli and LAG (fig. 11.10). Lamellar–zonal bone and peripheral LAG indicate a growth plateau and the attainment of adult size in *Apatosaurus*. These histological results parallel the sigmoidal, determinate growth curve for mass increase in extant mammals and birds (Case 1978a; Calder 1984) and contradict the traditional view of indeterminate growth in sauropods (Ricqlès 1968b, 1983; Rimblot-Baly et al. 1995). The predominance of laminar and plexiform bone throughout most of *Apatosaurus* ontogeny is most similar to that observed in actively growing extant mammals and birds (e.g. Ricqlès et al. 1991; Castanet et al. 1996, 2000). Amprino's Rule suggests that laminar tissue in sauropods grew at the same basic range of rates that occurs in extant mammals and birds. Bone tissue patterns thus suggest that birds and mammals are the most appropriate analogues for investigations of *Apatosaurus* growth.

These qualitative data are a useful starting point for understanding ages and longevity in *Apatosaurus* and serve as the basis for skeletochronologic estimates of age. Cycles similar

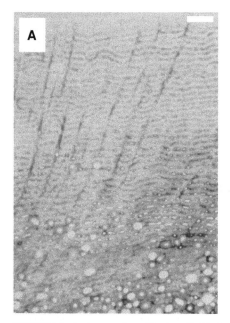
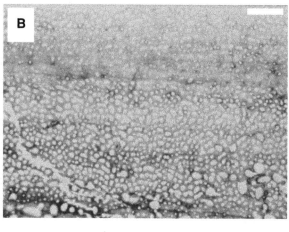

FIGURE 11.9. *Apatosaurus* tissue histology. Radii, ulnae, and scapulae each exhibit continuous, rapid deposition of highly vascularized, fibrolamellar bone for most of ontogeny. (A) Laminar bone maintains a consistently high rate of growth in radii and ulnae but is punctuated by areas of reduced vascular density and canal size in scapulae. Scale bar equals 100 μm. (B) Annuli and LAG do not mark vascular cycles observed in scapulae, thus attesting to the continuous, rapid deposition of primary bone in each element until late ontogeny. Scale bar equals 500 μm.

to those in *Apatosaurus* scapulae also occur in extant sirenians (Fawcett 1942), sea turtles (Rhodin 1985), some other sauropods (Ricqlès 1968b, 1983; Reid 1981; Rimblot-Baly et al. 1995; Sander 2000), and ichthyosaurs (Buffrénil and Mazin 1990). Just as in these taxa, *Apatosaurus* vascular cycles are assumed to represent annually periodic bone depositional events on the grounds of phylogenetic parsimony and tissue formation rates consistent with those in living vertebrates (Castanet and Smirina 1990; Erickson and Tumanova 2000). Endosteal reconstruction characterizes scapulae included in this analysis, but isolated patches of primary bone are retained in the deep cortex of the youngest scapula and allow gaps in the cortical bone record to be reconstructed. Lost growth rings can also be accounted for by sequentially superimposing subadult specimens on those from larger individuals. Data from the younger, less remodeled bones "fill in" missing data in older elements and provide a more complete history of growth. Maximum counts of 5 cycles in scapula 1 and 10 cycles in scapula 2 yielded conservative age estimates of 5 and 10 years, respectively. Because radii and ulnae did not have growth cycles or LAG, scapular cycle counts were applied to radius and ulna in the same relative age class to provide a testable hypothesis of absolute age for all bones in each age class (I–III). Age class I radius, ulna, and scapula were estimated at 5 years. Similarly, class II and III elements were estimated at between 8 and 10 years of age (table 11.3).

AMPRINO'S RULE, INFERENCES, AND BONE GROWTH RATE QUANTIFICATION

Laminar/reticular–plexiform bone tissue in a fibrolamellar context predominates for most of *Apatosaurus* ontogeny. Amprino's Rule provides a means for more precisely defining "rapid" bone growth for *Apatosaurus*. In this case, tissue growth rates for laminar–reticular bone are known in several modern taxa: ostrich (Castanet et al. 2000; Sander and Tückmantel 2003), emu (Castanet et al. 2000), mink (Ricqlès et al. 1991), duck (Castanet et al. 1996; Margerie et al. 2002),

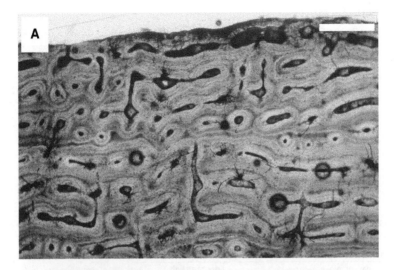

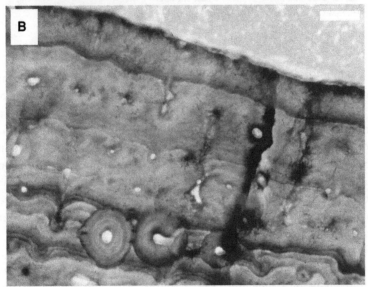

FIGURE 11.10. *Apatosaurus* exponential and stationary stages of growth. (A) Periosteal surface in juvenile *Apatosaurus* radius (50% of adult size) demonstrating fast-growth laminar, fibrolamellar bone in the external cortex. This continuous, rapid growth pattern is consistent with the exponential stage of growth. (B) Periosteal surface in adult *Apatosaurus* radius (90% of adult size). The first peripheral lines of arrested growth (LAG) appear in the external regions of *Apatosaurus* radius cortical bone. These LAG document the onset of the stationary phase of growth. Scale bars equal 100 μm.

and cow (Ricqlès et al. 1991). These experimental data indicate that the apposition rates of variably vascularized fibrolamellar bone can overlap and include a wide range of possible values. For example, in the duck *Anas platyrhynchos* (Margerie et al. 2002), the rates vary from less than 20 to 110 μm/day. Similarly, in ostriches and emus laminar–reticular fibrolamellar bone forms at rates of between 10 and 80 μm/day. Though tissue growth rates for a given typology vary among these taxa and can overlap and, at least in *Anas* (Margerie et al. 2002) the growth rates observed may not be directly linked to bone vascularity, application of any of these growth rates to the laminar tissue observed in *Apatosaurus* does have implications: namely,

TABLE 11.3
Apatosaurus *Ages Derived from Application of* Apatosaurus *Scapular Laminar Bone Growth Rates to Radius and Ulna*

	AGE (YEARS)	
ELEMENT	SCAPULA 1 RATE (10.1 µM/DAY)	SCAPULA 2 RATE (11.4 µM/DAY)
Radius 1	5.8	5.1
Radius 2	7.6	6.8
Radius 3	10.0	8.8
Ulna 1	5.5	5.1
Ulna 2	7.1	7.8
Ulna 3	10.5	9.3

NOTE: Scapula 1 and 2 laminar bone depositional rates are calculated by dividing the average amount of laminar bone deposited in each cycle by 378 (Wells, 1963). This rate is then applied to radius or ulna by:

$$\frac{X \, \mu m \text{ radial or ulnar laminar bone}}{X \text{ days of total growth}} = \frac{10.1/11.4 \, \mu m \text{ scapular laminar bone}}{1 \text{ day of growth}}$$

Converting days to years gives an answer for age for a particular relative age class.

providing a testable array of possible primary bone depositional rates for *Apatosaurus*. However, it is important to stress that this application of bone growth rates observed in living animals is merely a means of bounding the possibilities of growth rates in this dinosaur.

The second method for aging *Apatosaurus* outlined above provides just the data needed to determine how the speed of sauropod bone growth compares to the speed of bone growth in living vertebrates. Rather than relying on the application of bone appositional rates in living taxa, we turn to the bone of *Apatosaurus* itself. In the first step, we measure the thickness of laminar, fibrolamellar bone in scapular cycles and take an average of the growth rate around the cross section (thus negating the influences of the shape of the scapula). We divide this average amount of bone per cycle by the number of days in the Jurassic year (377 [Wells 1963]) and thus develop an estimate of average laminar bone growth rates directly from *Apatosaurus* bone. Average cycle thicknesses are consistent in the two included scapulae, and laminar tissue was deposited at comparable rates in each: for scapula 1 (~50% adult size), laminar bone accreted at a rate of ~10.1 µm/day. In scapula 2 (~60% adult size), the rate was slightly higher, at ~11.4 µm/day. We then apply this observed rate of bone deposition to other skeletal elements that lack cycles (radius and ulna) and arrive at an age for each *Apatosaurus* element under study. Measuring the minimum thickness of continuous laminar bone for each radius and ulna provides a conservative estimate of total bone growth. Conversion of days to years provides an age estimate for each *Apatosaurus* forelimb element. Age class I radius and ulna were between 4 and 6 years of age, class II radius and ulnae were between 6 and 9 years, and age class III elements were 8–11 years of age (table 11.3). Similar work conducted by Sander and Tückmantel (2003) yields comparable results for bone apposition rates in a variety of sauropods (between 1 and 20 µm/day, yielding comparable ages for the attainment of adult size). While significant, these quantified growth rates must be viewed with some caution because they document growth at very specific bone locations and only in a single direction (centripetal growth). Extending these rates to the entire organism is impossible without a complete understanding of the allometry of the skeleton. Because this type of data does not exist for any living taxon, it is unlikely that it will exist for a dinosaur in the near-future, and we must view these results with this in mind.

LONGEVITY ESTIMATES AND DEVELOPMENTAL MASS EXTRAPOLATION: APPLICATION AND INFERENCES

How do we move from the microscopic to the somatic level of growth rate interpretation in extinct taxa? In most extant animals, mass changes with respect to age show sigmoidal patterns (Sussman 1964). Standardized comparisons of maximum growth rates among phylogenetically and morphologically diverse taxa can be made using data from the exponential stage of development (Case 1978a,b; Calder 1984; Erickson and Tumanova 2000). To compare whole-body growth rates between dinosaurs and living vertebrates, similar quantified data are needed. This requires growth series that span the full range of dinosaur size, shape, and phylogeny, and accurate age and mass assessments at all ontogenetic stages. The merging of bone histological methods with scaling principles provided the requisite tools for the analysis of how diverse dinosaurs really grew (Erickson and Tumanova 2000; Erickson et al. 2001). Body mass estimates are obtained using long-bone circumferences and regression equations (Anderson et al. 1985). Because these equations do not work for all ontogenetic stages, the scaling principle called DME (Thompson 1943; Erickson and Tumanova 2000) was employed. The data for a diverse group of dinosaurs indicate that sigmoidal equations accurately fit the growth trajectories for Dinosauria and that dinosaurs as a whole did not show growth rates intermediate between those of birds/mammals and reptiles. Instead, our results indicate that Dinosauria was unique among major vertebrate clades in having members that show rates below, equivalent to, or above typical mammalian/avian rates. In addition, the distribution of these rates is related to the size of the organism under consideration: the larger the dinosaur, the faster it reaches a somatic maximum (Erickson et al. 2001). We included *Apatosaurus* ontohistogenetic data in our analysis of overall body growth rates (Erickson et al. 2001) and provided more detailed evidence on the relationship between bone tissue growth and mass increase during ontogeny in this taxon. Our analysis revealed that a sigmoidal equation accurately described the growth data for *Apatosaurus* (fig. 11.11). The exponential stage growth was approximately 6 years long, and the onset of somatic maturity occurred at about age 13. In terms of mass increase, *Apatosaurus* had growth rates similar to those of extant whales, and may have added about 14,460 g/day to their body mass. Interestingly, although *Apatosaurus* bone tissue is morphologically indistinguishable from that of altricial birds (i.e., both deposit abundant laminar, fibrolamellar bone), *Apatosaurus* accumulated body mass at rates about half that of a scaled-up altricial bird (123,025 g/day [Erickson et al. 2001]). Our regression analysis of dinosaur growth also allowed us to extrapolate to larger dinosaurs, and it is clear that the largest known dinosaur (*Argentinosaurus*; estimated adult body size of 100,000 kg [Appenzeller 1994]) would still have grown at 55,638 g/day, a rate absolutely slower than that estimated for a scaled-up altricial bird.

GROWTH RATE SYNTHESIS

All of the methods outlined above reveal that *Apatosaurus* grew rapidly throughout most of ontogeny. Histological typing identifies highly vascularized fibrolamellar bone tissue with no LAG or growth marks until late in ontogeny. The first evidence of a bone growth rate reduction occurs in subadults, and the largest adult *Apatosaurus* bones demonstrate continued diminished growth with attainment of adult body size. Categorization of bone tissue typologies is useful in that it provides a "first pass" on growth rates during ontogeny, as well as longevity estimates for the individuals under study.

Amprino's Rule allows direct comparison of quantified growth rates we might obtain directly from sauropod bones to those experimentally obtained in living vertebrates. In summary, Amprino's Rule tests the real possibility of our hypothesis: if rates we obtain are observed in

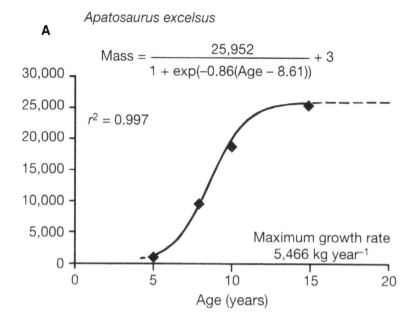
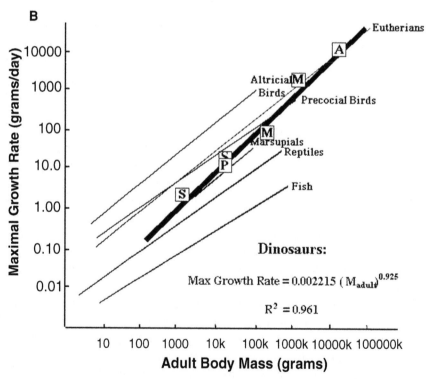

FIGURE 11.11. *Apatosaurus* growth rates. (A) Growth curve for *Apatosaurus*. The largest animal in the sample is among the largest specimens known. The growth curve for *Apatosaurus* is similar to that known in living vertebrates and includes an early-ontogeny lag stage, followed by an exponential stage, and, finally, a stationary stage. (B) Comparison of exponential stage growth rates in *Apatosaurus* (A), *Maiasaura* (M), *Massospondylus* (M), *Syntarsus*, (S), *Psittacosaurus* (P), and *Shuvuuia* (S) with typical values for extant vertebrates. Standardized comparisons are made using contrasts between animals of comparable adult mass to diminish signal from differences in shape and negate the effects of size. Growth rates for included dinosaurs are as follows: *Shuvuuia*, 3.4 g/day; *Psittacosaurus*, 12.5 g/day; *Syntarsus*, 23.9 g/day; *Massospondylus*, 90.3 g/day; *Maiasaura*, 2,793 g/day; and *Apatosaurus*, 14,460 g/day. (Modified from Erickson et al. 2001.)

living taxa for the same tissue typology, our argument for *Apatosaurus* is not disproven.

Application of DME and longevity estimates provide quantified measures of growth rates for all stages of whole-body growth, including the exponential stage. Because growth comparisons can be made with living taxa when the exponential stage whole-body growth rate is known and mass is standardized, DME provides one of the most groundbreaking advances in our ability to decipher whole-organism life history patterns in sauropods and other extinct taxa.

FUTURE DIRECTIONS

In this chapter we have demonstrated the utility of bone histological analysis for interpreting a wide range of paleobiological questions surrounding sauropods, which cannot be answered using traditional gross morphological studies. That said, we have only touched the surface on our way to a more comprehensive understanding of sauropod growth strategy on micro- and macroscopic levels. Bones hold the keys to addressing even more detailed questions of sauropod life history, overall growth strategy, evolution, and behavior.

As we have outlined in this chapter, the quality of available data varies with each step of an histological analysis, and all steps, from qualification of bone depositional rates to the whole-body approach of DME, are informative in some way. At the descriptive stage of an histological research project, we establish hypotheses of relative growth rates and age estimates based on histological characteristics and skeletochronology. These hypotheses are based on the comparisons drawn between the tissue of sauropods and that of living vertebrates, within a single ontogenetic series, and within the skeleton of a single individual. Thus, collecting as many data as possible, in terms of number of bones, number of individuals, and variety of comparative specimens, will significantly bolster our ability to interpret fossil tissues. Luckily, many sauropod taxa from around the world include growth series and will prove most valuable for life history studies. Similarly, comparisons among sauropods (e.g., Sander, 2000, 2003; Sander and Tückmantel 2003; Sander et al. 2004) highlight the potential of histology to inform about seemingly intractable questions such as sauropod sexual dimorphism.

As our understanding of histology, skeletochronology, and DME continues to grow, the phylogenetic significance of sauropod growth strategies can finally be addressed. Useful phylogenetic characters may be derived from growth curves, and it will be possible to compare overall body growth strategies employed by divergent sauropod taxa. For example, *Saltasaurus* and *Argentinosaurus* are both titanosaurs, but the end points of their growth trajectories vary dramatically. With a combined histological/DME approach, we can specifically evaluate the onset and length of exponential-stage growth rates to address the mechanisms of sauropods gigantism. With a deeper understanding of sauropod life history strategies, our knowledge will extend beyond the realm of sauropod biology to questions of ecology, resource allocation, and evolution.

ACKNOWLEDGMENTS

We wish to thank E. Schexnayder and C. Wong for assistance with illustrations. We are also grateful to Martin Sander and Anusuya Chinsamy, who provided helpful reviews of the manuscript. Of course, thanks would be incomplete without acknowledgment of all that Jack McIntosh has brought to the study of sauropods. This research is supported by the National Science Foundation, the Paleontology Department at the Science Museum of Minnesota, the Geology Department at Macalester College, and the Department of Biological Sciences at Florida State University.

LITERATURE CITED

Altman, P. L., and Ditmer, D. S. 1972. Biology Data Book, Vol. I. Federation of the American Society of Experimental Biology. Bethesda, MD.

Amprino, R. 1947. La structure du tissu osseux envisagée comme expression de différences dans la vitesse de l'accroissement. Arch. Biol. 58: 315–330.

———. 1948. A contribution to the functional meaning of the substitution of primary by secondary bone tissue. Acta Anat. 5(III): 291–300.

Anderson, J. F., Hall-Martin, A., and Russell, D. A. 1985. Long-bone circumference and weight in mammals, birds, and dinosaurs. J. Zool. A207: 53–61.

Appenzeller, T. 1994. Argentine dinos vie for heavyweight title. Science 266: 1805.

Breton, G., Fournier, R., and Watté, J.-P. 1985. Le lieu de ponte de dinosaures de Rennes-le-Chateau (Aude): premiers résultats de la campagne de fouilles 1984. Actes Coll. Dinosaures Chine France 1985: 127–140.

Buffrénil, V. de, and Mazin, J.-M. 1990. Bone histology of the ichthyosaurs: comparative data and functional interpretation. Paleobiology 16: 435–447.

Calder, W. A.,III. 1984. Size, Function, and Life History. Harvard University Press, Cambridge, MA.

Carrier, D., and Leon, L. R. 1990. Skeletal growth of the California gull (*Larus Californicus*). J. Zool. London 222: 375–389.

Carter, D. C., and Beaupré, G. S. 2001. Skeletal Function and Form, Mechanobiology of Skeletal Development, Aging, and Regeneration. Cambridge University Press, Cambridge. 318 pp.

Case, T. J. 1978a. On the evolution and adaptive significance of postnatal growth rates in the terrestrial vertebrates. Q. Rev. Biol. 53: 243–282.

———. 1978b. Speculations on the growth rate and reproduction of some dinosaurs. Paleobiology 4: 320–328.

Castanet, J., and Naulleau, G. 1985. La squelettochronologie chez les Reptiles. II. Résultats expérimentaux sur la signification des marques de croissance squelettiques utilisées comme critère d'âge chez les serpents. Remarques sur la croissance et la longévité de la Vipère Aspic. Ann. Sci. Nat. Zool. (Paris) 7: 41–62.

Castanet, J., and Smirina, E. 1990. Introduction to the skeltochronological method in amphibians and reptiles. Ann. Sci. Nat. Zool. (Paris) 11: 191–197.

Castanet, J., Francillon-Vieillot, H., Meunier, F. J. and Ricqlès, A. de 1993. Bone and individual aging. In: Hall, B. K. (ed.). Bone. Bone Growth—B. Vol. 7. CRC Press, Boca Raton, FL. Pp. 245–283.

Castanet, J., Grandin, A., Arbourachid, A., and Ricqlès, A. de 1996. Expression de l'dynamique de croissance dans la structure de los périostique chez *Anas platyrhynchous*. C. R. Acad. Sci. (Paris) Sci. Vie 319: 301–308.

Castanet, J., Curry Rogers, K., Cubo, J., and Boisard, J. J. 2000. Quantification of periosteal osteogenesis in ostrich and emu: implications for assessing growth in dinosaurs. C. R. Acad. Sci. (Paris) III 323: 543–550.

Chiappe, L. M., Coria, R. A., Dingus, L., Jackson, F., Chinsamy, A., and Fox, M. 1998. Sauropod dinosaur embryos from the Late Cretaceous of Patagonia. Nature 396: 258–261.

Chiappe, L. M., Salgado, L., and Coria, R. A. 2001. Embryonic skulls of titanosaur sauropod dinosaurs. science 293: 2444–2446.

Chinsamy, A. 1990. Physiological implications of the bone histology of *Syntarsus Rhodesiensis* (Saurischia: Theropoda). Palaeontol. Afr. 27: 77–82.

———. 1993. Bone histology and growth trajectory of the prosauropod dinosaur *Massospondylus carinatus* Owen. Modern Geology 18: 319–329.

Chinsamy, A., and Elzanowski, A. 2001. Evolution of growth patterns in birds. Nature 412: 402–403.

Coe, M. J., Dilcher, D. L., Farlow, J. O., Jarzen, D. M., and Russell, D. A. 1987. Dinosaurs and land plants. In: Friis, E. M., Chaloner, W. G., and Crane, P. R. (eds.). The Origins of Angiosperms and Their Biological Consequences. Cambridge University Press, Cambridge. Pp. 225–228.

Cooper, L. N., and Horner, J. R. 1999. Growth rate of *Hypacrosaurus stebingeri* as hypothesized from lines of arrested growth and whole femur circumference. J. Vertebr. Paleontol. 19: 39A.

Currey, J. D. 1987. The evolution of the mechanical properties of amniote bone. J. Biomech. 20: 1035–1044.

Curry, K. A. 1998. Histological quantification of growth rates in *Apatosaurus*. J. Vertebr. Paleontol. 18: 36A.

———. 1999. Ontogenetic histology of *Apatosaurus* (Dinosauria: Sauropoda): New insights on growth rates and longevity. J. Vertebr. Paleontol. 19: 654–665.

Curry Rogers, K. A., Erickson, G., and Norell, M. A. 2003. Dinosaurian life history strategies, growth rates, and character evolution: New insights garnered from bone histology and developmental mass extrapolation. J. Vertebr. Paleontol. 23(3): 90A.

Dodson, P. 1975. Functional and ecological significance of relative growth of *Alligator*. J. Zool. London 175: 315–355.

Enlow, D. H. 1963. Principles of Bone Remodeling. Charles C Thomas, Springfield, IL. 123 pp.

Enlow, D. H., and Brown, S. O. 1956. A comparative histological study of fossil and recent bone tissue (Part 1). Tex. J. Sci. 8: 405–443.

———. 1957. A comparative histological study of fossil and recent bone tissue (Part 2). Tex. J. Sci. 9: 186–214.

———. 1958. A comparative histological study of fossil and recent bone tissue (Part 3). Tex. J. Sci. 10: 187–230.

Erickson, G. M., and Brochu, C. 1999. How the "terror crocodile" grew so big. Nature 398: 205–206.

Erickson, G. M., and Tumanova, T. A. 2000. Growth curve of *Psittacosaurus mongoliensis* Osborn (Ceratopsia: Psittacosauridae) inferred from long bone histology. Zool. J. Linn. Soc. 130: 551–556.

Erickson, G. M., Curry Rogers, K., and Yerby, S. 2001. Dinosaurian growth patterns and rapid avian growth rates. Nature 412: 429–433.

Fawcett, D. W. 1942. Amedullary bones of the Florida manatee (*Trichechus Latirostris*). Am. J. Anat. 71: 271–303.

Foote, J. S. 1913. Th comparative histology of the femur. Smithsonian Miscellaneous Collection 61(2232).

Francillon-Vieillot, H., de Buffrénil, V., Castanet, J., Géraudie, J., Meunier, F. J., Sire, J. Y., Zylberberg, L., and Ricqlès, A. de 1990. Microstructure and mineralization of vertebrate skeletal tissues. In: Skeletal Biomineralization: Patterns, Processes, and Evolutionary Trends. Vol. 1. Van Nostrand Reinhold, New York. Pp. 473–530.

Gillette, D. D. 1991. *Seismosaurus halli* (n. gen., n. sp.) a new sauropod dinosaur from the Morrison Formation (Upper Jurassic–Lower Cretaceous) of New Mexico, U.S.A. J. Vertebr. Paleontol. 11: 417–433.

Horner, J. R., Ricqlès, A., de and Padian, K. 1999. Variation in skeletochronological indicators of the hadrosaurid dinosaur *Hypacrosaurus*: implications for age assessment in dinosaurs. Paleobiology 25: 295–304.

———. 2000. Long bone histology of the hadrosaurid dinosaur *Maiasaura peeblesorum*; growth dynamics and physiology based on an ontogenetic series of skeletal elements. J. Vertebr. Paleontol. 20: 115–129.

Jianu, C.-M., and Weishampel, D. B. 1999. The smallest of the largest, a new look at possible dwarfing in sauropod dinosaurs. Geol. Mijnbouw 78: 335–343.

Lacroix, P. 1971. The internal remodeling of bones. In: Bourne, G. H. (ed.). The Biochemistry and Physiology of Bone. Vol. 3. Academic Press, New York. Pp. 119–144.

Lanyon, L. E., and Rubin, C. T. 1984. Static vs. dynamic loads as an influence on bone remodelling. Journal of Biomechanics 17: 897–905.

Mantell, G. A. 1850. On the *Pelorosaurus*, an undescribed gigantic terrestrial reptile, whose remains are associated with those of the *Iguanodon* and the other saurians in the strata of the Tilgate Forest, in Sussex. Philos. Trans. Roy. Soc. London 140: 379–390.

Margerie, E. de, Cubo, J., and Castanet, J. 2002. Bone typology and growth rate: testing 'Amprino's Rule' in the mallard (*Anas platyrhynchos*). C. R. Biol. 325: 221–230.

Martin, R. B., and Burr, D. B. 1989. Structure, Function and Adaptation of Compact Bone. Raven Press, New York. 275 pp.

Owen, R. 1859. Monograph on the fossil Reptilia of the Wealden and Purbeck Formations. Supplement II. Crocodilia (*Streptospondylus* etc.). Palaeontogr. Soc. Monogr. 11: 20–44.

Padian, K., Ricqlès, A. J. de and Horner, J. R. 2001. Dinosaurian growth rates and bird origins. Nature 412: 405–408.

Powell, J. E. 1986. Revision de los titanosauridos de America del Sur. unpublished Ph.D. thesis, Universidad Nacional de Tucuman, Tucuman. 340 pp.

———. 1992. Osteologia de *Saltasaurus loricatus* (Sauropoda–Titanosauridae). In: Sanz, J. L., and Buscalioni, A. D. (eds.). Los Dinosaurios y su Entorno Biotico. Instituto "Juan de Valdes," Cuenca.

———. 2003. Revision of South American Titanosaurid dinosaurs: palaeobiological, palaeobiogeographical, and phylogenetic aspects. Rec. Queen Victoria Mus. Launceston 111: 1–173.

Purves, W. K., Orians, G. H., and Heller, H. C. 1992. Life: The Science of Biology. 3rd ed. Freeman, Salt Lake City, UT.

Queckett, J. T. 1855. Descriptive and Illustrated Catalogue of the Histological Series Contained in the Museum of the Royal College of Surgeons. II. Structure of the Skeleton of Vertebrate Animals. Royal College of Surgeons, London.

Reid, R. E. H. 1981. Lamellar-zonal bone with zones and annuli in the pelvis of a sauropod dinosaur. Nature 292: 49–51.

———. 1984. The histology of dinosaurian bone, and its possible bearing on dinosaurian physiology; pp. 629-663 in M. W. J. Ferguson (ed.), The Structure, Development and Evolution of Reptiles: Zoological Journal of London Symposia 52. Academic Press, London. Pp.629–663.

———. 1990. Zonal "growth rings" in dinosaurs. Modern Geol. 15: 19–48.

———. 1997. How dinosaurs grew. In: Farlow, J. O., and Brett-Surman, M. K. (eds.). The Complete

Dinosaur. Indiana University Press, Bloomington and Indianapolis. Pp. 403–413.

Rhodin, A. G. J. 1985. Comparative chondro-osseus development and growth of marine turtles. Copeia 1985: 752–771.

Ricqlès, A. de. 1968a. Quelques observations paléohistologiques sur les dinosaurien sauropode *Bothriospondylus*. Ann. Univ. Madagascar. 6: 157–209.

———. 1968b. Recherches paléohistologiques sur les os longs des tétrapodes. I—Origine du tissu osseux plexiforme des dinosauriens sauropodes. Ann. Paleontol. 54: 133–145.

———. 1977. Recherches paléohistologiques sur les os longs de Tétrapodes. VII (deuxième partie: fonctions) (suite). Ann. Paléontol. (Vertebr.) 62: 71–126.

———. 1980. Tissue structure of dinosaur bone. Functional significance and possible relation to dinosaur physiology. In: Thomas, R. D. K., and Olson, E. C. (eds.). A Cold Look at the Warm-Blooded Dinosaurs. AAAS Selected Symposium. Vol. 28. Westview Press, Boulder, CO. Pp. 103–139.

———. 1983. Cyclical growth in the long limb bones of a sauropod dinosaur. Acta Paleontologica 28: 225–232.

Ricqlès, A. de., Meunier, F. J., Castanet, J., and Francillon-Vieillot, H. 1991. Comparative microstructure of bone. In: Hall, B. K. (ed.). Bone, Bone Matrix, and Bone Specific Products. Vol. 3. CRC Press, Boca Raton, FL. Pp. 1–78.

Rimblot-Baly, F., Ricqlès, A. de, and Zylberberg, l. 1995. Analyse paléohistologique d'une série de croissance partielle chez *Lapparentosaurus madagascariensis* (Jurassique Moyen): Essai sur la dynamique de croissance d'un dinosaure sauropode. Ann. Paleontol. 81: 49–86.

Sander, P. M. 2000. Longbone histology of the Tendaguru sauropods: implications for growth and biology. Paleobiology 26: 466–488.

———. 2003. Long and girdle bone histology in sauropod dinosaurs: methods of study and implications for growth, life history, taxonomy, and evolution. J. Vertebr. Paleontol. 233 (Suppl.): 93A.

Sander, P. M., and Tückmantel, C. 2003. Bone lamina thickness, bone apposition rates, and age estimates in sauropod humeri and femora. Palaontol. Z. 77(1): 139–150.

Sander, P. M., Laven, T., Mateus, O., and Knoetschke, N. 2004. Insular dwarfism in a brachiosaurid sauropod from the Upper Jurassic of Germany. J. Vertebr. Paleontol. 243, (Suppl.): 108A.

Schweitzer, M. H., Watt, J. A., Avci, R. Forster, C. A., Krause, D. W., Knapp, L., Rogers, R. R., Beech, I., and Marshall, M. 1999. Keratin immunoreactivity in the Late Cretaceous bird *Rahonavis ostromi*. J. Vertebrate Paleontology 19: 712–722.

Schweitzer, M.H., Wittmeyer, J.L., Horner, J.R., and Janktoporski. 2005. Soft-tissue vessels and cellular preservation in Tyranosaurus Rex. Science 307: 1952–1955.

Seebacher, F. 2001. An new method to calculate allometric length-mass relationships of dinosaurs. J. Vertebr. Paleontol. 21: 51–60.

Smith, J. B., Lamanna, M. C., Lacovara, K. J., Dodson, P., Smith, J. R., Poole, J. C.,Giegengack, R., and Attia, Y. 2001. A giant sauropod dinosaur from an Upper Cretaceous mangrove deposit in Egypt. Science 292: 1704–1706.

Starck, J. M., and Chinsamy, A. 2002. Bone microstructure and developmental plasticity in birds and other dinosaurs. J. Morphol. 254: 232–246.

Sussman, M. 1964. Growth and Development. Prentice–Hall, Upper Saddle River, NJ.

Thompson, D. 1943. On Growth Form. Cambridge University Press. 1116 pp.

Varricchio, D. J. 1993. Bone microstructure of the upper Cretaceous theropod dinosaur *Troodon formosus*. J. Vertebr. Paleontol. 13: 99–104.

Weishampel, D. B., and Horner, J. R. 1994. Life history syndromes, heterochrony, and the evolution of Dinosauria. In: Carpenter, K., Hirsch, K. F., and Horner, J. R. (eds.). Dinosaur Eggs and Babies. Cambridge University Press, Cambridge. Pp. 229–243.

Wells, J. W. 1963. Coral growth and geochronometry. Nature 197: 948–950.

Witmer, L. M. 2001. Nostril positions in dinosaurs and other vertebrates and its significance for nasal function. science 293: 850–853.

Wykoff, R. W. G. 1975. Pleistocene and dinosaur gelatins. Comp. Biochem. Physiol. 55B: 95–97.

A CONVERSATION WITH JACK MCINTOSH

Jeffrey A. Wilson and Kristina Curry Rogers

John S. ("Jack") McIntosh has been a leading student of sauropod dinosaurs for well over half a century. During the course of his long career, Jack has been influenced by legendary paleontologists such as Barnum Brown, Richard Lull, Friedrich von Huene, and Alfred Romer, and he continues to influence young dinosaur paleontologists.

Jack's two main interests, sauropod dinosaurs and the history of North American paleontology, intersect in the badlands of the western United States during the last quarter of the nineteenth century, when O.C. Marsh and E.D. Cope discovered and described *Camarasaurus, Diplodocus, Allosaurus, Stegosaurus,* and many other dinosaurs. During long hours at museums studying bones and examining field notes, maps, and journals, Jack has reconstructed the events of many field seasons over decades of dinosaur collecting. His sleuthing has recovered lost details about the provenance and associations of many dinosaur skeletons and led to reconsideration of many of our old impressions of sauropods. The most famous recovered "detail" is Jack's revision of the skull of *Apatosaurus*. For the better part of a century, the famous Yale mount of an *Apatosaurus* skeleton bore a *Camarasaurus*-like skull because of a dotted-in sketch made in 1883 by Marsh. Through examination of the original quarry maps and shipping manifests, Jack discovered a "accounting error" and eventually recapitated *Apatosaurus* with the correct, *Diplodocus*-like skull.

In addition to his historical pursuits, Jack remains one of the foremost experts on sauropod dinosaurs. Jack was the first to summarize and synthesize skeletal, stratigraphic, taphonomic, and taxonomic data on the entirety of sauropod dinosaurs, a monumental effort that serves as a standard reference for any investigation into sauropods and also provides raw information for phylogenetic analyses of sauropod relationships. Incredibly, Jack managed this body of work while winning his bread as a theoretical physicist at Yale, Princeton, and Wesleyan. The conversation below recounts Jack's formative years and early encounters as a precocious student, his service in the Second World War, and paleontology in the badlands of the western United States. We trace Jack's interest in sauropods and delve into the some of the life experiences that solidified his position as one of our most celebrated colleagues.

The conversation was recorded and transcribed from discussions with Jack McIntosh on 3 April 2004 (Middletown, Connecticut) and 8 November 2004 (Denver, Colorado). Material from the two interviews has been woven together for continuity, and technical terms and colloquial phrasings have been edited to render them accessible to a broader readership. The recording was transcribed by Carole Goodyear and Amanda Kealey.

KCR: I love to hear the stories about how people get into paleontology. What sparked your interest in sauropods?

Mcintosh: Well, I think it was the same way as most people. I think I was about six. My father took me to the Carnegie Museum. I saw the *Diplodocus*, flipped. From there on, it continued to be part of my life.

JAW: So Jack, you wrote a famous letter at age thirteen or fourteen to [Richard] Lull. It seems you had already developed a very academic interest in dinosaurs. How did that emerge?

McIntosh: When I was a kid, I was sick most of the time. In the sixth grade I missed about one third of the year. When I was sick my uncle, who had gone to Yale and had taken a course under Lull, had the book *Organic Evolution*. He gave me that book to read, and that of course was very exciting.

JAW: *Your letter to Lull is pretty advanced for a 13 or 14 year old. You asked whether* Cardiodon *and* Cetiosaurus *should be the same thing," and "What about Apatosaurus?"*

KCR: *Had you already met Lull when you wrote to him, or did you write the letter without knowing him in advance?*

McIntosh: When you're that age, you're brash, you think you can do anything—you're crazy!

KCR: *Did Lull write back to you?*

McIntosh: Well, he wrote me back and said, "I really don't know that much about saurischians. Write to Gilmore."

JAW: *Do you still have the copy of Lull's letter?*

McIntosh: Oh of course, sure. And of course I have another one, which I cherish, that I got many, many years later from [Werner] Janensch. Well, it's in German of course.

KCR: *Did you get to meet these paleontologists as a young kid?*

McIntosh: I would go in every Saturday when I was in high school to Pittsburgh, to the Carnegie Library and Carnegie Museum, so I got to know all the people there of course. But I'd never even visited New York City until my father took me on to college.

After I graduated from high school in 1941, I started at Yale. I had never been to Connecticut before. So when my Father drove me to Yale, he took me up there and dumped me and went back home. They thought I was going to be terribly homesick and all that sort of thing. Well, I immediately went to the Sterling Library and was overwhelmed. I looked up von Huene because I had been looking for copies of the 1932 *Die fossil Reptil-Ordnung Saurischia*, and they didn't have it at the Carnegie Library. In fact I had written to the "Seven Book Hunters," who were advertised in *The New York Times*, and I made an absolutely silly statement. I said, "Can you possibly get me a copy of *Die fossil Reptil-Ordnung Saurischia*? I'll pay any price for it." Well, that was a mistake because I got back a letter a month or so later saying,

"We have found a copy and you can have it for $100." Well, of course back in the Depression days, I think I had seen a $20 bill, but the idea of $100 was totally out of the question. So I had to write them back sorrowfully saying I couldn't buy it. But when I got to the Yale Library and looked up von Huene, there was a copy of *Die fossil Reptil-Ordnung Saurischia*, so I immediately got it down. At the Carnegie Library they would put me in a room, where they would bring a book in. You could sit there and work on it but you could barely touch it. But here [Yale Library], I said, "Can I look at this?" and they said "Oh, you can take it out." Not only could you take it out, but you could take it out and renew it over and over again. So I took this thing out, brought it back to my room, and began copying large parts of it. I wasn't the least bit homesick. This was just absolutely wonderful.

So the second day I was there, I went up to the Peabody Museum. I went upstairs into the "secret" tower where Ed Lewis, who was the paleontologist there at the time, had his domain. I went in and talked to him, and told him that I wanted to work on dinosaurs, so he set me to work in the laboratory with the big *"Atlantosaurus"* (*Apatosaurus*) cervical vertebrae that were all broken. [Othniel C.] Marsh mounted two of these cervicals that were very good ones, but they had been mishandled, so Lewis put me to work getting those things back together. They were not too badly damaged; there were a number of pieces but you could put them back together. So the first thing I did was to get those big cervicals of *"Atlantosaurus"* (*Apatosaurus*) back together.

Freshman year I was working in one of the dining halls for a job, but from the second year I had a regular job putting dinosaur bones back together. What happened was I was working with a little *Coelurus*, which was very interesting. I had this femur and tibia and so forth, was studying them, and I felt somebody's breath on the back of my neck. It was Barnum Brown. He got all excited. "I have an animal," he said. "It has a nice tibia and fibula and the tibia looks just like yours except it's one and a half times as long but we don't have the femur." He said, "What I would like to ask you to do is to come down to the American Museum and take some measurements of our animal, then we will take a cast of your femur and we'll expand it." Well, up to then, I had not cut a single class. I was good about going to classes. But I'll tell you, the next Monday I was on the train for New York.

At the American Museum I rode on the elevator to the fifth floor, and I go in to see Rachel Nichols, who is running the place. She was the secretary; she knew everything. She sat me down and called Barnum Brown. He brought in a couple of trays, great big trays, and one of them had an animal that he called *"Daptosaurus,"* but he never did describe it. John Ostrom described another specimen of this animal as *Deinonychus* many years later. The other tray contained another very small theropod, which Brown called *"Macrodontosaurus."* Years later, Ostrom described this specimen under the name *Microvenator* since he determined that the big teeth found with the specimen belonged to a different animal. So Barnum Brown took me and he put me in a room with a great big long table. He put this box down and said, "Now why don't you study these things and you can measure them and draw them, and when you go back to Yale, we can decide what to do about this." When he brought these things in to show me, Rachel Nichols whispered, "He doesn't show those to anybody." So I had the afternoon to study *Deinonychus*—brand new from the Cloverly. I mean this was just absolutely out of this world. And with Barnum Brown of all people being the one that set me onto this thing. So that was great.

JAW: A brush with greatness. Did anything come of it?

McIntosh: I never went up to see Barnum Brown again. He retired right about that time anyway.

So then after two years at Yale I was in the Army Air Corps and I was sent to various colleges to learn meteorology and radar. I went to Brown and MIT and Harvard, and I went into the MCZ [Museum of Comparative Zoology, Harvard] to see what was in there. Here I was, this little insignificant shave-tail. I went up to the great door, knocked, and introduced myself to [Alfred S.] Romer. He was just absolutely delightful. He invited me in, and he began talking with me and pulled out these papers by [Chinese paleontologist] C. C. Young that had just come over from Chungking [as it was then spelled]. Chungking, of course, was the temporary capital of China at that time. The Japanese had taken over much of China, and Chungking was way back in Sichuan Province. Young had gone there, and he published some of these early papers on the *Lufengosaurus*, etc. The paper on which they were printed was of very poor quality. Somehow Young had gotten copies over to Romer, and Romer had papers on all of these different prosauropods. He brought them out and showed them to me and said, "Oh, why don't you borrow them over the weekend?" That was Romer. He probably had the only copies of those things in the United States, and he let me borrow them.

JAW: Were you in the war?

McIntosh: Oh, yes, sure. I had just started my junior year. What happened was I volunteered for this business [Army Air Corps], but they didn't take you immediately. I mean you were on call. So I was a month into my junior year when I finally got called.

I finally was sent over to Guam as a flight weather officer on a B-29. We flew 21 or 22 missions over Japan. We were in weather planes, so we would fly at high altitudes—32,000 feet, very high altitudes in those days—and we would fly over the different targets and then radio back to our group as to where the weather was the best and then they would pick them out. These missions would sometimes take 20 hours there and back, very long missions. Our group would fly a couple of hours behind us. We would pick out a target, and then they would go and bomb that target.

We were flying way up. The Japanese would shoot at us, and the stuff would rattle on the plane, but we were never badly hit. We had engines go out a couple of times, and we had to land on Iwo [Jima] on the way back or we would have gone into the ocean. Iwo [Jima] by that time looked like a huge aircraft carrier that had been completely macadamized over. It didn't look the way it did when the Marines were in there. But it was interesting seeing it, and it was very nice to have it there, because as I say, we would have been in the ocean. We lost two engines on the same side. I've forgotten whether this was due to ack-ack or to engine failure. But anyway, we did lose two engines. If you lose one on each side, that's all right, you can get back. But if you lose two on the same side, that's not good. And so Iwo [Jima] was there and we were able to get back to Guam.

About a week after they dropped the atom bombs, a Twentieth Air Force Colonel decided that he wanted to see what was going on and he came up and commandeered our plane. So we flew over both Hiroshima and Nagasaki in the same day, and that was a bit worrisome, too, because you're a single plane. But I don't think they even fired on us.

We flew another mission over Japan where we went out to supply materials for prisoners of war. We were assigned a particular camp to drop our stuff and we flew over in all directions. We couldn't find it, and we were very sad because this was great, you know, to be able to provision these guys that had been POWs for years. We turned around, and as we flew back home we saw one. I don't know if it was the one we were supposed to supply or not, but all these people were down there yelling and everything. We flew in at very low altitude and dropped all of this stuff, and they were just going mad. I mean that was exciting, to say the least.

The last mission over Japan was on V-J Day, when MacArthur had everybody flying over. However, we wanted to see what was going on. We were supposed to fly formation, but our particular group went in over the *U.S.S. Missouri* at low altitude and really saw what was going on down there.

JAW: This is when they were signing the surrender?

McIntosh: Yes. MacArthur was furious. He had us flying training missions for a week after that! There were a whole bunch of us that did this kind of thing.

After the war, I was at Tinker Field near Oklahoma City for a while. They were letting people out and they couldn't let everybody out at once. And so I was just there with absolutely nothing to do for about a month or so. I got a pass and went down to Norman to see Stovall and met Wann Langston. Wann showed me all around and we had an absolutely great time. Of course, there were lots of sauropod bones there, which Wann was responsible for putting together. Wann had been looking for *Die fosil Reptil-Ordnung Saurischia* too. I told him that I had access to it, and he was very envious. So when I got back to New Haven, I immediately copied all kinds of stuff and sent it to him. A few years ago he told me he still has it.

I got back about three years and one day after I went into the Air Corps—I didn't even go home. I went directly from where I was discharged at Indiantown Gap, Pennsylvania, out to New Haven, because the next term was starting on about the first of March. And I signed up for these courses, essentially the same ones that I had been taking when I left: same teachers, in the same rooms, and most of the same students with whom I had been in these courses three years earlier. The same thing had happened to them. They had started, and they had been called and so forth. It was incredible. It was as if three years had vanished. I can't even describe it. It was amazing, but anyway, then I got back into working in the laboratory.

I was majoring in physics, but I wasn't going into paleontology because you weren't allowed to "do dinosaurs" in those days. You did mammal teeth, and I was not interested in mammal teeth at all. But I could work on these dinosaurs in the laboratory, and that was my job. I worked on these things and I finished, got my degree, and then went on to get my Ph.D. in physics and go on to teach in various places: Princeton, Yale, and finally ended up at Wesleyan.

JAW: The years following the war were some lean years for vertebrate paleontology. You must have been very lonely for a while there—the only other person you could talk to about sauropods was José Bonaparte. Did you ever think for a minute about abandoning sauropods and working on a different group?

McIntosh: No. Oh, no. I continued, obviously, to have my interest in sauropods. Joe Gregory made me a research associate at Yale, so I would go down and work there any time I wanted to. And I did, as an avocation, in my spare time. Sometimes I had spare time; sometimes I didn't. But I could do it, particularly during the seven years that I taught at Yale. There are now things coming out all the time. I mean—How many new genera have been described in the last ten years?

KCR: Have you ever found a good intersection between your interest in dinosaurs and your interest in physics?

McIntosh: No, the things that interest me in paleontology don't overlap with my interests in physics.

JAW: You had another brush with greatness in your physics career too—hearing Einstein speak at Princeton. Did he have a real presence when he showed up?

McIntosh: Oh wow, when he walked into the room, you could have heard a pin drop anywhere. He shuffled into the room with the crazy old sweater that he always wore, and they pulled up a chair at the front of the room, and he sat down to lecture. He talked in a little voice that people in the back row could easily have heard. I was up near the front row. The room was so quiet that everybody heard every word he said. Another time, I was sitting in the theater to hear Linus Pauling, the famous chemist who won the Nobel Prize. He was going to give a lecture, and everybody was buzzing; it was a huge noise in the room. And all of a sudden every sound stopped. I mean it was just as if somebody had turned off a booming television or something. And I turned around and looked and Einstein was walking down the aisle.

Can you imagine? Every time you went anywhere to have the whole world stop. It must have been the most embarrassing thing for him. Of course, he didn't come out that often. That was the only other time that I saw him.

KCR: It really is hard to imagine that kind of awe. I wonder if there's anybody like that in paleontology. [All laugh.] What prompted your interest in sauropods? What about them was so exciting?

McIntosh: Oh, the Carnegie Museum! The *Diplodocus* and the *Camarasaurus* at the Carnegie Museum were the things that started it off. I'm five years old, and I'm taken into the Carnegie Museum by my father. It just never left. It's just that I wanted to find out as much about sauropods as I possibly could. Most people get over it in a couple of years, but I didn't. See, I'm not really that interested in this bird business. The whole thing has been proved to me. I know birds are dinosaurs, and they're going to find lots of them, and they're going to have feathers, and they're going to be wonderful. John Ostrom tells me that this [Yixian] formation is a mile high, with layer after layer, where they go in and find one after another, fossil birds. It's overwhelming. And it's all little stuff, and you can't really look at those vertebrae because [the animal] is so small and has little feathers coming out. I just don't have much interest in them. But if you collect a sauropod, and you can look at each vertebra, and see each lamina, and so on.

I got to the Carnegie Museum as a teenager and got to know "Pop" Kay [J. Leroy Kay], who was the curator at the time there. He was very nice and showed me all sorts of things. Much later, around 1960 or so, when I was getting older and when I was teaching, I wanted to check something. So I went down and met Craig Black, and he arranged for me to come in and do anything I wanted, which was extremely nice. And that's when I started going through the entire sauropod collection. I eventually went through all of the dinosaur collections. But with the sauropod collection I tried to identify everything.

JAW: And that's what led to that [1981] Bulletin of the Carnegie Museum?

McIntosh: Eventually. It didn't come out until the 1980s. When I started doing all that cataloging, I didn't really envision the catalog that came out later. But when I got through with it, I mean when I was well along with it, it seemed to be the obvious thing to do.

JAW: What initially got you interested in working with the discoveries of Cope and Marsh? Why did you start investigating the quarries and maps and all that?

McIntosh: Of course Marsh and Cope described many of the original sauropods of North America and their types are very important to separate out. The Cope bones of *Camarasaurus* in the American Museum—as you probably know—have mysterious numbers and letters on them. Osborn and Mook [who described the Cope collection] had no idea what these meant, and they don't even mention this in their monograph. The trouble was, Cope just didn't keep records, and he threw all his stuff away. Almost all the letters from his collectors have been lost, or somebody threw them away after he died, before the stuff was taken to the American Museum. But [Cope's collector] Lucas wrote all this stuff out in great detail—where these things were found and how. He drew pictures of a large number of these bones and lettered them, and a few of the drawings have survived—but only a few of them.

Cope described *Camarasaurus* and he had this drawing made of the animal, which is rather weird but good for a starter. He thought it all was one individual. It actually is more than one individual, at least two. Lucas found the second animal, which he thought was a single animal, and he called it *Camarasaurus* II. A few of his letters have survived, and it turned out to be two animals, too. The first one of those animals wasn't exactly articulated, but it was partially articulated and it was all together. What those mysterious numbers referred to are box numbers. But they [Osborn and Mook] didn't publish the box numbers at all because they didn't realize they were important. I was able to decode these things from the few surviving letters, which listed each lettered bone and the box it was shipped in.

When [the American Museum's W. D.] Matthew went to Philadelphia [to retrieve Cope's *Camarasaurus* bones], he put new numbers on all of these boxes for shipping purposes. So when the people in the American Museum prepared these bones, they put Matthew's box number on them, but they also put Lucas's box number and the letter assigned to each bone on them. These things were numbered in a certain order in the shipping manifest, and what I have figured out is that the order of the boxes corresponds to the way they were shipped. I now know what was in each box and the order in which those boxes came. There were 11 shipments in all, and there might be one or two more of a tooth or something like that, but there are essentially 11 shipments that were sent in by O. W. Lucas. Then there are about six or eight more that were sent in by Ira Lucas. So in the first set of boxes, shipment number 11 was sent in by O. W. Lucas and then the next one, number 12, is the one that's the key to everything—it tells me how to connect the Matthew numbers with the Lucas numbers. So I now know the order. So it turns out that shipment numbers 11, 12, and 13 are all one individual. They started collecting from the tail and they go forward as you get up to the cervicals and there is no overlap at all until you get to the very end. At the very end, you begin to get a second individual. I have pictures of the quarries that were taken at that time, and I'm trying to identify exactly which bone is which. I mean there's no duplication, and you get the right numbers of vertebrae and the right numbers of everything. When you go forward in the first individual of *Camarasaurus supremus*, there's a second individual that lies beyond that; and unfortunately it's the front end of that thing that's beginning to come in here, and so I can't quite separate them all exactly. I'm pretty sure how it goes.

JAW: So you're able to identify individuals?

McIntosh: Exactly. I hope they never do it, but if they actually wanted to mount a skeleton of *Camarasaurus supremus*, they can, because I can tell them exactly which individual each bone belongs to. This is the kind of thing that you can do if you have records.

One of the Carnegie Museum quarries where the quarry maps may be missing is the one on the Red Forks of Powder River, and there are two quarries there. We're not even sure if a map was ever made! The lower quarry, down by the river by the Red Fork, is the *Diplodocus hayi* that's out in Houston, and then there's another quarry that's up a bit higher where there are huge numbers of sauropod bones, interesting bones, too. Fortunately *Diplodocus hayi* is a unique individual; there's no mixture there. We know exactly which bones belong to that, and it came from a single quarry and then this other stuff. But this other stuff has all of these bones, and there is no indication at all as to what's what. I mean there are numbers on the bones, but there's no quarry map to tell you what those numbers correspond to, and so that is completely lost. Unless that quarry map should turn up, and I'm quite sure it won't now because Betty Hill [the former secretary at Carnegie Museum] and I have gone through absolutely everything. The trouble is, here you have this huge number of bones at the Carnegie Museum that are very interesting, and yet they're all mixed up. It may have been that even if you had the quarry map, you wouldn't be able to do much with it. But it may have been that at least you would know perhaps clusters of bones, you'd know something about it. But that quarry map is missing and so it's absolutely too bad that we don't know anything about that.

I have made a collection of quarry maps that I'm keeping—not to work on myself necessarily, but for the future. I mean because these things do get lost, and I want copies available, and I don't want what happened at the Carnegie Museum to happen again. Gilmore made huge quarry maps at Quarry C at the Sheep Creek, and those things have vanished too.

JAW: Have you spent a lot of time out west ground-truthing some of these localities?

McIntosh: Well, up until two or three years ago, I would make a trip out west every summer and I would go to various museums and so forth. In '73 and '74, I was in the field all summer with Bob Bakker, Peter Dodson, and Kay Behrensmeyer. That was absolutely marvelous—we went to most of the Jurassic dinosaur quarries, particularly the classic ones, that had ever been found, and I learned an awful lot there.

At that time, I had only known Jim Jensen for a couple of years. I stopped at Dinosaur National Monument every year, and one year I picked up a couple of hitchhikers on the way. They said that they had just heard on the radio on their last ride that an extraordinary dinosaur quarry had been found down in western Colorado, and I knew immediately this was Jim's site. So I got to Dinosaur National Monument, and instead of going up to the quarry area to look at things, I went to the telephone booth and called Delta, but I couldn't get a hold of the Joneses [Eddie & Vivian; friends of Jim's]. But the telephone operator said, "Oh, yes, I know all about this," and she told me about the Dry Mesa quarry and she told me exactly how to get there, and getting there was no simple problem. There were no numbers on those roads. She told me to go so many miles here and turn on this road and so forth. I hurried right down to Dry Mesa and I got up to this road that Jim had made to the quarry about a mile or two down. I was in my regular car—it wasn't four-wheel drive or anything. I decided, I don't want to try to go down this thing. So I put my pack on my back and walked down. When I got down there, there were cars and even trailers from some 24 states that had piled up. I mean the word had gotten out. Jim came rushing out and handed me a trowel and said, "Dig out that interesting *Diplodocus* caudal," and so that is the kind of thing that happened. Jim was just absolutely wonderful. It was just a beehive of activity, with everybody around there. It was extremely exciting, and they had already uncovered the big scapula.

A couple of years later I went down to see Jim and he decided that I needed to know more about the geology of Utah. So he got his truck out and he got Brooks Britt, who was about 16 at the time, and we got into the car and he drove us from one end of the San Rafael Swell to the other, showing us every single quarry. We even went to Gilmore's quarry where he found the *Alamosaurus*, and Jim showed me everything. I learned more Utah geology that day than I've ever learned [about anything] in that short a period of time in my life.

KCR: I think one of the big questions that kids have about dinosaurs is why "Brontosaurus" has a different name, and why the skulls were changed and what the whole situation is with that. Your work on debunking some of those old dinosaur legends is one of the most familiar examples of how our science works, for people all over the world. Can you tell us a little about how you got started on that, and how you dissected all these problems of sauropod heads and names?

McIntosh: Oh yes, at that time nobody knew what the skull of an *Apatosaurus* was. I was working through all of [Earl] Douglass's records and so forth. I have notebooks full of hundreds of letters like this. I don't have the letters themselves and I don't even have photocopies of them. But I copied out by hand everything that had to do with dinosaurs in those letters. One of the reasons I did that is because so many things have vanished over the years and I decided that everything that ought to be preserved should be preserved. And I have an enormous collection that people have allowed me to make copies of their records, so they're now preserved. A number of quarry maps have been lost or destroyed, and without maps, collections lose an enormous amount of their value, with no chance of determining in a multigenera or multi-individual quarry which thing might go with which.

KCR: You've been in the field of sauropod research for a long time, and you've seen a lot of students like us entering and gaining real interest in sauropods, and contributing to our changing view of the group. What do you think the future holds for sauropods? What questions are you most excited about finding answers to?

McIntosh: The main thing that I would like to have happen, which may never happen now, is to have them reopen the Marsh Quarry. There definitely is other *Diplodocus* material in that quarry [Felch Quarry, Garden Park, Colorado]. That quarry is not finished by any manner or means, and it's got all kinds of good stuff in it. I probably told you that I had Jensen ready to open it. The trouble is that the hill goes up, so you have to dig further and further into it. When Hatcher and Utterback collected *Haplocanthosaurus*, they had to dig back in to get those beautiful skeletons, some of the best stuff they took out of there aside from the *Ceratosaurus*. Jensen had already made arrangements to rent a bulldozer to get back in there, but on his way from Canyon City back to his home in Provo, he drove through Delta, Colorado, where friends of his, the Joneses, Vivian and Eddie, lived. They had found a huge claw [of *Torvosaurus*], and when they knew Jim was coming Vivian put this claw on the front table in the living room, and didn't say anything. Jim came in and sat down on the couch. They talked for about 15 minutes. All of a sudden Jim's eye landed on that claw.... "What is that!?" Vivian was just waiting, of course, for him to see it. She took him up to the Dry Mesa Quarry, and then all bets were off as far as opening the Garden Park quarry. He canceled everything back in Canyon City, and that was that. But somebody should open that again. There's still lots more there.

JAW: So what about places.... Where would you like to see more sauropods come from?

McIntosh: Well, of course I want the American Museum to collect that stuff from Mongolia. I'd also like to see all that stuff down in South America prepared and figured properly. You know, some of those skeletons hiding around there, and they even have some of the skulls!

JAW: What are you planning next?

McIntosh: Well, I'm 81 years old. I want to get all of my collections of pictures sorted out, and all the identifications written on them. I know what all the identifications are but some of those notes are very rough: I can read them but nobody else could.

JAW: Can you hire someone to help you with that?

McIntosh: Ah, I don't want anyone to help me with it. I want to do it myself. It keeps me busy. Lately I seem to be finding myself falling behind because I learn an awful lot of things at each SVP! There is just so much to learn about sauropods.

LIST OF CONTRIBUTORS

PAUL M. BARRETT, Department of Palaeontology, The Natural History Museum, Cromwell Road, London SW7 5BD, UK
p.barrett@nhm.ac.uk

MATTHEW T. CARRANO, Department of Paleobiology, Smithsonian Institution, P.O. Box 37012, MRC-121, Washington, DC 20013-7012
carranom@si.edu

LUIS M. CHIAPPE, Department of Vertebrate Paleontology, Natural History Museum of Los Angeles County, 900 Exposition Boulevard, Los Angeles, CA 90007
chiappe@nhm.org

RODOLFO A. CORIA, Museo Municipal Carmen Funes, Avenida Córdoba 55, Plaza Huincul 8318, Neuquén 8300, Argentina
coriarod@copelnet.com.ar

KRISTINA CURRY ROGERS, Paleontology Department, Science Museum of Minnesota, 120 West Kellogg Boulevard, St. Paul, MN 55102
krogers@smm.org

LOWELL DINGUS, Infoquest Foundation, 160 Cabrini Boulevard, No. 48, New York, NY 10033
ldingus@earthlink.net

GREGORY M. ERICKSON, Department of Biological Sciences, Florida State University, Conradi Building, Tallahassee, FL 32306-1100
gerickson@bio.fsu.edu

FRANKIE JACKSON, Department of Earth Sciences, Montana State University, Bozeman, MT 59717
frankiej@montana.edu

J. MICHAEL PARRISH, Department of Biological Sciences, Northern Illinois University, DeKalb, IL 60115
mparrish@niu.edu

PAUL C. SERENO, Department of Organismal Biology and Anatomy, University of Chicago, 1027 East 57th Street, Chicago, IL 60637
dinosaur@uchicago.edu

KENT A. STEVENS, Department of Computer and Information Science, 301 Deschutes Hall, University of Oregon, Eugene, OR 97403
kent@cs.uoregon.edu

PAUL UPCHURCH, Department of Earth Sciences, University College London, Gower Street, London WC1E 6BT, UK
p.upchurch@ucl.ac.uk

MATHEW J. WEDEL, University of California Museum of Paleontology and Department of Integrative Biology, 1101 Valley Life Sciences Building, Berkeley, CA 94720-4780
sauropod@berkeley.edu

JEFFREY A. WILSON, Museum of Paleontology and Department of Geological Sciences, University of Michigan, 1109 Geddes Road, Ann Arbor, MI 48109-1079
wilsonja@umich.edu

JOANNA L. WRIGHT, Department of Geology, University of Colorado at Denver, Campus Box 172, P.O. Box 173364, Denver, CO 80217-3364
jwright@carbon.cudenver.edu

INDEX

Aalenian, 105, 115–18, 120, 121, 122, 130–32
abelisaurids, 287
abundance, 137, 143, 146, 196
acetabulum, 180, 192
adaptation, 144, 148, 197, 201, 237
adaptive zone, 126
adductor fossa, 21, 36–37, 158
adipose tissue, 207, 208, 219, 223
Aegyptosaurus, 53, 68–69, 72, 73, 234, 246
Aeolosaurus, 53, 68–69, 72
Africa, 5, 10–11, 15–16, 274; dicraeosaurids in, 35, 172; Jurassic of, 35, 142, 145, 276; prosauropods in, 129, 140, 146; rebbachisaurids in, 143, 157; titanosaurs in, 20, 38–39, 143. *See also* names of specific countries
age and growth, 311, 313–14, 320; adult, 304, 312, 316–17, 319, 321–22; juvenile, 312, 316; subadult, 316–17, 321. *See also* longevity
air sacs: cervical, 216; pulmonary, 203, 217–20, 223
air volume in bones, 212–15, 217–21, 223
Alamosaurus: locomotion of, 240, 246; phylogeny of, 53–55, 56, 68–71, 72, 115, 296; *A. sanjuanensis*, 240, 246, 297
Albian: diplodocoids in, 136–37, 143; and diversity, 105, 112, 115–16, 119–20, 121, 130; feeding mechanisms in, 132–33, 134; teeth in, 135; titanosaurs in, 66, 136
alligators, 315
allometry, 312, 320
Allosaurus, 8
alveolar trough, 158, 159, 161–63, 165–66, 167–68
alveolus, 164
Amargasaurus, 37, 115, 172, 234, 243
Ampelosaurus, 53, 68–71, 72, 234, 245
Amphicoelias, 234
Amprino's rule, 313, 317–18, 321
Andesauridae, 51, 54, 71
Andesaurus: *A. delgadoi*, 38, 52; phylogeny of, 51, 56, 67, 71–72, 73; in trees and cladograms, 53–55, 68–69, 234, 296
angiosperms, 147, 196

angular, 59–60, 158–59, 160, 295
ankle, 21, 41
ankylosaurs, 242, 245
Antarctica, 10
Antarctosaurus: *A. giganteus*, 39; phylogeny of, 53, 55, 68–69, 71–72, 234; *A. septentrionalis*, 65; *A. wichmannianus*, 170
Antenonitrus, 2, 24, 27, 29, 31, 237
Apatosaurus, 17, 24, 115, 243; bones of, 180–81, 194, 242, 244, 307, 315–21; *A. excelsus*, 17, 240, 242, 317; feeding mechanism of, 128; feet of, 29, 31, 33, 240–41; growth of, 10, 315; limbs of, 24; *A. louisae*, 186, 190, 240–41, 244; neck of, 185–86, 195; phylogeny of, 234; and pneumaticity, 204, 205, 211, 214; skull of, 61, 327; specializations of, 37, 38; and tracks, 254–55, 260
apomorphies, 163
appendicular characters, 36
Aptian: and diversity, 105, 115–16, 119, 130, 135, 143; feeding mechanisms in, 132–33, 134; titanosaurs in, 66–67, 136
archosaurs (Archosauria), 167, 203, 208, 209, 239, 314
Argentina, 5, 119, 139, 287; eggs from, 10, 50, 285–86, 289; titanosaurs in, 51, 66–67, 296; tracks in, 265, 269
Argentinosaurus, 212, 244; growth of, 315, 323; phylogeny of, 53, 68–69, 71–72, 73; size of, 303–4
Argyrosaurus, 53, 68–69, 72, 234
Arizona, 259
Arkansas, 269
Asia, 5, 18, 142, 272, 276; titanosaurs in, 19, 39. *See also* names of specific countries
ASP (air-space proportion), 212–15, 220, 223
astralagus, 21, 33, 41
Atlantosauridae, 16, 51
Atlasaurus, 115, 117
Auca Mahuevo eggs. *See* eggs, Auca Mahuevo
Aucasaurus garridoi, 287
Augustinia, 53, 68–69, 72
Australia, 10, 268

337

Austrosaurus, 53, 115
autapomorphies, 37, 60, 108, 214

Bajocian, 105, 115–16, 120, 122, 130–32, 135
Barapasaurus, 2, 111, 115, 139, 234; feet of, 31–32; and pneumaticity, 202, 205–6, 207, 209, 211; quadrupedalism of, 237–39
Barosaurus, 8, 115, 172, 234, 256; *B. africanus*, 213, 246; bones of, 30, 213–14, 312, 315–16
Barremian: diplodocoids in, 136, 143; and diversity, 105, 115–16, 119, 120, 130, 136; feeding mechanisms in, 132–33, 134; teeth in, 135, 170, 174
Bathonian, 105, 115–16, 117–18, 130, 132, 134–36
behavior, 323. *See also* feeding; locomotion; rearing; reproduction
Bellusaurus, 115, 117
Berriasian: diplodocoids in, 136, 143; and diversity, 105, 115–16, 118–21, 130, 132, 135–36
bioenergetics, 194
biology, 109, 310, 312. *See also* paleobiology
biomass, 125, 140, 197. *See also* body size/mass
biomechanics, 215, 303–4, 310, 316
biorhythms, 305, 310, 312
biotic mixing, 121, 137
bipedalism, 23, 231–33, 237, 241, 245–46, 263
birds, 306, 330–31; altricial, 321–22; growth of, 307, 312, 313, 315, 317, 321–22; necks of, 182, 189–90, 191–92. *See also names of specific birds*
birds and pneumaticity, 203–4, 209–10, 212–13, 221–22, 223; adipose tissue in, 208, 219; air sacs in, 203, 220; diverticula of, 206, 216, 223
bison, 179
Blikanasaurus, 2, 27, 30–31, 115, 117, 140; limbs of, 23, 24, 29, 237
body plan, 1–2, 3, 20, 178
body size/mass, 6–7, 11, 144, 146, 174, 299; adult, 304, 312, 316–17, 319, 321–22; and bone growth rates, 313–15; decrease in, 233, 235–36, 246; differences in, 126, 138; evolution of, 230–36; increase in, 23, 173, 212, 231, 233, 235–37; and pneumaticity, 201, 212, 216–21; similarities in, 140, 144; and specializations, 23, 41–42; subadult, 312, 316–17; of titanosaurs, 39, 233, 244, 245, 303, 323
Bolivia, 269
bone growth, 303–23; annual, 310, 318; and annuli, 307, 310, 317; cyclicity/periodicity of, 307–10, 312, 313, 317, 320; and medullar expansion, 310, 311, 314; noncyclic, 313, 317; and polish lines, 312, 316; qualitative, 304, 312, 323; stages of, 313–15, 319, 321–23. *See also* LAG
bone growth rates, 303–23; and Amprino's rule, 313, 317, 321; of *Apatosaurus*, 316–23; average, 309, 313; maximum, 321–22; plateaus in, 309, 317; quantification of, 304–5, 312–13, 318–21; rapid, 305, 308, 311, 317–19; remodeling, 310; slow, 306, 307, 309–10,

312, 317; whole-body, 310–12, 313–15, 321, 323. *See also* growth curves
bones, 174, 273, 300; apneumatic, 204, 205, 219; cancellous, 204, 218–19; cortical, 307, 309–10, 311, 312, 315, 318–19; cut sections of, 222, 311, 316; embryonic, 285, 294; growth marks in, 307, 311, 314, 321; internal complexity of, 210–12, 214–15; limb, 217, 230; marrow-filled, 212, 222; morphology of, 304, 307; remodeling of, 306, 310–11, 312, 314, 316; texture of, 207, 209; with thin outer walls, 204, 209; volume of air in, 212–15, 217–21, 223. *See also* histology, bone; osteology; pneumaticity; reconstructions; skeletons; skulls; vertebrae; *names of specific bones; under names of specific genera*
bone tissue: fibrolamellar, 309–10, 311–12, 313, 315, 316–20, 321; Haversian, 306, 312; lamellar, 305–6, 307, 309; parallel-fibered, 305–6, 307, 312; secondary, 311; woven, 305–6, 307; zonal, 307, 309, 313, 317
bone vascularization: degree of, 309, 311–12, 313, 315, 317–18, 321; laminar, 306–9, 312, 313, 315, 316–17, 318–20; plexiform, 306–8, 312, 316–17, 318; reticular, 306–9, 312, 318–19; variable, 310, 319
"*Bothriospondylus*," 234, 315
Bothrosauropodidae, 16
brachiosaurids (Brachiosauridae), 212; feeding mechanism of, 128, 194, 196; locomotion of, 244–45; phylogeny of, 32, 38, 54
Brachiosaurus, 28, 115, 316; *B. altithorax*, 246; bones of, 194, 315; *B. brancai*, 38, 181, 182, 184–85, 186–89, 238, 240–42, 244, 246; feeding of, 196, 197; locomotion of, 238, 240–42, 244, 246; phylogeny of, 18, 38–39, 52, 68–71, 234, 296; and pneumaticity, 205–6, 207, 211, 214, 215; and tracks, 254–55
braincase, 297–98
brain size, 7
Brazil, 66, 168, 170, 268–69
Brazil Series B, 68–72
Breviparopus, 256, 260, 264–65, 266, 274; *B. taghbloutensis*, 257, 264
Brontopodus, 256, 265–66, 273–75; *B. birdi*, 257, 258, 260, 265
brontosaurs, 266
Brontosaurus, 16
Brown, Barnum, 328–29
browsing, 146, 194; by dorsiflexion (BD), 178, 179, 197; high, 128, 187, 196, 245; low, 127–28, 145, 179, 186, 188–89, 196–97, 245; maximum height of, 144, 195; medium-level, 127–28, 197; neutral (BN), 178–83, 197; by ventriflexion (BV), 178–79, 195, 197. *See also* feeding height
buoyancy, 221

caimans, 219
Callovian, 139; diplodocoids in, 134, 136; and diversity, 105, 115–16, 117, 130, 132, 135–36

camarasaurids (Camarasauridae), 32, 138, 139, 296; feeding of, 194, 197
camarasauromorphs (Camarasauromorpha), 254–56, 258, 260, 274
Camarasaurus, 16, 115, 169; bones of, 159, 181, 244, 246, 331; feeding of, 128, 196, 197; feet of, 30–32, 33, 240; *C. grandis*, 26, 240, 246; *C. lentus*, 160; limb bones of, 24, 26; phylogeny of, 38, 65, 68–71, 139, 234; and pneumaticity, 202, 204, 205; *C. supremus*, 17, 331; and tracks, 254–55; vertebrae of, 37–38, 186, 189, 211, 214
camels, 181, 182, 189–92
Campanian, 286; and diversity, 105, 115–17, 120, 130, 132–33, 135–36; feeding mechanisms in, 133, 134; titanosaurs in, 66–67, 131, 136–37
Canada, 144
Carnegie Museum, 327, 330–31, 332
Carnian, 2, 23, 139; and diversity, 105, 115–16, 117, 130–32; prosauropods in, 129–30
carnivores, 125–26
carpal, 21, 27, 33
carpus, 22, 41, 237, 245
cartilage, 26, 180, 245, 311
CCRs (candidate competitive replacements), 138–39, 141, 144–45, 147
Cedarosaurus, 115, 174
Cenomanian, 286; and diversity, 105, 115–16, 119, 130, 135–36; feeding mechanism diversity in, 132–33; titanosaurs in, 66–67, 131, 136, 143
ceratopsians (Ceratopsia), 231, 233, 237, 276
ceratopsids, 168, 173, 174
Ceratosaurus, 219, 333
cetaceans, 309
cetiosaurids (Cetiosauridae), 118, 127, 138, 139
Cetiosauriscus, 234
cetiosaurs, 237
Cetiosaurus: bones of, 16, 26; *C. mogrebiensis*, 246; and pneumaticity, 205, 216; range of, 113, 115; taxonomy of, 15, 112, 139, 234
chambers, vertebral, 201, 204–5, 209, 211, 214, 216
characters, 36, 52, 54, 65, 114, 304
carcharodontosaurids, 288
cheek embayment, 172
chickens, 203
Chile, 267–68
China: Jurassic of, 5, 18, 19, 145; Lower Lufeng Formation of, 140, 141, 144; prosauropods in, 129, 140; tracks in, 257, 269, 272
chirotheres, 263
Chubutisaurus, 234
clades, 23, 65, 67–68, 70–71, 141–42
cladistic analyses, 104, 108–9
cladistic hypotheses, 18–20
cladograms, 70, 106, 107, 113–15, 118, 128; podial consensus, 253–54; simplified, 296

classification, 5, 16–18. *See also* ingroups; nodes; outgroups; synonyms; trees, phylogenetic; taxonomy
claws: manual, 255–58, 260, 263, 274; pedal, 255–58, 262–65, 273–74
climate, 286–87. *See also* paleoclimate
cnemial crest, 230, 239, 244
Coelophysis, 216
Coelurus, 328
coevolution, 10, 147, 197
collagen, 305–6
Colorado, 253, 262, 264, 268, 332
competition, 140–43, 148
competitive exclusion, 140
competitive replacement, 138–39, 142, 143
condyle, 40, 41; and feeding, 181–84, 187, 193; and pneumaticity, 210–11, 214
Coniacian, 66, 105, 115–16, 119–21, 132
conifers, 146, 196, 197
convergence, 143
Coombs, W. C., 229
coracoid, 22, 40, 64
coronoid process, 158, 173, 174–75
cotyles, 181–84, 211, 214
cows, 315, 319
cranial remains, 19, 27, 36; embryonic, 285, 295–98. *See also* skulls
Cretaceous, 5, 7, 11, 32, 38, 51; diplodocoids in, 171; diversity in, 105, 115–16, 157; plants in, 195–96; radiation in, 104; tracks in, 260–61, 272. *See also names of specific periods*
Cretaceous, Early/Lower, 18, 106, 119, 145, 174; dicraeosaurids in, 35, 172; diplodocoids in, 137, 142–43; diversity in, 105, 119, 136, 138, 147; feeding mechanism diversity in, 133, 134, 142; rebbachisaurids in, 10, 157; titanosaurs in, 10, 137, 142–43; tracks in, 261, 268–73
Cretaceous, Late/Upper, 19, 58, 109, 145, 287; diplodocoids in, 143; diversity in, 105–6, 133, 134; eggs in, 285, 289, 294, 300; feeding mechanisms in, 134; NOOs for, 119–20; *T. madagascariensis* in, 56, 57, 59; teeth in, 170, 171, 173; titanosaurs in, 50, 137, 147, 261, 296, 299; tracks in, 261, 266, 269–71, 273
Croatia, 269
Crocodilians/crocodylians, 15, 39, 182, 208–9, 220
crown: broad (BC), 18, 21, 42, 127–29, 134–35, 141–42; embryonic, 295–96; narrow (NC), 16, 18, 42, 128–29, 134–35, 141–42, 170–72; of *Nigersaurus taqueti*, 159, 162–63, 167, 168–69; shape of, 21, 173–74. *See also* wear facets
CT (computed tomography), 159, 162, 206, 213, 222
cycadophytes, 146, 196, 197

data: on diversity, 113–14, 116; missing, 36, 72; problematic, 116, 139; qualitative, 317; quantitative, 321; skeletochronologic, 313, 314, 317. *See also* sampling

dating, radiometric, 111, 129
Datousaurus, 234
Deinonychus, 5
dentary, 21, 59–60, 160, 172, 173; embryonic, 295–96; of *Nigersaurus taqueti*, 158, 162, 163, 165–67, 168, 174–75
dentine, 166, 169, 170, 174
Depéret, Charles, 56–58
DGM Serie A, 212
diaphyses, 57, 212, 311, 314, 317
dicraeosaurids (Dicraeosauridae), 133, 134, 158, 171; feeding mechanism of, 128, 174; phylogeny of, 32, 35–36; teeth of, 36–37, 159, 167, 172
Dicraeosaurus, 115, 170, 172, 243, 315; *D. hansemanni*, 35, 186–87, 241, 246; locomotion of, 241, 246; neck of, 185–86, 187; phylogeny of, 69, 234; specializations of, 37, 38
diet, 138, 195. *See also* food; plants
digitigrady, 30, 34, 237–38, 239, 240, 258
digit: manual, 255–58, 265, 274–75; pedal, 255–56, 258, 262–63
DinoMorph, 185, 188, 190, 194
Dinosauria, The (Weishampel et al.), 2, 3
Dinosauriforms, 242
Dinosauromorpha, 114
dinosaurs (Dinosauria), 3, 4, 114, 243; body size of, 230–33; growth of, 309, 315, 322; locomotion of, 233, 243
diphyly, 141
diplodocids (Diplodocidae), 134, 297; compared to *Nigersaurus taqueti*, 159, 167, 169–72; and feeding, 174, 185–86, 194, 196–97; phylogeny of, 32, 35, 296; and pneumaticity, 205, 216; and tracks, 254–55, 264–65, 274; vertebrae of, 28, 192, 211. *See also* names of specific genera
diplodocines, 212, 223
diplodocoids (Diplodocoidea), 32, 119, 133; evolution of, 34–38, 243–44; feeding mechanisms of, 21, 134; locomotion of, 238–39, 242; phylogeny of, 5, 18, 19, 32; radiation of, 134–35; size of, 233, 235–36, 243, 246; specializations of, 8, 21–22, 36–38; TDEs of, 136–37; teeth of, 134, 171–72, 296; and titanosaurs, 55, 56, 67, 68, 70, 73, 142–43; and tracks, 258, 261; vertebrae in, 40–41, 51. *See also* dicraeosaurids; diplodocids; rebbachisaurids
Diplodocus, 7, 115, 128, 243; bones of, 37, 159, 180–81, 194, 304, 332; *D. carnegii*, 186, 193; head of, 28, 61, 160; *D. longus*, 1, 2, 32, 34–35, 38–39, 160; neck of, 185–86, 190, 195; phylogeny of, 69, 234; and pneumaticity, 202, 204, 205, 206, 215, 217–21; teeth of, 18, 166, 172, 196; and tracks, 254, 260
discoveries, 5–6, 7, 73, 106, 111, 266. *See also* fossils; trackways/ichnofossils
distribution: of clutches, 285; and gauge, 261; geographic, 104, 110, 144; global, 252, 270; paleoenvironmental, 271–72; paleolatitudinal, 271; of plants, 147, 196; of pneumaticity, 215–16, 223; spatiotemporal, 266–71;

stratigraphic, 41–42, 144, 146; temporal, 252; of tracks and body fossils, 272–73
divergence, 27, 117, 172
diversity, 9, 16, 104–24; absolute, 116, 134; and absolute age, 111–12; of amniotes, 125; and artifacts, 117–20; comparison of results on, 116–17; in Cretaceous, 115–16, 157; data on, 113–14, 116; decrease in, 109, 116–17, 118, 119, 120–21, 134, 138, 142; distortion in estimation of, 111, 113; in Early Cretaceous, 105, 119, 136, 138, 147; in Early Jurassic, 117, 129; and feeding mechanisms, 132, 137–38, 148; increase in, 104, 105, 110, 116–18, 120–21, 134; in Jurassic, 115–17; in Late Cretaceous, 105–6, 133; in Late Jurassic, 104, 105, 118, 121, 147; of locomotor morphology, 230, 247; low, 117, 143, 145; in Middle Jurassic, 105, 107, 116–18, 147; of ornithischians, 144; overestimation of, 112, 113, 117; patterns of, 117–20, 122; peak, 146; of plants, 147; and preservation biases, 109–11; previous work on, 105–7; processes affecting, 120–21, 122; of prosauropods, 140; relative, 134, 140, 142; results on, 114–20; and sampling biases, 109–10; of sauropodomorphs, 125–48; and stratigraphic ranges, 111–12; taxic approach to, 105; and taxonomic level, 112–13; and TDEs *vs.* PDEs, 107–9, 118, 119, 141, 145; of titanosaurs, 19; underestimation of, 116. *See also* NOOs; PDEs; radiation; tabefaction; TDEs
diversity curves, 109, 111, 114–16, 127
diverticula, 203; abdominal, 205; extraskeletal, 217–20, 223; intraosseous, 204, 220; vertebral, 202, 216, 221
DME (developmental mass extrapolation), 314–15, 321, 323
Dodson, P., 230
down-weighting, 111–12
dromaeosaurids, 288
duck, 309, 311, 318–19

edaphic processes, 291
egg-beds, 285–87, 289–93, 298. *See also* nests
eggs, Auca Mahuevo: abnormal, 285, 292; age of, 285–86, 291; clutches of, 273, 285, 286, 289–93, 298, 299, 300; laying of, 298, 299; megaloolithid, 299–300; morphology of, 285, 288, 292, 298–300; number of, 289, 291, 299; shells of, 285, 288–89, 292, 294, 298–99; titanosaur, 50, 297. *See also* embryos
Egypt, 66–67
Einstein, Albert, 330
elbow, 181, 245–46
Elephantopoides barkharusensis, 266
elephants, 229
embryos, 10, 19, 50, 294–300; development of, 285, 295, 297–98; titanosaur, 292, 294. *See also* eggs
emu, 309, 311, 318–19
enamel, 21, 170–71, 172, 173–74, 295; of *Nigersaurus taqueti*, 159, 162, 166–67, 168, 169
England, 15, 106, 112, 168

environment, 287, 310, 312. *See also* paleoenvironment
Eoraptor, 24, 31–32, 33, 37
Epachthosaurus, 52–55, 68–69, 71–72, 73, 204
epithelium, 213
erosional surfaces, 294–95
Erythrosuchus, 206–7
Euhelopodidae, 18–19
Euhelopus, 65, 115; locomotion of, 244, 246; neck of, 186, 187–89; phylogeny of, 18, 38, 55, 67, 68–71, 139, 234; and pneumaticity, 205, 212; *E. zdanskyi*, 187–89, 246
euornithopods, 9
Europe, 18, 111, 129, 142, 145, 146; rebbachisaurids in, 36, 157, 172; titanosaurs in, 20, 38, 51, 299; tracks in, 266. *See also names of specific countries*
eusauropods (Eusauropoda), 3, 19; basal, 118, 258; evolution of, 27–31, 297; feeding mechanism of, 21, 127–28; and pneumaticity, 223; posture of, 21; synamorphies of, 296; teeth of, 36–37; and tracks, 254–56, 258, 262–64, 274. *See also names of specific genera*
eutherians, 322
Eutitanosauria, 55
evolution, 8, 20–42, 118; of body size, 230–36; cranial, 121, 297; of Diplodocoidea, 35–38, 243–44; of Eusauropoda, 27–31, 297; lower-level patterns in, 243–46; macro-, 137, 138, 236; of Macronaria, 38–41; of Neosauropoda, 32–34, 243; of plants, 195; of posture, 23–27; of quadrupedalism, 230, 236–39; and synapomorphies, 20–22; theropod, 216; of tooth batteries, 171–75; of vertebrae, 211–12, 215. *See also* locomotion, evolution of; coevolution; specializations
extinction, 110, 111; of broad-crowned sauropods, 134; of cetiosauridae, 139; of dicraeosaurids, 133; of diplodocids, 142; at Jurassic-Cretaceous boundary, 105, 106, 145; mass, 109, 118–19, 120; of Nemegtosauridae, 143; of prosauropods, 129, 140, 144; of rebbachisaurids, 143; regional, 146; and sea level, 121

Farlow, J. O, 266
feeding, 41, 121; head-down, 187, 197; underwater, 190, 197. *See also* herbivory; jaw; teeth; vegetation
feeding height, 180, 185, 192, 194–95. *See also* browsing
feeding mechanisms, 142, 144; *Apatosaurus* type, 128, 132–33; brachiosaurid type, 128, 132–33, 134; *Camarasaurus* type, 128, 132–33; Cetiosaurid type, 127, 132–33, 134; dicraeosaurid type, 128, 132–33, 134; diplodocoid, 171; *Diplodocus* type, 128, 132–33, 134; and diversity, 137–38, 148; higher eusauropod type, 127–28, 132–33, 134; nemegtosaurid type, 128, 132, 133; *Nigersaurus* type, 128, 132, 134; prosauropod, 126, 141; sauropod, 127–29; *Shunosaurus* type, 127, 132–33, 134; titanosaur type, 128, 131–33, 134

feet. *See* forefeet; manus; metacarpals; metatarsals; pes; pes and tracks; phalanges; tarsals; trackways/ichnofossils
females, 298, 299, 312
femur: and body size, 231–33, 235–36; and classification, 23–24, 26; eccentricity of, 21–22, 230, 231, 243–44; and growth, 313, 314; and locomotion, 22, 40, 242–45; of titanosaurs, 38–39, 61, 63–64, 242, 256, 260
fenestra, 59–60, 160; embryonic, 295–96, 297; of *Nigersaurus taqueti*, 158–59, 163, 164, 169, 175
ferns, 146, 196–97
fibrolamellar complex, 306–7
fibula, 41, 64
fish, 322
Flagellicaudata, 22, 35, 37
flooding, 293
fodder. *See* plants
food, 143; processing of, 126, 144, 174, 194, 196. *See also* diet; mastication; plants
foot morphology, 253–58, 262, 264
footprints. *See* trackways/ichnofossils
foramina, 60, 160; of *Nigersaurus taqueti*, 158, 159, 161, 162–63, 164–66, 167–68; replacement, 162–63, 164–66, 167–68, 174; vertebral, 201–2, 203–5, 207, 209–10, 215, 219
forearm, 237
forefeet, 21, 33–34, 245. *See also* manus; metacarpals
forelimbs: and feeding, 180, 181; length of, 21, 179, 188, 231, 237, 243–46; and locomotion, 239, 245; of *Nigersaurus taqueti*, 61, 64, 158; and posture, 23–25, 40–41; of titanosaurs, 245–46
forests, 195–96, 276
fossa, vertebral, 201–2, 204, 211, 215–16, 218; blind, 207, 209, 215; pneumatic *vs.* nonpneumatic, 205–9; subfossae of, 206, 209, 215
fossil-bearing formations and sites, 109; Allen, 288; Anacleto, 286, 288–89, 293; Ardley Quarry, 256–58; Auca Mahuevo, 285–300; Barreales Escondido, 285–86, 287, 294, 298; Barreales Norte, 285–86, 287, 294, 298; Clarens, 144; Dry Mesa Quarry, 332, 333; Fatima, 256–57; Gadoufaoua beds, 167; Glen Rose, 252, 275–76; Kayenta, 144; Lower Elliot, 140; Lower Lufeng, 140, 141, 144; Maevarano, 59; Marsh Quarry, 333; McCoy Brook, 144; Morrison, 142, 195–96, 259, 261, 264; number of, 116, 119, 121, 267–69; Paluxy, 260, 275; Portland, 144; prosauropod, 129; Purbeck Limestone, 275; Purgatoire, 252–53, 258, 260, 264–65, 266, 275; Summerville, 259; Tendaguru Beds, 142, 196; Upper Elliot, 144
Fossil Reptil-Ordnung Saurischia, Die (Heune), 328
fossils, 2–4, 195, 287; distribution of, 272–73; earliest, 117; gaps in, 107–8; and growth, 303–4; quality of, 5, 10, 105, 110–11, 118, 121. *See also* preservation; trackways/ichnofossils

France, 66, 268, 294
frogs, 203
frontal, 59–60, 295–96, 297

gastric mill, 126
gastrolith, 174
gazelles, 179
genera, 53; and diversity estimation, 112–13, 114, 116, 127, 130; ichno-, 261–66, 274; number of, 125, 135–36. *See also names of specific genera*
geological age, absolute, 111–12, 114, 116
geological factors, 109, 120
Germany, 16, 265–66, 268, 274
ghost lineages, 2, 116–17, 133, 137
ginkgophytes, 196
giraffes, necks of, 179, 181, 183–84, 186, 193; flexibility of, 189, 191
glenoid, 180, 192
Gondwana, 146
Gondwanatitan, 53, 115, 246
Gongxianosaurus, 141; feet of, 27, 29, 30–31; phylogeny of, 23, 234; *G. shibeiensis*, 24
grazing, 179, 196
Greenland, 267, 274
gregariousness, 298
growth. *See* bone growth; bone growth rates
growth curves, 305, 314–15, 317, 322, 323
gymnosperms, 196, 197

habitats, 106, 138, 140, 146, 189; preferences of, 266, 271–72
hadrosaurids/hadrosaurs, 168, 173, 275, 276, 300
Haplocanthosaurus, 35, 55, 115, 139, 234, 333; and pneumaticity, 207–9, 215
hatchlings, 303–4
Hauterivian, 105, 115–16, 130, 132, 133, 135–36
head height, 178–79, 180, 183, 188, 194; low, 188, 195
herbivores, extant, 179, 183, 195
herbivory, 146; facultative, 125, 140; obligate, 125, 126, 140; and ornithischians, 157, 172; of sauropodomorphs, 125–26; and specializations, 21, 27–28, 36–37, 42. *See also* browsing
herds, 10, 276
Herrerasaurus, 28; bones of, 24, 26, 31–32, 33, 37, 209
heterodontosaurids, 144
heteropody, 265
Hettangian, 138; and diversity, 105, 111, 115–16, 118, 130–32, 135; prosauropods in, 129–30
hindfeet, 21
hindlimbs, 27, 29–30, 158; and locomotion, 40, 195, 239, 245, 276; and posture, 23–25; size of, 179, 180, 231, 243–44, 245; of titanosaurs, 61, 64, 245
hips, 239, 263, 273
histograms, 130–33, 135–36, 271, 275

histology, bone, 304–19, 321, 323; of *Apatosaurus*, 316–19; and growth, 307–15; qualitative, 315; and tissue organization, 305–6; variability in, 312; and vasculature, 306–7
holotypes, 158
Homalosauropodidae, 16
homology, 207–8
homoplasy, 143
horses, 179, 181, 182, 190
human factors, 109
humans, 309, 315
humeri, 203, 315; and locomotion, 22, 244–46; and posture, 26, 40–41, 181; size of, 232–33, 237; of titanosaurs, 56–57, 63, 64, 245–46
hydroxyapatite crystals, 305–6
Hypselosaurus, 299

ichnofossils. *See* trackways/ichnofossils
ichnogenera, 261–66, 274
ichnology, predictive, 252–58
ichthyosaurs, 6, 309, 318
iguanodontids, 275
ilium, 63, 64, 204–5; and locomotion, 41, 239, 241–42, 245
India, 5, 10, 111, 139; eggs in, 289, 299; titanosaurs in, 20, 38, 51, 56–57, 65, 66
ingroup, 52, 65, 68, 114
innovation, key, 126, 141
integument, 55, 285, 294. *See also* skin
Isanosaurus, 117, 140, 237; bones of, 26, 209; phylogeny of, 23, 234
ischium, 63, 64, 244–45
Isisaurus: colberti, 65, 69; phylogeny of, 20, 53, 56, 68, 70–72, 234. *See also* '*Titanosaurus*,' *colberti*
Isle of Wight, 56, 170, 174
Italy, 34–35, 257, 263, 264, 267–68, 274

Jabalpur Titanosaur indet., 65, 68–69, 72
Jainosaurus, 53, 67, 68–69, 72; *J. septentrionalis*, 65
Janensch, W., 16, 18
Janenschia, 72, 137, 316; bones of, 310, 315; locomotion of, 238, 240, 244, 246; phylogeny of, 38, 53, 68–69, 234; *J. robusta*, 238, 240, 246, 310
Japan, 329
jaw, 21, 36–37, 126, 127, 173; of *Nigersaurus taqueti*, 158, 159, 163, 167, 170, 172
Jensen, Jim, 332, 333
Jianu, C.-M., 107, 113, 116–17
Jingshanosaurus, 24, 31–32, 38
Jobaria, 115, 117; bones of, 21, 24, 32–34, 35, 169; *J. tiguidensis*, 240; vertebrae of, 209, 215, 216, 221
joints, 23–24, 34, 181, 246
jugal, 59–60, 295–96, 297
Jurassic, 4; diversity in, 105, 115–17; NOOs for, 120; plants in, 195–96; teeth in, 173; tracks in, 261. *See also names of specific periods*

Jurassic, Early/Lower, 2, 32, 125, 139, 276, 287; competition with prosauropods in, 140–41; diversity in, 117, 129, 133; feeding mechanisms in, 133, 134; prosauropods in, 129, 143–44, 146; tracks in, 34, 264, 267, 270–71, 273–74

Jurassic, Late/Upper, 11, 19, 32, 35, 139, 145; abundance in, 106; competition in, 142; diplodocids in, 36, 172; diversity in, 104, 105, 118, 121, 133, 147; feeding mechanisms in, 133, 134, 148; neosauropods in, 39, 233; radiation in, 137–38, 233; sympatry in, 194; titanosauriforms in, 260–61; titanosaurs in, 50, 137; tracks in, 258–59, 261, 266–68, 270–74

Jurassic, Middle, 2, 27, 104, 112, 139, 145; in China, 5, 18; diversity in, 105, 107, 116–18, 147; feeding mechanisms in, 134, 142, 148; neosauropods in, 10, 33; radiation in, 122; and titanosaurs, 38–39, 50, 106, 256, 260; tracks in, 256, 264, 267, 270–71, 273–74

Jurassic-Cretaceous boundary, 105, 106, 118–19, 145

juveniles, 63, 314, 316, 319. *See also* hatchlings

Kazakhstan, 119
Kimmeridgian: diplodocoids in, 134–36; and diversity, 105, 112, 115–18, 120, 129–30, 135; feeding mechanisms in, 132, 134; titanosaurs in, 66, 136–37, 138–39
knee, 40, 180, 239
Kotasaurus, 115, 234

lacrimal, 59–60, 295, 297
LAG (lines of arrested growth), 305, 307, 309–10, 313; in sauropods, 312, 315, 317, 319, 321
laminae, vertebral, 201–2, 205–9, 211, 223
Laplatasaurus, 57, 234
Lapparentosaurus, 234, 315
life history, 312, 323
ligament, 37, 179, 181, 187, 221
Limaysaurus, 20, 38
limbs, 9, 15, 239, 241; bones of, 217, 230; length of, 230, 243–44; proportions of, 126, 127–28, 299; scaling of, 231–33. *See also* forelimbs; hindlimbs; locomotion; posture
Lirainosaurus, 53, 65, 68–71, 72, 115, 234
lithology: carbonate, 267, 271–73, 275, 276; clastic, 267–69, 273, 275; limestone, 267–69, 272–73, 275; mudstone, 267, 269, 272, 286, 287–88, 292–93; sandstone, 267–69, 272–73, 285–86, 288, 293; siliciclastic, 271–73, 276; siltstone, 268–69, 286, 288
Lithostrotia, 52, 56, 67–71, 73
Lockley, M. G., 266, 271
locomotion, 39, 144; and gait, 39, 273; and tracks, 265, 273–76, 277; and weight distribution, 273, 275. *See also* bipedalism; posture; quadrupedalism; rearing
locomotion, evolution of, 229–47; and body size, 230–36, 246; lower-level patterns in, 243–46, 247; and quadrupedalism, 236–39, 246; and specializations, 230, 239–43, 247
longevity, 305, 310, 311, 321, 323
Losillasaurus, 143
Lourinhasaurus, 115
Lufengosaurus, 24–25, 31–32, 33, 37, 139
Lull, Richard, 327–28
lung, 203, 216, 217–20, 221, 223. *See also* air sacs, pulmonary
lycopods, 196–97

Maastrichtian, 270; and diversity, 105, 115–17, 120, 130, 132–33, 135–36; feeding mechanisms in, 132–33, 134; titanosaurs in, 66–67, 136–37
macronarians (Macronaria), 174; basal, 33, 134, 139, 236, 243, 244–45; derived, 233, 236; evolution of, 38–42; locomotion of, 238–39, 244–45; phylogeny of, 5, 8, 18, 32–33, 36, 38–39; posture of, 39–41; size of, 233, 235–36, 246, 265. *See also* brachiosaurids; Opisthocoelicaudiinae; saltasaurids; saltasaurines; Somphospondyli; Titanosauriformes; titanosaurs; *names of specific genera*
Madagascar, 5, 145; titanosaurs in, 38, 51, 56–59, 67, 296
Magyarosaurus, 53, 68–71, 72, 234–35, 303
Maiasaura, 309, 322
Malagasy Taxon B, 53, 68–69, 72
Malawi, 67
Malawisaurus, 72, 115; *M. dixeyi*, 52, 296; phylogeny of, 52–55, 56, 68–71, 73, 296
Mamenchisaurus, 19, 30, 115, 211–12; *M. constructus*, 246; locomotion of, 237, 246; phylogeny of, 19, 115, 234
mammals, extant, 168, 182, 189, 306; growth of, 311, 312–13, 315, 317; and locomotion, 229, 230, 231, 246; and pneumaticity, 212, 219, 221. *See also names of specific fauna*
mandibular symphysis, 158, 163, 175
manus, 35, 41, 64; and locomotion, 237–38, 239–40, 243
manus and tracks, 35, 39, 253–54; and claws, 255–57, 260, 263, 274–75; crescentic, 258, 263, 264, 265–66; depth of, 273–75; and digits, 255–57, 265, 275; semicircular/horseshoe-shaped, 258, 264, 265; size of, 259, 260, 265
manus-pes ratio, 260, 264, 265–66, 274
Marasuchus, 209
marginocephalians, 2, 3
marsupials, 322
mass. *See* body size/mass; DME
Massospondylidae, 138–39
Massospondylus, 139, 140, 322; bones of, 26, 33–34, 144; *M. carinatus*, 238, 241–42; locomotion of, 238, 241–42
mastication, 144, 169
maxilla, 59, 60, 160, 172, 173; embryonic, 295–96, 297; of *Nigersaurus taqueti*, 158, 162–64, 168, 169

McIntosh, John "Jack," 327–33
Meckel's canal, 158, 165
Megaloolithidae, 299–300
Megaloolithus, 289
Melanorosauridae, 138–39
Mesozoic, 10, 125, 145, 146, 266; plants in, 195, 196–97
metacarpal, 33–34, 41, 61, 64; evolution of, 33, 238; and locomotion, 237, 244; and tracks, 256–58, 261, 265
metacarpus, 21
metapodial, 34
metatarsal, 24, 27, 28, 29–30, 64; and locomotion, 21, 239, 240–41, 243
Mexico, 267
mink, 309, 318
Mongolia, 19, 56, 67, 73, 296
monophyly, 19, 125, 139; of prosauropods, 20, 22, 125; of Titanosauriformes, 65, 67, 70, 73; of titanosaurs, 52, 54–56, 65, 68, 69, 71
monospecificity, 112, 127
Montana, 196
Morocco, 34, 263, 264, 267, 274
MPTs (most parsimonious trees), 52, 55, 69, 73, 109, 114
muscles, 203, 207, 223; and growth, 304, 316; and locomotion, 239, 242–43, 245

nares, 59, 60–61, 171, 172; embryonic, 295, 297–98; of *Nigersaurus taqueti*, 158, 159, 161, 163, 169
narial fossa, 158, 159, 160, 161
neck, 221; curvature of, 172, 182, 185–89; dorsiflexion of, 185–86, 187–89, 190–92, 195; elongation of, 21, 28, 38; evolution of, 121; flexibility of, 180, 189–92; length of, 126, 127–28, 179, 195, 212, 245; ventriflexion of, 185–86, 189. *See also* vertebrae, cervical; vertebral column
nemegtosaurids (Nemegtosauridae), 55, 128, 133, 137, 143, 196
Nemegtosaurus, 72, 115; *N. mongoliensis*, 296, 297–98; phylogeny of, 19, 53, 55, 65, 68–71, 73; teeth of, 171, 297
neoceratopsians, 9
neosauropods (Neosauropoda), 19, 36, 231; bones of, 21, 38, 160; embryonic, 295–96; evolution of, 18, 32–35, 243; feet of, 30–31; locomotion of, 239, 240, 242, 243, 244, 245; and pneumaticity, 205, 206, 209, 216; radiation of, 107, 118, 233; and titanosaurs, 55, 62; and tracks, 254–56, 258, 264–65, 274. *See also* diplodocoids; macronarians; *names of specific genera*
neotheropods, 241
nests, 285, 292, 293, 298–300. *See also* egg-beds
Neuquensaurus, 72, 303; *N. australis*, 39; phylogeny of, 53–55, 56, 68–71, 234, 296
neural arches, 63–64, 202, 206, 209, 211, 218
neural canal, 202, 208
neural cavity, 202, 205
neural spine: caudal, 215, 218; cervical, 171, 173, 206, 218; dorsal, 63–64, 202, 205, 218; and pneumaticity,
201–2, 205–6, 209, 211, 215, 218, 220; presacral, 22, 37, 201–2
New Mexico, 256–57, 262
New York, 275
niche partitioning, 138, 142, 148, 194, 195
Niger, 157, 158, 168, 268
Nigersaurus, 9, 37, 38, 119, 128, 172
Nigersaurus taqueti, 157–72; skull of, 157–66, 169–70; teeth of, 158–60, 162–69, 170–75
nodes, 3, 20, 22, 52, 68–71
NOOs (number of opportunities to observe), 110, 117–18, 119–20, 121
Norian, 2, 23, 146, 216; and diversity, 105, 115–16, 117, 118, 130–32, 135; prosauropods in, 129, 140
North America, 23, 39, 111, 139, 146, 196; coexistence in, 142; diplodocids in, 35, 172; Late Jurassic of, 11, 35, 145; prosauropods in, 129; skeletons in, 15–16; tracks in, 34, 263, 266. *See also names of specific countries*

occlusion, 126, 174; development of, 141, 144; interdigitating, 127–28; and wear facets, 167, 169
olecranon, 21–22, 230, 239, 244
olecranon process, 41, 239
Omeisaurus, 115; bones of, 24–25, 28, 37; feet of, 30, 31–32, 33, 34, 258; locomotion of, 237–40, 246; phylogeny of, 18–19, 234; *O. tianfuensis*, 240, 246; and tracks, 254, 258
ontogeny, 16, 276, 297, 300; and growth, 303–4, 310–11, 312, 314; and pneumaticity, 206, 208, 216, 223
operational taxonomic units (OTUs), 109
Opisthocoelicaudia, 115; bones of, 24, 31–32, 40, 41, 254; locomotion of, 240, 244, 246; phylogeny of, 53–55, 68–72, 139, 234, 254; *O. skarzynskii*, 38, 52, 240, 244, 246; and titanosaurs, 56, 73
Opisthocoelicaudiinae, 38; 52, 67, 70–71, 73
opisthocoely, 181
orbit, 59, 159, 295, 297
origination times, 109, 118
ornithischians (Ornithischia), 4, 157, 239, 276; basal, 3, 34, 239; diversity of, 107, 121, 144; and sauropodomorphs, 143–46; teeth of, 172–74
ornithopods (Ornithopoda), 2, 3, 237; body size of, 231, 233; teeth of, 173, 174; tracks of, 40, 266, 271, 272
ossification, 41, 57, 316; reduced, 21, 26, 33
osteoblasts, 305
osteoclasts, 203
osteoderm, 56–57
osteology, 253; and feeding, 179, 181, 186, 189, 190–91, 193; and pneumaticity, 209–10. *See also* bone growth; bone growth rates; bones; bone tissue; bone vascularization
osteons, 306–7, 308
ostrich, 189, 206, 309, 311, 318–19
outgroups, 22, 29–30, 33, 36–37, 114; and titanosaurs, 68, 71

Owen, Richard, 15
Oxfordian, 138; diplodocoids in, 134, 136; and diversity, 105, 115–18, 120, 130, 132–33, 135–36

pachycephalosaurs, 37
pads, 263; heel/pedal, 255, 258–59, 263, 264, 265
paleobiology, 6, 185, 230; and pneumaticity, 201, 212, 216–17, 220–21; and reproduction, 10, 285, 292. *See also* biology
paleobotany, 195
paleoclimate, 272, 276
paleoecology, 121, 146, 178, 194–96
paleoenvironment, 137, 252–53, 316; coastal, 267–69, 271–73, 275, 299; floodplain, 268, 273; fluvial, 267–69, 271–73, 275, 286; forest, 268, 271; inland, 271, 273, 276; lacustrine, 268–69, 271–73; marine, 271–73, 276; plains, 272, 273, 286; subtropical, 266, 269; swamp, 268, 272
paleogeography, 270
paleolatitude, 253, 266
paleovertisols, 286
Pangaea, 272
Parabrontopodus, 256, 260, 273–74; *P. mcintoshi*, 257, 264–65
Paralititan, 53, 68–69, 72, 73, 303
paraoccipital process, 59–60, 158, 160
paraphyly, 19, 54, 139, 244, 300; of prosauropods, 22, 125, 141
parental care, 298
parietal, 59, 295–96, 297
Patagosaurus, 28, 115, 139, 205, 234
patristic distance, 235–36
PDEs (phylogenetic diversity estimates), 112, 121, 129, 137, 143, 147; disadvantages of, 109, 117, 127; and stratigraphic range, 111, 113; *vs.* TDEs, 107–9, 114, 116, 118–19
pectoral girdle, 40–41, 180, 188, 192, 244
Peirópolis form, 53
Pellegrinisaurus, 53, 115
pelvic girdle, 64, 244, 245
pelvis, 180, 276
Perú, 289
pes, 28–29, 30–32, 35, 39–40, 64; and locomotion, 21, 238–39, 240–41
pes and tracks, 253–57; and claws, 255–56, 258–59, 262–65, 273–74; digitigrade, 255, 258; and digits, 255–56, 262–63; semidigitigrade, 258. *See also* manus-pes ratio
phalanges, 28, 30, 31, 34, 41; and locomotion, 21–22, 239–40, 245
pharynx, 203
Phuwiangosaurus, 72, 115, 214–15; phylogeny of, 53, 54, 65, 67–71, 73, 234
phylogeny, 5–6, 8; diagrams of, 19, 20; and growth, 310, 312; and parsimony, 69, 73, 109, 114, 310, 318. *See also* cladograms; classification; diphyly; divergence; ingroups; lineages, ghost; monophyly; outgroups; PDEs; trees, phylogenetic; synapomorphies; *under specific clades and genera*
phylogeny of titanosaurs, 50–73; background on, 51–56; and interrelationships, 65–70; and *Rapetosaurus krausei*, 56–65
physiology, 221, 223, 299, 305
plants, 141, 145, 167, 195, 294; aquatic, 190, 196; cropping of, 127–28, 129, 174; evolution of, 107, 195; height of, 179, 194; and interactions with sauropodomorphs, 146–47; nipping of, 128, 129. *See also names of specific flora*
Plateosaurus, 24–25, 28, 30–31, 37; *P. engelhardti*, 22–23
plesiomorphy, 310
plesiosaurs, 6, 15
Pleurocoelus, 214–15, 265
Pliensbachian: and diversity, 105, 111, 115–16, 130–32; prosauropods in, 129, 140; tracks in, 259, 263
pneumaticity, 201–23; and air volume, 212–15; correlates of, 209–10; criteria for, 204–10; distribution of, 215–16, 223; extramural, 203, 204–5; and future research, 221–23; and internal complexity, 210–12; in living vertebrates, 201–4; and mass, 216–21; origin of, 207–9
Poland, 34–35, 267
pollex, 245
polytomies, 113–14, 116
polytomy, 65, 67–68, 69, 71, 72
Portugal, 256–57, 258, 260, 267–69, 274
postacetabulum, 242
postorbital, 158, 160, 296
posture, 16, 186, 195; columnar, 21, 23–27, 40–41, 230; digitigrade, 29–30, 34, 237–38, 240; forefoot, 33–34; graviportal, 40–41, 230, 273; hindfoot, 21, 28–32; and limb size, 232–33; neutral, 182; parasagittal, 39, 273; semidigitigrade hindfoot, 30; wide-gauge limb, 22, 39–41, 245. *See also* bipedalism; locomotion; quadrupedalism; semibipedalism
preacetabulum, 241–42, 244–45
predation, 179
predentary, 172
premaxilla, 24, 59, 160, 295; of *Nigersaurus taqueti*, 158, 159–63, 169
preservation: artifacts of, 257; and diversity estimates, 106, 109–11, 117, 119, 120–21, 129; poor, 141, 258, 275–76; and sea level, 137; and tracks, 257, 271–72, 276
primates, 203
proboscideans, 229, 239
prosauropods (Prosauropoda), 2–3, 10, 27–28, 117, 146; bones of, 33, 37–38, 209, 263; and competition with sauropods, 140–41; decline of, 143–44; distribution of, 4, 41; diversity of, 125–27, 129–48, 144; in Early Jurassic, 129, 143–44, 146; feet of, 31–32, 33, 263; limbs of, 24, 231; locomotion of, 237–40, 242; phylogeny of, 22–23, 234; TDEs for, 126–27, 129, 130, 140, 146. *See also names of specific genera*

Pseudotetrasauropus jaquesi, 262–63
Psittacosaurus, 322
PSP (postcranial skeletal pneumaticity), 203, 209, 216, 220–23. *See also* pneumaticity
pterosaurs, 204, 223
pterygoids, 59–62, 160, 295
pubis, 63, 64, 245, 315

quadrate, 59–60, 62, 158, 160, 295, 297–98
quadratojugal, 59, 158–59, 160, 295–96, 297
quadrupedalism, 21, 23–27, 42, 236–39, 241, 255; and body size, 231–33; obligate, 237–38
quadrupeds, extant/modern, 180, 181, 183. *See also* names of specific animals
Quaesitosaurus, 72, 115; phylogeny of, 19, 53, 55, 68–71, 73, 171
quail, 311

radiation, 110, 111, 120, 126, 141; of diplodocoids, 134–35; of heterodontosaurids, 144; in Late Jurassic, 137–38, 233; in Late Triassic, 137; of neosauropods, 107, 118, 233; of ornithischians, 145; of plants, 147; of prosauropods, 129; of sauropodomorphs, 148; of titanosaurs, 105, 137. *See also* diversity, increase in
radius, 22, 41, 61, 63, 64, 307; *Apatosaurus*, 316–20
ramus: alveolar, 159, 162–63; posterodorsal, 162, 169–70
range: geographical, 138, 140, 142–43, 144; ghost, 108, 112, 113–14, 116–17, 119
range, stratigraphic, 107–8, 109, 113–14; on charts/graphs, 105, 115–16, 130–32, 135–36, 261, 271; and competition, 138, 140, 142
Rapetosaurus, 170, 245; and embryos, 296, 297; *R. krausei*, 53, 55, 56–65, 67, 72, 73; phylogeny of, 68–71, 234
ratites, 189, 311
Rayosaurus, 115, 172, 234
rearing, 41, 195, 245
rebbachisaurids (Rebbachisauridae), 38, 119, 137; in Cretaceous, 10, 35, 171; phylogeny of, 32, 35, 42; teeth of, 37, 172, 173–74. *See also Nigersaurus taqueti*; names of specific genera
Rebbachisaurus, 172; *R. garasbae*, 35
reproduction, 10, 41, 212, 285, 292, 316. *See also* eggs, Auca Mahuevo
reptiles, 306, 312, 313, 322; modern, 298, 303, 315
respiratory system, 216, 223. *See also* air sacs, pulmonary; lungs; trachea
Rhaetian, 23, 139; and diversity, 105, 115–16, 117, 129–32, 135
rhinocerotids, 239
rhinos, 181, 190, 229
rhynchosaurs, 146
ribs, 61, 63–64, 173, 180, 188, 204–5
Rocasaurus, 53, 68–71, 72, 234
Romania, 66, 300, 303

Romer, Alfred S., 329
rostrum, 192, 297
Rotundichnus, 274; *R. muenchehagensis*, 257, 265–66

sacrum, 180, 245
Salgado, L., 52
saltasaurids (Saltasauridae), 216; phylogeny of, 38, 52, 55–56, 67, 69–71, 73; posture of, 22, 40–42
saltasaurines (Saltasaurinae), 233, 245, 297; phylogeny of, 39, 52, 65, 67–72, 73, 296
Saltasaurus, 115; bones of, 72, 214–15, 246; *S. loricatus*, 22–23, 27, 32, 35, 38–39, 52, 246; phylogeny of, 53–55, 56, 68–71, 234, 296; and pneumaticity, 202, 204–5; size of, 303–4, 323
sampling, 107, 113; biases in, 109–10, 117, 120, 122, 145; differences in, 116, 121; errors in, 118; insufficient, 119, 222
Santa Rosa indet., 53, 68–69, 72
Santonian, 105, 115–16, 119–20, 130, 132–33, 135–36
saurischians (Saurischia), 2–4, 18, 22, 26, 42, 237; locomotion of, 237, 239, 246; and pneumaticity, 209, 216. *See also* prosauropods; theropods
Sauropodichnus, 274; *giganteus*, 257, 265–66
sauropodomorphs (Sauropodomorpha), 4, 22, 107, 125, 237, 254; and interactions with ornithischians, 143–46; and interactions with plants, 146–47; and pneumaticity, 215–16; radiation of, 148. *See also* diversity; prosauropods; sauropods
Sauropodopus antiquus, 262–63
sauropods, 23; adult, 303, 304; competition with, 140–41, 148; derived, 216, 240, 260; fossil record of, 2–6, 10–11, 120; juvenile, 204, 210, 211, 299; size of, 1, 6–7, 230, 235, 303–4, 312. *See also* eusauropods; neosauropods
sauropods, basal/primitive, 5, 134, 139; and body size, 230, 236; bones of, 30, 33–34, 207; locomotion of, 237–39, 240, 243; and tracks, 255, 258. *See also* names of specific genera
Sauroposeidon, 206, 207, 209–11, 214, 222, 244
scapulae, 22, 61, 63, 64, 180–81; *Apatosaurus*, 316–18, 320; of *Nigersaurus taqueti*, 158, 159
scapulocoracoids, 180
sea level, 106, 120–21, 137
Seismosaurus, 243, 303
semibipedalism, 231–33, 237
semidigitigrady, 30, 258, 264
Senonian, 66
septa, 173; of *Nigersaurus taqueti*, 163, 168; vertebral, 202, 207, 209, 210, 211
Sereno, P. C., 18–19, 106–7
sexual dimorphism, 312, 316, 323
sexual maturity, 303, 316
SG (specific gravity), 217, 220, 221
shoulder, 192, 195, 239, 245, 246
shoulder girdle, 40–41, 181

Shunosaurus, 2, 115, 127; bones of, 24–25, 29; feet of, 31–32, 33–34; *S. lii*, 27, 238, 240–41, 246; locomotion of, 237–42, 246; phylogeny of, 18, 234; and tracks, 254–55; vertebrae of, 28, 37, 209, 211, 216
Shuvuuia, 322
Simpson, George Gaylord, 1, 2
simulations, 109, 113, 194
Sinemurian, 259; and diversity, 105, 111, 115–16, 130–32, 135; prosauropods in, 129–30
sirenians, 310, 318
sister-taxa, 54, 55, 65, 72, 108
sites. *See* formations and sites
skeletochronology, 310, 313, 314, 317
skeletons, 9, 15, 40, 185, 276, 311; appendicular, 180–81, 237, 243; axial, 180, 181–83; complete, 16, 27; juvenile, 56, 59–61; neutral pose of, 178, 180–83; theropod, 287; weight of, 220–22. *See also* bones; pneumaticity; skulls
skin, 203, 275, 295. *See also* integument
skulls, 19, 56; embryonic, 295–98; evolution of, 36–37, 121; holotypic, 59–60; juvenile, 60–63; *Nigersaurus taqueti*, 158–60, 169–70; of titanosaurs, 59–63, 295. *See also* cranial remains
slickensides, 286, 290, 291
snakes, 203
snout, 28, 36, 172, 296; of *Nigersaurus taqueti*, 159
soft tissues, 304. *See also* adipose tissue; cartilage; ligaments
soil, 286
Somphospondyli, 38, 55, 62, 65, 70–71, 73
south, the, 157
South Africa, 117, 144, 263
South America, 5, 19, 142, 143, 146, 157; dicraeosaurids in, 35–36, 172; Late Cretaceous of, 145, 170; titanosaurs in, 38, 54, 57; tracks in, 34, 274. *See also names of specific countries*
South Korea, 261, 266, 269, 271–72
Spain, 66, 268–69
specializations: appendicular, 26; herbivorous, 21, 27–28, 36–37, 42; locomotor, 230, 239–43, 247; presacral, 22, 37; tail, 22, 38
speciation, allopatric, 106, 121, 137
species: and diversity estimation, 107, 109, 110, 112–13, 127; number of, 1, 6
spongiosa, 205, 311
squamates, 39, 220
squamosal, 59, 295–96, 298
squirrels, 314
sternal plates, 22, 40
stratigraphic stages: on charts/graphs, 105, 115–16, 130–32, 135–36, 261, 271. *See also* Cretaceous; Jurassic; Mesozoic; Triassic; *names of specific stages*
stress, 306, 310
stride length, 231, 239, 243, 244, 257, 265
subadults, 61, 316–17, 321
subclades, 16, 52

subunguligrady, 239
success rate, 9
Supersaurus, 205, 303
surangular, 59–60, 62, 158–59, 160, 163, 170
sutures, 202, 203, 208
swimming, 275
Switzerland, 261, 263, 267–68, 274
sympatry, 138, 148, 194, 197
symphysis, 158, 165, 170
symplesiomorphies, 244
synapomorphies, 3–4, 16, 20–23, 32–33, 41; appendicular, 39, 246; cranial, 36, 62; and diversity estimation, 107, 113, 127; and embryos, 295, 296; of Eusauropoda, 27, 296; of titanosaurs, 52, 54, 55, 65; and tracks, 252, 254, 255
synapsids, 203
synonyms, 264, 266, 289
synovial capsule, 181, 185, 189–90, 191
Syntarsus, 309, 322
systematics, 104, 139, 157

tabefaction, 141, 142, 143, 146
Tadjikistan, 267, 269
tails, 22, 37–38, 41, 252. *See also* vertebrae, caudal
Tanzania, 66, 137
taphonomic effect, 106, 120–21, 137
taphonomy, 196, 299
tarsal, 21, 27, 33
tarsus, 41
taxa, 109; extant/modern, 309, 312, 318–19, 323; number of, 105, 115–16, 142; sister, 54, 55, 65, 72, 108; terminal, 66–67. *See also* TDEs; *names of specific taxa*
taxonomy, 7, 51, 139. *See also* clades; cladograms; classification
TDEs (taxic diversity estimates): for diplodocoids, 136–37, 143; and feeding, 133, 138; and herbivory, 125–26; and plants, 147; for prosauropods, 126–27, 129, 130, 140, 146; for sauropods, 110, 117–20, 127, 130, 133; and stratigraphic range, 111, 113; and teeth, 134–35; for titanosaurs, 136–37. *vs.* PDEs, 107–9, 114, 116, 118–19
tectonics, 137
teeth, 126, 287–88; abrasion of, 168, 172; batteries of, 162, 167–69, 171–75; and classification, 18; columns of, 163, 164, 168, 169, 172, 173; dentary, 162, 166, 167, 171, 172; of diplodocoids, 37; and diversity, 127–29; embryonic, 167–68; of Eusauropoda, 28, 36; lower, 166–67, 168, 172, 196; maxillary, 162–63, 164, 166–67, 168–69, 172; of *Nigersaurus taqueti*, 158–60, 162–69, 170–75; number of, 158, 166–68, 171, 172; peglike, 194; pencil-shaped, 295–96, 297; premaxillary, 163, 166–67, 169; replacement, 21, 159, 163, 164, 172, 173; rows of, 21, 27, 168, 170; spatulate, 16, 194–95; of titanosaurs, 60–62; upper, 166–67, 168, 172, 196. *See also* crowns; dentary; occlusion; wear facets

Tehuelchesaurus, 115, 234, 246
Telmatosaurus transylvanicus, 300
temperature, body, 221
Tendaguria, 206
tetrapods, 138, 178, 203, 240
Tetrasauropus, 2, 23, 31, 35; tracks of, 262–63, 274; *T. unguiferous*, 262–63
Texas, 265, 269, 275–76
Thailand, 23, 67, 117, 268
Thecodontosaurus caducus, 215–16
theropods (Theropoda), 24, 27–28, 57; basal/primitive, 32, 37, 239; bones of, 24, 182, 241; diversity of, 2, 121; feet of, 31–32, 33, 34; locomotion of, 239, 240, 241–42, 243; phylogeny of, 4, 234; and pneumaticity, 204, 210, 214, 216, 219, 223; tracks of, 40, 272, 275. *See also* birds
thin sections, 311, 316
thorax, 41, 180, 203
thyreophorans, 2, 144, 231, 237
Thyreophora, 3
tibia, 24, 29, 40, 41, 64; and locomotion, 21–22, 239
Titanosauridae, 138–39, 254–55, 274, 296
Titanosauriformes: monophyly of, 54–56, 65, 67, 70, 73; phylogeny of, 38, 52, 62, 67–71, 296; possible, 139, 265–66; and tracks, 254–56, 258, 259, 260, 265–66, 274; vertebrae of, 211, 219
Titanosaurinae, 51
Titanosaurinae indet., 53
Titanosauroidea, 51, 54, 55
titanosaurs/titanosaurians (Titanosauria), 50–73; background on, 51–56; basal, 55, 73, 244; derived, 56, 137, 240; and diplodocoids, 55–56, 67, 68, 70, 73, 142–43; eggs and embryos of, 285, 288, 292–97; feeding mechanism of, 128, 131; interrelationships of, 65–70; locations of, 38–39, 50, 51; locomotion of, 10, 239, 240, 241–42, 243, 244–46; missing data on, 72; monophyly of, 52, 54–56, 65, 68, 69, 71; phylogeny of, 9, 19–20, 32, 38–39, 42, 67–71; and plants, 147; and pneumaticity, 205, 212; radiation of, 105, 137; *Rapetosaurus krausei*, 53, 55, 56–65, 67, 72, 73; reproduction of, 298; size of, 39, 233, 244, 245, 303, 323; TDEs of, 136–37; teeth of, 18, 134, 167, 170, 171, 194, 296; time period of, 38, 50, 133, 139; and tracks, 39–40, 106, 118, 253–54, 256–57, 260, 266. *See also* Lithostrotia; Opisthocoelicaudiinae; saltasaurids; saltasaurines; *names of specific genera*
'Titanosaurus', 54–55, 113, 115; *T. colberti*, 53, 55, 246; *T. indicus*, 50–51, 73; *T. madagascariensis*, 56–57, 59. *See also Isisaurus*
Tithonian: diplodocoids in, 136, 171; and diversity, 105, 112, 115–18, 120, 130, 135–36; feeding mechanisms in, 132, 134
Toarcian: and diversity, 105, 115–18, 120–21, 130–32, 135; prosauropods in, 130–31, 140
toes, 264
tooth combs, 126

topotypes, 264
trachea, 203, 216, 217–20, 221
tracks, pneumatic, 204, 209
trackways/ichnofossils, 10; cladistic approach to, 254–55; and classification, 258–66; depth of, 273; and diversity, 106; earliest, 30; and eggs, 286; gauge of, 276; ideal, 254, 258; juvenile, 259, 263, 272–73; and locomotion, 39, 239, 273–76; multiple, 275–76; paleoenvironmental distribution of, 271–72; paleolatitude of, 253; parallel, 275; prdicted, 252–58; size of, 258–59, 263, 266; spatiotemporal distribution of, 266–71; and speed, 265, 277; of titanosaurs, 38, 50, 106, 118, 245, 261; Triassic, 19, 23, 31, 34–35, 267, 273–74. *See also* fossil formations and sites
trackways/ichnofossils, narrow-gauge, 39, 255–61, 264, 273–75; Cretaceous, 268–69, 273; Jurassic, 106, 261, 267–68, 273–74; Triassic, 263, 267, 273–74
trackways/ichnofossils, wide-gauge, 39–40, 255–61, 263, 265–69, 273–75; titanosaur, 50, 106, 245, 261; titanosauriform, 258–61, 265–66
transverse process, 205, 222
trees, 196. *See also* forests
trees, phylogenetic, 19–20, 139, 234; Adams consensus, 55, 67–71, 72; composite, 107; most parsimonious (MPTs), 52, 55, 69, 73, 109, 114; reduced consensus, 114; strict consensus, 65, 70, 72, 73
Triassic, 4, 10, 125, 256; diversity in, 105, 115–16. *See also names of specific periods*
Triassic, Late/Upper, 27, 30, 109, 134, 146; diversity in, 117, 137; origin in, 2, 104, 125; prosauropods in, 19, 140; tracks in, 23, 32, 34–35, 261–64, 266–67, 270–71, 273–74
tripodality, 128, 195
trochanter, lesser, 239, 242
trunk, 180, 192, 195, 243–44, 245
turkeys, 182, 190
Turonian, 105, 115–16, 119–21, 131, 132, 135
turtles, 318
Tyrannosaurus, 222

ulna, 41, 63, 64, 316–18, 320
unguals, 21, 31, 237, 240, 264
ungulates, 189, 219
United Kingdom, 118, 139, 256–58, 263, 267–68, 274. *See also* England; Isle of Wight; Wales
United States, 16, 66, 144, 267–69, 274. *See also names of specific states*
Upchurch, P., 18, 19–20
Utah, 259, 262, 267
Uzbekistan, 119

Valanginian, 105, 115–16, 119–20, 130, 132, 135–36
variation, ontogenetic, 16
vascular canals, 158, 165, 306–7, 308–9, 317, 318. *See also* bone tissue, vascularization of

vegetation. See plants
vertebrae: apneumatic, 205, 210, 218, 221; camellate, 202, 205, 210–12; complexity of, 204, 211–12, 214–15; evolution of, 211–12, 215; fused, 204; number of, 21–22, 28, 37, 171, 172, 189; polycamerate, 211–12; presacral, 63, 204, 205, 211, 215–16; sacral, 37, 62, 64, 204, 215–16, 218; semicamellate, 211–12; somphospondylous, 211; thoracic, 183–84, 216; types of, 211. See also chambers, vertebral; diverticula, vertebral; foramina, vertebral; fossae, vertebral; laminae, vertebral; neural spines; septa, vertebral
vertebrae, caudal, 15, 24, 215; number of, 37–38, 41, 62; and pneumaticity, 204, 215–16, 218–19, 223; of titanosaurs, 51, 52, 54, 56–57, 59, 62–63
vertebrae, cervical, 37, 61, 189, 193; articulation of, 184–85, 187; diplodocoid, 171–73; number of, 21–22, 62, 171–72; opisthocoelous, 181, 184–85; and pneumaticity, 206, 207–8, 210, 213, 216, 218; wedge-shaped, 183–84, 186–87
vertebrae, dorsal, 37, 64, 180; and head height, 183, 192; number of, 21–22, 62; and pneumaticity, 202, 206–7, 213, 216, 218
vertebral centra, 181, 183, 186–87; and pneumaticity, 201–2, 204, 206, 207, 210, 215; volume of air in, 218
vertebral column: arching/curvature of, 180–81, 182, 183, 185, 188, 192; caudal, 195; cervical, 182–92; dorsal, 181, 182, 188; dorsiflexion of, 181, 184, 193–94; neutral, 182; pneumaticity along, 215–16; presacral, 181, 192; ventriflexion of, 181, 184, 194
vertebrates, 230, 254, 306, 310
vertebrates, extant/modern: growth of, 304–5, 311, 313, 318, 320–22; necks of, 181, 182, 183, 189–90, 191; and pneumaticity, 201, 222. See also names of specific fauna
vigilance, 179–80
viscera, 203
Volkheimeria, 234
vomer, 160
Vulcanodon, 2, 115, 140, 231; bones of, 24–26, 28, 29–30; hindfeet of, 30–31; *V. karibaensis*, 241–42; locomotion of, 237–39, 241–42; phylogeny of, 23, 234, 296; and tracks, 254–55
Vulcanodontidae, 138–39

Wales, 263, 267, 274
water, 196, 271–73, 275, 276, 286. See also sea level
wear facets, 174, 196; buccal, 172; distal, 127–28; external, 166, 170–71, 172; internal, 166, 170, 171; mesial, 127–28; of *Nigersaurus taqueti*, 127–28, 166, 167, 168–69
Weishampel, D. B., 107, 113, 116–17
whales, 315, 321

xenarthrans, 241

zygapophyses, 206, 210; and feeding, 181, 182–85, 187, 189–93